GANGLIOSIDE STRUCTURE, FUNCTION, AND BIOMEDICAL POTENTIAL

ADVANCES IN EXPERIMENTAL MEDICINE AND BIOLOGY

Editorial Board:

NATHAN BACK, *State University of New York at Buffalo*
NICHOLAS R. DI LUZIO, *Tulane University School of Medicine*
EPHRAIM KATCHALSKI-KATZIR, *The Weizmann Institute of Science*
DAVID KRITCHEVSKY, *Wistar Institute*
ABEL LAJTHA, *Rockland Research Institute*
RODOLFO PAOLETTI, *University of Milan*

Recent Volumes in this Series

Volume 166
BIOLOGICAL RESPONSE MODIFIERS IN HUMAN ONCOLOGY AND IMMUNOLOGY
Edited by Thomas Klein, Steven Specter, Herman Friedman, and Andor Szentivanyi

Volume 167
PROTEASES: Potential Role in Health and Disease
Edited by Walter H. Hörl and August Heidland

Volume 168
THE HEALING AND SCARRING OF ATHEROMA
Edited by Moshe Wolman

Volume 169
OXYGEN TRANSPORT TO TISSUE – V
Edited by D. W. Lübbers, H. Acker, E. Leniger-Follert, and T. K. Goldstick

Volume 170
CONTRACTILE MECHANISMS IN MUSCLE
Edited by Gerald H. Pollack and Haruo Sugi

Volume 171
GLUCOCORTICOID EFFECTS AND THEIR BIOLOGICAL CONSEQUENCES
Edited by Louis V. Avioli, Carlo Gennari, and Bruno Imbimbo

Volume 172
EUKARYOTIC CELL CULTURES: Basics and Applications
Edited by Ronald T. Acton and J. Daniel Lynn

Volume 173
MOLECULAR BIOLOGY AND PATHOGENESIS OF CORONAVIRUSES
Edited by P. J. M. Rottier, B. A. M. van der Zeijst,
W. J. M. Spaan, and M. C. Horzinek

Volume 174
GANGLIOSIDE STRUCTURE, FUNCTION AND BIOMEDICAL POTENTIAL
Edited by Robert W. Ledeen, Robert K. Yu, Maurice M. Rapport,
and Kunihiko Suzuki

A Continuation Order Plan is available for this series. A continuation order will bring delivery of each new volume immediately upon publication. Volumes are billed only upon actual shipment. For further information please contact the publisher.

GANGLIOSIDE STRUCTURE, FUNCTION, AND BIOMEDICAL POTENTIAL

Edited by

Robert W. Ledeen
Albert Einstein College of Medicine
Bronx, New York

Robert K. Yu
Yale University School of Medicine
New Haven, Connecticut

Maurice M. Rapport
New York State Psychiatric Institute
New York, New York

Kunihiko Suzuki
Albert Einstein College of Medicine
Bronx, New York

PLENUM PRESS • NEW YORK AND LONDON

Library of Congress Cataloging in Publication Data

Main entry under title:

Ganglioside structure, function, and biomedical potential.

(Advances in experimental medicine and biology; v. 174)
Proceedings of a symposium held at Parksville, Vancouver Island, British Columbia, Canada, July 6-10, 1983.
Bibliography: p.
Includes index.
1. Gangliosides—Congresses. 2. Gangliosides—Therapeutic use—Testing—Congresses. 3. Nerves, Peripheral—Diseases—Treatment—Congresses. I. Ledeen, Robert W. II. Series.
QP752.G3G35   1984   599'.0188   84-8232
ISBN 0-306-41707-3

Fidia Research
Frontiers in Neuroscience

Proceedings of the symposium on Ganglioside Structure, Function, and Biomedical Potential, held July 6-10, 1983, at Parksville, Vancouver Island, British Columbia, Canada. This was a satellite to the ninth meeting of the International Society for Neurochemistry held July 10-15, 1983 at Vancouver, British Columbia. The symposium program was organized by Robert W. Ledeen, Robert K. Yu, Maurice M. Rapport, Kunihiko Suzuki, and Guido Tettamanti. Financial support was provided by Fidia Research Laboratories and the National Tay-Sachs and Allied Diseases Association.

©1984 Plenum Press, New York
A Division of Plenum Publishing Corporation
233 Spring Street, New York, N.Y. 10013

All rights reserved

No part of this book may be reproduced, stored in a retrieval system, or transmitted, in any form or by any means, electronic, mechanical, photocopying, microfilming, recording, or otherwise, without written permission from the Publisher

Printed in the United States of America

PREFACE

This volume contains the proceedings of the symposium on "Ganglioside Structure, Function and Biomedical Potential" which was held at Parksville, Vancouver Island, B.C., Canada on July 6-10, 1983. The symposium was organized as a satellite to the ninth meeting of the International Society for Neurochemistry, held immediately afterward in Vancouver City, B.C. Close to 50 speakers from 9 countries presented papers on a wide range of topics on the ganglioside theme. These encompassed the many aspects of basic research that have evolved over the past half-century, as well as some newer topics relating to the biomedical potential of gangliosides as therapeutic agents. One of the purposes of the meeting was to encourage dialogue between investigators in these seemingly diverse areas in the hope that each would come away with fresh perspective toward his own work. Judging from the many spirited, informed discussions that took place throughout the four-day meeting, this goal was achieved.

The charm and beauty of Vancouver Island undoubtedly contributed to the congenial atmosphere which quickly developed among the 120 or so participants. Drawn from 13 countries of Europe (East and West), America (North and South), Asia and the Middle East, this gathering reflected the broad geographical scale on which ganglioside research is now conducted. That this field is still in its "logarithmic" stage of growth is indicated by the increasing number of citations appearing each year in the various abstracts (e.g. fig. next page).

This volume is organized along the lines of the meeting itself. The three plenary lectures, reflecting an historical perspective, were delivered by investigators who have themselves made history in the ganglioside/glycolipid field. Most of the sessions that followed were opened with a background lecture by an invited speaker who presented a brief overview of the cumulative progress in the area, followed by his own recent findings. We are indebted to Drs. Yoshitaka Nagai, Guido Tettamanti, Sen-itiroh Hakomori, and Alfredo Gorio for contributing additional chapters to this volume summarizing these background materials which they presented at the meeting (review chapters were also provided by two of the editors: Drs. Robert Yu and Kunihiko Suzuki).

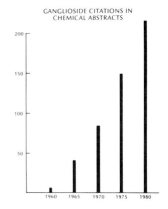

GANGLIOSIDE CITATIONS IN CHEMICAL ABSTRACTS

The first few sessions were devoted to the "conventional" topics of structure, metabolism, and distribution. Although these represent the more established areas of ganglioside research, this meeting again demonstrated that they retain novelty and excitement despite their early origins. The same may be said of neurological diseases, which led to the discovery of these substances. Receptor function remains a subject of keen interest, following the discovery of specific binding to cholera toxin a decade ago; a new perspective now emerging, that gangliosides may function in some cases as receptor modulators, was described in further detail.

The most lively discussions occurred during the remainder of the program dealing with neuritogenesis, regeneration, and potential clinical applications. Earlier reports on the ability of exogenous gangliosides to induce neurite growth in cell cultures received additional support from several of the speakers. Also heard were reports on the use of various animal models to study gangliosides as agents for neurological regeneration and repair. While we tend to view these as newly discovered phenomena, it is well to recall the studies of Henry McIlwain and his colleagues two decades ago demonstrating a restorative effect of exogenous gangliosides on "injured" neurons in their response to electrical stimulation. A similar theme - neuronal survival as facilitated by gangliosides - was developed by some of the speakers at this meeting. Further confirmation of these effects and elucidation of their mechanism could mark a turning point in ganglioside research. This effort will also assist in evaluating a proposed rationale for the employment of gangliosides as pharmaceutical agents in treatment of certain nervous system disorders. The final portion of the book contains several reports on clinical trials of this type.

Finally, it is our pleasure to acknowledge the generous support for this meeting provided by the Fidia Research Laboratories. Contributions by the National Tay-Sachs and Allied Diseases Association as well as the Fidia Laboratories provided a special scholarship fund that subsidized the attendance of several graduate students and postdoctoral fellows. The ability to attract talented young investigators portends well for the future of the ganglioside field.

Robert W. Ledeen
Robert K. Yu
Maurice M. Rapport
Kunihiko Suzuki

CONTENTS

PLENARY LECTURES

Wonders in Glycolipids - A Historical View 3
   T. Yamakawa

Present Status of the Immunology of Gangliosides 15
   M. M. Rapport and Y.-Y. Huang

Use of Cell Cultures in Ganglioside Research 27
   P. Mandel, H. Dreyfus, Y. Matsui, and G. Rebel

STRUCTURE ANALYSIS AND PHYSICAL PROPERTIES

Gangliosides: Structure and Analysis 39
   R. K. Yu

Structure Analysis of Glycosphingolipids Using
   Fast Atom Bombardment (FAB) Techniques 55
   H. Egge, J. Peter-Katalinic, and P. Hanfland

Application of Field Desorption and Secondary Ion
   Mass Spectrometry for Glycolipid Analysis 65
   S. Handa and Y. Kushi

New Techniques for the Investigation of Structure
   and Metabolism of Sialic Acids 75
   R. Schauer, C. Schröder, and A. K. Shukla

Recent Advances in Structural Analysis of Gangliosides:
   Primary and Secondary Structures 87
   R. K. Yu, T. A. W. Koerner, Jr., P. C. Demou,
   J. N. Scarsdale, and J. H. Prestegard

Adrenal Medulla Gangliosides 103
   M. Sekine, T. Ariga, and T. Miyatake

Glycosphingolipids of Equine Erythrocytes Membranes:
    Complete Characterization of a Fucoganglioside 111
    S. Gasa, A. Makita, K. Yanagisawa, and M. Nakamura

Physical Behaviour of Glycolipids in Bilayer Membranes:
    Distribution and Accessibility 119
    M. W. Peters and C. W. M. Grant

DISTRIBUTION AND TRANSPORT

Ganglioside Distribution at Different Levels of
    Organization and its Biological Implications 135
    Y. Nagai and M. Iwamori

The Labeling of the Retina and Optic Tectum
    Gangliosides and Glycoproteins of Chickens
    in Darkness or Exposed to Light 147
    R. Caputto

Biosynthesis and Transport of Gangliosides in
    Peripheral Nerve 155
    A. J. Yates, U. R. Tipnis, J. H. Hofteig, and
    J. K. Warner

Cellular Localization of Gangliosides in the Mouse
    Cerebellum: Analysis Using Neurological Mutants 169
    T. N. Seyfried and R. K. Yu

Gangliosides and Related Compounds as Biological
    Response Modifiers 183
    Y. Nagai, S. Tsuji, and Y. Sanai

METABOLISM

An Outline of Ganglioside Metabolism 197
    G. Tettamanti

Activator Proteins for the Catabolism of
    Glycosphingolipids 213
    Y.-T. Li and S.-C. Li

Ganglioside Biosynthesis in Rat Liver Golgi Apparatus:
    Stimulation by Phosphatidylglycerol and
    Inhibition by Tunicamycin 227
    H. K. M. Yusuf, G. Pohlentz, G. Schwarzmann, and
    K. Sandhoff

CONTENTS

Incorporation Rate of GM1 Ganglioside Into Mouse
    Brain Myelin: Effect of Aging and Modification
    by Hormones and Other Compounds 241
  S. Ando, Y. Tanaka, Y. Ono, and K. Kon

Biosynthesis In Vitro of Gangliosides Containing
    Gg- and Lc-Cores 249
  S. Basu, M. Basu, J. W. Kyle, and H.-C. Chon

Genetic Regulation of GM2(NeuGc) Expression in
    Liver of Mouse 263
  A. Suzuki, Y. Hashimoto, M. Abe, Y. Kiuchi, and
  T. Yamakawa

New Approaches in the Study of Ganglioside Metabolism 273
  G. Tettamanti, R. Ghidoni, S. Sonnino, V. Chigorno,
  B. Venerando, A. Giuliani, and A. Fiorilli

MUTANTS AND DEVELOPMENT

Glycosphingolipids of Chicken Skeletal Muscle in Early
    Development and Genetic Dystrophy 287
  E. L. Hogan, J.-L. Chien, and S. Dasgupta

Ganglioside Alterations in the Genetically-Determined
    Hypertrophic Neuropathy of the Murine
    Neurological Mutant Trembler 297
  M.-L. Harpin, J. Portoukalian, B. Zalc, and
  N. Baumann

Comparative and Developmental Behavior of Alkali
    Labile Gangliosides in the Brain 307
  R. Ghidoni, S. Sonnino, V. Chigorno, A. Malesci, and
  G. Tettamanti

Phylogeny and Ontogeny of Vertebrate Brain
    Gangliosides 319
  L. N. Irwin

RECEPTORS AND FUNCTION

Ganglioside Receptors: A Brief Overview and
    Introductory Remarks 333
  S.-I. Hakomori

Gangliosides as Modulators of the Coupling of Neuro-
    transmitters to Adenylate Cyclase 341
  G. Dawson and E. Berry-Kravis

Gangliosides, The Thyrotropin Receptor, and Autoimmune
    Thyroid Disease 355
  P. Lacetti, D. Tombaccini, S. Aloj, E. F. Grollman and
  L. D. Kohn

Specific Gangliosides are Receptors for Sendai Virus 369
  M. A. K. Markwell, P. Fredman, and L. Svennerholm

Gangliosides as Receptor Modulators 381
  E. G. Bremer and S.-I. Hakomori

Brain Gangliosides and Thermal Adaptation in
    Vertebrates 395
  H. Rahmann, R. Hilbig, W. Probst, and M. Muhleisen

GANGLIOSIDES AND NEUROLOGICAL DISEASES

Gangliosides and Disease: A Review 407
  K. Suzuki

Cerebral and Visceral Organ Gangliosides and Related
    Glycolipids in GM1-Gangliosidosis Type 1,
    Type 2 and Chronic Type 419
  T. Taketomi, A. Hara, and T. Kasama

Canine GM_2-Gangliosidosis: Chemical and Enzymatic
    Features 431
  Y. Eto, L. Autilio-Gambetti, and J. T. McGrath

Immunological Expression of Gangliosides in Multiple
    Sclerosis and in a Demyelinating Model Disease
    in Rabbits 441
  B.-A. Sela, H. Offner, G. Konat, V. Lev-Ram,
  O. Cohen, and I. R. Cohen

Antibodies to Glycosphingolipids in Patients with
    Multiple Sclerosis and SLE 455
  T. Endo, D. D. Scott, S. S. Stewart, S. K. Kundu, and
  D. M. Marcus

NEURITOGENESIS AND REGENERATION

Neuritogenesis and Regeneration in the Nervous System:
    An Overview of the Problem and on the Promoting
    Action of Gangliosides 465
  A. Gorio, D. Janigro, and R. Zanoni

CONTENTS

Effects of Gangliosides on the Functional Recovery
  of Damaged Brain 475
  G. Toffano, G. Savoini, C. Aldinio, G. Valenti,
  R. Dal Toso, A. Leon, L. Calza, I. Zini,
  L. F. Agnati, and K. Fuxe

Exogenous Gangliosides Enhance Recovery From CNS
  Injury 489
  S. E. Karpiak

Ganglioside Induced Surface Activity and Neurite
  Formation of Neuro-2A Neuroblastoma Cells 499
  F. J. Roisen, D. A. Spero, S. J. Held, G. Yorke, and
  H. Bartfeld

Effect of Exogenous Gangliosides on the Morphology
  and Biochemistry of Cultured Neurons 513
  H. Dreyfus, B. Ferret, S. Harth, A. Gorio,
  L. Freysz, and R. Massarelli

Studies of Gangliosides in Diverse Nerve Cell Cultures 525
  W. Dimpfel, and U. Otten

In-Vitro and In-Vivo Studies on Gangliosides in the
  Developing and Regenerating Hippocampus of
  the Rat 535
  W. Seifert and H.-J. Fink

BIOMEDICAL POTENTIAL OF GANGLIOSIDES

Ganglioside Treatment of Genetic and Alloxan-Induced
  Diabetic Neuropathy 549
  A. Gorio, F. Aporti, F. Di Gregorio, A. Schiavinato,
  R. Siliprandi, and M. Vitadello

Double-Blind Controlled Trial of Purified Brain Gangliosides in Amyotrophic Lateral Sclerosis and Experience
  with Peripheral Neuropathies 565
  W. G. Bradley

Trials of Ganglioside Therapy for Amyotrophic Lateral
  Sclerosis and Diabetic Neuropathy 575
  M. Hallett, H. Harrington, H. R. Tyler, T. Flood, and
  N. Slater

Treatment of Painful Diabetic Polyneuropathy with Mixed
  Gangliosides 581
  A. Naarden, J. Davidson, L. Harris, J. Moore, and
  S. DeFelice

Ganglioside (Cronassial) Therapy in Diabetic Neuropathy 593
    S. H. Horowitz

Multicentre Trial on Gangliosides in Diabetic Peripheral
    Neuropathy 601
    D. Fedele, G. Crepaldi, and L. Battistin

A Double Blind Placebo Controlled Trial of Mixed
    Gangliosides in Diabetic Peripheral and Autonomic
    Neuropathy 607
    R. R. Abraham, R. M. Abraham, and V. Wynn

Gangliosides--Clinical Overview 625
    S. L. DeFelice and M. Ellenberg

Participants 629

Author Index 635

Subject Index 639

Structure Analysis and Physical Properties

WONDERS IN GLYCOLIPIDS - A HISTORICAL VIEW

Tamio Yamakawa

Tokyo Metropolitan Institute of Medical Science
Honkomagome, Bunkyo-ku, Tokyo

It is my great privilege to present the opening lecture at this very exciting meeting. The organizer, Dr. Ledeen, suggested a historical outline of glycolipid research as an introduction, and I hope you will allow me to talk freely as I outline some of the thoughts and questions that have occured to me during my 30 years of research in this field.

The first naturally occurring glycolipid that was found was cerebroside, or galactosyl ceramide (Fig. 1). It was isolated by Thudichum about 100 years ago and described in his textbook.[1] Because of the infamous "protagon" controversy, his finding was only accepted after his death. At that time, galactose was also called cerebrose because of its presence in cerebroside. The monohexosyl ceramide most commonly found in various animal tissues is glucosyl ceramide. Galactosyl ceramide occurs abundantly in white matter of brain, and is also present in considerable amount in kidney along with glucosyl ceramide.

We may ask the question, why is the distribution of galactosyl ceramide and glucosyl ceramide so specific? The localization of galactosyl ceramide in brain and kidney cells was visualized by Nagai and his co-workers in 1981 by means of the immunofluorescence technique. Oligodendrocytes, myelin sheath and choroid plexus were definitely stained.[2] Furthermore, they showed positive staining at the microtubule-like cytoskeleton structure in cultured monkey kidney cells with anti-galactosyl ceramide.[3] The localization of galactosyl ceramide in renal tissue raises an interesting question, and may be a clue to solving the differing roles of glucosyl ceramide and galactosyl ceramide. For the biosynthesis of more complex oligoglycosylceramides, glucosyl ceramide seems to be convenient for elongation of carbohydrate chains, but this is not

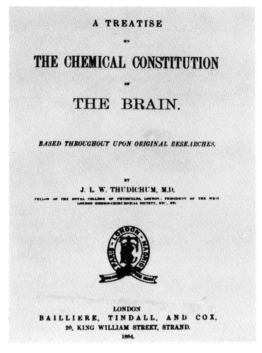

Fig. 1 Title page of Thudichum monograph.

the case for galactosyl ceramide. Abrahamsson[4] pointed out that in oligoglycosylceramides the oxygen at position 4 of hexose is usually involved in glycosidic linkage to the next sugar. The conformation of galactose in cerebroside would not allow addition of another sugar residue. It is therefore of interest to note that almost all known oligoglycosylceramides have glucose instead of galactose in proximal linkage to ceramide. So, a possible explanation may be that a membrane site which requires restriction to a monohexosyl ceramide, for example in the myelin sheath, selects galactosyl ceramide.

The second glycolipid which appeared was sulfatide or cerebroside sulfuric ester. The presence of sulfate in brain lipids had been suggested by Thudichum. Koch subsequently separated a substance in which cerbroside and phosphatide seemed to be esterified with sulfuric acid[5] (Fig. 2). When Blix isolated sulfatide in 1933, he arbitrarily suggested the formula of cerebroside-6'-sulfate as its probable structure.[6] In 1951, Nakayama presented evidence that sulfatide is cerebroside-6'-sulfate because he could not get the triphenylmethyl derivative of sulfatide by reaction with trityl chloride.[7] In 1955, Thannhauser and Schmidt reported that they were unable to tritylate cerebroside, and therefore regarded Nakayama's evidence as inconclusive. However, they also stated that sulfatide is cerebroside-6'-sulfate based on paper chromatography of partially methylated galactose obtained after hydrolysis of permethylated sulfatide.[8]

During our structural study of glycolipids in 1962 we used permethylation and subsequent methanolysis. We tried to get some standard samples of partially methylated methylglucosides and methylgalactosides for gas chromatography. In order to obtain methyl 2,3,4-tri-0-methyl galactoside, we permethylated sulfatide and prepared the partially methylated methyl galactoside. The gas chromatogram of this material resembled closely that of methyl 2,4,6-tri-0-methyl galactoside but was definitely different from that of methyl 2,3,4-tri-0-methyl galactoside, which was obtained by chemical synthesis (Fig. 3). The result indicated the galactose in sulfatide to be substituted at position 3. Subsequent examination of sulfatide by periodate oxidation indicated the galactose in sulfatide was not degraded by periodate, which presented further evidence that the brain sulfatide is cerebroside-3'-sulfate.[9] Somewhat later, Stoffyn in Folch's laboratory attained the same conclusion by the resistance of sulfatide towards alkali.[10] The reason that Thannhauser et al. reached an erroneous conclusion was that the R_G-value of methyl 2,3,4-tri-0-methyl-galactose (0.64) is similar to that of methyl 2,4,6-tri-0-methylgalactose (0.67) with the solvent system they employed.

Returning to sulfated glycolipids, we purified a glycolipid from boar testis in 1973 which proved to be a sulfated glycoglycerolipid. The structure was 1-0-palmityl-2-0-palmitoyl-3-(β-3'-sulfogalactosyl)-glycerol (Fig. 4). We named it Seminolipid because it was present in high concentration in mammalian spermatozoa. This evidence suggested that this glycolipid originated only in germinal cells in the testis.[11]

Another sulfated glycoglycerolipid was previously obtained by Kates[12] from extremely halophilic bacteria, Halobacterium cutiru-

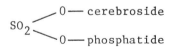

Fig. 2. The structure proposed for a substance containing sulfuric acid ester by Koch in 1907.

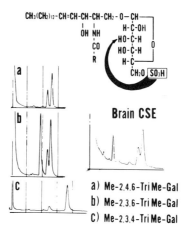

Fig. 3. Determination of sulfate position in sulfatide by gas liquid chromatography.[9]

brum (Fig. 4). In this case, too, the sulfate was attached to the 3-position of the terminal galactose residue. These three examples might indicate the structural specificity of the position of the sulfate group, and it suggests that the enzyme transfers the sulfate group only to the 3rd OH of galactose of glycolipids. On the other hand, in the case of acid mucopolysaccharides, the sulfate group is attached to either the 4th or 6th OH group of N-acetylgalactosamine as in the chondroitin sulfates. Characterization of the sulfotransferases may give us a reasonable explanation for the structural specificity.

The existence of a glycoglycerolipid, such as Seminolipid in mammalian cells is rather unusual. Generally glycoglycerolipids are present in higher plants and microorganisms as the structural unit of membranes. Glycolipids in animal tissues or cells on the other hand exist in the form of sphingolipids. Possibly an early stage of phylogeny is expressed in the early stage of ontogeny.

The next big event was the discovery of ganglioside by Klenk in the late 1930's.[13] Earlier, in 1927, Walz in Tübingen[14] and simultaneously Levine and Landsteiner in New York[15] found glycolipids in bovine spleen and in horse kidney, respectively, which gave a purple color when heated with Bial's orcinol reagent. Klenk noticed the increased intensity of the color reaction in the glycolipid obtained from a patient's brain with Tay-Sachs disease in 1939.[16] Blix found a similar glycolipid with the same color reaction in normal bovine brain in 1938.[17] Klenk reported that the

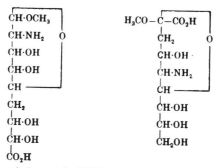

Fig. 4. Chemical structure of sulfated glycolipids.

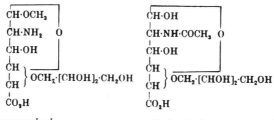

Yamakawa and Suzuki 1952   Gottschalk 1955

Neuraminic   N-Acetyl-neuraminic acid
Acid-methylglycoside

Klenk, Faillard, Weygand and Schöne. (1956)

Fig. 5. Proposed structures for sialic acid.

content of this glycolipid was much higher in grey matter than in white matter of brain and named it Ganglioside.[18] Klenk isolated neuraminic acid, which later proved to be a methoxyl derivative, in a crystalline form from ganglioside and demonstrated it to be the principal chromogen in 1942.[19]

In 1936, Blix obtained a crystalline sugar acid from bovine submaxillary mucin which exhibited a purple color when heated with Ehrlich's p-dimethylaminobenzaldehyde reagent.[20] As for the identity of these two substances, hot disputes continued for several years. Finally, both materials proved to be closely related. Now neuraminic acid is accepted to be the core substance of Blix's substance which later was named sialic acid. The chemical structure of sialic acid was proposed by us,[21] Klenk and others[22] but the formula proposed by Gottschalk in 1955[23] proved to be correct (Fig. 5).

The structural study of gangliosides has also been a subject of research. In the early 1950's, ganglioside was considered to be a single compound. In 1954, Svennerholm separated ganglioside into two components by cellulose column chromatography.[24] However, many workers continued to study the chemical structure of ganglioside with mixed materials. In 1963, Kuhn and Wiegandt established the structures with gangliosides purified by chromatography[25] (Fig. 6).

In 1950, we obtained a glycolipid from equine erythrocyte membranes which exhibited purple color with Bial's orcinol reagent. I named it hematoside (Fig. 6). It was the first observation of a sialic acid residue at the surface of cell membranes and also of a ganglioside in extraneural tissues.[26] In the same year, Klenk found a neutral glycolipid containing galactosamine in human

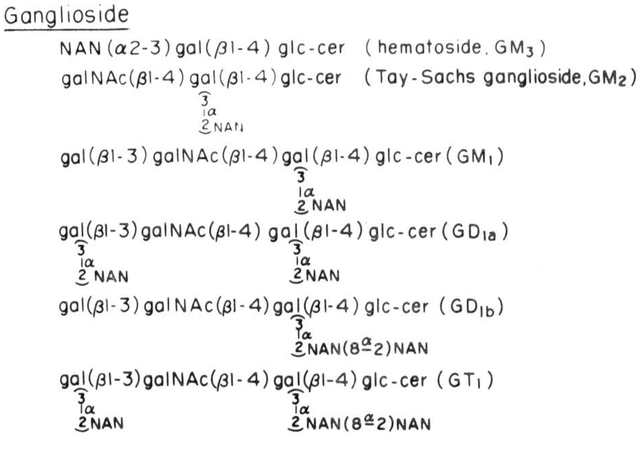

Ganglioside
NAN(α2-3)gal(β1-4)glc-cer (hematoside, GM$_3$)
galNAc(β1-4)gal(β1-4)glc-cer (Tay-Sachs ganglioside, GM$_2$)
 $\overset{3}{\underset{2\,NAN}{|\alpha}}$

gal(β1-3)galNAc(β1-4)gal(β1-4)glc-cer (GM$_1$)
 $\overset{3}{\underset{2\,NAN}{|\alpha}}$

gal(β1-3)galNAc(β1-4) gal(β1-4)glc-cer (GD$_{1a}$)
 $\overset{3}{\underset{2\,NAN}{|\alpha}}$ $\overset{3}{\underset{2\,NAN}{|\alpha}}$

gal(β1-3)galNAc(β1-4)gal(β1-4)glc-cer (GD$_{1b}$)
 $\overset{3}{\underset{2\,NAN(8\underline{\alpha}2)NAN}{|\alpha}}$

gal(β1-3)galNAc(β1-4)gal(β1-4)glc-cer (GT$_1$)
 $\overset{3}{\underset{2\,NAN}{|\alpha}}$ $\overset{3}{\underset{2\,NAN(8\underline{\alpha}2)NAN}{|\alpha}}$

Fig. 6. Structures of gangliosides.

Structure	Name	Amount (μg/100 ml)
Glc-Cer	Glucosylceramide	279
Gal-Glc-Cer	Lactosylceramide	2,211
Gal-Gal-Glc-Cer	Trihexosylceramide	1,021
GlcNAc-Gal-Glc-Cer	Amino-CTH	47
GalNAc-Gal-Gal-Glc-Cer	Globoside I	9,600
GalNAc-GalNAc-Gal-Gal-Glc-Cer	Para-Forssman antigen	7
Gal-GlcNAc-Gal-Glc-Cer	Paragloboside	553
Gal-Gal-GlcNAc-Gal-Glc-Cer	P_1 antigen	5-10
GalNAc-Gal(Fuc)-GlcNAc-Gal-Glc-Cer	A antigen	124
or Gal-Gal(Fuc)-GlcNAc-Gal-Glc-Cer	B antigen	120
NeuAc-Gal-Glc-Cer	Hematoside	355
NeuAc-Gal-GlcNAc-Gal-Glc-Cer	Sialosylparagloboside	426

Fig. 7. Glycolipids in human erythrocytes.

erythrocytes.[27] We confirmed his result and named it globoside.[28] Since then, my interest has focused on the species-specificity of erythrocyte glycolipids.

Actually, mammalian erythrocyte glycolipids are divided into three classes (globoside type, hematoside type and ganglioside type) based on the presence or absence of either hexosamine or sialic acid in the major glycolipid.[29] Lipid composition of erythrocytes is similar to that of brain tissue in several respects. In both tissues, triglyceride and cholesterol ester are absent, phospholipids and free cholesterol are predominant, and moreover glycolipid is present in a considerable amount. Of course, erythrocytes are homogeneous cells with the major function of oxygen exchange, but brain is an organ composed of a variety of cells with very complex functions. However, both are good sources of glycolipids and many kinds of acidic and neutral glycolipids have been obtained and characterized. Based on these fundamental results, many glycolipids have been isolated from various tissues and organs.

These glycolipids with many varieties of carbohydrate structure must have a function in recognition and certain other roles. In human erythrocyte membranes, the amount of glycolipid is only 5% of the total lipid, but the glycolipids can be separated into more than ten kinds of minor glycolipids.[30] In 1953, we found ABO blood group-specific activity in the crude "globoside" from human erythrocytes.[31] This was the first finding of the biological activity of naturally occurring sphingoglycolipids exhibiting a kind of recognition. The ABO group active glycolipids were purified by chromatography and amounted to only 0.1% of the total glycolipids; the activities are determined by the terminal carbohydrate structures (Fig. 7).

The real meaning of glycolipid at the surface of erythrocytes is obscure, but at least some glycolipids have definite antigenic properties. For example, Forssman hapten of sheep and goat erythrocytes resides in the glycolipid, and the activity is expressed in the terminal N-acetylgalactosaminyl-α1-3N-acetylgalactosaminyl residues of glycolipid elucidated by Siddiqui and Hakomori.[32] Moreover, human P blood group activity revealed by glycolipids was elucidated by Marcus and Naiki,[33] and Feizi[34] found that Ii antigens or determinants of cold hemagglutinins are also due to the carbohydrate structure of erythrocyte glycolipids.

On the other hand, we are still seeking a correspondingly prominent example of such an antigenicity in glycoprotein. Of course several reports suggest the activity is present at the carbohydrate in glycoprotein of erythrocytes, but I think most of them lack definite chemical characterization.

Carbohydrate structures of glycoproteins have been determined in these past ten years and the mechanism of biosynthesis and processing of carbohydrate chains of the so-called N-glycosidic type glycoproteins is elucidated. In contrast to this, the biosynthesis of the sugar chain in O-glycosidic type glycoproteins is, in a manner similar to glycolipids, thought to be conducted by successive elongation with monosaccharides supplied from sugar nucleotides. The carbohydrate sequence of N-glycosidic type glycoproteins seems not so diverse, but relatively simple. It seems to me that the sugar sequence responsible for antigenicity or recognition may take place preferably in O-glycosidic type glycoproteins and glycolipids.

As the sugar structures of glycolipids are determined strictly by genetic regulation, our present interest is to study the genetic expression of glycolipids, as will be presented by my co-worker Akemi Suzuki.

I have presented a brief history of glycolipid research, especially limited to structural studies up to 1970. The progress of research in ten years since then is brilliant, and the history is now being developed by the workers attending this symposium, step by step. Each new finding produces more questions. I expect many of these questions and riddles will be solved by your efforts and those of your successors.

REFERENCES

1. J. L. W. Thudichum, A treatise on the chemical constitution of the brain, Bailliere, Tindall and Cox, London (1884).
2. T. Uchida, K. Takahashi, H. Yamaguchi, and Y. Nagai, Localization of galactocerebroside in oligodendrocytes, myelin sheath and choroid plexus, Japan J. Exp. Med. 51:29 (1981).
3. K. Sakakibara, T. Momoi, T. Uchida, and Y. Nagai, Evidence for association of glycosphingolipid with a colchicine-sensitive microtubule-like cytoskeletal structure of cultured cells, Nature 293:76 (1981).
4. S. Abrahamsson, B. Dahlen, H. Lofgren, I. Pascher, and S. Sundell, Molecular arrangement and conformation of lipids of relevance to membrane structure, in: "Structure of Biological Membranes", S. Abrahamsson and I. Pascher, eds., Plenum Press, New York and London (1977).
5. W. Koch, Zur kenntnis der schwefelverbindugen des nerven systems, Z. Physiol. Chem. 53:496 (1907).
6. G. Blix, Zur kenntnis der schwefelhaltigen lipoidstoffe des gehirns. Uber cerebronschwefelsaure, Z. Physiol. Chem. 219:82 (1933).
7. T. Nakayama, Studies on the conjugated lipids. II. On cerebron sulphuric acid, J. Biochem. 38:157 (1951).
8. S. J. Thannhauser, J. Fellig, and G. Schmidt, The structure of cerebroside sulfuric ester of beef brain, J. Biol. Chem. 215:211 (1955).
9. T. Yamakawa, N. Kiso, S. Handa, A. Makita, and S. Yokoyama, On the structure of brain cerebroside sulfuric ester and ceramide dihexoside of erythrocytes, J. Biochem. 52:226 (1962).
10. P. J. Stoffyn and A. Stoffyn, Structure of sulfatides, Biochim. Biophys. Acta 70:218 (1963).
11. I. Ishizuka, A. Suzuki, and T. Yamakawa, Isolation and characterization of a novel sulfoglycolipid, "Seminolipid", from boar testis and spermatozoa, J. Biochem. 73:77 (1973).
12. M. Kates, B. Palamenta, M. B. Perry, and G. A. Adams, A new glycolipid sulfate ester in Halobacterium cutirubrum, Biochim. Biophys. Acta 137:213 (1967).
13. E. Klenk, Uber die natur der phosphatide und anderer lipoide des gehirns und der leber bei der Niemann-Pickschen Krankheit (12. Mitteilung uber phosphatide), Z. Physiol. Chem. 235:24 (1935).
14. E. Walz, Uber das vorkommen von kerasin in der normalen rindermilz, Z. Physiol. Chem. 166:210 (1927).
15. P. A. Levene and K. Landsteiner, On some new lipids, J. Biol. Chem. 75:607 (1927).
16. E. Klenk, Beitrage zur chemie der lipoidosen (3. Mitteilung) Niemann-Picksche Krankheit und amaurotische Idiotie, Z. Physiol. Chem. 262:128 (1939).

17. G. Blix, Einige beobachtungen uber eine hexosaminhaltige substanz in der protagonfraktion des gehirns, Skand. Arch. Physiol. 80:46 (1938).
18. E. Klenk, Uber die ganglioside, eine neue gruppe von zukerhaltigen gehirnlipoiden, Z. Physiol. Chem. 273:76 (1941).
19. E. Klenk, Neuraminsaure, das spaltprodukt eines neuen gehirnlipoids, Z. Physiol. Chem. 268:50 (1942).
20. G. Blix, Uber die kohlenhydratgruppen des submaxillarismucins, Z. Physiol. Chem. 240:43 (1936).
21. T. Yamakawa and S. Suzuki, The chemistry of the lipids of posthemolytic residue or stroma of erythrocytes. III. On the structure of hemataminic acid, J. Biochem. 39:175 (1952).
22. E. Klenk, H. Faillard, F. Weygand, and H. H. Schone, Untersuchungen uber die konstitution der neuraminsaure, Z. Physiol. Chem. 304:35 (1956).
23. A. Gottshalk, Structural relationship between sialic acid, neuraminic acid and 2-carboxy-pyrrole, Nature (London) 176:881 (1955).
24. L. Svennerholm, Partition chromatography of brain in gangliosides on cellulose, Acta Chem. Scand. 8:1108 (1954).
25. R. Kuhn and H. Wiegandt, Die konstitution der ganglioside G_{II}, G_{III} und G_{IV}, Z. Naturforsch. 18b:541 (1963).
26. T. Yamakawa and S. Suzuki, The chemistry of the lipids of posthemolytic residue or stroma of erythrocytes. I. Concerning the ether-insoluble lipids of lyophilized horse blood stroma, J. Biochem. 38:199 (1951).
27. E. Klenk and K. Lauenstein, Uber die zuckerhaltigen lipoid der formbestandteile des menschlichen blutes, Z. Physiol. Chem. 288:220 (1951).
28. T. Yamakawa and S. Suzuki, The chemistry of the lipids of posthemolytic residue or stroma of erythrocytes. III. Globosides, the sugar-containing lipid of human blood stroma, J. Biochem. 39:393 (1952).
29. T. Yamakawa, Glycolipids of mammalian red blood cells, in: "Lipoide" Shutte, ed., 16. Colloquium der Gesell. Physiol. Chem. in Mosbach. (1965).
30. T. Yamakawa and Y. Nagai, Glycolipid at the cell surface and their biological functions, Trends Biochem. Sci. 3:128 (1978).
31. T. Yamakawa and T. Iida, Immunological study on the red blood cells. I. Globoside as the agglutinogen of the ABO system on erythrocytes, Japan J. Exp. Med. 23:327 (1953).
32. B. Siddiqui and S. Hakomori, A revised structure for the Forssman glycolipid hapten, J. Biol. Chem. 246:5766 (1971).
33. M. Naiki and D. M. Marcus, An immunochemical study of the human blood group Il, P, and P^k glycosphingolipid antigens, Biochemistry 14:4837 (1975).

34. T. Feizi, E. A. Kabat, G. Vicari, B. Anderson, and W. L. Marsh, Immunological studies on blood groups, XIX. The I antigen complex: Specificity differences among anti-I sera revealed by quantitative precipitin studies; Partial structure of the I determinant specific for one anti-I serum, J. Immunol. 106:1578 (1971).

PRESENT STATUS OF THE IMMUNOLOGY OF GANGLIOSIDES

Maurice M. Rapport and Yung-yu Huang

Division of Neuroscience, New York State Psychiatric
Institute, and Department of Biochemistry, Columbia
University College of Physicians and Surgeons, New York
N. Y. 10032

INTRODUCTION

Since the reaction of neurochemists to immunology, and specifically to the immunology of gangliosides, may involve a time-delayed reversal of enthusiasm, we hope this survey will be an aid in tempering this reaction. There is a story told that on one occasion when ships of the British Navy were passing in review, one ship had the misfortune to commit six successive blunders. All eyes turned to the flagship anticipating the admiral's horrendous reaction and were hopelessly puzzled when his signal flashed "Good". A full twenty seconds later the message was completed by the word "God". Although the interval in working in the field of immunology is much longer than twenty seconds, the complete message is often the same.

Immunological experience with glycosphingolipids is much greater with neutral glycosphingolipids than with the acidic gangliosides since the immunogenic activity of gangliosides is considerably weaker. This weak immunogenic activity results in considerable softness in the generalizations derivable from published reports. We may therefore anticipate that many novel immunological observations will continue to be described, and that these may be difficult to reproduce.

The four questions most frequently asked concerning antibodies to gangliosides are: 1. How do you get them? 2. How specific are they? 3. What can they be used for? and 4. Are they involved in disease? This brief survey will be addressed to aspects of the first three of these questions, since Dr. Marcus and Dr. Sela will be discussing the fourth.

ANTIBODIES TO GANGLIOSIDES - HOW DO YOU GET THEM AND HOW SPECIFIC ARE THEY?

The most dependable method, which is also the most demanding, is to raise them. An alternative is to "beg, borrow, or steal" them from investigators who have already been lucky. One such antibody, a monoclonal type called GQ (more appropriately A2B5), is now available commercially (at a handsome price).

From the standpoint of control, both of quality and supply, there is no substitute for raising antiganglioside antibodies in one's own laboratory. In raising such antibodies, both the type of immunogen and the animal species are important. Although the more systematic and conventional procedure is to use pure gangliosides or ganglioside derivatives as immunogens, greater novelty and success have been recorded by the use of cells, cell fragments, and isolated membranes. Several examples may be cited. Antibodies to "chol-1", a ganglioside whose structure is still not established, were obtained by injecting sheep with a synaptic plasma membrane fraction from the electric organ of the Torpedo.[1] The monoclonal antibody A2B5 which reacts with ganglioside GQ1C (and others) was obtained by injecting mice with chick retina cells,[2] and monoclonal antibodies to GD3 were obtained by several groups[3,4] by injecting mice with cultured human melanoma cells. An unexpected source of such antibodies was found in human lymphocytes that preserved their antibody-producing capacity after transformation. Two such human lymphocytes lines from a melanoma patient were found to produce antibodies to GM2 and GD2.[5,6] The injection of mice with human colon carcinoma cells produced a monoclonal antibody to a complex ganglioside.[7] The novelty of these observations indicates that spectacular achievements in acquiring antibodies to gangliosides may still be expected from such random efforts.

The most direct method is to use isolated gangliosides or suitable derivatives as immunogens. Since a decade ago pure individual gangliosides were not usually obtainable in reasonable quantity, much of the early work on raising antibodies to gangliosides was done with the ganglioside mixture from brain.[8] These mixtures contained 4 major gangliosides (GM1, GD1a, GD1b, GT1b) and half a dozen minor components in varying amounts. However, it is now possible to proceed more systematically and to use isolated gangliosides or suitable derivatives as immunogens and to inject them with bovine serum albumin or methylated albumin as adjuvants.[9-16] A summary of the responses obtained by injecting rabbits and chickens with different gangliosides is presented in Table 1. Injecting rabbits with total brain ganglioside or pure GM1 induced polyclonal antibodies with similar degrees of cross-reaction, namely to GD1b and asialo GM1.[8,9] With GT1b in rabbits an even greater degree of cross-reaction was observed.[16] On the other hand, specific polyclonal antibodies to GM3, GM4 and GD3 were raised.[11,14,15] In the chicken,

Table 1. Specificity of Polyclonal Antibodies Raised Against Pure Gangliosides and Derivatives.

Antibody to:	Species	Test Method	Cross Reactions	Ref
GM1	rabbit	complement-fixation	GD1b, asialo GM1	9,10
	rabbit	complement-fixation	GD1b, asialo GM1, GM2	11
	chicken	solid phase-ELISA	asialo GM1	12
GM2	rabbit	gel diffusion	?	13
	chicken	solid phase-ELISA	most gangliosides	12
GM3	rabbit	complement fixation, gel diffusion, inhibition of passive agglutination	none	14
	chicken	solid phase-ELISA	all gangliosides	12
GM4	rabbit	solid phase-RIA with protein A	none	15
GD3	rabbit	complement-fixation	none	11
GT1b	rabbit	complement-fixation	GD1a, GD1b, GM1, asialo GM1	16
Sialosyl para globoside	chicken	solid phase-ELISA	none	12

No antibody response in chicken to GD1a, GD1b, or GT1b   12
No antibody response in mouse to GM1, GM2, or GM3   17

specific antibodies were obtained to sialosyl paragloboside, whereas GM2 and GM3 antibodies showed extensive cross-reactions.[12] Antibody responses in the chicken to GD1a, GD1b, or GT1b were not seen.[12] Efforts to produce antibodies to GM1, GM2 and GM3 gangliosides in the mouse have not been successful.[17]

Gangliosides or derivatives conjugated to proteins or other polymers through a carboxyl group have been useful in eliciting antibodies in the rabbit to GD3 and GM3.[11,14] Reports have not yet appeared on the use of ganglioside derivatives conjugated through the amino group after deacylation.

It may be noted in passing that in Waldenstrom's macroglobulinemia, myeloma proteins have been detected that react with six or more different antigens containing sialic acid groups. These "antibodies" probably do not discriminate between sialic acid on glycoproteins and gangliosides (reviewed in [17]).

The specificity of antiganglioside antibodies is difficult to establish rigorously. It depends on the sensitivity of the particular immunological test employed and is limited by the availability of individual gangliosides, glycoproteins, and neutral glycolipids that can be tested. The principal immunochemical methods presently in use for studying gangliosides are complement-fixation, precipitation, cytotoxicity, liposome lysis, liposome aggregation, indirect ELISA, and autoradiography of thin layer chromatograms. Each of these methods may have a different level of sensitivity and may thus show a particular antibody to have different degrees of cross-reactivity with various ganglioside molecules. Rabbit antibodies to GM3, GM4 and GD3 and chicken antibodies to sialosyl paragloboside have been reported to show a high degree of specificity (Table 1), whereas antibodies to GM1 and GT1b are much less specific. Transformed human lymphoblast antibodies to GM2 and GD2,[5,6] and mouse monoclonal antibodies to GQ1c,[18] GD3,[3,4] and sialyl Le^a [7] show a high degree of specificity against known gangliosides.

Some of the complications one faces in establishing specificity can be seen by comparing two recent reports on the specificity of the commercially available monoclonal reagent A2B5. Yu and Kasai,[18] using a solid phase radio-immunoassay, found this antibody to react with GQ1c but not with GQ1b, GT1a, GT1b, GD1b or GM1. On the other hand, Fredman et al.[19] using autoradiography of thin layer chromatograms, detected reactions with many (perhaps 10) different gangliosides, none corresponding in mobility to the major gangliosides in mammalian brain. These results, which, it should be pointed out, are not discordant, emphasize the general problem of trying to define specificity rigorously when the sensitivities of different methods may differ by orders of magnitude, when our knowledge of naturally occurring substances of related structure is still incomplete, and when many of these substances are still unavailable for testing. The recent studies of Hunter and Dunn[20] show the presence of 120 ganglioside species in mouse and chick brain, and most of these are not yet characterized structurally.

Other considerations with regard to specificity of antibodies include the class of antibody (IgG and IgM molecules react differently) and improvements in specificity of polyclonal antibodies that can be achieved by affinity chromatography. In our laboratory an enormous effort was required to get rabbit anti-GM1 ganglioside antibodies pure by affinity chromatography, since the procedures described in the literature resulted in the concentration of a contaminant (100K protein) from rabbit serum that could not be readily removed (S. P. Mahadik and V. Ciccarone, unpublished).

Solid phase methods are now the most convenient for studies of antibodies to gangliosides, but all the variables associated with adsorption of gangliosides and immunoglobulins to a plastic surface are still not well understood. Using the ELISA method, our rabbit antisera raised recently against total brain ganglioside or GM1 showed relatively poor specificity, but affinity purified antibodies eluted from GM1 liposomes reacted only with GM1, GD1b and asialo GM1 (Table 2), reproducing the specificity previously found for native antisera by complement-fixation.[8] Attention has been called to the fact that when measuring antibody using ELISA (and presumably other solid phase) techniques antibody affinity may be more important than antibody concentration.[21]

ANTIBODIES TO GANGLIOSIDES - WHAT CAN THEY BE USED FOR?

Limitations arising from imperfect specificity of antibodies to gangliosides differ depending on the purposes for which the antibodies are to be used and the questions asked. Although this meeting has centered its attention on gangliosides, most of us agree that the function of these molecules is still more the object of a search than a recognizable fact. We are therefore making progress if we can obtain physical evidence implicating gangliosides in some biological process. The effects of antiganglioside antibodies provide such evidence even when the antibodies lack the ability to differentiate among the various molecular species.

Of several possible applications of antibodies to gangliosides (Table 3), our laboratory has had most success in showing that antibodies to GM1 act as interventive agents in blocking some CNS functions when injected intracerebrally.[22] A surprising amount of selectivity was found in vivo. Thus antibodies to gangliosides induced EEG spiking,[23,24] inhibited morphine analgesia,[25] blocked reserpine sedation, and interfered with passive avoidance learning[26] but did not noticeably affect pattern discrimination,

Table 2. Specificity of Affinity Purified[a] Antibodies to GM1 Ganglioside By Solid Phase Assay.[b]

Test Antigen[c]	Specimen 1[d]		Specimen 2[e]	
	RIA[f]	ELISA[g]	RIA	ELISA
	% Reaction		% Reaction	
GM1	100[h]	100[i]	100[j]	100[k]
GD1b	49	38	25	42
Asialo GM1	39	32	26	31
GD1a	<1	<1	<1	<1
GM2	<1	<2	<1	<1
GD3	<1	<1	<1	<1
Cytolipin H	4	<1	3	<1
Cytolipin K	2	<2	<1	<1

a. by elution from GM1 liposomes[28] b. Dynatech microtiter plates were superior to Falcon microtiter plates c. 1 µg test antigen in 100 µl ethanol was placed in each well, evaporated to dryness, and rinsed with PBS containing 0.3% gelatin d. from rabbit antiserum against total ganglioside e. from rabbit antiserum against GM1 f. using iodinated antibody in 100 µl PBS with 0.3% gelatin, 120 min., 25° g. using antibody in 100 µl PBS with 0.3% gelatin, 120 min., 25°, followed by alkaline phosphatase-conjugated goat anti-rabbit IgG, 120 min., 25° h. 100% = 870 cpm/73 ng antibody i. 100% = absorbance of 0.73 in 10 min. with 100 ng antibody j. 100% = 1180 cpm/73 ng antibody k. 100% = absorbance of 1.13 in 10 min with 100 ng antibody.

activity levels, fixed-ratio conditioning, self-stimulation, or pain threshold. In vitro, these antibodies enhanced depolarization-induced release of GABA from brain slices without significantly affecting spontaneous release of GABA or depolarization-induced release of serotonin or norepinephrine.[27] More recently, antibodies to GM1 ganglioside have been shown to inhibit sprouting induced in dorsal root ganglia by NGF[28] and to inhibit the elevation of ornithine decarboxylase activity induced by NGF in cell cultures.[29] These results implicate GM1 ganglioside in several types of functions, since the action of the antibody is presumably to block specific sites containing this ganglioside.

Several instances have already been cited in which both polyclonal and monoclonal antibodies have led to the detection of novel ganglioside structures.[1,7] More recently several additional struc-

Table 3. What Antibodies to Gangliosides Can Be Used For.

In Practice

1. To detect possible ganglioside functions
   a. In vivo: as interventive agents in study of CNS functions
   b. In vitro: by determining functional consequences of blocking specific membrane sites
   c. To elucidate the mechanism of action of biologically active agents such as toxins, hormones, viruses, growth factors, interferon, etc.

2. To detect novel ganglioside structures and follow their isolation

3. To localize gangliosides immunocytochemically

In Prospect

4. To diagnose cancer

5. To destroy melanoma and other types of cancer cells
   a. directly
   b. as carriers of anticancer drugs

6. To identify cell types
   a. cell sorting

7. To purify gangliosides by affinity chromatography

---

tures have been detected with monoclonal antibodies directed against human melanoma cells.[30] For studies of this type, lack of antibody specificity is not a serious obstacle.

Antibodies have one of their most important applications at present in immunocytochemical localization, and they have been used for this purpose to study gangliosides.[31-33] In this area of research, specificity may be critical, since very small amounts of particular ganglioside structures may be concentrated in some unusual cells. For localization of GM1 ganglioside, cholera toxin or its B subunit are probably far more specific than antibodies.[34,35,36] Antibodies were used recently to show that chol-1 (or cross-reacting gangliosides) is localized to cholinergic terminals of rat cortex and hippocampus based on complement-mediated release of neurotransmitters.[37]

Some progress is now being made suggesting that antibodies to gangliosides will become useful in the diagnosis of cancer. A monoclonal antibody against a colorectal adenocarcinoma cell line has been used to detect a tumor-associated ganglioside in the serum of patients with several different types of carcinoma.[38] The use of antiganglioside antibodies in the treatment of melanoma is suggested by studies indicating that immunotherapy with OFA-I (GM2 and GD2) enriched tumor cell vaccine induced tumor regression and prolonged survival in patients with disseminated melanoma.[39] The application of monoclonal antibodies to gangliosides and other glycoconjugates in both detection and treatment of tumor cells has been concisely presented.[17]

In concluding this brief survey, we would like to pay homage to the chemists who have developed and are developing the methods for detecting gangliosides, for determining their structures, and for isolating them in workable amounts. Although substantial progress has been made in developing immunological tools that can help us to relate ganglioside structure to biological function or to understand pathological processes, these tools depend on a solid foundation of chemical information. Without this foundation, our immunological observations, applied to such a rapidly expanding field of closely related structures, might only bedevil us.

SUMMARY

A brief survey has been presented addressed to 3 questions frequently asked about antibodies to gangliosides. 1) How do you get them? 2) How specific are they? 3) What can they be used for?

ACKNOWLEDGMENT

We wish to thank Dr. S. Hakomori for his courtesy in making his manuscript[17] available in advance of its publication.

REFERENCES

1. P. J. Richardson, J. H. Walker, R. T. Jones, and V. P. Whittaker, Identification of a cholinergic-specific antigen chol-1 as a ganglioside, J. Neurochem. 38:1605 (1982).
2. G. S. Eisenbarth, F. S. Walsh, and M. Nirenberg, Monoclonal antibody to a plasma membrane antigen of neurons, Proc. Natl. Acad. Sci. USA 76:4913 (1979).
3. C. S. Pukel, K. O. Lloyd, L. R. Trabassos, W. G. Dippold, H. F. Oettgen, and L. F. Old. GD3, A prominent ganglioside of human melanoma: detection and characterization by a mouse monoclonal antibody, J. Exp. Med. 155:1133 (1982).

4. E. Nudelman, S. Hakomori, R. Kannagi, S. Levery, M. Y. Yeh, K. E. Hellstrom, and I. Hellstrom, Characterization of a human melanoma-associated ganglioside antigen defined by a monoclonal antibody, 4.2, J. Biol. Chem. 257:12752 (1982).
5. L. D. Cahan, R. F. Irie, R. Singh, A. Cassidenti, and J. C. Paulson, Identification of a human neuroectodermal tumor antigen (OFA-I-2) as ganglioside GD2, Proc. Natl. Acad. Sci. USA 79:7629 (1982).
6. T. Tai, J. C. Paulson, L. D. Cahan, and R. F. Irie, Ganglioside GM2 as a human tumor antigen (OFA-I-1), Proc. Natl. Acad. Sci. USA 80:5392 (1983).
7. J. L. Magnani, B. Nilsson, M. Brockhaus, D. Zopf, Z. Steplewski, H. Koprowski, and V. Ginsburg, A monoclonal antibody-defined antigen associated with gastrointestinal cancer is a ganglioside containing sialylated lacto-N-fucopentaose II, J. Biol. Chem. 257:14365 (1982).
8. M. M. Rapport, L. Graf, Y. L. Huang, W. Brunner, and R. K. Yu, Antibodies to total brain gangliosides: titer and specificity of antisera, in: "Structure and Function of Gangliosides", L. Svennerholm, P. Mandel, H. Dreyfus, and P. F. Urban, eds., Plenum Press, New York, p. 327 (1980).
9. M. Naiki, D. M. Marcus, and R. Ledeen, Properties of antisera to ganglioside GM1 and asialo GM1, J. Immunol. 113:84 (1974).
10. N. A. Gregson and C. T. Hammer, Antibodies against defined nerve cell components: gangliosides, J. Roy. Soc. Med. 73:501 (1980).
11. S. K. Kundu, D. M. Marcus, and R. W. Veh, Preparation and properties of antibodies to GD3 and GM1 gangliosides, J. Neurochem. 34:184 (1980).
12. I. Mioshi, Y. Fujii, and M. Naiki, Avian antisera to various gangliosides: detection by enzyme immunoassay, J. Biochem. 92:89 (1982).
13. T. A. Pascal, A. Saifer, and J. Gitlin, Immunochemical studies on normal and Tay-Sachs brain gangliosides, Proc. Soc. Exp. Biol. Med. 121:739 (1966).
14. R. A. Laine, G. Yogeeswaran, and S. Hakomori, Glycosphingolipids covalently linked to agarose gel or glass beads, J. Biol. Chem. 249:4460 (1974).
15. R. I. Jacobson, N. Kasai, F. F. Richards, and R. K. Yu, Preparation of anti-GM4 antiserum and its assay by a solid phase radioimmunoassay, J. Neuroimmunol. 3:225 (1982).
16. N. A. Gregson and C. T. Hammer, Some immunological properties of antisera raised against the trisialoganglioside GT1b, Molec. Immunol. 19:543 (1982).
17. S. Hakomori, Monoclonal antibodies directed to cell surface carbohydrates, in: "Hybridoma and Functional Cell Lines", R. H. Kennett, K. B. Bechtol, and J. McKearn, eds., Plenum Press, New York (1984).
18. R. K. Yu and N. Kasai, The monoclonal antibody A2B5 is specific to ganglioside GQ1c, J. Neurochem. 41:S 122B (1983).

19. P. Fredman, J. L. Magnani, G. B. Grunwald, G. D. Trisler, M. Nirenberg, and V. Ginsberg, Monoclonal antibodies A2B5(105) and 18B8 react with many gangliosides in neuronal tissue, in: "Glycoconjugates, Proc. 7th Internat. Symp.", M. A. Chester, D. Heinegard, A. Lundblad, and S. Svensson, eds., Lund-Ronneby, Sweden, p. 242 (1983).
20. G. D. Hunter and A. J. Dunn, Ganglioside patterns of the embryonic chick optic tectum: molecular heterogeneity and developmental time course, J. Neurochem. 41:S 151B (1983).
21. J. E. Butler, T. L. Feldbush, P. L. McGirern, and N. Stewart, The enzyme-linked immunosorbent assay (ELISA): a measure of antibody concentration or affinity?, Immunochem. 15:131 (1978).
22. M. M. Rapport, S. E. Karpiak, and S. P. Mahadik, Biological activities of antibodies injected into the brain, Fed. Proc. 38:2391 (1979).
23. S. E. Karpiak, S. P. Mahadik, L. Graf, and M. M. Rapport, An immunological model of epilepsy: seizures induced by antibodies to GM1 ganglioside, Epilepsia 22:189 (1981).
24. S. E. Karpiak, Y. L. Huang, and M. M. Rapport, Immunological model of epilepsy. Epileptiform activity induced by fragments of antibody to GM1 ganglioside, J. Neuroimmunol. 3:15 (1982).
25. S. E. Karpiak, Antibodies to GM1 ganglioside inhibit morphine analgesia, Pharm. Biochem. Behav. 16:611 (1982).
26. S. E. Karpiak and M. M. Rapport, Inhibition of consolidation and retrieval stages of passive-avoidance learning by antibodies to gangliosides, Behav. Neur. Biol. 27:146 (1979).
27. B. Frieder and M. M. Rapport, Enhancement of depolarization-induced release of GABA from brain slices by antibodies to ganglioside, J. Neurochem. 37:634 (1981).
28. M. Schwartz and N. Spirman, Sprouting from chicken embryo dorsal root ganglia induced by nerve growth factor is specifically inhibited by affinity-purified antiganglioside antibodies, Proc. Natl. Acad. Sci. USA 79:6080 (1982).
29. F. J. Roisen, M. M. Rapport, Y. Huang, and G. Yorke, Ganglioside mediation of NGF enhanced neuronal maturation, Soc. Neurosci. Abstr. 9:615 (1983).
30. J. Holgersson, K. -A. Karlsson, P. Karlsson, M. Ringqvist, N. Stromberg, J. Thurin, M. Blaszczyk, M. Herlyn, Z. Steplewski, and H. Koprowski, Mouse monoclonal antibodies with specificity for human melanoma cells are directed against hitherto unknown disialogangliosides, in: "Glycoconjugates, Proc. 7th Internat. Symp.", M. A. Chester, D. Heinegard, A. Lundblad, and S. Svensson, eds., Lund-Ronneby, Sweden, p. 856 (1983).
31. C. De Baecque, A. B. Johnson, M. Naiki, G. Schwarting, and D. M. Marcus, Ganglioside localization in cerebellar cortex: an immunoperoxidase study with antibody to GM1 ganglioside, Brain Res. 114:117 (1976).

32. H. Laev, M. M. Rapport, S. P. Mahadik, and A. J. Silverman, Immunohistological localization of ganglioside in rat cerebellum, Brain Res. 157:136 (1978).
33. M. C. Raff, K. L. Fields, S. Hakomori, R. Mirsky, R. M. Pruss, and J. Winter, Cell-type specific markers for distinguishing and studying neurons and the major classes of glial cells in culture, Brain Res. 174:283 (1979).
34. H. -A. Hansson, J. Holmgren, and L. Svennerholm, Ultrastructural localization of cell membrane GM1 ganglioside by cholera toxin, Proc. Natl. Acad. Sci. USA 74:3782 (1977).
35. M. Willinger and M. Schachner, GM1 ganglioside as a marker for neuronal differentiation in mouse cerebellum, Dev. Biol. 74:101 (1980).
36. M. Willinger, The expression of GM1 ganglioside during neuronal differentiation, in: "Gangliosides in Neurological and Neuromuscular Function, Development and Repair", M. M. Rapport and A. Gorio, eds., Raven Press, New York, p. 17 (1981).
37. P. J. Richardson, Presynaptic distribution of the cholinergic specific antigen chol-1 and 5'-nucleotidase in rat brain, as determined by complement-mediated release of neurotransmitters, J. Neurochem. 41:640 (1983).
38. J. Holmgren, L. Lindholm, O. Nilsson, T. Lagergard, B. Persson, and L. Svennerholm, Detection of CA-50, a monoclonal antibody-defined cancer-associated antigen, in serum from tumor patients, in: "Glycoconjugates, Proc. 7th Internat. Symp.", M. A. Chester, D. Heinegard, A. Lundblad, and S. Svensson, eds., Lund-Ronneby, Sweden, p. 850 (1983).
39. S. S. Ahn, R. F. Irie, T. H. Weisenburger, P. C. Jones, G. Juillard, D. J. Roe, and D. L. Morton, Humoral immune response to intralymphatic immunotherapy for disseminated melanoma: correlation with clinical response, Surgery 92: 362 (1982).

USE OF CELL CULTURES IN GANGLIOSIDE RESEARCH

Paul Mandel, Henri Dreyfus, Yoshiki Matsui, and
Gerard Rebel

Centre de Neurochimie du CNRS
5, rue Blaise Pascal, 67084 Strasbourg Cedex, France

INTRODUCTION

Gangliosides are present in a variety of cell types: one might thus expect that they are involved in some common functions. In brain the existence of a great variety of gangliosides may suggest specific functions due to their occurrence in different cell populations and structures. Moreover, gangliosides are membrane constituents, partly embedded in the bilayer of the membrane and partly exposed to the external environment with negatively charged reactive polysaccharide chains.

In view of the wide distribution and membrane localization, several hypotheses were elaborated suggesting the involvement of gangliosides in:

1) cell surface phenomena: receptor functions, cell-cell interaction, structural plasticity and binding properties;

2) cell differentiation and communication through exogenous signals received;

3) some pathological phenomena including metabolic alterations and tumoral growth.

In the central nervous system (CNS), in addition to functions common to other organs, attention has been mainly focused on the role of gangliosides in cell surface phenomena including receptor functions,[1-3] in cation binding such as Ca^{++} transport and release processes,[4] in neurotransmission[4,5] in view of the abundance of gangliosides in plasma and synaptic membranes,[6] and in storage diseases due to alterations of ganglioside metabolism in some genetic disorders.[7]

Nevertheless, the molecular mechanisms involving ganglioside functions and the control of their metabolism remain often hypothetical or under critical evaluation (see for review ref. 8). In fact, it is difficult to achieve in any organ, and particularly so in the CNS, which is morphologically and functionally heterogeneous, an unequivocal image of the metabolism and functions of different types of ganglioside. Cell cultures may provide a model system for studying molecular mechanisms of potential ganglioside functions on a rather homogeneous cell population in a well defined medium. Moreover, monolayer cell cultures offer a unique advantage for investigating the surface components, for instance the ectoenzymes.[9] In this paper, we will discuss some aspects of ganglioside pattern and metabolism (enzymology). Attention will be focused on the involvement in neuronal specificity, neuron-glial interaction and cell transformation. The potential effects of gangliosides on nerve cell cultures will be discussed in this volume.

Three types of nerve cell cultures will be considered:

1) Primary neuronal cultures (homogeneous or with neuronal predominance)

2) Primary astrocytic and oligodendroglial cultures (homogeneous or with predominance of one species)

3) Clonal cell lines, tumoral or transformed cells.

GANGLIOSIDE PATTERNS IN PRIMARY CULTURES

In neurons

Ganglioside pattern was investigated in pure neuronal cultures obtained from chick,[10] rat[11,12] or human embryonic brain.[13]

Chick neurons could be maintained in pure culture for 7 days. The quantity of gangliosides increased 10 fold on DNA basis from the 1st to the 7th day (5.6 fold from day 3 to day 7). Rather low values of neuraminic acid (NeuAc) per DNA unit were attained. It is likely that it may be due to the absence of glial cells and/ or to the lack of functional activity. Nevertheless, similar low amounts of lipid NeuAc were found in cell cultures from embryonic rat brain enriched in neurons where some astrocytes were present.[12]

All common species of gangliosides described in brain homogenates were present in the cultured neurons, especially di-, tri-, tetra-sialogangliosides and pentasialogangliosides (Table 1). The main qualitative changes observed in ganglioside pattern during chick neuron differentiation in culture concerned a decrease of GD3, which is the major ganglioside before the appearance of synaptic junctions, and the increase of GD1a, tri-, tetra- and penta-sialogangliosides. This phenomenon might be considered as a major change from "embryonic" to "mature" cell. One may assume that increase of gangliosides from 3 to 7 day old chick neuron cultures correlates with modifications of the plasma membrane. This assumption was confirmed by the development of the activities of glycosyl- and sialyl-transferases involved in ganglioside biosynthesis.[14] A striking increase of trisialyl and tetrasialyl derivatives to about 40% of the cellular gangliosides was also observed after 3 weeks in neuron-enriched cultures from rat embryonic brain.[11] The pattern of major gangliosides found in cultured embryonic rat brain at 10 days of culture was very close to that of the 5 day chick-neuron cultures.[15]

In glial cells

Glial cells survived in culture for much longer periods of time than neurons. However, their differentiation was also much slower and depended strongly on the culture medium and probably on the presence of some growth factors. Astrocytes from newborn animals were easily maintained in cultures. After a few days, neurons disappeared and a pure glial cell culture could be obtained. In a 20-day culture of chick embryo glial cells the

Table 1. Ganglioside Pattern of Neurons Derived From Chick Embryo Brain During Maturation in Culture.

Gangliosides	Days in culture		
	3	5	7
GM3	2.8	2.4	3.4
GM2	--	0.6	2.8
GM1	1.2	1.3	3.7
GD3	36.6	26.6	20.0
GD1a	15.9	15.6	23.9
GT1a + GD2	6.3	5.0	3.3
GD1b	16.5	13.9	13.2
GT1L + GT1b	14.1	19.9	17.3
GQ1	5.2	11.0	8.0
GP	1.4	3.7	4.4
Total: nmol/mg DNA	43.8	184.7	350.5

Results are expressed as the percentage of total ganglioside-NeuAc.

variety of gangliosides detected was similar to that of neurons. However, the amount of GD1a and of trisialo and tetrasialo-gangliosides was much lower while GM3 occurred in a much higher amount.[15] Only traces of tetrasialogangliosides could be detected. The presence of tri- and tetra-sialogangliosides in glial cells was not due to neuronal contamination since no neurons could be detected in these cultures. The ganglioside pattern of rat astrocytes obtained from newborn rats after 16 days of culture was rather immature with a low amount of tri- and tetra-sialogangliosides and rather high amounts of GM3. It is also likely that the relatively high amount of GM3 found in some neuronal cultures could be related to the presence of glial cells. Nevertheless, in subcultures much higher amounts of GD1a and some trisialogangliosides could be found.[12] It is noteworthy that depending on the origin of cultured astroblasts (embryonic, newborn, young, or adult brain) the ganglioside pattern may differ, suggesting again that several factors or regulatory mechanisms which control ganglioside biosynthesis may be missing in the culture medium. The ganglioside pattern of oligodendrocytes isolated from white matter of three- to six-month old lamb brains and maintained in culture was characterized by the presence as major gangliosides of GM3, GM1, GD3 and GD1a. During the first ten days there was little incorporation of [^{3}H] galactose in gangliosides, but this was followed by a stimulation of uptake for the major gangliosides reaching a maximum after approximately 25-30 days in vitro. There was little incorporation into GM2 or trisialogangliosides through the life of the cultures.[16]

As compared to neurons, the distribution of gangliosides in cultured mouse glial cells included much less trisialo- and tetrasialogangliosides and higher amounts of disialogangliosides. In cocultures of neurons and glial cells from embryonic mouse brain there was a slight increase of GD1a which was higher than in either neurons or glial cells.[15]

GANGLIOSIDE PATTERN IN CLONAL CELL LINES

Neuroblastoma and transformed neuronal cells

As reported by several authors,[15,17-20] the ganglioside pattern of neuronal clonal cell lines from either tumors or transformed nerve cells differs strikingly from that of neurons isolated from brain by cellular fractionation or maintained in primary cultures. In all clonal cell lines except the rat pheochromocytoma clone PC 12,[2] the tri- and tetra-sialogangliosides are missing.[15,17-19]

Another difference between neurons and neuroblastoma lies in the high percentage of the two monosialogangliosides, GM3 and GM2;[15,17] sometimes high amounts of GD1a could be found.[15] The decrease among total gangliosides and the simplification of the pattern was ascribed to the malignant transformation.[20,21] The disappearance of tri- and tetra-sialogangliosides may be due either to a decrease of synthesis or an increase of degradation.[20-23] One cannot exclude that some clones or tumoral cells have retained the ability to synthesize trisialogangliosides but that these lipids cannot be incorporated in cell membranes, being rapidly catabolized.

One may expect that the morphological differentiation of neuroblastoma cells induced either by withdrawal of the serum from the culture medium or by cyclic AMP or bromodeoxy-uridine (BrDU) would produce more neuronal type distribution of gangliosides. This is not the case (ref. 17, Table 2), although sometimes these types of differentiation are accompanied in some clones by development of numerous long processes and great increases in the synthesis of neurotransmitters and in cellular respiration and excitability.[17,24,25] It seems that dibutyryl cyclic AMP produces more extensive biochemical and morphological differentiation than serum withdrawal or BrDU; the amount of gangliosides is increased to a greater extent. Moreover, several changes were observed in the mono- and di-sialoganglioside distribution after treatment: for instance an increase of GM1 and sometimes in GD_{1a}. However, trisialogangliosides only exceptionally appear.[2,15] Apparently,

Table 2. Gangliosides of Proliferating and Serum Deprived Clones.

Clone	NIE 115		M1			
	Prol.	Diff.	Prol.	Diff.		
				S.D.	DBcAMP	BrDU
Tot. lip. S.A.[a] μg/g dry weight	282 ± 18	413 ± 25	498 ± 18	630 ± 21	736 ± 21	564 ± 23
		% of total gangliosides[a]				
GM3	4.7±0.5	–	5.7±0.9	1.1±0.3	20.5±2.1	1.0±0.7
GM2	63.9±4.1	23.6±3.3	23.9±3.1	24.7±3.9	26.3±1.9	17.3±2.0
GM1	8.7±2.5	14.2±3.7	12.6±4.1	17.3±2.9	13.4±3.2	13.7±3.6
GD3	traces	15.9±2.2	–	1.5±0.5	traces	2.8±1.0
GD1a	22.7±2.3	46.3±2.6	53.1±3.9	51.7±4.2	39.1±2.8	57.4±5.4
GD1b	–	–	4.7±0.3	3.7±0.5	0.5±0.3	7.8±2.2
GT1	–	–	–	–	–	–
GQ1	–	–	–	–	–	–

[a]Each value is the mean of three experiments ± standard deviation
Prol.: proliferation; Diff.: differentiation; S.D.: serum deprived.

only the neuroblastoma-glioma hybrid binds just a discernable amount of labelled tetanos toxin[26] which characterizes neurons. In cocultures of neuroblastoma or transformed neuronal cells with transformed glial cells trisialogangliosides sometimes appear. It is also noteworthy that reisolated neuroblastoma cells may retain the capacity of synthezising small quantities of trisialogangliosides.[15]

Glioma and transformed glial cells

Tumoral or transformed glial cells from different origin, (mouse,[15] rat,[12] hamster[27,28] and human[29]) were used. Two clones of glial cells were most frequently investigated: spontaneously transformed hamster astrocytes, the clone NN, and glioblasts originating from a rat glioma clonal line C6. They did not undergo a morphological differentiation when treated either with dibutyryl cyclic AMP or by serum removal.[27] When NN cells were grown in the presence of BrDU the morphology became closer to that of mature glioblasts and notable changes in ganglioside distribution were observed; namely an increase of GM1 and GD1a and a decrease of GM3. High GM3 levels were reported in tissue cultures of rat and mouse glial tumors, in hamster astroglia and rat astroblasts. Synthesis of large amounts of GM3 and GD3, but not of GM1 and GD1a, was reported in C6 glioblastoma and G2620 oligodendroglioma.[19,27,28]

ECTOGLYCOSYL TRANSFERASE ACTIVITIES AT THE SURFACE OF CULTURED NEURONS

The properties of enzyme systems that catalyze the formation and degradation of gangliosides were largely described.[3,16,19,20,30-34] Nevertheless, our knowledge of regulatory mechanisms involved in the functional activities of gangliosides is rather limited.

An increasing amount of evidence suggests that glycoconjugates, in particular surface sialoconjugate glycosyl transferases,[3,30] may play an important role in cell to cell recognition and interaction and in contact inhibition. Plasma membrane gangliosides may also be involved in cell differentiation and in carcinogenesis as suggested by the observed loss of polysialogangliosides in tumoral cells.[8,15] We investigated glycosyl transferase activities: ectogalactosyl, ectofucosyl and ectosialyl transferases at the external surface of exclusively neural or glial (astrocyte) cultures. An appropriate methodology was developed to eliminate potential sources of error put forward in investigations on ectoenzymes. The sources of error due to the hydrolysis of nucleotide-sugar substrates or due to cellular uptake of free sugars were eliminated. Ovomucoid and asialofetuin coupled to Sepharose and Ultrogel beads were used as exogenous substrates to avoid a possible substrate pinocytosis. Ectoglycosyl transferase activities were studied as functions of protein concentration, incubation time and amount of bead-coupled exogenous acceptors. The effect of an eventual release of cellular enzymes was taken into consideration. The data showed that galactosyl, fucosyl, and sialyl transferase activities were present at the external surface of the neuronal membrane.[9]

Until now, in contrast to the results obtained with neurons, the ectoenzyme activities of cultured astrocytes appeared to be rather weak.[34] It may be related to an insufficiency of differentiation of the cells used. The problem is under investigation.

DISCUSSION

The investigations on nerve cell cultures provided much useful information. However, we have to be aware of some weaknesses of this approach. It is of great interest to know the specific ganglioside patterns of pure neuronal and pure glial cell cultures. Nevertheless, the question can be raised whether pure neuronal or glial cell cultures provide the whole desirable information. Actually, homogeneous cell cultures lack the neuron-glia interaction which probably plays an important role; the ganglioside pattern of neuron-glia cocultures differs from that of individual cultures. Moreover, the metabolism may differ depending on the culture medium and the presence or absence of growth and differentiation factors not yet identified. This seems particularly the case for astrocytes in primary cultures. Thus, the observations in tissue culture have to be critically controlled.

The enzymes involved in the metabolism of gangliosides are quite well known. We need more information on the specific regulatory mechanisms which occur in neurons and glial cells and which underlie the biosynthesis and degradation at the transcription and translational levels. Here again, nerve cell cultures may provide advantages.

But except for the case of cholera toxin, we ignore the sequence of events between the reception of an exogenous signal by plasma membrane gangliosides and the effector sites. The cultures can be used to obtain further information, for instance concerning the problems of the effects of the toxin on both cell types, neurons and glia, possessing GM1.

The discovery of ectoenzymes which may be involved in ganglioside synthesis raises several questions: the source of the precursor nucleotide sugars, the mechanisms which initiate and stop the reaction, the possibility that ectoenzymes are involved in plasma membrane repair mechanisms, the nature of the signals which start the repair.

One of the most exciting questions is the involvement of gangliosides in carcinogenesis and specifically in contact inhibition. Apparently the simplification of the ganglioside pattern seems related to tumorigenesis or cell transformation. Nevertheless, we should not forget that some tumoral cells possess trisialogangliosides (PC 12). There are some data in favor of an alteration of biosynthesis, some other of the degradation or alteration of the integration processes in the membrane.[8,21] Here again, investigations using nerve cell cultures may be very useful.

During the last few years unexpected effects of exogenous gangliosides or ganglioside antibodies on cultured nerve cells were reported (see for reviews papers in this volume); the mechanisms involved are rather obscure.

Finally, several pieces of evidence suggest the involvement of gangliosides in neurotransmitter functions[4,14,35] and in drug (opiate) effects[36] which could also be investigated in tissue culture.

REFERENCES

1. J. Holmgren, H. Helwing, P. Fredman, O. Strannegard, and L. Svennerholm, Gangliosides as receptors for bacterial toxins and Sendai virus, in: "Structure and Function of Gangliosides," L. Svennerholm, P. Mandel, H. Dreyfus, and P. F. Urban, eds., Plenum Press, New York (1980).
2. W. Seifert, Gangliosides in nerve cell cultures, in: "Gangliosides in Neurological and Neuromuscular Function, Development and Repair," M. M. Rapport and A. Gorio, eds., Raven Press, New York (1981).
3. B. D. Schur and S. Roth, Cell surface glycosyltransferases, Biochim. Biophys. Acta 415:473 (1975).
4. L. Svennerholm, Gangliosides and synaptic transmission, in: "Structure and Function of Gangliosides," L. Svennerholm, P. Mandel, H. Dreyfus, and P. F. Urban, eds., Plenum Press, New York (1980).
5. H. Rahmann, Functional implication of gangliosides in synaptic transmission, Neurochem. Int. 5:549 (1983).
6. R. W. Ledeen, Ganglioside structures and distribution: are they localized at the nerve ending?, J. Supramol. Struct. 8:1 (1978).
7. K. Sandhoff, K. Harzer, W. Waessle, and H. Jatzkewitz, Enzyme alterations and lipid storage in three variants of Tay-Sachs disease, J. Neurochem. 18:2469 (1971).
8. P. H. Fishman and R. O. Brady, Biosynthesis and function of gangliosides, Science 194:906 (1976).
9. Y. Matsui, D. Lombard, B. Hoflack, S. Harth, R. Massarelli, P. Mandel, and H. Dreyfus, Ectoglycosyltransferase activities at the surface of cultured neurons, Biochem. Biophys. Res.

Commun. 113:446 (1983).
10. H. Dreyfus, J. C. Louis, S. Harth, and P. Mandel, Gangliosides in cultured neurons, Neuroscience 5:1647 (1980).
11. E. Yavin and Z. Yavin, Ganglioside profiles during neural tissue development, Develop. Neurosci. 2:25 (1979).
12. G. Rebel, M. Mersel, and P. Mandel, in preparation.
13. L. M. Hoffman, S. E. Brooks, and L. Schneck, Human fetal brain cells in culture, Biochim. Biophys. Acta 665:359 (1981).
14. H. Dreyfus, S. Harth, R. Massarelli, and J. C. Louis, Mechanisms of differentiation in cultured neurons: involvement of gangliosides, in: "Gangliosides in Neurological and Neuromuscular Function, Development and Repair," M. M. Rapport and A. Gorio, eds., Raven Press, New York (1981).
15. P. Mandel, H. Dreyfus, A. N. K. Yusufi, L. Sarlieve, J. Robert, N. Neskovic, S. Harth, and G. Rebel, Neuronal and glial cell cultures, a tool for investigation of ganglioside function, in: "Structure and Function of Gangliosides," L. Svennerholm, R. Mandel, H. Dreyfus, and P. F. Urban, eds., Plenum Press, New York (1980).
16. S. R. Mack, S. Szuchet, and G. Dawson, Synthesis of gangliosides by cultured oligodendrocytes, J. Neurosci. Res. 6:361 (1981).
17. J. Ciesielski-Treska, J. Robert, G. Rebel, and P. Mandel, Gangliosides of active and inactive neuroblastoma clones, Differentiation 8:31 (1977).
18. G. Yogeeswaran, R. K. Murray, M. L. Pearson, B. D. Sanwal, F. A. McMorris, and F. H. Ruddle, Glycosphingolipids of clonal lines of mouse neuroblastoma and neuroblastoma X L cell hybrids, J. Biol. Chem. 248:1231 (1972).
19. G. Dawson, S. F. Kemp, A. C. Stoolmiller, and A. Dorfman, Biosynthesis of glycosphingolipids by mouse neuroblastoma (NB41A), rat glia (RGC-6) and human glia (CHB-4) in cell culture, Biochem. Biophys. Res. Commun. 44:687 (1971).
20. S. F. Kemp and A. C. Stoolmiller, Studies on the biosynthesis of glycosphingolipids in cultured mouse neuroblastoma cells: characterization and acceptor specificities of N-acetylneuraminyl and N-acetylgalactosaminyltransferases, J. Neurochem. 27:723 (1976).
21. S. Hakomori, Glycosphingolipids in cellular interaction, differentiation and oncogenesis, Ann. Rev. Biochem. 50:733 (1981).
22. F. A. Cumar, R. O. Brady, E. H. Kolodny, V. W. McFarland, and P. T. Mora, Enzymatic block in the synthesis of gangliosides in DNA virus-transformed tumorigenic mouse cell lines, Proc. Natl. Acad. Sci. USA 67:757 (1970).
23. P. H. Fishman, R. O. Brady, R. M. Bradley, S. A. Aaronson, and G. J. Torado, Absence of a specific ganglioside galactosyltransferase in mouse cells transformed by murine sarcoma virus, Proc. Natl. Acad. Sci. USA 71:298 (1974).
24. W. Dimpfel, W. Möller, and U. Mengs, Ganglioside-induced neurite formation in cultured neuroblastoma cells, in: "Gangliosides in Neurological and Neuromuscular Function, Development and

Repair," M. M. Rapport and A. Gorio, eds., Raven Press, New York (1981).
25. P. Mandel, J. Ciesieski-Treska, and M. Sensenbrenner, Neurons in vitro, in: "Molecular and Functional Neurobiology," W. H. Gispen, ed., Elsevier, Amsterdam (1976).
26. W. Dimpfel, R. T. C. Huang, and E. Habermann, Gangliosides in nervous tissue cultures and binding of ^{125}I-labelled tetanus toxin, a neuronal marker, J. Neurochem. 29:329 (1977).
27. J. Robert, G. Rebel, and P. Mandel, Glycosphingolipids from cultured astroblasts, J. Lipid Res. 18:517 (1977).
28. G. Rebel, J. Robert, and P. Mandel, Glycolipids and cell differentiation, in: "Structure and Function of Gangliosides," L. Svennerholm, P. Mandel, H. Dreyfus, and P. F. Urban, eds., Plenum Press, New York (1980).
29. W. Seifert, Gangliosides in nerve cell cultures, in: "Gangliosides in Neurological and Neuromuscular Function, Development and Repair," M. M. Rapport and A. Gorio, eds., Raven Press, New York (1981).
30. S. Roth, E. J. McGuire, and S. Roseman, Evidence for cell-surface glycosyltransferases, J. Cell Biol. 51:536 (1971).
31. P. Stoffyn and A. Stoffyn, Biosynthesis in vitro of mono- and di-sialogangliosides from gangliotetraosylceramide by cultured cell lines and young rat brain, Carbohydrate Res. 78:327 (1980).
32. P. W. Robbins and I. MacPherson, Control of glycolipid synthesis in a cultured hamster cell line, Nature 229:569 (1971).
33. J. R. Moskal, D. A. Gardner, and S. Basu, Changes in glycolipid glycosyltransferases and glutamate decarboxylase and their relationship to differentiation in neuroblastoma cells, Biochem. Biophys. Res. Commun. 61:751 (1974).
34. H. Dreyfus, Y. Matsui, and P. Mandel, Ectogylcosyltransferase activities of astrocytes and gliomas cells, in preparation.
35. J. C. Louis, B. Pettman, J. Courageot, J. F. Rumigny, P. Mandel, and M. Sensenbrenner, Developmental changes in cultured neurons from chick embryo cerebral hemispheres, Exp. Brain Res. 42:63 (1981).
36. G. Dawson, R. McLawhon, and R. J. Miller, Opiates and enkephalins inhibit synthesis of gangliosides and membrane glycoproteins in mouse neuroblastoma cell line N4TG1, Proc. Natl. Acad. Sci. USA 76:605 (1979).

Plenary Lectures

GANGLIOSIDES: STRUCTURE AND ANALYSIS

Robert K. Yu

Department of Neurology
Yale University School of Medicine
New Haven, CT 06510 U.S.A.

INTRODUCTION

It was nearly half a century ago that Professor E. Klenk first isolated gangliosides from the brain of a Niemann-Pick patient[1] and later from the brain of a Tay-Sachs patient.[2] Subsequently, these substances were isolated from normal brains,[3,4] and the name "gangliosides" was proposed by Professor Klenk to reflect their glycosidic nature and apparent localization in the ganglion cells (Ganglienzellen) of the brain. Early structural studies by Klenk and other investigators also established the presence of sialic acid in the oligosaccharide chain. However, the detailed chemical structure of a ganglioside was not determined until the early 50's and 60's when Yamakawa, Klenk and their co-workers isolated hematoside from horse erythrocytes and elucidated its structure.[5,6] In 1963, Kuhn and Wiegandt[7,8] determined the structure of the four major mammalian brain gangliosides, namely G_I, G_{II}, G_{III} and G_{IV}, now more widely known as G_{M1}, G_{D1a}, G_{D1b} and G_{T1b}, respectively.[9] Shortly thereafter, the correct structure of the major ganglioside in Tay-Sachs brain, G_{M2}, was proposed by Makita and Yamakawa[10] (the asialo form) and proved by Ledeen and Salsman.[11] Finally, the α-D ketosidic configuration of the sialic acid moiety in gangliosides was established by Yu and Ledeen in 1969.[12]

With improved isolation and purification procedures, many new ganglioside structures were discovered in neural and extraneural tissues and body fluids. To-date, more than 60 different gangliosides have been characterized. For a complete compilation of these structures, the readers can refer to several comprehensive reviews[13-17] and monographs[18-20] that have appeared in recent years.

With few exceptions, all gangliosides isolated from vertebrate tissues can be classified into the following series according to the oligosaccharide structures shown in Table 1.

The characterization of the diverse oligosaccharide structures in various gangliosides still remains a challenge to biochemists. In order to characterize their structures, it is necessary to determine their (A) carbohydrate composition, (B) carbohydrate sequence, (C) glycosidic substitution sites, and (D) anomeric configuration and conformation. This information can be obtained by application of various standard procedures employed by carbohydrate chemists on the intact glycolipids or on the liberated oligosaccharides. Several discussions of these procedures have

Table 1. Classification of Gangliosides

Series	Example
Gala	NeuAcα2-3Galβ1-1´Cer
	G_{M4}
Hemato	NeuAcα2-3Galβ1-4Glcβ1-1´Cer
	G_{M3}
Neolacto	NeuAcα2-3Galβ1-4GlcNAcβ1-3Galβ1-4Glcβ1-1´Cer
	L_{M1} or SPG
Globo*	NeuAcα2-3Galβ1-3GalNAcβ1-3Galα1-4Galβ1-4Glcβ1-1´Cer
Ganglio	Galβ1-3GalNAcβ1-4Galβ1-4Glcβ1-1´Cer 3 α| 2 NeuAc
	G_{M1}

*Globo-series gangliosides were characterized only recently.[21,22] A simpler globo-series ganglioside with the structure NeuAc(α2-3)GalNAc(β1-3)Gal(α1-4)Gal(β1-4)Glc-Cer was tentatively identified in human teratocarcinoma cells.[23]

appeared recently.[14,17,24] Some of the newer procedures will be outlined in this chapter.

CARBOHYDRATE COMPOSITION

For determination of sialic acid, the identifying sugar of all gangliosides, the thiobarbituric acid[25,26] and resorcinol-HCl[27] methods are probably the most widely used procedures. Several modifications that provide either enhanced sensitivity or specificity of the two colorimetric methods have been described and reviewed recently.[14,28-30]

Sialic acid, along with other sugar constituents, can also be conveniently analyzed by gas-liquid chromatography (GLC). The original method developed by Sweeley and co-workers[31-33] still enjoys wide application. The procedure employs vigorous methanolysis in 0.5-1.5 N hydrochloric acid to convert the sialic acid residues and other sugar constituents to their methyl glycosides, followed by formation of the trimethylsilyl ether derivatives for GLC analysis. Alternatively they can be analyzed as their pertrifluoroacetyl derivatives.[34,35] The major advantage of these procedures is that they afford the entire carbohydrate compositional data simultaneously. However, the vigorous acid treatment employed in cleavage of individual sugar residues also removes all O-acyl and N-acyl groups, thus destroying the identity of the original sialic acid. Yu and Ledeen[36] found that under mild acid methanolic conditions (0.05 N HCl in methanol, $80°$, 1 hr.), the sialic acid is converted to its methyl ketoside methyl esters with minimal cleavage of the N-acyl group. This observation forms the basis for a GLC method that simultaneously determines the two major forms of sialic acid, namely, NeuAc and NeuGc. The sensitivity of this procedure can be increased considerably by using mass-fragmentography.[37] Several variations of the GLC procedure for sialic acid analysis have been reviewed by Schauer.[29,30]

The discovery of gangliosides which contain O-acetylated sialic acids[38-41] requires special caution in identifying the sialic acid residue. Characterization of these sialic acids can be effected by isolation of the O-acetylated sialic acid after neuraminidase or mild acid treatment, followed by GLC or GLC-mass spectrometric analysis.[30,42] Some of the more recent advances in elucidating the various types of sialic acid by capillary column GLC, high-performance liquid chromatography, and mass spectrometry are discussed in this volume by Schauer.[43]

Neutral and aminosugars can also be analyzed by GLC as alditol and hexosaminitol acetates following borohydride reduction of the liberated monosaccharides.[44,45] The major advantage of this method is that each alditol gives rise to a single peak and is more easily resolved and quantitated than methyl glycosides. Some improvements in sample preparation and resolution have been introduced recently.[46,47] Alternatively, the carbohydrate composition can be analyzed by GLC-mass spectrometry of the oligosaccharides following their release from glycosphingolipids by trifluoroacetolysis.[48,49]. The procedure, however, is applicable only to sphingolipids containing long-chain bases with a double bond at the $\Delta^{4,5}$ position.

A novel method for oligosaccharide structure determination has been developed that simultaneously establishes the linkage and the ring form (pyranose and furanose) of the monosaccharide residue.[50] The method involves the regiospecific reductive cleavage of the glycosidic carbon-oxygen bonds in the fully methylated polysaccharide. The reductive cleavage is achieved by ionic hydrogenation, employing triethylsilane as the reducing agent and boron trifluoride etherate or trimethylsilyl trifluoromethanesulfonate as the acid catalyst in dichloromethane as solvent. Pyranosyl and furanosyl residues are converted to 1,5- and 1,4-anhydroalditols, respectively, which are subsequently acetylated in situ and analyzed by GC-mass spectrometry.

CARBOHYDRATE SEQUENCE

The most widely used method for establishing the carbohydrate sequence of a ganglioside is by partial hydrolysis of the glycolipid and analysis of the liberated sugars or the unhydrolyzed glycolipid remnants. This can be effected by a mild acetolysis procedure used originally by Kuhn and Wiegant[7,8] in their determination of the structures of the four major brain gangliosides. Alternatively, mild acid hydrolysis (5.6 mM formic acid, 80°) which partially removes sialic acid residues proves useful in assessing the arrangement of sialic acid residues in polysialogangliosides.[51-54] More rigorous treatment (0.1 to 1 M formic acid at 100°) removes all sialic acid residues with only partial degradation of the remaining oligosaccharide structure.[55-57] The ability of neuraminidase from Arthrobacter ureafaciens to remove all sialic acid residues from gangliosides may prove useful for obtaining the neutral oligosaccharide core structure.[58]

With the availability of various highly specific glycosidases, it is feasible to sequence the oligosaccharide chain by sequential application of these enzymes. This technique has been outlined in detail by Li and Li[59] and its application to ganglioside structural analysis has been reviewed by Ledeen and Yu.[14] The major advantage of this approach is that the enzyme not only establishes the identity of the terminal glycosyl unit, but also its anomeric configuration.

First applied to glycosphingolipids by Sweeley and Dawson[60,61] and Samuelsson and Samuelsson,[62] mass spectrometry has emerged as a powerful tool in furnishing detailed structural information regarding the carbohydrate composition and sequence, and in some instances, the lipophilic constituents of gangliosides. Earlier investigations in this field have relied upon either the electron impact (EI) or the chemical ionization (CI) mode for the analysis of glycolipids or the liberated oligosaccharides as their pertrimethylsilylated, peracetylated, permethylated, or pertrifluoroacetylated derivatives.[14,17,48,63-65] With the advent of new ionization methods, particularly the desorption ionization (DI) techniques,[66] it is possible to obtain mass spectra directly from nonderivatized, large, and nonvolatile biomolecules of low thermal stability. Among these procedures, the recently developed fast-atom bombardment (FAB) and field-desorption (FD) techniques are particularly useful. Their effectiveness in providing sequence information as well as molecular weights of underivatized glycosphingolipids has been clearly demonstrated.[67-73] Some of the salient features of these techniques have been discussed in detail in this volume by Egge et al.[74] and Handa and Kushi.[75] It is expected that these newer ionization techniques will rapidly gain popularity in glycosphingolipid analysis.

In recent years, many highly specific monoclonal antibodies which recognize well-defined carbohydrate structures have become available. These antibodies have already proven useful as an alternative in providing detailed structural information. For example, the monoclonal antibody which detects the stage-specific embryonic antigen 3 (SSEA-3) recognizes a number of glycolipids with the sequence R-3GalNAcβ1-3Galα1-R´.[22] The structure that is recognized by anti-SSEA-1 has the sequence Galβ1-4[Fucα1-3]GlcNAcβ1-R.[76,77] The potential of using specific monoclonal antibodies to supplement structural analysis by chemical methods has recently been advocated by Hakomori.[78]

GLYCOSIDIC SUBSTITUTION

Permethylation and periodate oxidation have been the standard chemical techniques for establishing the glycosidic substitution sites in an oligosaccharide chain. In the permethylation analysis, the most widely used procedure is that of Hakomori.[79] The procedure involves reacting glycolipids with an excess of methyl iodide in the presence of dimethylsulfinyl carbanion, generated from sodium hydride and dimethyl sulfoxide. Alternative methylation reagents have been described employing sodium hydride and methyl iodide in dimethylformamide[80,81] and potassium tert-butoxide in dimethyl sulfoxide.[82] A modification of

Hakomori's procedure has recently been described.[83] These procedures lead to complete methylation of all free hydroxyl and acetamido groups. Following hydrolysis of the permethylated glycolipids, the substituted monosaccharides are converted to partially methylated alditol acetates which are identified by GLC or GLC-mass spectrometry.[84,85] Hexosamines require a modified procedure of glycosidic cleavage, involving sequential acetolysis and hydrolysis, to ensure yield.[86] An alternative procedure utilizing methyl glycoside derivatives has succeeded in revealing substitution sites on galactosamine and glucosamine.[87,88]

Periodate oxidation is also useful in delineating the glycosidic substitution sites. The reagent specifically cleaves the carbon-carbon bonds bearing vicinal diol or similar structures. The aldehydes thus generated are frequently reduced with sodium borohydride. Following acid hydrolysis, the liberated fragments can be identified by GLC or other chromatographic means.[89] Thus, glucose substituted at the C-4 hydroxyl, but not at the C-2 or C-3 positions, gives rise to erythritol, whereas galactose similarly substituted generates threitol. These can be differentiated by GLC of the tetraacetates.[11] Either sugar substituted at the C-2 hydroxyl yields glycerol, as do unsubstituted sugars at a terminal position, whereas substitution at C-3 hydroxyl blocks oxidation. Unsubstituted fucose yields dihydroxypropane and glyceraldehyde.[90] Sialic acid with a substitution at the C-8 hydroxyl remains intact, but the terminal sialic acid residue yields the 7-carbon analog (N-acylheptulosaminic acid). Hence, periodate oxidation is particularly useful for identifying the NeuAc2-8NeuAc linkage.[52,53]

Information on glycosidic substitution sites can also be obtained by ^{1}H- or ^{13}C-NMR spectroscopy. This can be deduced from the chemical shift difference that occurs at a particular proton or carbon, due to the deshielding effect following the formation of a new anomeric linkage (glycosidation shift, $\Delta\delta$).[24,91,92] Thus, formation of NeuAc2-3Gal glycosidic linkages causes a predictable 0.2-0.6 ppm downfield shift of the Gal (H-3) proton in the ^{1}H-NMR spectra of several asialogangliosides.[92,93] However, formation of GalNAc1-4Gal and Gal1-3GalNAc glycosidic linkages causes irregular effects on the proton directly involved, which may be due to secondary structural effects. Care, therefore, must be exercised in using the ^{1}H glycosidic shift data for assignment of glycosidic substitution sites. The glycosidic shifts calculated from ^{13}C resonance spectra are far more reliable. Thus, for neutral hexopyranoside linkages the glycosidic shifts typically fall between 7-10 ppm, and for sialic acids, the values are between 2-4 ppm.[91,94,95] Confirmation of the linkage sites can also be achieved by nuclear Overhauser enhancement spectroscopy[93] or by ^{13}C-^{13}C connectivity plots.[96] The advantage of these procedures is that they are applicable to the intact oligosaccharide moiety.

ANOMERIC CONFIGURATION AND CONFORMATION

The standard method for establishing anomeric configurations of gangliosides, and of glycosphingolipids in general, has been the use of specific glycosidases as described above for sequencing. Alternatively, it can be determined by polarimetry or circular dichroism.[97] Among the various chemical methods, chromium trioxide oxidation is still in wide use. This method is based on the original observation by Angyal and James[98] that chromium trioxide in acetic acid rapidly oxidizes peracetylated hexopyranosides in which the aglycon occupies an equatorial position (e.g., β-pyranosides). The corresponding anomers in which the aglycon occupies an axial position (e.g., α-pyranosides) are oxidized only slowly, yielding 1-0-acyl derivatives which can be quantitated by GLC. On the other hand, acetylated β-furanosides are more resistant to oxidation than α-furanosides.[99] Application of this method for determination of the anomeric configuration of sugar residues in acetylated oligosaccharides has been described by Hoffman et al.[100] and Laine and Renkonen.[101]

^{1}H-NMR has fast become a popular method for determining anomeric configuration of the sugar residues because this procedure is rapid, sensitive, quantitative, and non-destructive. In most cases, the anomeric configuration of the sugar residues in glycosphingolipids can be determined from their $J_{1,2}$ coupling constants. Thus a large (>6 Hz) coupling indicates β-D linkage, and a small (<4 Hz) coupling indicates α-D linkage.[24,92,93] This applies to most D-aldohexopyranosides set in a $^4C_1$ chair conformation except D-mannose whose H-2 is in an equatorial configuration. In general, the chemical shift of the anomeric proton of α-D-mannose is downfield to that of the β-anomer. L-Fucose can be easily recognized by a distinct 3-proton doublet at 1.04 ppm due to its H-6 protons. The combination of changes in conformation ($^4C_1$ to $^1C_4$) and absolute stereochemistry (D to L) cancel each other, with the net result of a small coupling constant (<4 Hz) indicating α-L-fucose.

Elucidation of the anomeric configuration by ^{13}C-NMR is relatively difficult because of the low natural abundance of this isotope. The best approach is from the one-bond coupling for the anomeric carbon ($^1J_{C1,H1}$). The difference in the couplings for the two anomeric configurations is generally about 10 Hz with the value for the equatorial ^{13}C-^{1}H coupling being the larger. The ^{13}C-NMR spectra of several gangliosides and related glycosphingolipids have been assigned.[91,94,95,102-106] The β-pyranosyl conformation of D-glucosylceramide has also been demonstrated by this technique.[103] For a detailed discussion of the use of ^{1}H- and ^{13}C-NMR to determine residue configuration and conformation, several recent reviews should be consulted.[24,93,107-110]

REFERENCES

1. E. Klenk, Über der Natur der phosphatide und anderer Lipoide des Gehirns und der Leber in Niemann-Pickscher Krankheit, Hoppe-Seyler's Z. Physiol. Chem. 235:25 (1935).
2. E. Klenk, Beiträge zur Chemie der Lipoidosen, III: Neimann-Picksche Krankheit und amaurotische Idiotie, Hoppe-Seyler's Z. Physiol. Chem. 262:128 (1939).
3. G. Blix, Einige Beobachtungen über eine hexosaminehaltinge Substanz in der Protagonfraktion des Gehirns, Skand. Arch. Physiol. 80:46 (1938).
4. E. Klenk, Uber die Ganglioside, eine neue Gruppe von zuckerhaltigen Gehirnlipoiden, Hoppe-Seyler's Z. Physiol. Chem. 273:76 (1942).
5. T. Yamakawa and S. Suzuki, The chemistry of posthemolytic residue or stroma of erythrocytes. I. Concerning the ether-insoluble lipids of lyophilized horse blood stroma, J. Biochem. 38:199 (1951).
6. E. Klenk and G. Padberg, Uber die Ganglioside von Pferdeerythrocyten, Hoppe-Seyler's Z. Physiol. Chem. 327:249 (1962).
7. R. Kuhn and H. Wiegandt, Die Konstitution der Ganglio-N-tetraose und des Gangliosides G_I, Chem. Ber. 96:866 (1963).
8. R. Kuhn and H. Weigandt, Die Konstitution der Ganglioside G_{II}, G_{III} und G_{IV}, Z. Naturforsch. 18b:541 (1963).
9. L. Svennerholm, Chromatographic separation of human brain gangliosides, J. Neurochem. 10:613 (1963).
10. A. Makita and T. Yamakawa, The glycolipids of the brain of Tay-Sachs' disease - the chemical structure of a globoside and main ganglioside, Japan J. Exp. Med. 33:361 (1963).
11. R. Ledeen and K. Salsman, Structure of the Tay-Sachs' ganglioside. I., Biochemistry 4:2225 (1965).
12. R. K. Yu and R. W. Ledeen, The glycosidic linkage of sialic acid, J. Biol. Chem. 244:1306 (1969).
13. S. -i. Hakomori, Glycosphingolipids in cellular interaction, differentiation, and oncogenesis, Ann. Rev. Biochem. 50:733 (1981).
14. R. W. Ledeen and R. K. Yu, Gangliosides: structure,

isolation, and analysis, Methods Enz. 83:140 (1982).
15. R. W. Ledeen, Gangliosides, in: Handbook of Neurochemistry, A. Lajtha, ed., Vol. 3, 2nd Ed., pp. 41-90, Plenum, New York (1983).
16. H. Wiegandt, The gangliosides, Adv. Neurochem. 4:149 (1982).
17. S. Ando, Gangliosides in the nervous system, Neurochem. Internat. 5:507 (1983).
18. E. Brunngraber, Neurochemistry of Aminosugars. Neurochemistry and Neuropathology of the Complex Carbohydrates, Charles C. Thomas, Springfield, IL (1979).
19. L. Svennerholm, P. Mandel, H. Dreyfus, and P. -F. Urban, eds., Structure and Function of Gangliosides, Plenum, N.Y. (1980).
20. A. Makita, S. Handa, T. Taketomi, and Y. Nagai, eds. New Vistas in Glycolipid Research, Plenum, N.Y. (1982).
21. J. -L. Chien and E. L. Hogan, Novel pentahexosyl ganglioside of the globo-series purified from chicken muscle, J. Biol. Chem. 258:10727 (1983).
22. R. Kannagi, S. B. Levery, F. Ishigami, S. -i. Hakomori, L. H. Shevinsky, B. B. Knowles, and D. Solter, New globo-series glycosphingolipids in human teratocarcinoma reactive with the monoclonal antibody directed to a developmentally regulated antigen, stage-specific embryonic antigen 3, J. Biol. Chem. 258:8934 (1983).
23. G. A. Schwarting, P. C. Carroll, and W. C. DeWolf, Fucosyl-globoside and sialosyl-globoside are new glycolipids isolated from human teratocarcinoma cells, Biochem. Biophys. Res. Comm. 113:935 (1983).
24. J. Dabrowski, P. Hanfland, and H. Egge, Analysis of glycosphingolipids by high-resolution proton nuclear magnetic resonance spectroscopy, Methods Enz. 83:69 (1982).
25. L. Warren, The thiobarbituric acid assay of sialic acids, J. Biol. Chem. 234:1971 (1959).
26. D. Aminoff, Method for quantitative estimation of N-acetylneuraminic acid and their application to hydrolysates of sialomucoids, Biochem. J. 81:384 (1961).
27. L. Svennerholm, Quantitative estimation of sialic acids. II. A colorimetric resorcinol-hydrochloric acid method, Biochim. Biophys. Acta 24:604 (1957).
28. R. W. Ledeen and R. K. Yu, Chemistry and analysis of sialic acids, in: Biological Roles of Sialic Acid, A. Rosenberg and C. -L. Schengrund, eds., pp. 1-57, Plenum, N.Y. (1976).
29. R. Schauer, Sialic Acids, Chemistry, Metabolism and Function, Springer-Verlag, New York (1982).
30. R. Schauer, Chemistry, metabolism, and biological functions of sialic acids, Adv. Carbohyd. Chem. Biochem. 40:131 (1982).

31. C. C. Sweeley and B. Walker, Determination of carbohydrates in glycolipids and gangliosides by gas chromatography, Anal. Chem. 36:1461 (1964).
32. D. E. Vance and C. C. Sweeley, Quantitative determination of neutral glycosylceramides in human blood, J. Lipid Res. 8:621 (1967).
33. S. J. Rickart and C. C. Sweeley, Quantitative analysis of carbohydrate residues of glycoproteins and glycolipids by gas-liquid chromatography. An appraisal of experimentals, J. Chromatogr. 147:317 (1978).
34. S. Ando and T. Yamakawa, Application of trifluoroacetyl derivatives to sugar and lipid chemistry. I. Gas chromatographic analysis of common constituents of glycolipids. J. Biochem. 70:335 (1971).
35. J. P. Zanetta, W. C. Breckenridge, and G. Vincendon, Analysis of monosaccharides by gas-liquid chromatography of the O-methyl glycosides as trifluoracetate derivatives. Application to glycoproteins and glycolipids, J. Chromatog. 69:291 (1972).
36. R. K. Yu and R. W. Ledeen, Gas-liquid chromatographic assay of lipid-bound sialic acid: measurement of gangliosides in brain of several species, J. Lipid Res. 11:506 (1970).
37. J. Ashraf, D. A. Butterfield, J. Jarnefelt, and R. A. Laine, Enhancement of the Yu and Ledeen gas-liquid chromatographic method for sialic acid estimation: use of methane chemical ionization mass fragmentography, J. Lipid Res. 21:1137 (1980).
38. S. Hakomori and T. Saito, Isolation and characterization of a glycosphingolipid having a new sialic acid, Biochemistry 8:5082 (1969).
39. R. Ghidoni, S. Sonnino, G. Tettamanti, N. Baumann, G. Reuter, and R. Schauer, Isolation and characterization of trisialoganglioside from mouse brain, containing 9-O-acetyl-neuraminic acid, J. Biol. Chem. 255:6990 (1980).
40. S. Sonnino, R. Ghidoni, V. Chigorno, and G. Tettamanti, Chemistry of gangliosides carrying O-acetylated sialic acid, Adv. Exp. Med. Biol. 152:55 (1982).
41. V. Chigorno, S. Sonnino, R. Ghidoni, and G. Tettamanti, Isolation and characterization of a tetrasialoganglioside from mouse brain, containing 9-O-acetyl, N-acetylneuraminic acid, Neurochem. Intern. 4:531 (1982).
42. J. P. Kamerling, J. F. G. Vliegenthart, C. Versluis, and R. Schauer, Identification of O-acetylated N-acyl-neuraminic acids by mass spectrometry, Carbohyd. Res. 41:7 (1975).
43. R. Schauer, C. Schroder, and A. K. Shukla, New techniques for the investigation of structure and metabolism of

sialic acids, This volume (1984).
44. J. S. Sawardeker, H. J. Sloneker, and A. Jeanes, Quantitative determination of monosaccharides as their alditol acetates by gas-liquid chromatography, Anal. Biochem. 37:1602 (1965).
45. G. G. S. Dutton, Application of gas-liquid chromatography to carbohydrates: Part I, Adv. Carbohyd. Chem. Biochem. 28:11 (1973).
46. L. A. Torello, A. Y. Yates, and D. K. Thompson, Critical study of the alditol acetate method for quantitating small quantities of hexoses and hexosamines in gangliosides, J. Chromatogr. 202:195 (1980).
47. A. B. Blakeney, P. J. Harris, R. J. Henry, and B. A. Stone, A simple and rapid preparation of alditol acetates for monosaccharide analysis, Carbohyd. Res. 113:291 (1983).
48. B. Nilsson and D. Zopf, Gas chromatography and mass spectrometry of hexosamine-containing oligosaccharide alditols as their permethylated N-trifluoroacetyl derivatives, Methods Enz. 83:46 (1982).
49. B. Nilsson and D. Zopf, Oligosaccharides released from glycolipids by trifluoroacetolysis can be analyzed by gas chromatography-mass spectrometry, Arch. Biochem. Biophys. 222:628 (1983).
50. D. Rolf and G. R. Gray, Reductive cleavage of glycosides, J. Am. Chem. Soc. 104:3539 (1982).
51. I. Ishizuka and H. Wiegandt, An isomer of trisialoganglioside and the structure of tetra- and pentasialogangliosides from fish brain, Biochim. Biophys. Acta 260:279 (1972).
52. S. Ando and R. K. Yu, Isolation and characterization of a novel trisialoganglioside, G_{T1a}, from human brain, J. Biol. Chem. 252:6247 (1977).
53. S. Ando and R. K. Yu, Isolation and characterization of two isomers of brain tetrasialogangliosides, J. Biol. Chem. 254:12224 (1979).
54. R. K. Yu and S. Ando, Structures of some new complex gangliosides of fish brain, Adv. Exp. Med. Biol. 125:33 (1980).
55. L. Svennerholm, J.-E. Mansson, and Y.-T. Li, Isolation and structural determination of a novel ganglioside, a disialosylpentahexosylceramide from human brain, J. Biol. Chem. 248:740 (1973).
56. T. Itoh, Y.-T. Li, S.-C. Li, and R. K. Yu, Isolation and characterization of a novel monosialosylpentahexosyl ceramide from Tay-Sachs' brain, J. Biol. Chem. 256:165 (1981).
57. N. Kasai, L. O. Sillerud, and R. K. Yu, A convenient method for the preparation of asialo-G_{M1}, Lipids 17:107 (1982).

58. M. Saito, K. Sugano, and Y. Nagai, Action of <u>Arthrobacter ureafaciens</u> sialidase on sialoglycolipid substrates, <u>J. Biol. Chem.</u> 254:7845 (1979).
59. Y. -T. Li and S. -C. Li, Utilization of glycosidases for the structural studies of complex carbohydrate chains, in: <u>CNRS International Symposium on the Structure and Methodology of Glycoconjugates,</u> Vol. 1, pp. 339-350 (1973).
60. C. C. Sweeley and G. Dawson, Determination of glycosphingolipid structures by mass spectrometry, <u>Biochem. Biophys. Res. Commun.</u> 37:6 (1969).
61. G. Dawson and C. C. Sweeley, Mass spectrometry of neutral, mono- and disialoglycosphingolipids, <u>J. Lipid Res.</u> 12:56 (1971).
62. K. Samuelsson and B. Samuelsson, Gas-liquid chromatography-mass spectrometry of cerebrosides as trimethylsilyl ether derivatives, <u>Biochem. Biophys. Res. Commun.</u> 37:15 (1969).
63. K. -A. Karlsson, Structural fingerprinting of gangliosides and other glycoconjugates by mass spectrometry, <u>Adv. Exp. Med. Biol.</u> 125:47 (1980).
64. T. Ariga, R. K. Yu, M. Suzuki, S. Ando, and T. Miyatake, Characterization of G_{M1} ganglioside by direct inlet chemical ionization mass spectrometry, <u>J. Lipid Res.</u> 23:437 (1982).
65. M. McNeil, A. G. Darvill, P. Aman, L. -E. Franzen, and P. Albersheim. Structural analysis of complex carbohydrates using high-performance liquid chromatography, gas chromatography, and mass spectrometry, <u>Methods Enz.</u> 83:3 (1982).
66. K. L. Busch and R. G. Cook, Mass spectrometry of large, fragile, and involatile molecules, <u>Science</u> 218:247 (1982).
67. K. L. Rinehart, Jr., Fast atom bombardment mass spectrometry, <u>Science</u> 218:254 (1982).
68. M. Arita, M. Iwamori, T. Higuchi, and Y. Nagai, 1,1,3,3,-tetramethylurea and triethanolamine as a new useful matrix for fast atom bombardment mass spectrometry of gangliosides and neutral glycosphingolipids, <u>J. Biochem.</u> 93:319 (1983).
69. M. Arita, M. Iwamori, T. Higuchi, and Y. Nagai, Negative ion fast atom bombardment mass spectrometry of gangliosides and asialogangliosides: A useful method for the structural elucidation of gangliosides and related neutral glycosphingolipids, <u>J. Biochem.</u> 94:249 (1983).
70. A. Dell, H. R. Morris, H. Egge, H. von Nicolai, and G. Strecker, Fast-atom-bombardment mass-spectrometry for carbohydrate-structure determination, <u>Carbohyd. Res.</u> 115:41 (1983).

71. Y. Kushi and S. Handa, Application of field desorption mass spectrometry for the analysis of sphingoglycolipids, J. Biochem. 91:923 (1982).
72. S. Handa and Y. Kushi, High-performance liquid chromatography and structural analysis by field desorption mass spectrometry of underivatized glycolipids, Adv. Exp. Med. Biol. 152:23 (1982).
73. S. Handa, Y. Kushi, H. Kambara, and K. Shizukushi, Secondary ion mass spectra of neutral sphingoglycolipids, J. Biochem. 93:315 (1983).
74. H. Egge, J. Peter-Katalinic, and P. Hanfland, Structure analysis of glycosphingolipids using fast atom bombardment (FAB) techniques, in: This volume (1984).
75. S. Handa and Y. Kushi, Application of field desorption and secondary ion mass spectrometry for glycolipid analysis, in: This volume (1984).
76. T. Feizi, The antigens Ii, SSEA-1, and ABH are in an interrelated system of carbohydrate differentiation antigens expressed on glycosphingolipids and glycoproteins, Adv. Exp. Med. Biol. 152:167 (1982).
77. R. Kannagi, E. Nudelman, S. B. Levery, and S. Hakomori, A series of human erythrocyte glycosphingolipids reacting to the monoclonal antibody directed to a developmentally regulated antigen, SSEA-1, J. Biol. Chem. 257:14865 (1982).
78. S.-i. Hakomori, Monoclonal antibodies directed to cell surface carbohydrates, in: Monoclonal Antibodies and Functional Cell Lines, R. H. Kennett, K. D. Bechtol and T. J. McDearn, eds., (in press) Plenum Publishing Corp., New York (1984).
79. S.-i. Hakomori, A rapid permethylation of glycolipid and polysaccharide catalyzed by methylsufinyl carbanion in dimethyl sulfoxide, J. Biochem. 55:205 (1964).
80. T. Imanari and Z. Tamura, Gas chromatography of glucuronides, Chem. Phar. Bull. 15:1677 (1967).
81. S. Ando, K. Kon, Y. Nagai, and T. Murata, Chemical ionization and electron impact mass spectra of oligosaccharides derived from sphingoglycolipids, J. Biochem. 82:1623 (1977).
82. J. Finne, T. Krusius, and H. Rauvala, Use of potassium tert-butoxide in the methylation of carbohydrates, Carbohyd. Res. 80:336 (1980).
83. T. Narui, K. Takahashi, M. Kobayashi, and S. Shibata, Permethylation of polysaccharides by a modified Hakomori method, Carbohyd. Res. 103:293 (1982).
84. H. Bjorndal, C. G. Hellerqvist, B. Lindberg, and S. Svensson, Gas-liquid chromatography and mass spectrometry in methylation analysis of polysaccharides, Angew. Chem. Int. Ed. Engl. 9:610 (1970).

85. C. G. Hellerqvist, B. Lindberg, A. Pilotti, and A. A. Lindberg, Structural studies of the O-specific side-chains of the cell-wall lipopolysaccharide from Salmonella senftenberg, Carbohyd. Res. 16:297 (1971).
86. K. Steller, H. Saito, and S. -i. Hakomori, Determination of aminosugar linkage in glycolipids by methylation: aminosugar linkages of ceramide pentasaccharides of rabbit erthrocytes and of Forssman antigen, Arch. Biochem. Biophys. 155:464 (1973).
87. S. K. Kundu, R. W. Ledeen, and P. A. J. Gorin, Determination of position of substitution on 2-acetamido-2-deoxy-D-galactosyl residues in glycolipids, Carbohyd. Res. 39:179 (1975).
88. S. K. Kundu, R. W. Ledeen, and P. A. J. Gorin, Determination of position of substitution on 2-acetamido-2-deoxy-glucosyl residues in glycolipids, Carbohyd. Res. 38:329 (1975).
89. H. Yamaguchi, T. Ikenaka, and Y. Matsushima, An improved method for gas-liquid chromatographic analysis of Smith degradation products from oligosaccharides, J. Biochem. 68:253 (1970).
90. S. Ando and R. K. Yu, Isolation and structural study of a novel fucose-containing disialoganglioside from human brain, Glycoconjugates Res. 1:79 (1979).
91. L. O. Sillerud, R. K. Yu, and D. E. Schafer, Assignment of the carbon-13 nuclear magnetic resonance spectra of gangliosides G_{M4}, G_{M3}, G_{M2}, G_{M1}, G_{D1a}, G_{D1b}, and G_{T1b}, Biochemistry 21:1260 (1982).
92. T. A. W. Koerner, Jr., J. H. Prestegard, P. C. Demou, and R. K. Yu, High-resolution proton NMR studies of gangliosides. I. Use of homonuclear two-dimensional spin-echo J-correlated spectroscopy for determination of residue composition and anomeric configurations, Biochemistry 22:2676 (1983).
93. R. K. Yu, T. A. W. Koerner, Jr., P. C. Demou, J. N. Scarsdale, and J. H. Prestegard, Recent advances in structural analysis of gangliosides: Primary and secondary structures, This volume (1984).
94. L. O. Sillerud and R. K. Yu, Comparison of the ^{13}C-N.M.R. spectra of ganglioside G_{M1} with those of G_{M1}-oligosaccharide and asialo-G_{M1}, Carbohyd. Res. 113:173 (1983).
95. R. K. Yu and L. O. Sillerud, Carbon-13 nuclear magnetic resonance studies of hematoside and globoside, Adv. Exp. Med. Biol. 152:41 (1982).
96. S. L. Patt, F. Sauriol, and A. S. Perlin, Determination of the positions of glycosidic linkages from $^{13}C-^{13}C$ connectivity plots, Carbohyd. Res. 107:C1 (1982).
97. R. W. Ledeen, New developments in the study of ganglioside structures, Chem. Phys. Lipids 5:205 (1970).

98. S. J. Angyal and K. James, Oxidation of carbohydrates with chromium trioxide in acetic acid, Aust. J. Chem. 23:1209 (1970).
99. M. Oshima and T. Ariga, Analysis of the anomeric configuration of a galactofuranose containing glycolipid from an extreme thermophile, FEBS Lett. 64:440 (1976).
100. J. Hoffman, B. Lindberg, and S. Svensson, Determination of the anomeric configuration of sugar residues in acetylated oligo- and polysaccharides by oxidation with chromium trioxide in acetic acid, Acta Chem. Scand. 26:661 (1972).
101. R. A. Laine and O. Renkonen, Analysis of anomeric configurations in glyceroglycolipids and glycosphingolipids by chromium trioxide oxidation, J. Lipid Res. 16:102 (1975).
102. L. O. Sillerud, J. H. Prestegard, R. K. Yu, D. E. Schafer, and W. H. Konigsberg, Assignments of the ^{13}C NMR spectrum of aqueous gangliosides G_{M1} micelles, Biochemistry 17:2619 (1978).
103. T. A. W. Koerner, Jr., L. W. Cary, S. -C. Li, and Y. -T. Li, Carbon-13 NMR spectroscopy of a cerebroside. Proof of the β-pyranosyl structure of D-glucosylceramide, J. Biol. Chem. 254:2325 (1979).
104. T. A. W. Koerner, Jr., L. W. Cary, S. -C. Li, and Y. -T. Li, Carbon-13 NMR spectroscopy of Forssman hapten, Biochem. J. 195:529 (1981).
105. H. A. Nunez and C. C. Sweeley, Carbon-13 nuclear magnetic resonance spectrometry of globotriaosylceramide, J. Lipid Res. 23:863 (1982).
106. P. L. Harris and E. R. Thornton, Carbon-13 and proton nuclear magnetic resonance studies of gangliosides, J. Am. Chem. Soc. 100:6738 (1978).
107. L. D. Hall, High-resolution nuclear magnetic resonance spectroscopy, in: The Carbohydrates, Chemistry and Biochemistry, W. Pigman, D. Horton and J. D. Wander, eds., Vol. IB, pp. 1299-1326, Academic Press, N.Y. (1981).
108. P. A. J. Gorin, Carbon-13 nuclear magnetic resonance spectroscopy of polysaccharides, Adv. Carbohyd. Chem. Biochem. 38:13 (1981).
109. K. Bock and H. Thogersen, Nuclear magnetic resonance spectroscopy in the study of mono- and oligosaccharides, in: Annual Reports on NMR Spectroscopy, G. A. Webb, ed., Vol. 13, pp. 1-57, Academic Press, New York, (1982).
110. R. Barker, H. A. Nunez, P. R. Rosevear, and A. S. Serianni, ^{13}C NMR analysis of complex carbohydrates, Methods Enz. 83:58 (1982).

STRUCTURE ANALYSIS OF GLYCOSPHINGOLIPIDS USING FAST ATOM BOMBARDMENT (FAB) TECHNIQUES

Heinz Egge*, Jasna Peter-Katalinic* and Peter Hanfland**

*Institute for Physiological Chemistry
**Institute for Experimental Hematology and Bloodtransfusion
University of Bonn G.F.R.

INTRODUCTION

The development of fast atom bombardment mass spectrometry[1] (FAB MS) and high field and high resolution magnetic sector mass spectrometry[2] have disclosed new dimensions in the analysis and sequencing of polar nonvolatile compounds of vital biological importance such as oligonucleotides,[3] peptides,[4] and glycoconjugates[5] with molecular weights up to 8000 daltons.

MATERIALS

The gangliosides were prepared 20 years ago in R. Kuhn's laboratory.[6] The preparation of the neutral GSL's from erythrocyte membranes followed procedures already described.[7]

METHOD PRINCIPLE

In a cold cathode discharge ion source, rare gas ions (Ar,Xe) of controlled energy are produced. The focussed emergent beam passes through a collision chamber containing rare gas of 10^{-3}-10^{-4} mbar pressure where charge exchange without loss of kinetic energy takes place. Outside the chamber, rare gas ions are "cleaned out" by electrostatic deflector plates. The material to be analysed, applied to the target in a matrix of a high boiling liquid, interferes with the fast atom beam followed by desorption of molecular ions and molecular debris. A collisional cascade ("spike") desorption mechanism has been postulated to explain the relatively high percentage of intact molecular versus fragment ions. The energy of the emitted molecular ions is so low, that this process can be described as "cold desorption". Due to their thermodynamic stability the molecular ion intensities can be increased by addition of inorganic and organic salts and acids, yielding the pseudo-molecular $[M+Me]^+$, $[M+H]^+$ in the positive and $[M-H]^-$ ions in the negative mode.

FAB spectra were recorded on a ZAB HF mass spectrometer with reversed geometry giving a mass range of 8000 at 3 kV acceleration voltage (VG Analytical, Altrincham, U.K.). Rare gas xenon atoms were used having a kinetic energy equivalent to 9 keV. The mass marker was calibrated with CsI giving signals $[Cs_{n+1}I_n]^+$ in the positive and $[Cs_nI_{n+1}]^-$ in the negative ion mode. Native glycolipids were dissolved in MeOH/glacial acetic acid 7/3 to a concentration of 5 µg/µl. Glycerol or 1-mercapto-2,3-propanediol (thioglycerol) dosed with glacial acetic acid up to 5% was used as a matrix when measuring in the negative ion mode.

In the positive ion mode, the stainless steel target was first loaded with 0.1% solution of sodium acetate in MeOH, dried and loaded again with thioglycerol. Peracetylated or permethylated GSL dissolved in MeOH to a concentration of 5 µg/µl were added to the matrix. Spectra up to 4000 amu were evaluated by counting the spectral lines. The mass numbers are therefore smaller by about one unit in the mass range of 2000 amu than those calculated on the basis of the exact atomic weights. Signals above 4000 amu were determined by interpolation between adjacent CsI or KI signals. In this case, nearest whole numbers of the exact physical masses are presented.

RESULTS AND DISCUSSION

Native gangliosides and neutral GSL´s are preferably analysed in the negative ion mode. In the positive ion mode and in the presence of Na^+, clusters of ions with a low relative intensity are produced by the addition of n+1 sodium ions in exchange of n protons. This is exemplified in Fig. 1, representing the molecular ion region of GD1a.

In the negative ion mode intense M-1 signals are observed. Di-, tri- and tetrasialo-gangliosides added as neutral salts to the target may, depending on the pH of the matrix, retain one or more cations in exchange of a proton. This is especially the case during upfield scans of longer duration after evaporation of the low boiling acetic acid. As an example the FAB spectrum of GD1a, (Fig. 2c) is presented. In addition to the pseudomolecular ions $[M-1]^-$, a series of fragment ions can be observed that allows

STRUCTURE ANALYSIS OF GLYCOSPHINGOLIPIDS

sequencing, especially regarding the points of attachment of the sialic acid residues. Prominent fragment ions found in the negative ion FAB spectra of gangliotetraosylceramide, GM1, GD1a and GT1b are presented in Fig. 1 and Scheme 1. Fragment ions derived from the "nonreducing" end that contains NeuAc residues are mostly stabilized by the elimination of two protons. These "doublets" can easily be recognized in the spectra. On the other hand fragments containing the ceramide residue normally reflect the specific distribution of the sphingosine and fatty acid pattern that is also exhibited in the M-1 signals. Thus negative ion FAB spectra of native gangliosides provide valuable data, esecially concerning the points of attachment of the sialic acid residues, which are not easily obtained by other methods such as EI-MS or CI-MS.

The results of negative ion FAB MS can be corroborated by the positive ion FAB MS of permethylated gangliosides. In the presence of Na^+ ions very intense pseudomolecular ions $[M+Na]^+$ are observed. These can favorably be used for the detection of higher or lower homologs present in a preparation because they are easily discriminated from other fragment ions that normally do not contain the Na^+. In addition, highly specific fragment ions are formed starting from the "nonreducing" end that are com-

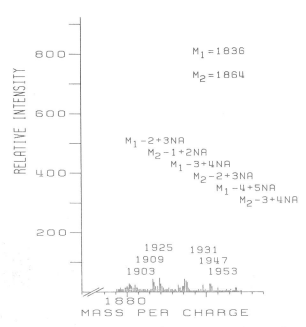

Fig. 1. Molecular ion region of GD1a measured in thioglycerol matrix in the presence of sodium acetate by positive ion FAB.

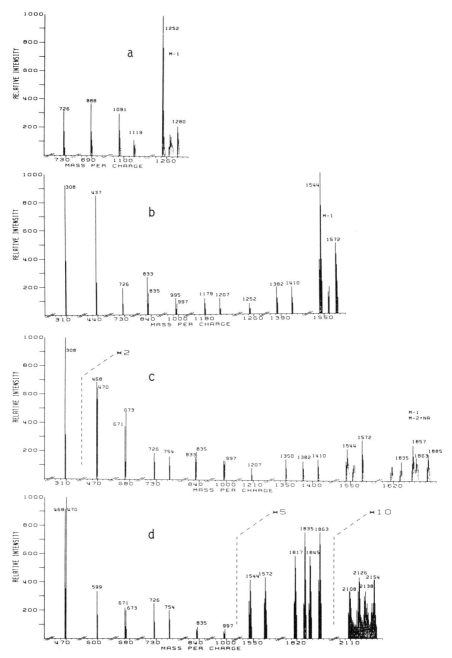

Fig. 2. Negative ion FAB spectra of native GgOse$_4$Cer (a), GM1 (b), GD1a (c) and GT1b (d). Peaks produced by the matrix were omitted.

plementary to those observed in the negative ion mode. This is exemplified by the FAB spectra obtained from a series of GSL's derived from rabbit and human B erythrocytes. These GSL's with up to 25 sugar components, exhibit a similar but not identical principle of construction. Several N-acetyl-lactosamine residues (LacNAc) are linked to each other by either β1-3 or β1-6 bonds, thus forming long branched carbohydrate chains that are terminated by the blood group specific α-Gal and α-Fuc residues. In the lower homologs sequence determination is rather straight forward whereas in the higher homologs containing up to nine LacNAc residues many possible alignments are feasible. High resolution NMR furnished data on the number of antennae, the types (α,β) and sites of linkage of LacNAc residues, but no clues on their overall arrangement. Here FAB MS furnished the decisive information due to a highly selective mode of fragmentation. This is characterized as follows:

The terminal constituents such as ceramide and the "non-reducing" monosaccharide units furnish characteristic ions like m/e 219/187 for hexose, m/e 189/157 for fucose, m/e 260/228 for hexosamine and e.g. 658 for the ceramide. Further fragmentation proceeds almost exclusively starting from the "nonreducing" end of the GSL by cleavage of the glycosidic bonds of the GlcNAc residues. The results are summarized in Scheme 2. The terminal Gal-Gal-GlcNAc- residues are arranged as antennae along a backbone chain formed by 1-3 linked LacNAc units. "Tree-like" structures could be safely ruled out. In the undegraded GSL these sequence ions follow a pentasaccharide rhythm with a difference of 1102 amu. The α-galactosidase treated GSL's accordingly exhibit a tetrasaccharide rhythm with a difference of 898 amu and after additional Smith degradation a trisaccharide rhythm is observed with a difference of 694 amu between the most prominent sequence ions as shown in Fig. 3.

A more complex type of architecture is found in human blood group B-active GSL. Here either the "core" neolacto-N-tetraose or one of the antennae may be elongated by one LacNAc residue. The FAB MS of the permethylated ceramide hexadecasaccharide (B-V) shows that in this case both the "core" as well as one of the antennae are elongated by one LacNAc residue (Fig. 4). This can be clearly deduced from the $[M+Na]^+$ ion and the sequence ions at 842; 1291; 2567 and 3016 amu (Scheme 3).

SUMMARY

The results presented show, that with the aid of negative ion FAB MS native glycosphingolipids and especially gangliosides are amenable to sequence analysis. The preferred formation of pseudomolecular ions M-1 and of sialic acid containing fragments

gives conclusive information on the number of sialic acids
present and the sites of their attachment to the oligosaccharide
backbone. Positive ion FAB MS of branched permethylated glycosphingolipids with up to 25 sugar units yielded pseudomolecular
ions [M+Na]$^+$ in excess of 6000 daltons, that allowed an exact
calculation of carbohydrate constituents. Furthermore highly

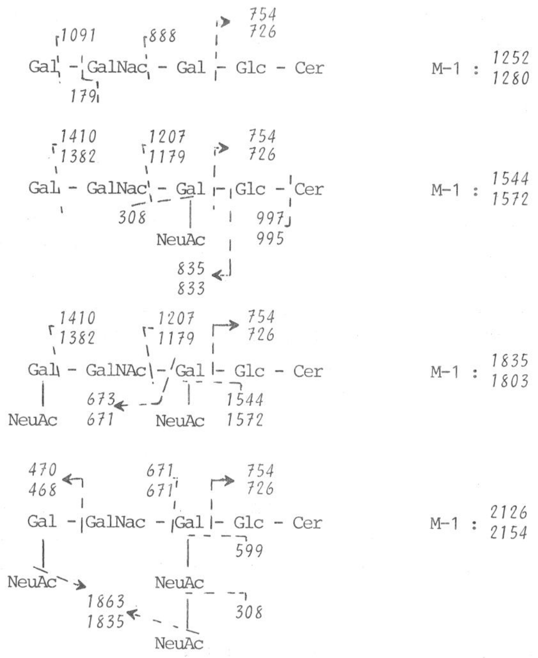

```
        1091    888      → 754
                           726
     Gal -|GalNac| - Gal |- Glc - Cer       M-1 : 1252
          |                                        1280
         179|


      1410    1207       → 754
      1382    1179         726
     Gal - GalNac| - Gal |- Glc -|Cer       M-1 : 1544
                 |       |       |                 1572
                308      |  | 997
                       NeuAc  995
                         835
                         833


      1410    1207      → 754
      1382    1179        726
     Gal - GalNAc| -/Gal |- Glc - Cer       M-1 : 1835
       |    673     |     1544                     1803
     NeuAc  671   NeuAc  1572


      470      671      → 754
      468      671        726
     Gal -|GalNac -|Gal |- Glc - Cer        M-1 : 2126
       |           |     599                       2154
     NeuAc       NeuAc
            1863         308
            1835
                       NeuAc
```

Scheme 1. Fragmentation pattern of gangliosides analysed by negative ion FAB MS. Only major fragments are indicated.

specific fragmentation patterns furnished information on number
and positions of branching points as well as on the ceramide
moiety. It can be anticipated, that FAB MS will be very useful
in the analysis of more complex gangliosides carrying additional
fucose or acyl residues and of even larger molecules with molecular weights up to 15,000 daltons.

STRUCTURE ANALYSIS OF GLYCOSPHINGOLIPIDS

```
                    Gal - Gal - GlcNAc                    M+Na⁺ : 2222 amu
                              668 -- \
                                     Gal - GlcNAc - Gal - Glc ─ Cer
                                                1117 ─'        '658
```

```
Gal - Gal - GlcNAc /
               /   1770 --                                        M+Na⁺ : 5083 dalton
           668 -\   Gal - GlcNAc     2872 -
Gal - Gal - GlcNAc \  668 -\  Gal - GlcNAc     3974 --
           Gal - Gal - GlcNAc \         \    Gal - GlcNAc - Gal - Glc - Cer
                          668 -\  Gal - GlcNAc                    '658
                          Gal - Gal - GlcNAc
```

```
Gal - Gal - GlcNac /
               /   1770 -                                         M+Na⁺ : 6184 dalton
           668   Gal - GlcNAc     2872 --
Gal - Gal - GlcNAc\ 668 --  Gal - GlcNAc
           Gal - Gal - GlcNAc\       \    3974 --
                          668 -\  Gal - GlcNAc
                          Gal - Gal - GlcNAc\    \  Gal - GlcNAc - Gal - Glc - Cer
                                        668 -\  Gal - GlcNAc           658
                                        Gal - Gal - GlcNAc
```

Scheme 2.

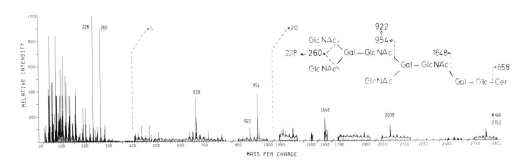

Fig. 3.   Positive ion FAB spectrum of a permethylated ceramide nonasaccharide obtained from ceramide pentadecasaccharide of rabbit erythrocytes after α-galactosidase treatment and Smith degradation.

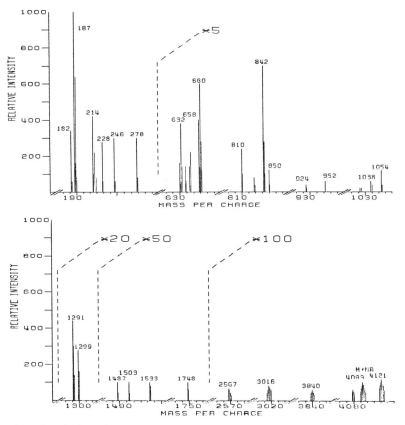

Fig. 4. Positive ion FAB mass spectrum of permethylated glycosphingolipid B-V from human B erythrocytes. Matrix derived peaks are omitted.

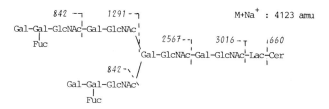

Scheme 3. Simplified formula of glycolipid B-V.

REFERENCES

1. M. Barber, R. S. Bordoli, R. D. Sedgwick and A. N. Tyler, Fast atom bombardment of solids as an ion source in mass spectrometry, Nature 293:270 (1981).
2. A. Dell and H. R. Morris, Fast atom bombardment-high field magnet mass spectrometry of 6000 dalton polypeptides, Biochem. Biophys. Res. Commun. 106:1456 (1982).
3. D. H. Williams, C. Bradley, G. Bojesen, S. Santikarn and L.C.E. Taylor, Fast atom bombardment mass spectrometry: a powerful technique for the study of polar molecules, J. Am. Chem. Soc. 103:5700 (1981).
4. M. Barber, R. S. Bordoli, G. J. Elliott, N. J. Horoch and B. N. Green, Fast atom bombardment mass spectrometry of human pro-insulin, Biochem. Biophys. Res. Commun. 110:753 (1983).
5. A. Dell, J. E. Oaks, H. R. Morris and H. Egge, Structure determination of carbohydrates and glycosphingolipids by fast atom bombardment mass spectrometry, Int. J. Mass Spectrom. & Ion Phys. 46:415 (1983).
6. R. Kuhn and H. Egge, Uber Ergenbnisse der Permethylierung der Ganglioside G_I and G_{II}, Chem. Ber. 96:3338 (1963).
7. P. Hanfland, H. Egge, U. Dabrowski, S. Kuhn, D. Roelcke and J. Dabrowski, Isolation and characterization of an I-active ceramide decasaccharide from rabbit erythrocyte membranes, Biochemistry 20:5310 (1981).

The work was supported by the Deutsche Forshungsgemeinschaft.

APPLICATION OF FIELD DESORPTION AND SECONDARY ION MASS SPECTROMETRY

FOR GLYCOLIPID ANALYSIS

Shizuo Handa and Yasunori Kushi

Department of Biochemistry, Faculty of Medicine
Tokyo Medical and Dental University
Yushima, Bunkyo-ku, Tokyo 113

INTRODUCTION

Glycolipids are well known cell membrane components and play a role in several biological phenomena on the cell surface.[1] As glycolipids are relatively minor components of cells, it is necessary to develop analytical methods for small amounts of samples. Mass spectrometry have proved to be powerful tools for this purpose. Electron impact and chemical ionization mass spectrometry provide good information on the structure of permethylated or permethylated and reduced glycolipids.[2-4] However, these ionization methods require the derivatization of the sample, and the ion intensities of the high mass region are very weak and difficult to detect continuously during analysis. The recently introduced "soft ionization techniques" are successfully applied for the analyses of non-volatile and thermolabile compounds. These include field desorption (FD), fast atom bombardment (FAB) and secondary ion mass spectrometry (SI-MS). However only a few reports on the application of these new techniques for the analysis of glycolipids[5-11] have been published.

In this report, FD and SI-MS have been applied to the analyses of several glycolipids. These methods are compared to each other and shown to be useful for the elucidation of chemical structures and molecular species distribution.

This investigation was supported in part by a grant from the Ministry of Education, Science and Culture of Japan. ABBREVIATIONS: FD, field desorption; SI, secondary ion; MS, mass spectrometry.

EXPERIMENTAL PROCEDURE

Materials

Gangliosides and sulfatide were prepared from bovine brain as described previously.[7] Seminolipid was a generous gift from Dr. Ineo Ishizuka, Teikyo University. Methylation of glycolipids was performed according to Hakomori.[11]

Mass spectrometry

Mass spectra were recorded on a double-focusing instrument (Hitachi M-80) equipped with a FD or SI ion source and interfaced with a Hitachi M-003 computer system. The conditions for the analyses of glycolipids by FD were the same as those reported previously.[6] The sample solution in chloroform-methanol (2/1,v/v) was loaded on to the carbon emitter with a microsyringe. The accelerating voltage of 3 KV or 1.5 KV and cathode voltage of -4 KV to -6 KV were applied. The emitter current was increased slowly at a rate of 1.25 mA/min from 0 to 40 mA. All spectra shown in figures were obtained by averaging 20 or more spectra which were continuously taken during the period when sufficient total ions were being produced. The conditions for SI analysis were the same as those reported previously.[9] The accelerating voltage was 3 KV or 1.5 KV. Xenon ions were used as the primary ion with an energy of 5 KeV. The Xe^+ ion current was 5×10^{-8} A. One or two μg of each sample dissolved in 1 μl of a mixture of chloroform-methanol (2/1,v/v) was mixed with 1 μl of glycerol on the silver plate.

RESULTS AND DISCUSSION

Neutral glycolipids

FD and SI-MS spectra of neutral glycolipids were presented previously.[6,7,9] In the FD spectra of underivatized neutral glycolipids,[6,7] the ions $(M+H)^+$ and $(M+H-H_2O)^+$ decrease and instead, $(M+Na)^+$ becomes the predominant ion. These ions provide good information on the distribution of molecular species. On increasing emitter current, the intensities of the fragment ions are increased. These were almost all due to the loss of sugar moieties by sequential cleavage at the glycosidic linkages. This makes FD-MS analysis very useful for the determination of the sugar sequence of complex glycolipids. Previously, it had been thought that the FD spectra are transient and variable depending on the change of the emitter current, making it difficult to get reproducible spectra. However, as shown in the previous papers, reproducible spectra were obtained by increasing the emitter current at the same rate and integrating the spectra during the analyses.[6,7]

In SI spectra of underivatized neutral glycolipids, ion peaks are assigned by comparison with the spectra of several glycolipids with different ceramide or different sugar compositions, and also by referring to the data obtained by FD-MS.[9] As in FD spectra, $(M+H)^+$ and $(M+H-H_2O)^+$ ions are observed as relatively intense ions. As the sugar chain length becomes longer, the intensities of these ions decreased and instead, $(M+Na)^+$ become the predominant ions. In all spectra the ion derived from the ceramide is observed as an intense peak comparable to those of quasimolecular ions, and none of the cationized ions of these fragments are observed. The ions derived from the fragments cleaved at each of the glycosidic linkages are also detected, but the intensities of these ions are lower than those obtained by FD. In the low mass region, there are many ion peaks, including those of ions arising from the glycerol matrix and the assignment of these ions is difficult.

Ganglioside

Because brain gangliosides contain both C20 and C18 sphingosine bases, ions containing sphingosine appeared as pairs which were separated from each other by m/z 28. In the FD spectrum of GM_1, (Fig. 1.) fragments due to asialo-moieties at m/z 1305 and 1277 gave intense peaks, and the other fragments due to the sequential cleavage at the glycosidic linkages were clearly detected as in the case of neutral glycolipids (m/z, 1143 and 1115, 940 and 912, and 778 and 750). However, it was very hard to detect the ions in the molecular region, and only the cationized and dehydrated molecular ions were detected at m/z 1578 and 1550. Fragment ions due to ceramide were also difficult to detect. These observations were confirmed by the FD spectra of other gangliosides (GM_3, GM_2 and GD_{1a}) obtained in our laboratory. The ions at m/z 267 and 239 originated from the cleavage of the sphingosine base between C2 and C3, and the ion at m/z 163 was due to galactose.

In the SI spectra, cationized molecular ions were clearly detected. The other fragments due to asialo-moieties at m/z 1305 and 1277, and asialo- agalacto-moieties at m/z 1143 and 1115, were detected, although the relative intensities of these ions were not as great as those in FD spectra. Fragments due to ceramide were observed at m/z 576 and 548. These fragment ions were detected in SI spectra of all gangliosides but not in FD spectra, except for GM_3.

Permethylated gangliosides

Spectra of permethylated GD_{1a} are shown in Fig. 2. Although ions in the molecular ion region were very difficult to detect in the FD spectra of underivatized gangliosides, cationized molecular ions at m/z 2237 and 2209 $(M+Na)^+$ were the base peaks in the

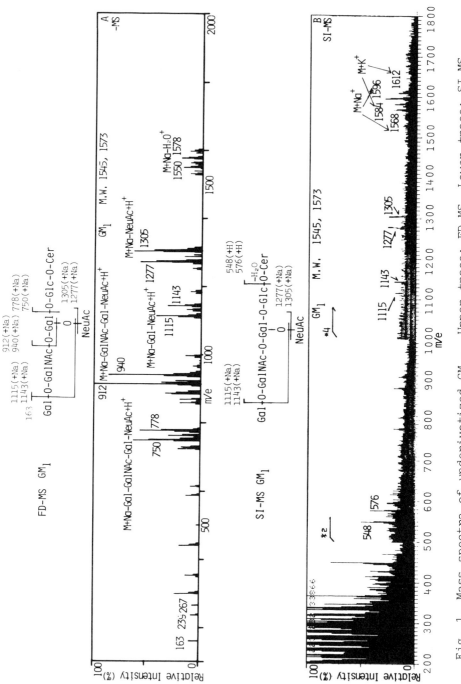

Fig. 1. Mass spectra of underivatized GM_1. Upper trace: FD-MS, Lower trace: SI-MS.

FD spectra of permethylated samples. Fragments derived from the loss of sialic acid at m/z 1879 and 1848 were also prominent, but the other ions were not.

In the SI spectra, protonated molecular ions at m/z 2215 and 2187 as well as ions $(M+H-32)^+$ at m/z 2183 and 2155 were clearly detected. Intensities of fragment ions due to the loss of one and two sialic acids were relatively low. Fragment ions due to ceramide were at m/z 604 and 576. In the SI spectra of permethylated samples, ion peaks in the low mass region were clearly assigned with low background, in contrast to those of underivatized samples. These ions were due to hexosamine at m/z 228, sialic acid at m/z 376 and 344 and sialosylgalactosyl-N-acetylgalactosamine at m/z 825.

Sulfatide

In the FD spectra (not shown), the prominent ion peaks were due to ceramide containing non-hydroxy and hydroxy fatty acids. Ions due to loss of the sulfate group showed relatively high intensities. However no ions were detected in the molecular ion region.

In contrast to the FD spectra, the SI spectra showed the cationized molecular ions clearly. The other prominent ions were due to loss of the sulfate group from the cationized molecular ions and ions due to ceramide, but the intensities of these ions were not so intense as in the FD spectrum.

Seminolipid

Seminolipid was presented as an example of glyceroglycolipid containing sulfate ester. Its structure is 1-O-alkyl-2-O-acyl-3 (β-3´-sulfogalactosyl)-glycerol. Almost all of the constituent fatty acids were palmitic, and glyceryl ethers were composed of chimyl alcohol (approximately 80%) and its C14 analog (15%).[13]

In the FD spectra of seminolipid molecular ions were hard to detect, but the cationized ion due to alkylacylglycerol was the base peak and the cationized ion due to the loss of sulfate group was predominant. (Dr. I. Ishizuka, personal communication.)

In the SI spectra of seminolipid (Fig. 3) cationized molecular ions were clearly detected and ions corresponding to desulfo- and desulfo-degalacto-derivatives were prominent. Ions due to the different alkyl chain lengths were differentiated from each other by m/z 28. The assignments of the ion peaks were confirmed by comparison with the spectra of deacyl-seminolipid (Fig. 3).

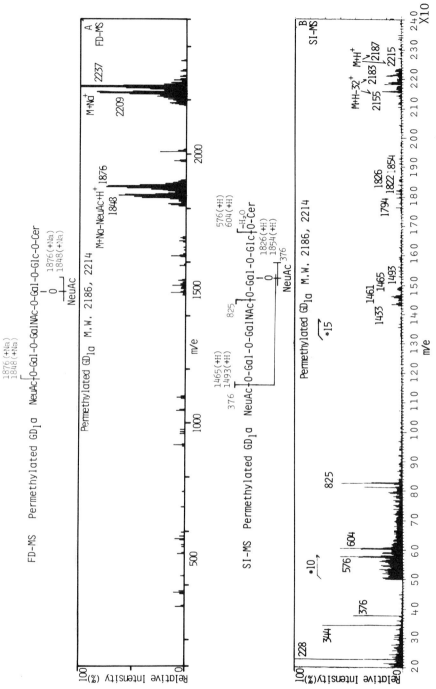

Fig. 2. Mass spectra of permethylated GD$_{1a}$. Upper trace: FD-MS, Lower trace: SI-MS.

CONCLUSION

FD spectra of underivatized neutral glycolipids can provide useful information on molecular species distribution, ceramide composition and sugar sequences. FD spectra of underivatized acidic glycolipids, such as sulfatide and ganglioside, also give similar information but yield scarcely any information on molecular weight. On the other hand, in SI spectra of both underivatized neutral and acidic glycolipids, molecular and ceramide ions are prominent but intensities of ions due to the sequential loss of sugars are relatively low.

Thus FD and SI-MS can provide complementary data and the combination of these methods is useful for glycolipid analysis. The fact that SI-MS does not require a special emitter and produces stable ion patterns during analysis is a positive feature of SI-MS in comparison with FD-MS. Even though derivatization of the large molecular weight glycolipids may impose instrumental limitations, it gives useful information on molecular weight in both FD and SI-MS. SI spectra of permethylated glycolipids are similar to EI-spectra. The signal to noise ratio is improved in low mass regions and this makes it possible to elucidate the oligosaccharide structures. Both ionization techniques will provide powerful tools for glycolipid analysis.

SUMMARY

Field desorption (FD) and secondary ion mass spectrometry (SI-MS) mass spectra of several glycolipids are presented to demonstrate their potential for the analysis of glycolipids. FD and SI-MS give useful information on molecular weight, ceramide structure and sugar sequence. In general, FD provides clearer fragment ion peaks for the analysis of sugar sequence than SI-MS. For underivatized acidic glycolipids such as gangliosides, sulfatide and seminolipid, SI-MS provides quasimolecular ions which are hardly produced by FD. In contrast to underivatized gangliosides, permethylated samples give molecular ion species of high intensity in both FD and SI-MS, but no fragment ions pertinent to carbohydrate sequence could be observed in FD spectra. SI-MS spectra of permethylated samples provide good information on sugar chain structure. Thus FD and SI-MS mass spectra complement each other, and the combination of these ionization methods will provide powerful tools for glycolipid analysis.

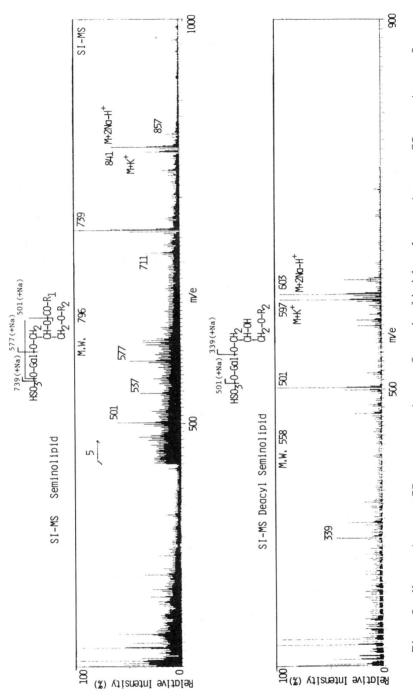

Fig. 3. Upper trace: SI mass spectrum of seminolipid, Lower trace: SI spectrum of deacyl-seminolipid.

REFERENCES

1. A. Makita, S. Handa, T. Taketomi, and Y. Nagai, eds., "New Vistas in Glycolipid Research," Plenum Publishing Corp., New York (1982).
2. R. W. Ledeen, S. K. Kundu, H. C. Price, and J. W. Fong, Mass spectra of permethyl derivatives of glycosphingolipids, Chem. Phys. Lipid 13:429 (1974).
3. K.-A. Karlsson, I. Pascher, W. Pimlott, and B. E. Samuelsson, Use of mass spectrometry for the carbohydrate composition and sequence analysis of glycosphingolipids, Biomed. Mass Spectrom 1:49 (1974).
4. T. Ariga, R. K. Yu, M. Suzuki, S. Ando, and T. Miyatake, Characterization of GM_1 ganglioside by direct inlet chemical ionization mass spectrometry, J. Lipid Res. 23:437 (1982).
5. C. E. Costello, B. W. Wilson, K. Biemann, and J. V. Reinhold, Analysis of glycosphingolipids by field desorption mass spectrometry, in: "Cell Surface Glycolipids," C. C. Sweeley, ed., ACS Symp. Series 128 p. 36 (1980).
6. Y. Kushi and S. Handa, Application of field desorption mass spectrometry for the analysis of sphingoglycolipids, J. Biochem. 91:923 (1982).
7. S. Handa and Y. Kushi, High performance liquid chromatography and structural analysis by field desorption mass spectrometry of underivatized glycolipid, in: "New Vistas in Glycolipid Research," A. Makita, T. Taketomi, S. Handa, and Y. Nagai, eds., Plenum Publishing Corp., New York (1982).
8. H. Egge, J. Dabrowski, P. Hanfland, A. Dell, and U. Dabrowski, High resolution ^1H nuclear magnetic resonance spectroscopy and "Soft ionization" mass spectrometry of glycosphingolipids, in: "New Vistas in Glycolipid Research," A. Makita, T. Taketomi, S. Handa, and Y. Nagai, eds., Plenum Publishing Corp., New York (1982).
9. S. Handa, Y. Kushi, H. Kambara, and K. Shizukuishi, Secondary ion mass spectra of neutral sphingolipids, J. Biochem. 93:315 (1983).
10. M. Arita, M. Iwamori, T. Higuchi, and Y. Nagai, 1,1,3,3-Tetramethylurea and triethanolamine as a new useful matrix for fast atom bombardment mass spectrometry of gangliosides and neutral glycosphingolipids, J. Biochem. 93:319 (1983).
11. K. L. Rinehart, Fast atom bombardment mass spectrometry, Science 218:254 (1982).
12. S. Hakomori, A rapid permethylation of glycolipids and polysaccharides catalyzed by methylsulfinyl carbanion in dimethyl sulfoxide, J. Biochem. 55:205 (1964).
13. I. Ishizuka, M. Suzuki, and T. Yamakawa, Isolation and characterization of a novel sulfoglycolipids, "Seminolipid", from boar testis and spermatozoa, J. Biochem. 73:77

NEW TECHNIQUES FOR THE INVESTIGATION OF STRUCTURE AND METABOLISM OF SIALIC ACIDS

Roland Schauer, Cornelia Schröder and Ashok K. Shukla

Biochemisches Institut, Christian-Albrechts-Universität
Kiel, Olshausenstrasse 40, D-2300 Kiel, West Germany

INTRODUCTION

Sialic acids are derivatives of neuraminic acid with either an acetyl or glycolyl residue at the amino function and frequently one or more O-acetyl group(s) at C-4, C-7, C-8 and preferably at C-9. C-9 can also be O-lactylated and C-8 O-methylated.[1,2] Sialic acids are usually determined by colorimetry,[3] fluorimetry,[4] thin-layer and gas-liquid chromatography.[3,5] Structural analysis is performed by gas-liquid chromatography combined with mass spectrometry (GLC-MS)[5] or by ^{1}H-nuclear magnetic resonance spectroscopy.[6] Most of these methods require purification of the neuraminic acid derivatives, which is time-consuming and may lead to loss of sialic acids and hydrolysis or migration of O-acetyl groups.[7] In GLC analysis the derivatization procedure used so far[5] has sometimes failed to give the expected derivatives, or these were not stable. Furthermore, due to insufficient separation on packed columns, quantitative determination was not satisfactory, either.

Therefore, gas chromatographic analysis of sialic acids has been improved using capillary columns and a more effective derivatization technique,[8] and high performance liquid chromatography was adopted for rapid sialic acid determination.[9]

METHODS AND RESULTS

Capillary Gas-Liquid Chromatography of Sialic Acids

For gas-chromatographic analysis the purified[10] sialic acids are freeze-dried, suspended in dry pyridine and derivatized using N-methyl-N-trimethylsilyltrifluoracetamide (MSTFA, Macherey-Nagel & Co., Düren, West Germany).[8] Depending on the grade of purification of sialic acids and the nature of their substituent at the amino group, trimethylsilyl (TMS) esters, per-O-TMS ethers and/or TMS esters, per-O-TMS ethers, N-TMS derivatives are formed, which are stable for at least two weeks at room temperature. The silylation mixture can be injected directly into the gas chromatograph and no extraction procedure, which may cause loss of sialic acids, is needed. The trimethylsilylated sialic acid derivatives are well separated on fused silica capillary columns (OV-101 or OV-17, Macherey-Nagel & Co., Düren, West Germany). For example, a gas chromatogram of a fraction of free sialic acids from rat urine is shown in Fig. 1. Here the difficulties in gas-chromatographic analysis become obvious: samples obtained from biological material in many cases remain complex mixtures, even after purification by column chromatography on different gels, ion-exchange resins and cellulose.[10] Therefore, a strong silylation reagent like MSTFA is needed to achieve complete derivatization, and for separation of substances having very similar properties the use of capillary columns is recommended. Due to the nearly ideal peak shape, the detection limit of this method is 0.5 ng sialic acid when using a flame ionization detector.

As can be seen in the Table, combination of this sensitive GLC method with mass spectrometry described below had led to the identification of various sialic acid species in the different organisms investigated. The great sialic acid variety in mammals, expecially in cow, is remarkable. In man, six sialic acids were identified, although N-acetylneuraminic acid predominates. This number includes N-glycolylneuraminic acid which has been found in very small amounts (about 0.01% of the total sialic acid content) in human tissues.[10] Some unusual "experiments" with the neuraminic acid molecule have been made by the starfish, belonging to the Echinodermata, which are supposed to be the "inventors" of sialic acid.[1] Substitution at C-8, even by a methyl group, is uncommon in the animal kingdom, as is the combination of a methyl and an acetyl group in one molecule. Figure 1 and the Table further show the occurrence of 2,7-anhydroneuraminic acid derivatives in the sialic acid preparations from rat and starfish, as well as the occurrence of different N- and O-acyl or O-methyl derivatives of 2-deoxy-2,3-dehydroneuraminic acid in various animals including man.

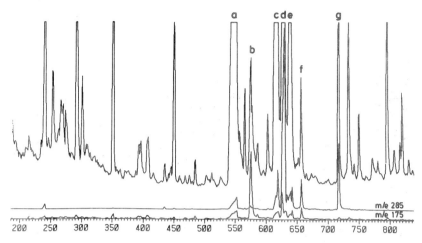

Fig. 1. Gas chromatogram with mass chromatography (m/e 285), indicating the presence of 2,3-unsaturated sialic acids; m/e 175, indicating the presence of an O-acetyl group at C-9) of sialic acids from rat urine, analyzed as trimethylsilyl esters, per-O-trimethylsilyl ethers. a, Neu2,7anhydro5Ac; b, Neu2,7anhydro5,9Ac$_2$ (?); c, Neu5Ac; d, Neu2en5Ac and traces of Neu5,9Ac$_2$; e, Neu2,7anhydro5Gc; f, Neu2en5,9Ac$_2$; g, Neu2en5Gc and traces of Neu5Gc. (For explanation of the abbreviations see the Table.)

High Performance Liquid Chromatography of Sialic Acids

Fast, accurate and sensitive analysis of sialic acids is possible by HPLC using a column of 40 x 4.6 mm filled with a strong basic anion-exchange resin (Aminex A-29, Bio-Rad, München, West Germany). Sialic acids are applied to the column without derivatization and frequently without prior purification, and are eluted isocratically with 0.75 mM sodium sulfate at a flow rate of 0.5 ml/min, 15 bar and detection at 200 nm wavelength. In the system used, not only saturated sialic acids are separated from each other, but also unsaturated sialic acids, different sialyllactoses and the C-7 and C-8 analogues or CMP-glycosides of sialic acid (Fig. 2). Minimum amounts of 200 pg (0.6 pmol) can be determined. For further details about the technique and the retention times of the different neuraminic acid derivatives see ref. 9.

This HPLC method can also be applied to the study of enzyme reactions, where an aliquot of the enzyme assay mixture can directly be injected into the column. (Only in the case of crude enzyme preparations, purification of the sialic acids prior to

Table 1. Naturally Occurring N,O-substituted Saturated, Unsaturated and Anhydro Sialic Acids.

○, sialic acid described in literature (1-3); △, sialic acid newly discovered (unpublished); ☐, sialic acid with known structure, but with so far unknown occurrence (unpublished). The sialic acid nomenclature corresponds to a proposal made at the 5th International Symposium on Glycoconjugates, held at Kiel-Damp in 1979.
Neu, neuraminic acid. Substituents (R): Ac, acetyl; Gc, glycolyl; Lt, lactyl; Me, methyl. Neu2en, 2-deoxy-2,3-dehydroneuraminic acid. Neu2,7anhydro, 2,7-anhydroneuraminic acid. The numbers of abbreviations represent the positions of substituents, for example: Neu7,9Ac$_2$5Gc means 7,9-di-O-acetyl-N-glycolylneuraminic acid.

	Neu5Ac	Neu4,5Ac$_2$	Neu5,7Ac$_2$	Neu5,8Ac$_2$	Neu5,9Ac$_2$	Neu4,5,9Ac$_3$	Neu5,7,9Ac$_3$	Neu5,8,9Ac$_3$	Neu5,7,8,9Ac$_4$	Neu5Ac9Lt	Neu5Ac8Me	Neu4,5Ac$_2$9Lt	Neu5Gc	Neu4Ac5Gc	Neu7Ac5Gc	Neu8Ac5Gc	Neu9Ac5Gc	Neu9Ac5Gc8Me	Neu7,9Ac$_2$5Gc	Neu8,9Ac$_2$5Gc	Neu7,8,9Ac$_3$5Gc	Neu5Gc9Lt	Neu5Gc8Me	Neu2,7anhydro5Ac	Neu2,7anhydro5Gc	Neu2,7anhydro5Gc8Me	Neu2enSAc	Neu2en5,9Ac$_2$	Neu2en5Ac9Lt	Neu2en5Gc	Neu2en5Gc9Lt	Neu2en5Gc8Me
Man	○				○								○														○					
Cow	○		○	○	○		○	○	○	☐			○		○△		○		○	○	○						☐	△		○		
Horse													○														☐					
Pig				○	○					☐			○	○		○	○										☐		△	○	△	
Rat					○					☐	○	○	○				○					△		△	△		☐	△		☐	△	
Chicken	○																															
Starfish													○					△					○			△						△

analysis is necessary.) When compared with other methods,[2,3] only very small amounts of substrate and enzyme are required, and the technique is especially suited for kinetic measurements. A few examples are given in the following.

In the case of sialidase (E.C. 3.2.1.18) information about all the components of interest which may participate in the reaction can be obtained in the same run: e.g. sialyllactose (substrate), sialic acid (product) and 2-deoxy-2,3-dehydro-N-acetylneuraminic acid (inhibitor) (Fig. 2).

The activity of N-acetylneuraminate lyase (EC 4.1.3.3) is determined by measuring the decrease of sialic acid concentration or the formation of acylmannosamines in the course of reaction (Fig. 3). Furthermore, the lyase can be of help for the unequivocal identification of sialic acids, as only HPLC peaks corresponding to sialic acids disappear after treatment by the lyase.

The activity of acylneuraminate cytidylyltransferase (EC 2.7.7.43) is determined by measuring the rate of disappearance of free sialic acids or of appearance of CMP-sialic acids. The retention times of substrates and products can easily be distinguished. The nature of the sialic acids of the CMP-glycosides can be established by hydrolysis of the CMP-glycosides at pH 3-4. It

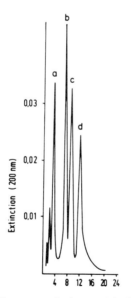

Fig. 2. HPLC analysis of neuraminic acid derivatives on Aminex A-29 [9]. a, (2-3)sialyllactose; b, Neu5Ac; c, Neu5Gc; d, Neu2en5Ac.

was described earlier[11] that only CMP-N-acetylneuraminic acid and not CMP-N-acetyl-9-O-acetylneuraminic acid can be prepared using a partially purified, highly active CMP-sialic acid synthase preparation from frog liver, which was believed to be an expression of the substrate specificity of the synthase from this tissue. However, this phenomenon can now be explained by the presence of a sialate O-acetyl esterase in frog liver (see below) which contaminated the CMP-sialic acid synthase preparation and hydrolyzed the 9-O-acetyl groups in the course of the synthase reaction. This problem has been overcome by the use of diisopropylfluorophosphate, which is known to be an inhibitor of esterases but not of the CMP-sialic acid synthase. Thus, CMP-Neu5,9Ac$_2$ can now be prepared in good yield.

Similar to the acylneuraminate cytidylyltransferase reaction, the activity of CMP-N-acylneuraminate phosphodiesterase (EC 3.1.4.40) is measured by the formation of sialic acid or the disappearance of the corresponding CMP-glycoside.

Using the HPLC method described, carboxyl esterase (EC 3.1.1.?) activities in the liver of horse, cow and frog have been discovered which hydrolyze O-acetyl groups of sialic acids. These esterases seem to be specific with regard to sialic acid and the position of the O-acetyl group, as frog liver esterase mainly hydrolyzes O-acetyl groups at C-9 of sialic acid,[12] while 4-O-acetyl groups are not attacked at a significant rate under identical conditions. In contrast, equine liver carboxyl esterase hydrolyzed 4-O-acetyl groups, the rate, however, being 30% lower than that for 9-O-acetyl groups. As equine tissues contain 4-O-acetylated sialic acids, which cannot be released by sialidase,[2] the availability of a specific esterase activity involved in the degradation of 4-O-acetylated sialo-glycoconjugates appears necessary. It is not yet clear whether the equine esterase activity consists of two different enzymes or whether the same esterase hydrolyzes both 4-O-acetyl and 9-O-acetyl groups. The esterase from equine liver does not hydrolyze 7-O-acetyl groups at a significant rate. Since O-acetyl groups can migrate from C-7 to C-9 under neutral or alkaline conditions (Fig. 3), as known from in vitro experiments and followed by several analytical techniques including NMR-spectroscopy,[7] it is assumed that also in vivo the 7-O-acetyl group migrates from C-7 to C-9 followed by hydrolysis by the esterase. This facilitates the actions of sialidase and sialic acid lyase on the resulting non-esterified sialic acid. In Fig. 3 an example for the study by HPLC of these isomerization and enzymic reactions is shown.

<u>Capillary Gas-Liquid Chromatography and High Performance Liquid Chromatography in Combination with Mass Spectrometry</u>

Most of the GLC-MS data for sialic acids described in the

STRUCTURE AND METABOLISM OF SIALIC ACIDS

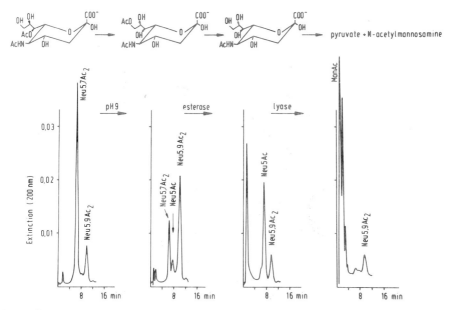

Fig. 3. Study of enzyme reactions in sialic acid metabolism with HPLC. For technical details see text and ref. 9.

literature have been obtained from methyl esters, per-O-TMS ethers, which were prepared by diazomethane and hexamethyldisilazane/ trimethylchlorosilane/pyridine. Based on these mass spectra a fragmentation scheme for sialic acids has been set up.[5] This has proved to be also valid for the TMS esters, per-O-TMS ethers, N-TMS derivatives, as is shown in Fig. 4 for this corresponding derivative of Neu5Ac. However, there are some changes when compared to a mass spectrum of the methyl ester, per-O-TMS ether of Neu5Ac (not shown). In the spectra of TMS esters, per-O-TMS ethers, N-TMS derivatives, fragment G becomes the base peak instead of fragment D. Fragment B has a relatively higher intensity than A, and E-TMSOH is more intense than E. The N-TMS group is easily split off from the fragments, A, B, C and D, resulting in the fragments A', B', C' and D' (Fig. 4). In some cases, the molecular ion can be observed in the mass spectra of the TMS esters, per-O-TMS ethers, N-TMS derivatives, which has never been found for other types of derivatives of sialic acid, e.g. methyl esters, per-O-TMS ethers or TMS esters, per-O-TMS ethers.

Figure 5 shows the mass spectrum of underivatized N-acetyl-neuraminic acid, which has not been recorded before. It has been obtained by a HPLC-MS interface technique, using a direct liquid introduction interface. The elution solvent (acetonitrile:water,

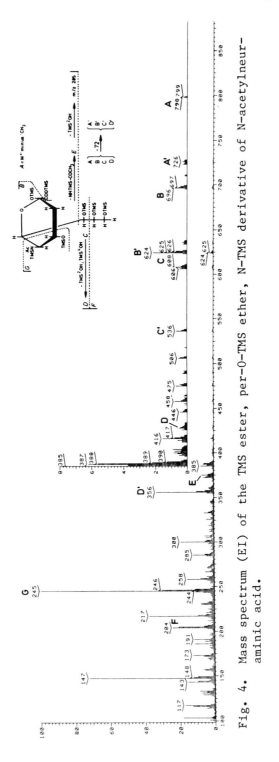

Fig. 4. Mass spectrum (EI) of the TMS ester, per-O-TMS ether, N-TMS derivative of N-acetylneuraminic acid.

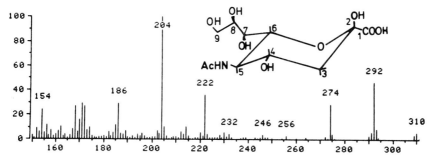

Fig. 5. Mass spectrum (CI) of underivatized N-acetylneuraminic acid.

60:40) was taken for chemical ionization (CI). The fragment with the m/e value 310 represents the $M^+ + H$ ion, which by the loss of H_2O shifts to m/e 292. A second elimination of H_2O leads to m/e 274. Loss of the C_1-C_3 part of the molecule from m/e 310, m/e 292 and m/e 274 gives the fragments m/e 222, m/e 204 and m/e 186, respectively. HPLC-MS can also be applied for the analysis of O-acetylated and unsaturated sialic acids. We thank Prof. Dr. E. Th. Rietschel, Forschungsinstitut Borstel, for giving us the opportunity to perform the HPLC-MS experiments.

DISCUSSION

Capillary GLC and HPLC, especially in combination with mass spectrometry, have been shown to be valuable tools for the analysis of sialic acids, both in the structural elucidation of glycoconjugates and in the study of metabolic reactions. HPLC allows the analysis of sialic acids in most cases without prior purification or derivatization. Therefore, this method will be more commonly used. However, HPLC-MS is less sensitive than capillary GLC-MS, and preference is given to the latter method when determination of trace amounts of sialic acids is necessary.

The use of both capillary GLC and HPLC markedly contributed to our knowledge of the great variety of sialic acids existing in animals and in man and deepened our understanding of the enzyme reactions of sialic acid metabolism. For instance, application of HPLC led to the discovery of sialate O-acetyl esterase, which was a "missing link" in sialic acid metabolism. This enzyme facilitates the action of the catabolic enzymes sialidase and N-acetylneuraminate lyase, as the reaction rates of these are decreased depending on the presence and position of O-acetyl groups on the sialic acid molecule. While the 4-O-acetylated derivatives of neuraminic acid are completely resistant towards the action of sialidases and the lyase, the O-acetyl groups of the sialic acid side chain markedly hinder the activity of these enzymes.[2]

The wide occurrence (Table) of 2,3-unsaturated sialic acids in animals from starfish to mammals is striking and has, with the exception of man,[2] not been reported before. With regard to their origin, earlier studies have shown the formation of Neu2en5Ac from CMP-Neu5Ac by a non-enzymic elimination reaction which can take place under physiological conditions.[2] The formation of Neu2en5Gc and of the 2-eno-sialic acids with O-substituents is assumed to occur in a similar way, as the CMP-glycosides of e.g. Neu5Gc[13] and Neu5,9Ac$_2$[14] have been detected in porcine and bovine submandibular glands, respectively. Although the 2,3-unsaturated sialic acids are potent inhibitors of most sialidases in vitro,[2,15] it is not known whether they have such a function in vivo.

The origin – metabolic or by chemical degradation – of several derivatives of 2,7-anhydroneuraminic acid (Table) also reported here for the first time to occur in biological materials, cannot be explained at present. The formation of N-acetyl-2,7-anhydroneuraminic acid under strong acid conditions has been described by Lifely and Cottee.[16] However, at least in rat urine Neu2,7anhydro5Ac and Neu2,7anhydro5Gc seem not to be artefacts formed during the purification procedure, as the sialic acids were not treated by acids or temperatures higher than 37°C.

The analytical methods described will help answer these questions and further elucidate pathways in sialic acid metabolism or clarify the biological role of sialic acid, which is the only carbohydrate in nature with such manifold chemical modifications and biological functions.

SUMMARY

Sialic acid analysis in biological material including gangliosides is often confronted with the necessity to determine trace amounts of various N,O-substituted species. Therefore, techniques of high sensitivity and resolution are required, such as capillary gas-liquid chromatography (GLC) and high performance liquid chromatography (HPLC). Both methods in combination with mass spectrometry allow structural analysis of the different neuraminic acid derivatives. Thus, the number of natural sialic acids known so far has increased to more than 30, including not only saturated, but also 2,3-unsaturated and 2,7-anhydro-sialic acids. Furthermore, HPLC has proved to be especially useful for the study of enzyme reactions, as the sialic acids of enzyme assay mixtures in most cases can be analyzed without prior extensive purification or derivatization.

REFERENCES

1. A. P. Corfield and R. Schauer, Occurrence of sialic acids, in: "Sialic Acids - Chemistry, Metabolism and Function", R. Schauer, ed., Cell Biology Monographs, Vol. 10, Springer-Verlag Wien, New York, p. 5 (1982).
2. R. Schauer, Chemistry, metabolism and biological functions of sialic acids, in: "Adv. Carbohydr. Chem. Biochem.", R. S. Tipson and D. Horton, eds., Vol. 40, Academic Press, New York, p. 131 (1982).
3. R. Schauer, Characterization of sialic acids, Methods Enzymol. 50C:64 (1978).
4. A. K. Shukla and R. Schauer, Fluorimetric determination of unsubstituted and 9(8)-O-acetylated sialic acids in erythrocyte membranes, Z. Physiol. Chem. 363:255 (1982).
5. J. P. Kamerling and J. F. G. Vliegenthart, Gas-liquid chromatography and mass spectrometry of sialic acids, in: "Sialic Acids - Chemistry, Metabolism and Function", R. Schauer, ed., Cell Biology Monography, Vol. 10, Springer-Verlag Wien, New York, p. 95 (1982).
6. J. Haverkamp, H. van Halbeek, L. Dorland, J. F. G. Vliegenthart, R. Pfeil, and R. Schauer, High-resolution ^{1}H-NMR spectroscopy of free and glycosidically linked O-acetylated sialic acids, Eur. J. Biochem. 122:305 (1982).
7. J. P. Kamerling, H. van Halbeek, J. F. G. Vliegenthart, R. Pfeil, A. K. Shukla, and R. Schauer, Migration of O-acetyl groups in sialic acids, in: "Proc. VIIth Int. Symp. Glycoconjugates", Ronneby/Sweden, 160 (1983).
8. C. Schröder and R. Schauer, Qualitative und quantitative Bestimmung von Acylneuraminsäuren mit Hilfe der Capillarsaulen-Gas-Chromatographie, Fresenius Z. Anal. Chem. 311:385 (1982).
9. A. K. Shukla and R. Schauer, Analysis of N,O-acylated neuraminic acids by high performance liquid chromatography, J. Chromatogr. 244:81 (1982).
10. C. Schröder, U. Nöhle, A. K. Shukla, and R. Schauer, Improved methods for the isolation and structural analysis of trace amounts of new sialic acids - identification of N-glycolylneuraminic acid in man, in: "Proc. VIIth Int. Symp. Glycoconjugates", Ronneby/Sweden, 162 (1983).
11. A. P. Corfield, R. Schauer, and M. Wember, The preparation of CMP-sialic acids using CMP-acylneuraminate synthase from frog liver immobilized on Sepharose 4B, Biochem. J. 177:1 (1979).
12. A. K. Shukla and R. Schauer, High performance liquid chromatography assay of enzymes of the sialic acid metabolism, A. Physiol. Chem. 363:1039 (1982).
13. H.-P. Buscher, J. Casals-Stenzel, P. Mestres-Ventura, and R. Schauer, Biosynthesis of N-glycolylneuraminic acid in porcine submandibular glands, Eur. J. Biochem. 77:297 (1977).

14. A. P. Corfield, C. Ferreira do Amaral, M. Wember, and R. Schauer, The metabolism of O-acyl-N-acylneuraminic acids, Eur. J. Biochem. 68:597 (1976).
15. U. Nöhle, A. K. Shukla, C. Schröder, G. Reuter, J. P. Kamerling, J. F. G. Vliegenthart, and R. Schauer, Synthesis and natural occurrence of 2-deoxy-2,3-didehydro-N-glycolyl-neuraminic acid, Z. Physiol. Chem. 363:1036 (1982).
16. M. R. Lifely and F. H. Cottee, Formation and identification of two novel anhydro compounds obtained by methanolysis of N-acetylneuraminic acid and carboxyl-reduced, meningococcal B polysaccharide, Carbohydrate Res. 107:187 (1982).

RECENT ADVANCES IN STRUCTURAL ANALYSIS OF GANGLIOSIDES: PRIMARY AND SECONDARY STRUCTURES

Robert K. Yu, Theodore A. W. Koerner*, Jr., Peter C. Demou, J. Neel Scarsdale, and James H. Prestegard

Yale University
New Haven, CT 06510 U.S.A.

INTRODUCTION

The growing interest in the biological function of cell surface glycosphingolipids (GSLs) has stimulated the constant search for new methods for analyzing their primary and secondary structures. To determine the primary structure of the oligosaccharide moiety of a GSL, it is necessary to establish the composition and configuration of its sugar residues, and the sequence and linkage sites of the oligosaccharide chain. This information has traditionally been obtained by application of a combination of such procedures as compositional analysis by gas-liquid chromatography, mass spectrometry, permethylation studies, Smith degradation, partial acid or enzyme hydrolysis, optical rotation measurements, etc. However, these techniques are relatively time-consuming, frequently require elaborate derivatization of the intact molecules, and access to many different instruments. Furthermore, these procedures do not afford secondary structural information (conformation) which is important in determining the biological activities of these compounds. Although x-ray crystallography has been used to provide accurate information about the conformation of a carbohydrate by measurement of bond lengths, bond angles, and interatomic distances, it is not certain whether the conformation that exists in a crystalline state is the one that is preferred in solution.

---

*The present address of TAWK is Dept. of Pathology, Tulane University School of Medicine, New Orleans, LA 70112.

Proton nuclear magnetic resonance (NMR) spectroscopy is well-suited for providing the primary and secondary structural information of the oligosaccharide moiety of a GSL because this method is rapid, quantitative, sensitive and nondestructive. However, previous utilization of proton NMR has yielded only fragmentary data concerning the primary structure of GSLs.[1-15] Its potential in providing complete structural information has not been fully realized. The major obstacle lies in the severe resolution problem encountered when most of the sugar ring resonances, except anomeric resonances, fall within a one-ppm chemical shift range that often includes solvent resonances. Previous solutions to these resolution problems have relied upon a variety of double resonance methods.[10-12,14,15] However, these procedures are relatively time-consuming when dealing with complex GSLs where many discrete irradiation frequencies are required. Moreover, while they allow the chemical shifts of most ring protons to be assigned, many of the coupling constants have not been determined.[15]

Two recent advances in NMR spectroscopy have greatly improved the analytical potential of NMR methods. These are the introduction of very high field (500 MHz) NMR spectrometers and the development of 2-dimensional NMR methods. The most useful of the latter methods include 2-D J-correlated spectroscopy (SECSY or COSY), 2-D J-resolved spectroscopy, and 2-D nuclear Overhauser effect (NOE) spectroscopy. Originally proposed by Jeener and developed by Ernst and coworkers,[17-20] these procedures offer the advantages, compared to 1-D experiments, of enhancing the resolution of closely spaced resonances and in providing a direct manifestation of scalar couplings and through-space couplings in a macromolecule. Although these procedures have proven effective for the analysis of proteins, their potential in providing structural information for the primary and secondary structures of oligosaccharides has only been exploited recently. Thus, 2-D SECSY has been used to completely assign the proton NMR spectrum of a ceramide trisaccharide,[21] mono- and di-saccharides,[22-23] and to partially assign the spectrum of a ceramide pentadeca-saccharide.[15] Employing a combination of 1-D and 2-D NMR techniques, we have been able to determine the complete structures of a series of complex GSLs, independent of other methods of structural analysis.[21,24,25,26] In the first part of this paper, we illustrate the stepwise determination of G_{M2} structure by proton NMR.

DETERMINATION OF THE PRIMARY STRUCTURE OF G_{M2} VIA 1-D AND 2-D NMR SPECTROMETRY

Ganglioside G_{M2} (Fig. 1), 1 mg, was dissolved in 0.5 ml of $Me_2SO-d_6-D_2O$ (98:2, v/v) containing 4 mM of Me_4Si as a reference. NMR spectra were recorded on a Bruker WM-500 spectrometer equipped with an Aspect 2000 computer, operating in the Fourier-transform mode with quadrature detection. The probe temperature was kept at 30°C.

Two types of 2-D NMR experiments were performed. The first experiment, 2-D SECSY, which establishes scalar coupling (J) connectivities between peaks, was executed with two 90° pulses separated by a time 1/2 t_1. The first time domain was formed by incrementing t_1. Free induction decays (FIDs) acquired at the end of time t_1 provide the second time domain. The data are displayed as a contour plot after Fourier-transformation in both domains. Except for small displacements due to J-coupling, the central horizontal region corresponds to a normal 1-D spectrum, and off-axis peaks occur at a vertical position corresponding to 1/2 of the chemical shift distance to a spin-coupled resonance. Sequential construction of vertical, 135°, and vertical lines identifies coupled resonances. Each spectrum required a total of 32 pulses in a 256 x 2048 data set, which took about 2 h to acquire. Processing and plotting time for each spectrum was approximately 2 h.

The second experiment (2-D NOE), which establishes through-space connectivities due to dipole-dipole cross relaxation of protons, was executed by using two 90° pulses separated by 1/2 t_1 to selectively invert magnetization. Acquisition after a mixing delay, a third 90° pulse, and an additional delay of 1/2 t_1 is used to establish a second time domain, t_2. Mixing and pulse times were 0.5 s and 11 μs, respectively. Zero filling

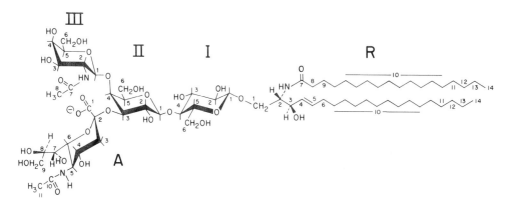

Fig. 1. Structure and numbering system for G_{M2}.

and a window function of $\cos^2\Theta$ (phase shifted by $\pi/4$) were used in both dimensions. Phase cycling and a random increment added to the mixing delay were used to suppress J-peaks. Processing and graphical identification of NOE-coupled resonances are similar to those in the 2-D SECSY experiment. The spectrum required a total of 88 scans in a 256 x 2048 data set, which took 20 h to acquire. Processing and plotting required another 2 h.

Fig. 2 shows a 1-D NMR spectrum of G_{M2}. Except for the region between 3 and 4 ppm in which most of the sugar ring proton resonances reside, several classes of resonances can be identified. In the up-field alkyl region, a 6-proton alkyl methyl multiplet can be seen at 0.85 ppm, which corresponds to the terminal methyl groups of sphingosine and fatty acid. Two 3-proton acetamido methyl singlets are found at 1.88 and 1.78 ppm, characteristics of A-11 and III-8, respectively. The axial and equatorial methylene resonances of sialic acid (A-3a and A-3e) are found at 1.62 and 2.56 ppm, respectively. In the down-field

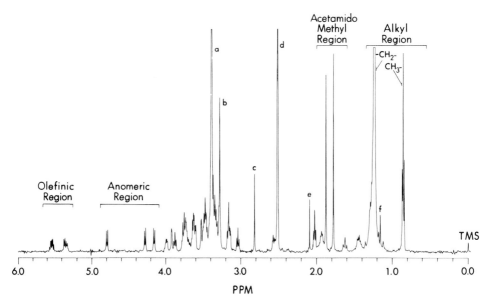

Fig. 2. Proton NMR spectrum of G_{M2}. The following peaks are due to (a) HOD, (b and c) EDTA methylenes, (d) Me_2SO-d_6 methyls, (e) acetone methyls, and (f) stopcock grease methylenes. TMS, Me_4Si reference. Reproduced with permission from ref. 25.

anomeric region (4.0 - 5.0 ppm), three well-resolved one-proton doublets are detected at 4.15, 4.27, and 4.79 ppm. Consideration of the chemical shifts and coupling constants of these anomeric protons indicates that they correspond to β-D-glucopyranosyl, β-D-galactopyranosyl and 2-acetamido-2-deoxy-β-D-galactosyl residues. In the olefinic region (5.2 - 5.6 ppm) are multiplets due to the tran double bond methine protons of the sphingosine moiety. Integration of the above resonance peaks immediately establishes the composition of G_{M2} as NeuAc:Gal:Glc:GalNAc:ceramide (1:1:1:1:1).

In order to assign the sugar ring proton resonances in the crowded region (3-4 ppm) and further confirm identification of composition, the 2-D SECSY experiment proves advantageous as all the ring protons in a sugar residue are circuitally coupled. In addition to chemical shift data, it provides better resolution of spin-spin coupling constants which can be more accurately determined by J-resolved spectroscopy and used to extract configurational data. Fig. 3b shows the 2-D SECSY spectrum of G_{M2} in the 3-5 ppm region. The corresponding segment of the 1-D spectrum is shown in Fig. 3a for comparison. Spectral assignments begin with the anomeric resonances, well resolved in the 4-5 ppm region. Three series of J-connectivities can be established; each series corresponds to a single aldopyranoside ring. With these connectivities, integration data and the process of elimination as guides, all resonances in each of the three aldopyranosides can be resolved and assigned. Table 1 lists the proton chemical shifts and coupling constants for G_{M2}.

The configuration of residues I, II, and III can be independently assigned by considering the coupling constants manifested by their ring protons (H-1 through H-5). This configurational analysis is possible because aldopyranoside rings typically present in gangliosides are known to exist in rigid $^4C_1$ chair conformation and the Karplus relationships between the ring protons of such rings have been well studied.[27,28] Thus, for residue I the series of coupling constants $J_{1,2}$, $J_{2,3}$, $J_{3,4}$, and $J_{4,5}$ are all greater than 6 Hz, indicating that all ring protons are axial. Hence residue I must assume a β-gluco configuration. Similar analysis for residues II and III reveals that both residues are of β-galacto configuration. Since the chemical shifts for H-1 and H-2 of residue III are further downfield compared with those for residues I and II, III should then correspond to the 2-acetamido-2-deoxy-β-galacto-pyranosyl residue.

The sequence and sites of glycosidic linkage of the neutral oligosaccharide backbone of G_{M2} are revealed by the 2-D NOE spectrum shown in Fig. 4. Inspection of the anomeric region

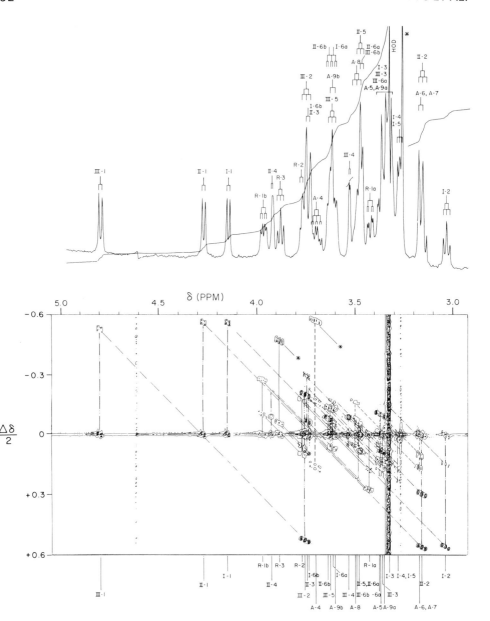

Fig. 3. Top : Integrated, 1-D NMR spectrum of G_{M2} between 3 and 5 ppm. Bottom : 2-D SECSY spectrum of G_{M2}. J-conectivities are labeled for sugar residues as follows: I, β-glucopyranosyl (-·-); II, β-galactopyranosyl (-··-); III, 2-acetamido-2-dexoy-β-galactopyranosyl (-···-); A, N-acetylneuraminosyl (--); R, ceramide. A asterisk denotes J-connectivities to protons outside observed window. Reproduced with permission from ref. 25.

Table 1. Chemical Shifts and Coupling Constants for G_{M2}.[a]

proton chemical shifts, δ (ppm)

residue	H-1	H-2	H-3	H-4	H-5	H-6a	H-6b	H-7	H-8	H-9a	H-9b	H-10[b]
I	4.150	3.035	3.325	3.277	3.289	3.613	3.744	–	–	–	–	–
II	4.274	3.158	3.744	3.920	3.470	3.463	3.613	–	–	–	–	–
III	4.794	3.753	3.39	3.524	3.642	3.39	3.463	–	1.777	–	–	–
A	–	–	2.558(e) 1.621(a)	3.700	3.39	3.148		3.161	3.485	3.340	3.613	–
R	3.421(a) 3.987(b)	3.774	3.875	5.346	5.534	1.932		–	2.023	–	–	1.235

proton-proton coupling constant J (Hz)

residue	$J_{1,2}$	$J_{2,3}$	$J_{3,4}$	$J_{4,5}$	$J_{5,6a}$	$J_{5,6b}$	$J_{6a,6b}$	$J_{6,7}$	$J_{7,8}$	$J_{8,9a}$	$J_{8,9b}$	$J_{9a,9b}$
I	7.9	8.2	9.6	9.6	6.0	1.5	–11.5	–	–	–	–	–
II	9.6	8.7	<1.5	<1.5	6.0	5.6	–12.1	–	–	–	–	–
III	8.8	9.6	1.5	<1.5	5.9	6.0	–12.0	–	–	–	–	–
A	–12.4 (3e,3a)	5.1(3e,4) 10.6(3a,4)		10.6	9.1		–	1.9	9.8	6.0	1.5	–10.8
R	3.3(1a,2) –9.6 4.1(1b,2) (1a,1b)	7.7	7.4	15.7	6.0		–	–	–	7.4	7.4	–

[a] The estimated error for δ is ± 0.001 ppm and that for J is ± 0.6 Hz.
[b] Includes R-9, R-11, R-12 and R-13.

reveals two types of through-space couplings for each of the anomeric protons. The first type is due to the intra-residue 1,3 and 1,5 diaxial couplings, which are useful for confirming configurational assignments. The second type is due to inter-residue couplings between the anomeric proton and the transglycosidic proton. The latter information is used to establish the site of glycosidation. Thus, I-1 is coupled to R-1a, II-1 to I-4, and III-1 to II-4. Combination of these data then establishes the sequence of the neutral hexoses as III(1-4)II(1-4)I(1-1´)R. The structure is therefore 2-acetamido-2-deoxy-galactopyranosyl(β1-4)galactopyranosyl(β1-4)glucopyranosyl (β1-1´)ceramide.

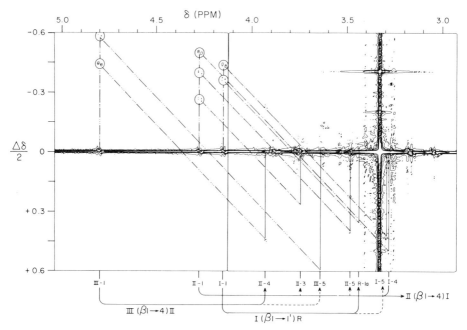

Fig. 4. 2-D NOE spectrum of G_{M2} between 3 and 5 ppm. NOE couplings are labeled for each of the anomeric protons. At the bottom, <u>inter-residue</u> couplings are labeled with solid lines and <u>intra-residue</u> couplings with dashed lines. The glycosidic linkages between residues are revealed by inter-residue couplings. Reproduced with permission from ref. 26.

The final step in establishing the primary structure of G_{M2} is to locate the site of substitution of the sialic acid residue. Examination of the proton resonances of the core oligosaccharide in G_{M2} and those in asialo-G_{M2}[25,26] reveals that II-3 proton resonance in G_{M2} is shifted downfield by 0.23 ppm compared with the II-3 proton in asialo-G_{M2}. Hence the sialic acid residue is linked to II-3. When this information is combined with the structure deduced for the neutral oligosaccharide core, the complete structure of G_{M2} can be established as GalNAc(β1-4)[NeuAc(α2-3)]Gal(β1-4)Glc(β1-1´) ceramide. The structure is identical to that assigned by chemical and enzymatic methods.[29,30]

DETERMINATION OF THE SECONDARY STRUCTURE OF GLOBOSIDE

Since GSLs are involved in cell-cell recognition, enzyme substrate interaction, antigen-antibody interaction, and other molecular events that may involve recognition processes, it is reasonable to assume that these events are necessarily governed not only by the primary structures but also by the secondary structures of the oligosaccharide moieties. NMR has been proven to be particularly useful in this regard. Studies from Barker's laboratory on a number of oligosaccharides employing ^{13}C-^{13}C and ^{13}C-^1H couplings to estimate the Φ and Ψ dihedral angles about a glycosidic bond have indicated that a specific conformation exists between glycosidic residues.[31-34] However, the measurement of natural abundance ^{13}C-^{13}C and ^{13}C-^1H couplings is difficult. Specific ^{13}C enrichment is necessary but is difficult to achieve. An alternative uses hard sphere exo anomeric (HESA) calculations to locate a potential minimum. The conformation corresponding to this minimum can be experimentally verified by NOE difference spectroscopy or proton-proton spin-lattice relaxation rate measurements.[35,36]

In our own work using 2-D NOE spectroscopy to measure the proton-proton dipolar cross-relaxation network in various GSLs,[21,25,26] we frequently found that the magnitude of the NOE between trans-glycosidic pair of protons were equal to or greater than that between intra-residue 1,3 or 1,5-diaxial couplings. Since the cross-relaxation rate of protons (NOE) is inversely proportional to the distance to the sixth power, this suggests that the trans-glycosidic pairs of protons are close in space and conformers which bring these pairs close must be highly populated. Since a rigorous treatment of equilibrium NOEs in groups of dipolar coupled protons requires explicit consideration of all cross relaxation pathways and determination of a sufficient number of enhancements to yield a unique set of interprotonic distances for this entire set,[37] we have employed time-dependent 1-D NOE spectroscopy which allows one to analyze interactions on a pair-wise basis. This allows us to estimate the <u>average</u> (weighed by $1/\underline{r}^6$) interprotonic distances, $\underline{r}$, across the glycosidic linkages of several oligosaccharides. This is illustrated using globoside as an example.

The structure of globoside, GalNAc(β1-3)Gal(α1-4)Gal(β1-4)Glc (β1-1')ceramide, is shown in Fig. 5. Time dependent NOE data were generated for the four anomeric protons whose chemical shifts have been previously assigned[38] on the basis of a 2-D NOE experiment and a 2-D SECSY spectrum (J.N. Scarsdale, T.A.W. Koerner, J.H. Prestegard, and R.K. Yu, unpublished data). Spectra were obtained in the difference mode[39] with varying irradiation times for each of the anomeric protons. For such a series of experiments the time dependent NOE is expressed as $\eta_j(t) = \sigma_{ij}/\rho_j(1-e^{-\rho_j t})$, where $\eta_j(t)$ is the NOE, σ_{ij} the cross-relaxation rate between protons i (the irradiated proton) and j, ρ_j the total relaxation of proton j, and t the time of irradiation.[40] The term σ_{ij}, which is proportional to $1/\underline{r}^6$, can be determined for both intra-residue and inter-residue due pairs. Since the inter-residue distance is known, the ratio of inter- and intra-residue σs can be used to estimate the inter-residue interprotonic distance r_{ij}. The results of such a study are shown in Table 2.

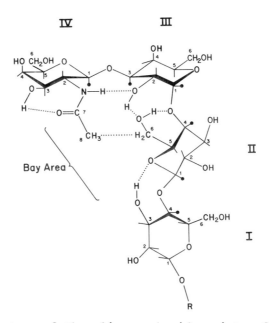

Fig. 5. Structure of the oligosaccharide moiety of globoside. The secondary structure shown is the one that could be stabilized by an extensive network of H-bonds and a van der Waals interaction.

Table 2. Interprotonic Distances of Globoside Terminal Residues

proton pair	σ	r (A)
IV(1)-IV(3)	-.16	(2.5)*
IV(1)-IV(5)	-.16	(2.5)*
IV(1)-III(3)	-.15	2.5
IV(1)-III(4)	-.086	2.8
III(1)-III(2)	-.30	(2.5)*
III(1)-II(4)	-.30	2.5
III(1)-II(5)	-.05	3.2
III(1)-III(3)	-.10	**

*The intra-residue distances, known from x-ray crystallographic data,[41,42] are used as calibration standards.
**This intensity could be the result of strong coupling between III(2) and III(3), hence no distances are calculated.

Construction of Dreiding models of the glycosidic linkages of globoside with fixed interprotonic distances (Table 2) and systematic inspection of allowed dihedral angles Φ and Ψ (+60°, 0°, and -60°) reveal a likely conformation to be that shown in Figs. 5 and 6.

While it is possible that such conformations are highly solvent dependent, several interesting structural features are apparent for this conformation: 1) An extended H-bond network involving ten atoms and four H-bonds is possible. This begins with the IV-3 hydroxyl and ends with the II-4 linkage oxygen. The sequence of this extended H-bond network is O(II-4)...H-O(II-6)...H-O(III-2)...H-N-C=O(IV-2)...H-O(IV-3), proceeding from H-bond acceptor terminus to the H-bond donor terminus. Such extended, vectoral H-bond networks could result in stabilized secondary structures for oligosaccharides, as has been noted by Jeffrey and Takagi;[43] 2) A van der Waals interaction between the IV-2-acetamido methyl and the II-6 methylene can exist; 3) An overall "L-shape" for the oligosaccharide with a hydrophobic inner-side or "bay area" (containing the van der Waal interaction and extended H-bond network) and a hydrophilic outer-side (presenting all non-H-bonded hydroxyl groups) is apparent. We are currently extending the above study to aqueous solvents and to other GSLs with the hope of relating conformational properties to membrane surface interaction.

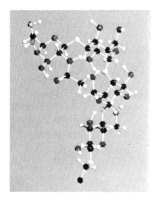 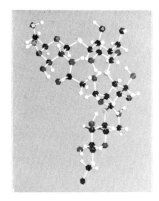

Fig. 6. A 3-D stereo representation of the oligosaccharide moiety of globoside.

SUMMARY

High-field (500 MHz) proton NMR has been used to elucidate the primary and secondary structures of glycosphingolipids (GSLs). Using 2-D J-correlated spectroscopy (2-D SECSY) which establishes scalar couplings of protons, the monosaccharide composition, anomeric configuration and aglycon structures of a GSL can be established. 2-D nuclear Overhauser effect spectroscopy (2-D NOE) then establishes through-space intra- and inter-residue couplings of cross-relaxing protons. We have found that each anomeric proton is involved in NOE couplings with inter- and intra-residue protons. The inter-residue coupling, resulting from interaction of protons across the glycosidic linkage, establishes the n-1 sugar residue and specific glycosidation site to which the n-residue is linked. When such information is known for each residue and is combined, the sequence of the core oligosaccharide is obtained. The sialylation-induced glycosidation shift is then used to establish the site of sialic acid residue attachment in a ganglioside molecule. We have also observed that the anomeric proton inter-residue NOE couplings can be used to suggest the preferred conformation of an oligosaccharide. We have found that the oligosaccharide residue of globoside exists in a unique and rather rigid conformation which could be stabilized by hydrogen bonds and van der Waals interactions. Since GSLs are known to have a receptor role and are implicated in cell-cell recognition, enzyme-substrate interaction and antigen-antibody interaction, the determination of their conformation should be useful in understanding their biological functions.

ACKNOWLEDGMENT

This research was supported by USPHS grants HL06442, NS 11853, NMSS grant RG 1289-B-3, and NSF grant CHE-7916210.

REFERENCES

1. J. Kawanami and T. Tsuji, Structure of the oligosaccharide from mammalian glycolipids, Chem. Phys. Lipids 7:49 (1971).
2. S. Handa, T. Ariga, T. Miyatuke, and T. Yamakawa, Presence of α-anomeric glycosidic configuration in the glycolipids accumulated in kidney with Fabry's disease, J. Biochem. 69:625 (1971).
3. S. Ando, M. Kon, M. Isobe, Y. Nagai, and T. Yamakawa, Existence of glucosaminyl lactosyl ceramide (amino CTH-I) in human erythrocyte membranes as a possible precursor of blood group-active glycolipid, J. Biochem. 79:625 (1976).
4. P. L. Harris and E. R. Thornton, Carbon-13 and proton nuclear magnetic resonance studies of gangliosides, J. Amer. Chem. Soc. 100:6738 (1980).
5. K.-E. Falk, K.-A. Karlsson, and B. E. Samuelsson, Proton nuclear magnetic resonance analysis of anomeric structure of glycosphingolipids: the globo-series (one to five sugars), Arch. Biochem. Biophys. 192:164 (1979).
6. K.-E. Falk, K.-A. Karlsson, and B. E. Samuelsson, Proton nuclear magnetic resonance analysis of anomeric structure of glycosphingolipids: blood group ABH-active substance, Arch. Biochem. Biophys. 192:177 (1979).
7. K.-E. Falk, K.-A. Karlsson, and B. E. Samuelsson, Proton nuclear magnetic resonance analysis of anomeric structure of glycosphingolipids: Lewis-active and Lewis-like substance, Arch. Biochem. Biophys. 192:191 (1979).
8. J. F. G. Vliegenthart, High resolution ^{1}H-NMR spectroscopy of carbohydrate structure, Adv. Exp. Med. Biol. 125:77 (1980).
9. S. Gasa, T. Mitsuyama, and A. Makita, Proton nuclear magnetic resonance of neutral and acidic glycosphingolipids, J. Lipid Res. 24:174 (1983).
10. J. Dabrowski, H. Egge, and P. Hanfland, High resolution nuclear magnetic resonance spectroscopy of glycosphingolipids. I. 360 MHz ^{1}H and 90.5 MHz ^{13}C NMR analysis of galactosylceramide, Chem. Phys. Lipids 26:187 (1980).
11. J. Dabrowski, P. Hanfland, and H. Egge, Structural analysis of glycosphingolipids by high-resolution ^{1}H nuclear magnetic resonance spectroscopy, Biochemistry 19:5652 (1980).

12. P. Hanfland, H. Egge, U. Dabrowski, S. Kulin, D. Roelcke, and J. Dabrowski, Isolation and characterization of an I-active ceramide decasaccharide from rabbit erythrocyte membrane, Biochemistry 20:5310 (1981).
13. J. Dabrowski, P. Hanfland, H. Egge, and U. Dabrowski, Immunochemistry of the Lewis-blood-group system: proton nuclear magnetic resonance study of plasmatic Lewis-blood-group-active glycosphingolipids and related substances, Arch. Biochem. Biophys. 210:405 (1981).
14. J. Dabrowski and P. Hanfland, Structure determination of a ceramide pentadecasaccharide by two-dimensional J-resolved and J-correlated NMR spectroscopy, FEBS Lett. 142:138 (1982).
15. J. Dabrowski, P. Hanfland, and H. Egge, Analysis of glycosphingolipids by high-resolution proton nuclear magnetic resonance spectroscopy, Methods Enz. 83:69 (1982).
16. A. Yamada, J. Dabrowski, P. Hanfland, and H. Egge, Preliminary results of J-resolved, two-dimensional ^{1}H-NMR studies on glycosphingolipids, Biochim. Biophys. Acta 618:473 (1980).
17. W. P. Aue, E. Bartholdi, and R. R. Ernst, Two-dimensional spectrscopy. Application to nuclear magnetic resonance, J. Chem. Phys. 64:2229 (1976).
18. K. Nagayama, A. Kumar, K. Wuthrich, and R. R. Ernst, Two-dimensional spin-echo correlated spectroscopy (SECSY) for ^{1}H NMR studies of biological molecules, J. Mag. Res. 40:321 (1980).
19. J. Jeener, B. H. Meier, P. Bachmann, and R. R. Ernst, Investigation of exchange processes by two-dimensional NMR spectroscopy, J. Chem. Phys. 71:4546 (1979).
20. A. Kumar, R. R. Ernst, and K. Wuthrich, A two-dimensional nuclear Overhauser enhancement (2D NOE) experiment for the elucidation of complete proton-proton cross-relaxation networks in biological macromolecules, Biochem. Biophys. Res. Commun. 95:1 (1980).
21. J. H. Prestegard, T. A. W. Koerner, P. C. Demou, and R. K. Yu, Complete analysis of oligosaccharide primary structure using two-dimensional high field proton NMR, J. Amer. Chem. Soc. 104:4993 (1982).
22. M. A. Bernstein and L. D. Hall, De novo sequencing of oligosaccharides by proton NMR spectroscopy, J. Am. Chem. Soc. 104:5553 (1982).
23. M. A. Bernstein, L. D. Hall, and S. Sukumar, Assignment of proton n.m.r. spectra of carbohydrates, using two-dimensional techniques: COSY and SECSY, Carbohydrate Res. 103:C1 (1982).
24. T. A. W. Koerner, J. H. Prestegard, and R. K. Yu, Analysis of

ganglioside structure via high resolution proton NMR spectroscopy, Fed. Proc. 41:1170 (1982).
25. T. A. W. Koerner, J. H. Prestegard, P. C. Demou, and R. K. Yu, High-resolution proton NMR studies of gangliosides. 1. Use of homonuclear two-dimensional spin-echo J-correlated spectroscopy for determination of residue composition and anomeric configurations, Biochemistry 22:2676 (1983).
26. T. A. W. Koerner, J. H. Prestegard, P. C. Demou, and R. K. Yu, High-resolution proton NMR studies of gangliosides. 2. Use of two-dimensional nuclear Overhauser effect spectroscopy and sialylation shifts for determination of oligosaccharide sequence and linkage sites, Biochemistry 22:2687 (1983).
27. L. D. Hall, High-resolution nuclear magnetic resonance spectroscopy, in: The Carbohydrates, Chemistry and Biochemistry (W. Pigman, D. Horton, and J.D. Wander, eds.) Vol. IB, pp. 1299-1326, Academic Press, N.Y.
28. C. Altona and C. A. G. Haasnoot, Prediction of anti and gauche vicinal proton-proton coupling constants in carbohydrates: a simple additivity rule for pyranose rings, Org. Mag. Resonance 13:417 (1980).
29. A. Makita and T. Yamakawa, The glycolipids of the brain of Tay-Sachs' disease - the chemical structure of a globoside and main ganglioside, Japan J. Exp. Med. 33:361 (1963).
30. R. Ledeen and K. Salsman, Structure of the Tay-Sachs' ganglioside. I., Biochemistry 4:2225 (1965).
31. H. A. Nunez and R. Barker, Enzymatic synthesis and carbon-13 nuclear magnetic resonance conformational studies of disaccharides containing β-D-galactopyranosyl and β-D-[1-^{13}C]galactopyranosyl residue, Biochemistry 19:489 (1980).
32. M. L. Hayes, A. S. Serianni, and R. Barker, Methyl β-lactoside: 600-MHz ^{1}H and 75-MHz ^{13}C-N.M.R. Studies of ^{2}H-and ^{13}C-enriched compounds, Carbohydrate Res. 100:87 (1982).
33. P. R. Rosevear, H. A. Nunez, and R. Barker, Synthesis and solution conformation of the type 2 blood group oligo-saccharide αL Fuc (1→2)βDGal(1→4)βDGlcNAc, Biochemistry 21:1421 (1982).
34. R. Barker, H. A. Nunez, P. R. Rosevear, and A. S. Serianni, ^{13}C NMR analysis of complex carbohydrates, Methods Enzymol. 83:58 (1982).
35. R. U. Lemieux, K. Bock, L. T. J. Delbaere, S. Koto, and V. S. Rao, The conformations of oligosaccharides related to the ABH and Lewis human blood group determinants, Can. J. Chem. 58:631 (1980).

36. K. Bock and R. U. Lemieux, The conformational properties of sucrose in aqueous solution: Intramolecular hydrogen bonding, Carbohydrate Res. 100:63 (1982).
37. A. A. Bothner-By and J. H. Noggle, Time development of nuclear Overhauser effects in multispin systems, J. Am. Chem. Soc. 101:5152 (1979).
38. T. A. W. Koerner, J. H. Prestegard, and R. K. Yu, Determination of the sequence and trans-glycosidic interprotonic distances of globoside using two-dimensional nuclear Overhauser effect proton NMR spectroscopy, Biochem. Biophys. Res. Commun. (submitted).
39. R. Richarz and K. Wuthrich, NOE difference spectroscopy: A novel method for observing individual multiplets in proton NMR spectra of biological macromolecules, J. Magn. Reson. 30:147 (1978).
40. J. H. Noggle and R. E. Shirmer, The nuclear Overhauser effect, Academic Press, N.Y.
41. F. Longchambon, J. Ohannessian, D. Avenel, and A. Neuman, Structure cristalline due β-D-galactose et de 1′α-L-fucose, Acta Crystallog. 31B:2623 (1975).
42. B. Sheldrick, The crystal structures of the α- and β-anomers of D-galactose, Acta Crystallogr. 32B:1016 (1976).
43. G. A. Jeffrey and S. Takagi, Hydrogen-bond structure in carbohydrate crystals, Accounts Chem. Res. 11:264 (1978).

ADRENAL MEDULLA GANGLIOSIDES

Michiko Sekine, Toshio Ariga and Tadashi Miyatake*

Department of Biochemistry and Metabolism
Tokyo Metropolitan Institute of Medical Science
Honkomagome, Bunkyo-ku, Tokyo 113, Japan
*Department of Neurology
Brain Research Institute
Niigata University
Asahimachi 1, Niigata 951, Japan

INTRODUCTION

The adrenal medulla, which belongs to the autonomic nervous system, contains high amounts of gangliosides.[1,2] Ledeen and co-workers isolated three monosialogangliosides and two disialo-gangliosides from bovine adrenal medulla and determined their structures as N-acetyl GM3, N-glycolyl GM3, GM1, GD1a and GD1a containing 1 mole each of N-acetyl- and N-glycolylneuraminic acid (Table 1).[1,3] Additionally, three disialogangliosides and four trisialogangliosides were recently found in bovine adrenal medulla (Table 1).[4,5]

Gangliosides have been recognized as constituents of cell membranes and are implicated in cell fusion and recognition phenomena as receptors. The adrenal medulla contains chromaffin granules which store the catecholamine and release it into the bloodstream by exocytosis. The gangliosides in adrenal medulla have been studied to better understand the molecular mechanism implicated in the exocytosis process.

Abbreviations: GM3(A), NeuNAc-Gal-Glc-Cer; GM3(G), NeuNGly-Gal-Glc-Cer; GD3(A,A), NeuNAc-NeuNAc-Gal-Glc-Cer; GD3(G,G), NeuNGly-NeuNGly-Gal-Glc-Cer.

Table 1. Bovine Adrenal Medulla Gangliosides.

(1) NeuNAc(2→3)Gal(1→4)Glc(1→1')Cer

(2) NeuNGly(2→3)Gal(1→4)Glc(1→1')Cer

(3) Gal(1→3)GalNAc(1→4)Gal(1→4)Glc(1→1')Cer
                              (3←2)α
                             NeuNAc

(4) Gal(1→3)GalNAc(1→4)Gal(1→4)Glc(1→1')Cer
    (3←2)α   (3←2)α
    NeuNAc   NeuNAc

(5) Gal(1→3)GalNAc(1→4)Gal(1→4)Glc(1→1')Cer
    (3←2)   (3←2)α
    NeuNAc   NeuNGly

(6) NeuNAc(α2→8)NeuNGly(α2→3)Gal(β1→4)Glc(β1→1')Cer

(7) NeuNGly(α2→8)NeuNGly(α2→3)Gal(β1→4)Glc(β1→1')Cer

(8) Gal(β1→3)GalNAc(β1→4)Gal(β1→4)Glc(β1→1')Cer
    (3←2)α
    NeuNGly   NeuNGly

(9) Gal(β1→3)GalNAc(β1→4)Gal(β1→4)Glc(β1→1')Cer
                              (3←2)α
                             NeuNAc
                              (8←2)α
                             NeuNAc

(10) Gal(β1→3)GalNAc(β1→4)Gal(β1→4)Glc(β1→1')Cer
     (3←2)α   (3←2)α
     NeuNAc   NeuNGly
                           (8←2)α
                           NeuNAc

(11) Gal(β1→3)GalNAc(β1→4)Gal(β1→4)Glc(β1→1')Cer
     (3←2)α   (3←2)α
     NeuNGly   NeuNAc
                           (8←2)α
                           NeuNAc

(12) Gal(β1→3)GalNAc(β1→4)Gal(β1→4)Glc(β1→1')Cer
     (3←2)α   (3←2)α
     NeuNAc   NeuNGly
                           (8←2)α
                           NeuNGly

Ganglioside (1)-(5): Ledeen, et al.[1] and Price, et al.[3]
Ganglioside (6)-(12): Ariga, et al.[4,5]

In the present study we have examined gangliosides of adrenal glands from several animals and separated chromaffin granules to clarify whether specific gangliosides exist in this organ and organelle.

MATERIALS AND METHODS

Extraction and preparation of gangliosides from adrenal glands

Adrenal glands of mouse, rat, guinea pig, rabbit, monkey, pig, ox, and chicken were obtained shortly after death and lyophilized. Gangliosides were prepared by the method of Ando et al.[6] with a slight modification. Total lipids were extracted from the lyophilized tissues with chloroform/methanol (2:1), (1:1) and chloroform/methanol/water (30:60:4.5). The lipids, dissolved in chloroform/methanol/water (30:60:4.5), were applied to a DEAE-Sephadex A-25 column. After removing neutral lipid with chloroform/methanol (1:2) and methanol, the acidic lipid fraction was eluted with 0.2M sodium acetate in methanol. After mild alkaline treatment and neutralization, the acidic lipid fraction was desalted by Sephadex LH-20 column chromatography with methanol as the eluting solvent to yield crude gangliosides.

Determination of ganglioside composition and sugar sequence

Lipid-bound sialic acid was measured by the resorcinol method.[7] The ganglioside composition was examined by TLC with the solvent systems of (a) chloroform/methanol/water (55:45:10,v/v/v) containing 0.02% $CaCl_2 \cdot 2H_2O$ and (b) chloroform/methanol/5N NH_4OH/0.4% $CaCl_2 \cdot 2H_2O$ (60:40:4:5). The gangliosides were visualized by heating the plates at 95°C with the resorcinol reagent and the chromatogram was scanned at 580 nm using a TLC densitometer.

The sugar compositions of isolated gangliosides were determined by gas liquid chromatography (GLC). Neutral sugars were determined as their trimethylsilyl derivatives by GLC.[8] Sialic acid was determined by the method of Yu and Ledeen.[9]

Subcellular fractionation

Chromaffin granules, mitochondria and microsomes were prepared from bovine adrenal medulla by the method of Smith and Winkler[10] with a slight modification. Each fraction, thus obtained, was suspended in 0.005M Tris-succinate buffer (pH 5.9), freeze-thawed and centrifuged at 100,000g for 45 min. This procedure was repeated five times to yield the purified membranes.

RESULTS AND DISCUSSION

Gangliosides of adrenal glands

Ganglioside contents in adrenal glands of the examined eight animals were from 19.4 µg sialic acid/g wet tissue in guinea pig to 186.2 in pig. Price and Yu[2] have analyzed gangliosides from adrenal glands of human and four domestic mammals including ox and pig. Our data for ox and pig agreed with their results.

As demonstrated in Fig. 1, the major gangliosides were GM3(G) in mouse and GM3(A) in rat and guinea pig. Rabbit had GM3(A) and GD3(A,A), which were similar to the ganglioside pattern for human.[11] Chicken had also GM3(A) and GD3(A,A) as the major gangliosides. Monkey had GM3(G) and GD3(G,G). In guinea pig and chicken, a ganglioside which co-migrated on TLC with human brain GM4 was detected. Gangliosides were separated and purified from 25g of guinea pig adrenal glands and the ganglioside which showed the same Rf on TLC as GM4 was obtained. The sugar composition of this ganglioside was determined to be galactose and N-acetylneuraminic acid in the molar ratio of 1:1, and this ganglioside was converted to galactosylceramide by neuraminidase treatment. By these analytical data, this ganglioside was determined to be GM4. Originally, GM4 was found in human brain[12] and then in some organs of chicken. From these results, the major adrenal gland gangliosides were determined to be GM3 and GD3 with N-acetyl- or N-glycolylneuraminic acid.

Distribution of ganglioside

We have examined the ganglioside concentrations of bovine adrenal medulla and cortex. In adrenal medulla, the ganglioside-sialic acid was about 3 times as much as in adrenal cortex.

Three subcellular fractions (chromaffin granule, microsomal and mitochondrial) were separated from bovine adrenal medulla. Electron micrographic examinations of chromaffin granules showed well-preserved structures characteristic of the granules in the cells of bovine adrenal medulla. Mitochondria were not found in this fraction. Two marker enzymes, NADPH-cytochrome reductase for microsomes and cytochrome oxidase for mitochondria, were very low in the fraction of chromaffin granules. Therefore, the contamination by microsomes or mitochondria was considered to be minor. The distribution of gangliosides in these subcellular fractions is shown in Table 2. The ganglioside concentration of chromaffin granule was about 3 and 7 times as much as microsomal and mitochondrial fractions, respectively. Two gangliosides, GM3(A) and GM3(G) were found in these three fractions and the

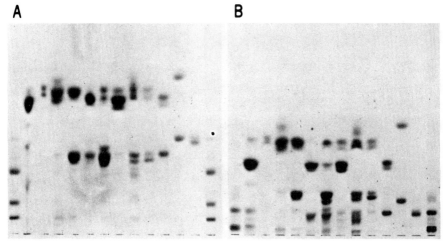

Fig. 1. Thin layer chromatogram of gangliosides from adrenal glands. Lanes are as follows: 1, human grey matter gangliosides; 2, mouse; 3, rat; 4, guinea pig; 5, rabbit; 6, monkey; 7, pig; 8 ox; 9, chicken; 10, GM3 (A) and GD3(A,A); 11, GM3(G) and GD3(G,G); 12 GM4 and GM1; 13, N-glycolyl GM1; 14, human gray matter gangliosides. The solvent systems were (A) chloroform/methanol/water (55:45:10), containing 0.02% $CaCl_2 \cdot 2H_2O$ and (B) chloroform/methanol/5N NH_4OH/0.4% $CaCl_2 \cdot 2H_2O$ (60:40:4:5). Gangliosides were visualized after resorcinol spray.

Table 2. Distribution of Gangliosides in Subcellular Membranes of Bovine Adrenal Medulla.

	Lipid-bound sialic acid	NeuNAC	NeuNGly
	nmol/mg protein	%	
Chromaffin granules (N=7)	39.2 ± 4.9	47.6	52.4
Mitochondria (N=7)	5.5 ± 1.6	38.7	61.3
Microsomes (N=7)	14.2 ± 1.9	40.8	59.2

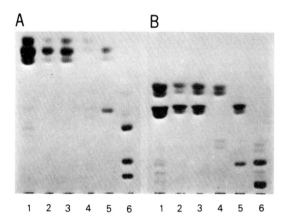

Fig. 2. Thin layer chromatogram of gangliosides from subcellular fractions. Lanes are as follows; 1, microsomal fraction; 2, mitochondrial fraction; 3, chromaffin granules; 4 GM3(A) and GD3(A,A); 5, GM3(G) and GD3(G,G); 6, human gray matter gangliosides. The solvent systems were (A) chloroform/methanol/water (55:45:10,v/v/v) containing 0.02% $CaCl_2 \cdot 2H_2O$ and (b) chloroform/methanol/5N NH_4OH/ $CaCl_2 \cdot 2H_2O$ (60:40:4:5). Gangliosides were visualized after resorcinol spray.

ratio of GM3(A) to GM3(G) was very similar in these three fractions (Fig. 2).

Ganglioside compositions of bovine chromaffin granules have been reported.[13,14] Dreyfus et al. found only GM3(A) in chromaffin granules, obtained by their preparation.[13] Geissler et al., however, demonstrated two hematosides, GM3(A) and GM3(G), in their chromaffin granules, which agrees with our results described above. Recently, the topography of gangliosides of the chromaffin granule has been studied[15] and at least 75% of the gangliosides are located on the inner leaflet of the membrane. However, the biological function of these gangliosides in chromaffin granules remains obscure.

SUMMARY

The gangliosides were examined in adrenal glands of mouse, rat, guinea pig, rabbit, monkey, pig, ox and chicken. GM3 ganglioside was predominant in all examined animals except pig. In pig GD3 ganglioside was the major one. GM4 ganglioside was found in guinea pig and chicken. The distribution of sialic acid varied in each species. NeuNGly containing gangliosides were not detected in rat, guinea pig, rabbit and chicken. The other animals have both NeuNAc and NeuNGly containing gangliosides.

Chromaffin granules from bovine adrenal medulla contain gangliosides at concentrations 3 and 7 times as great as microsomal and mitochondrial fractions, respectively. These gangliosides were NeuNAc-containing GM3 and NeuNGly containing GM3 in the same amount.

REFERENCES

1. R. Ledeen, K. Salsman, and M. Cabrera, Gangliosides of bovine adrenal medulla, Biochemistry 7:2287 (1968).
2. H. C. Price and R. K. Yu, Adrenal medulla gangliosides. A comparative study of some mammals, Comp. Biochem. Physiol. 54B:451 (1976).
3. H. Price, S. Kundu, and R. Ledeen, Structures of gangliosides from bovine adrenal medulla, Biochemistry 14:1512 (1975).
4. T. Ariga, M. Sekine, R. K. Yu, and T. Miyatake, Disialogangliosides in bovine adrenal medulla, J. Biol. Chem. 257:2230 (1982).
5. T. Ariga, M. Sekine, R. K. Yu, and T. Miyatake, Isolation and characterization of the trisialogangliosides from bovine adrenal medulla, J. Lipid Res. 24:737 (1983).
6. S. Ando, N.-C Chang, and R. K. Yu, High performance thin-layer chromatography and densitometric determination of brain ganglioside composition of several species, Anal. Biochem. 89:437 (1978).
7. L. Svennerholm, Quantitative estimation of sialic acids. II. A colorimetric resorcinol-hydrochloric acid method, Biochim. Biophys. Acta 24:604 (1957).
8. C. C. Sweeley, R. Bently, M. Makita, and W. W. Wells, Gas-liquid chromatography of trimethylsilyl derivatives of sugars and related substances, J. Am. Chem. Soc. 85:2497 (1963).
9. R. K. Yu and R. W. Ledeen, Gas-liquid chromatographic assay of lipid-bound sialic acid: measurement of gangliosides in brain of several species, J. Lipid Res. 11:506 (1970).
10. A. D. Smith and H. Winkler, A simple method for the isolation of adrenal chromaffin granules on a large scale, J. Biochem. 103:480 (1967).
11. T. Ariga, S. Ando, A. Takahashi, and T. Miyatake, Gangliosides and neutral glycolipids of human adrenal medulla, Biochim. Biophys. Acta 613:480 (1980).
12. R. Kuhn and H. Wiegandt, Weitere Ganglioside aus Menschenhirn, Z. Naturforsch. 19b:256 (1964).
13. H. Dreyfus, D. Aunis, S. Harth, and P. Mandel, Gangliosides and phospholipids of the membranes from bovine adrenal medullary chromaffin granules, Biochim. Biophys. Acta

489:89 (1977).
14. D. Geissler, A. Martinek, R. U. Margolis, R. K. Margolis, J. A. Skrivanek, R. Ledeen, P. König, and H. Winkler, Composition and biogenesis of complex carbohydrates of ox adrenal chromaffin granules, Neuroscience 2:685 (1977).
15. E. W. Westhead and H. Winkler, The topography of gangliosides in the membrane of the chromaffin granule of bovine adrenal medulla, Neuroscience 7:1611 (1982).

GLYCOSPHINGOLIPIDS OF EQUINE ERYTHROCYTES MEMBRANES:

COMPLETE CHARACTERIZATION OF A FUCOGANGLIOSIDE

Shinsei Gasa, Akira Makita, Ken Yanagisawa and
Mitsuru Nakamura

Biochemistry Laboratory, Cancer Institute, Hokkaido
University School of Medicine, Sapporo 060, Japan

INTRODUCTION

The major glycosphingolipids of mammalian erythrocytes have been demonstrated to be characteristic of the species from which they are derived.[1] In equine erythrocytes the predominant glycolipids are hematoside (II^3NeuGc-LacCer)[2] and its derivative which contains O-acetyl ester at the N-glycolylneuraminyl moiety.[3-5] In connection with the previous study,[5] it was found that equine erythrocytes have additional minor ganglioside components. A fucoganglioside purified from the minor ganglioside components was studied for its chemical structure by means of nuclear magnetic resonance (MNR) spectrometry, and destructive methods.

GLYCOSPHINGOLIPID COMPOSITION OF EQUINE ERYTHROCYTES

Glycosphingolipid composition of equine erythrocyte membranes is relatively simple as follows:

Neutral glycosphingolipids	(500 mg/100 ml of packed cells)
GlcCer	trace
LacCer	99%
Triglycosylceramide	trace
Gangliosides	(170 mg/100 ml of packed cells)
II^3NeuGc(4-Ac)-LacCr	72%
II^3NeuGc-LacCer	24%
Minor gangliosides	4%

Thus, the major glycosphingolipids of equine erythrocytes are composed of lactosylceramide and its sialylated derivatives. Of the minor gangliosides which have a core structure of the ganglio type, one contained fucose which is described below.

ISOLATION OF FUCOGANGLIOSIDE

The solvent mixture ratio is expressed by volume. The ganglioside fraction was prepared as described previously.[5] Briefly, erythrocyte membranes prepared from two horses (thoroughbred breed) were extracted with acetone and then twice with ethanol at 65° for 15 min each time. After evaporation of the solvent, the combined ethanol extracts were suspended in diethyl ether and centrifuged. The ether-insoluble lipids were subjected to DEAE-Sephadex A-25 (Pharmacia) chromatography[6] for separation into neutral glycosphingolipids and gangliosides. The gangliosides were chromatographed on a silica gel (Iatrobeads 6RS) column with a linear gradient elution of chloroform-methanol-water (60:40:2) to (20:80:4) using a modification of the procedure described previously.[7] A ganglioside fraction which contained II^3NeuGc-LacCer and minor gangliosides was eluted from the column after a large amount of a hematoside containing 4-O-acetyl-N-glycolylneuraminic acid (II^3NeuGc(4-Ac)-LacCer).[5] Rechromatography of this ganglioside fraction on a silica gel column in a manner similar to that described above yielded a minor fucose-containing ganglioside. Final purification of the fucoganglioside was accomplished by preparative thin layer chromatography.

The fucoganglioside had a mobility of 0.27 relative to II^3NeuGc-LacCer on a thin layer chromatogram developed with $CHCl_3$/CH_3OH/2.5 M NH_4OH (60:40:9).

FATTY ACID AND LONG CHAIN BASE COMPOSITION

The equine fucoganglioside contained a higher amount of unsaturated fatty acids (Table 1) than the major ganglioside, II^3NeuGc(4-Ac)-LacCer:[5] 33% of the total acid compared to 12% in the major ganglioside. On the other hand, the constituent long chain base of the minor ganglioside was almost exclusively sphingosine (99%), similar to that of the major ganglioside.

ANALYSIS OF COMPOSITION AND ANOMERIC STRUCTURE BY NMR SPECTROMETRY

The NMR spectrum of the equine fucoganglioside, from upfield to downfield, demonstrates the presence of methyl protons of L-fucose (Fuc C^6H_3 in Fig. 1) and acetylamide (N-Ac), anomeric protons of various monosaccharide linkages, olefinic protons present

Table 1. Fatty Acid Composition of Fucoganglioside (%).

16 : 0	2.3	22 : 0	3.1
16 : 1	2.3	22 : 1	trace
18 : 0	15.4	24 : 0	25.8
18 : 1	24.6	24 : 1	11.1
20 : 0	10.2	26 : 0	trace
20 : 1	5.1		

Table 2. NMR Spectrometry Data of Fucoganglioside.

Protons	Cer	β-Glc	β-Gal	β-Gal[a]	β-GalNAc	α-Fuc	NeuGc
Amide[b]							
Chemical shift (ppm)	7.48 7.52				7.02		7.71
Coupling constant (Hz)	7.8				7.6		8.2
Anomer[c]							
Chemical shift (ppm)		4.20	4.27	4.48	4.72	5.11	
Coupling constant (Hz)		7.8	7.8	7.9	8.0	3.5	
Molar ratio measured by intensity	1.00	0.99	0.95	0.95	0.89[d] (1.10)[e]	0.97	1.13

a) β-Galactoside to which fucose is linked.
b) Measured at 25°.   c) Measured at 110°.
d) From amide proton.   e) From anomeric proton.

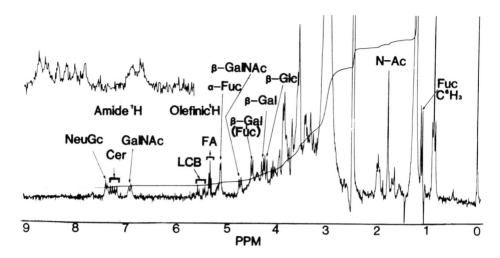
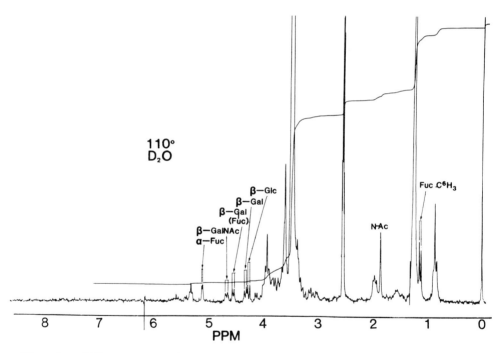

Fig. 1. NMR spectra of equine fucoganglioside. The spectra were obtained at 110° in dimethylsulfoxide-d_6 without (upper) or with D_2O (lower), as described previously.[8] The inlet in the upper figure is an expanded spectrum of the amide proton region.

Table 3. Glycosidase Treatment of Fucoganglioside.

Enzyme treatment	4Glc	2Gal	3,4Gal	3Gal	0Gal	3GalNAc	0GalNAc	0Fuc
None	1.0	0.80	0.72			0.91		0.69
I. αFuc'ase	1.0		0.68		1.01	0.90		
II.I βGal'ase	1.0		0.67				0.87	
III.II βHex'ase	1.0				0.98			
IV.III SA'ase	1.0				0.99			

Prefixed number in monosaccharide indicates the position unsubstituted by methoxyl group in methylation analysis. Values indicate molar ratio relative to 4Glc (2,3,6-tri-0-methyl-1,4,5-tri-0-acetyl-glucitol).
Abbreviations: αFuc'ase, α-fucosidase; βGal'ase, β-galactosidase; β-Hex'ase, β-N-acetylhexosaminidase; SA'ase, sialidase from Clostridium perfringens.

in long chain base (LCB) and fatty acid (FA) and amide protons (Fig. 1, upper). NMR analysis of a number of intact glycosphingolipids established the regularity of the resonance positions of amide protons in N-acetylhexosamine, sialic acid and ceramide moieties.[8] Thus, N-glycolylneuraminic acid (NeuGc in Fig. 1), whose amide proton resonates farther upfield than does that of N-acetylneuraminic acid, was identified in this ganglioside. The O-acetyl group was not detected in this ganglioside based on the absence of methyl protons due to O-acetyl. When D_2O was introduced, all the amide protons completely disappeared due to H-D exchange (Fig. 1, lower). As for anomeric protons, signals due to β-glucoside, β-galactosides, β-N-acetylgalactosaminide and α-fucoside were apparent in the spectra of this ganglioside. It was found that the anomeric proton of a β-galactoside moiety (β-Gal-(Fuc) in Fig. 1) which is linked to fucose is shifted downfield as compared to that of another β-galactoside.

Integration of the intensities of amide and anomeric proton peaks allows the determination of the composition of ceramide[8] and carbohydrates in glycosphingolipids.[8,9] The molar composition of

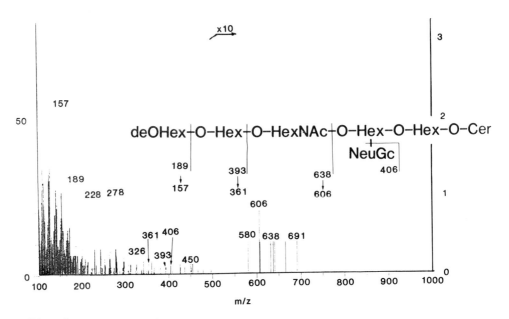

Fig. 2. Direct inlet mass spectrum of permethylated fucoganglioside.

the constituents of the equine fucoganglioside calculated by this method is shown in Table 2, along with the data on the proton signals.

SEQUENCE AND POSITION DETERMINATION OF GLYCOSIDIC LINKAGES

The permethylated ganglioside was subjected to direct inlet mass spectrometry, as described previously.[5] As shown in Fig. 2, mass fragments from the fucoganglioside derivative demonstrated the presence of non-reducing terminal deoxyhexose (m/z, 189 → 157) and N-glycolylneuraminic acid (m/z, 406). In addition to these signals, the sequences of deoxyhexose-hexose (m/z, 393 → 361) and deoxyhexose-(hexose)$_2$-acetylhexosamine (m/z, 638 → 606) were recognized.

In addition, the ganglioside was examined by a coupled glycosidase treatment[10] and methylation analysis. In this procedure, after each treatment with an exoglycosidase, the glycolipid product was permethylated, derivatized to a partially methylated alditol acetate and subjected to combined gas chromatography and mass spectrometry. The results are shown in Table 3.

On the basis of the above results, the equine erythrocyte fucoganglioside was concluded to have the following chemical structure:

Fuc(α1-2)Gal(β1-3)GalNAc(β1-4)Gal[3-2αNeuGc](β1-4)Glc(β1-1)Cer

The equine fucoganglioside did not react with chicken antisera (kindly provided by Drs. M. Naiki and N. Kasai) against II^{3}NeuGc-LacCer, when examined by immunodiffusion in agar. The fucoganglioside also did not exhibit human blood group H antigenicity at 250 μg/ml, when examined by hemagglutination inhibition test of group O erythrocytes with eel serum (8 units) under the conditions that human H-antigenic glycoprotein (2 μg/ml) exhibited the H antigenicity.

A compound similar to this fucoganglioside and an oligosaccharide therefrom were demonstrated in pig adipose tissue[11] and bovine liver,[12] respectively. A fucoganglioside which contains N-acetylneuraminic acid instead of N-glycolylneuraminic acid was characterized in bovine testis,[13] brain,[14] and thyroid gland.[15] The anomeric configuration of these gangliosides however has not been determined.

REFERENCES

1. T. Yamakawa and Y. Nagai, Trends Biochem. Sci. 3:128 (1978).
2. T. Yamakawa and S. Suzuki, J. Biochem. 38:199 (1951).
3. S. Hakomori and T. Saito, Biochemistry 8:5082 (1969).
4. R. Schauer, R. Veh, M. Sander, A. P. Corfield, and H. Wiegandt, in: "Structure and Function of Gangliosides", L. Svennerholm, P. Mandel, H. Drefus, and P. -F. Urban, eds., p. 283, Plenum Press, New York (1980).
5. S. Gasa, A. Makita, and Y. Kinoshita, J. Biol. Chem. 258:876 (1983).
6. R. W. Ledeen, R. K. Yu, and L. F. Eng, J. Neurochem. 21:829 (1973).
7. T. Momoi, S. Ando, and Y. Nagai, Biochim. Biophys. Acta 441:488 (1976).
8. S. Gasa, T. Mitsuyama, and A. Makita, J. Lipid Res. 24:174 (1983).
9. T. A. W. Koerner, J. H. Prestegard, P. C. Demou, and R. K. Yu, Biochemistry 22:2676 (1983).
10. Y. Yoda, A. Makita, and S. Gasa, J. Biochem. 91:627 (1982).
11. M. Ohashi and T. Yamakawa, J. Biochem. 81:1675 (1977).
12. H. Wiegandt, Hoppe-Seyler's Z. Physiol. Chem. 354:1049 (1973).
13. A. Suzuki, I. Ishizuka, and T. Yamakawa, J. Biochem. 78:947 (1975).
14. R. Ghidoni, S. Sonnino, G. Tettamanti, H. Wiegandt, and V. Zambotti, J. Neurochem. 27:511 (1976).
15. B. Macher, T. Pacuszka, B. R. Mullin, C. C. Sweeley, R. O. Brady, and P. H. Fishman, Biochim. Biophys. Acta 588:35 (1979).

PHYSICAL BEHAVIOUR OF GLYCOLIPIDS IN BILAYER MEMBRANES:

DISTRIBUTION AND ACCESSIBILITY

Mark W. Peters and Chris W. M. Grant

Department of Biochemistry
University of Western Ontario
London, Ontario, Canada N6A 5C1

INTRODUCTION

A major concern of our laboratory has been to fit glycolipids into a realistic picture of the eukaryote cell membrane. This problem actually consists of a number of smaller problems including: behaviour of the ceramide backbone in a lipid bilayer, headgroup behaviour, interaction with other membrane components, distribution, and function as receptors. Here we describe our recent attempt to isolate 2 such problems: glycolipid distribution in bilayers, and the source of glycolipid crypticity. The reader is referred to Figure 1 for an illustration of the concepts involved.

An early question with regard to how glycosphingolipids fit into membranes has been whether their ceramide backbones obey the principles established for phosphatidylcholines. We demonstrated using spin labels covalently attached to the fatty acid chain of galactosyl ceramide that there are indeed great similarities,[3] and that the concepts of fatty acid flexibility gradient[4] and order parameter[5] are applicable. Unfortunately, to date many such experiments have been performed only with the simple glycolipid, galactosyl ceramide, since no others have been probe-labelled on acyl chains.

Basic headgroup dynamics of various glycolipids have been studied by several laboratories utilizing spectroscopic techniques.[6-9] Whether in model or real membranes, the picture which has emerged is one in which carbohydrate residues move with high freedom of motion in a totally hydrophilic environment. This conclusion has been the same for both neutral and sialic

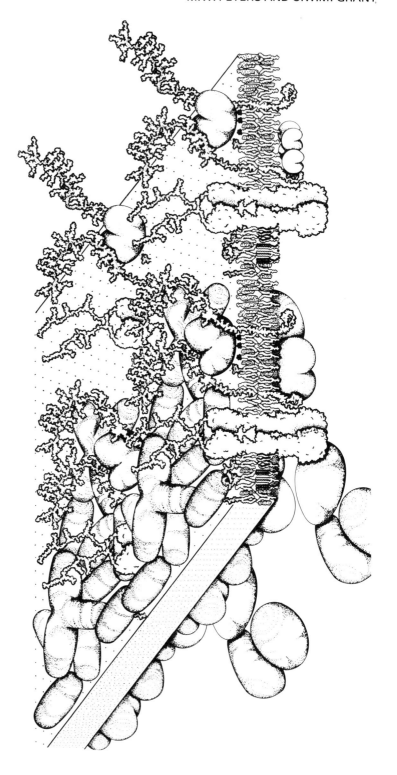

GLYCOLIPIDS AND BILAYER MEMBRANES

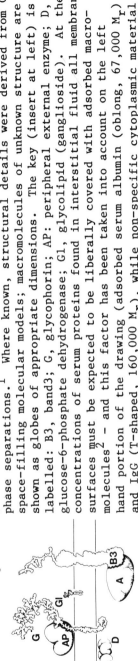

Fig. 1. Scale drawing. An attempt to show glycoplipids in a setting appropriate to a eukaryote membrane. The protein features, arrangements, and lipid/protein ratio illustrated have been chosen to accurately correspond to the human erythrocyte. Details of the lipid and protein arrangements have been described elsewhere with regard to lipid phase separations.[1] Where known, structural details were derived from CPK space-filling molecular models; macromolecules of unknown structure are shown as globes of appropriate dimensions. The key (insert at left) is labelled: B3, band3; G, glycophorin; AP: peripheral external enzyme; D, glucose-6-phosphate dehydrogenase; Gl, glycolipid (ganglioside). At the concentrations of serum proteins found in interstitial fluid all membrane surfaces must be expected to be liberally covered with adsorbed macromolecules[2] - and this factor has been taken into account on the left hand portion of the drawing (adsorbed serum albumin (oblong, 67,000 M_r) and IgG (T-shaped, 160,000 M_r), while non-specific cytoplasmic material has been indicated at the lower surface).

acid-bearing complex glycolipids.[3] Having said this, however, it is important to note that good evidence has been presented for restriction to rotation about anomeric linkages in glycolipid oligosaccharides at 23°C (ref. 10 and references therein).

The question of how small quantities of glycolipids distribute in cell membranes is extraordinarily complex. In fact very little is known about the details of lateral distribution for any lipid in membranes of 2 or more components. Significant advances in this area have resulted from the original efforts of H. M. McConnell and D. Chapman which permitted the prediction of gross phospholipid composition of coexisting phase-separated domains in 2-component phospholipid bilayers (reviewed in ref. 11). For glycolipids there is the problem of micro-distribution (e.g., possible clustering,[3] possible specific associations[12-14]), and gross distribution (e.g., are gangliosides concentrated at synapses?)[15] We have suggested that a tendency for glycolipids to form small clusters in phospholipid bilayer membranes could be used to account for several of our observations with spin labelled species (ref. 3 and citations therein). Delmelle et al. have felt that such clustering was only seen in rigid bilayers that tended to exclude glycosphingolipids.[16] Bunow and Bunow have interpreted their calorimetric results as not supporting clustering.[17] Here we record an attempt to adapt freeze-etch EM to localization of gangliosides in bilayer membranes of controlled composition and physical properties. This was achieved by using the lectin, wheat germ agglutinin (WGA), and also by using specifically bound lipid vesicles.

An important aspect of Fig. 1 that we would like to address in this article is glycolipid headgroup crowding and shielding by other membrane surface structures. It is apparent that glycolipid headgroups are dwarfed by components normally thought of as being associated with even the relatively simple erythrocyte membrane. Moreover, the outer surface of all types of cells must be expected to be thickly covered with adsorbed macromolecules ranging from albumin to fibronectin. This consideration brings to mind some well-known literature observations that glycolipids can be "cryptic" (reviewed in ref. 18); for instance, globoside is relatively better bound by specific antisera in fetal erythrocytes than in mature cells, and glycolipids are typically more susceptible to enzyme attack in transformed and trypsinized cells. Such observations are consistent with the idea that cell surface material, including proteins and oligosaccharides, can physically shield glycolipids from macrostructures that might otherwise contact them.

We have tested this concept using protein and oligosaccharide macrostructures which could be forced to occupy the lipid/water interface of bilayer membranes. Serum albumin provided a convenient 67000 M_r globular protein; while Dextrans offered suitable massive polysaccharides. Previous workers who have wished to strongly associate Dextrans[19] or albumin[20] with lipid bilayers for other reasons have suggested derivatizing them with stearic acid. We have also found this highly satisfactory, although we have had technically cleaner results using the more fluid oleic acid.

MATERIALS & METHODS

Sources of phospholipids, gangliosides and other materials were as described previously.[9,22] Lipid purity was checked by thin-layer chromatography on silica gel GF_{254} (Stahl). <u>Vibrio cholerae</u> neuraminidase (1 I.U./ml) was purchased from Calbiochem.

<u>Preparation of model membranes and assay for N-acetylneuraminic acid release by neuraminidase</u>

Vesicles of egg phosphatidylcholine containing 5 mol% of bovine brain ganglioside were prepared by hydration of thin films dried down in a test tube from chloroform/methanol (1:1), as described previously.[22] Vesicles prepared in this way have a high percentage of single bilayer structures. Each sample consisted of 30 μl of the above suspension of lipid vesicles, to which was added either 10 μl of saline (blank) or 10 μl of saline containing 0.5 mg of Dextran or albumin surface-coat material. The samples were incubated with surface-coat material for at least 30 min at 22°C prior to assaying for sialic acid release by neuraminidase. Subsequent to incubation with surface-coat material, to each sample was added 10 μl of neuraminidase (0.01 I.U.) or 10 μl of saline (control). The mixture was allowed to react at 37°C for 15 min.

Enzymatically released N-acetylneuraminic acid was quantitated using a scaled-down version of the thiobarbiturate assay described by Warren.[21] Interfering chromophores absorbing maximally at 232 nm were controlled for by simultaneously assaying coated and uncoated liposomes that had not been treated with neuraminidase.

<u>Surface coat material</u>

Serum albumin was derivatized with fatty acids following the general procedure of Lapidot et al.,[20] while the basic method employed for linking fatty acids to Dextran was that described by Wolf et al.[19] for stearic acid. The details of our synthetic procedures have been reported previously,[22] along with a detailed characterization of the lipid membrane attaching properties of the surface coat material so produced. Bound surface coat material remained attached even in the face of extensive washing over a 24 hr period.

Wolf et al.[19] have reported essentially complete surface coverage of bilayer membranes by stearic acid-derivatized Dextran at 0.04 mg/ml. Experiments described in this paper have typically been carried out using 10 mg/ml of conjugated albumin or Dextran - a concentration which at a very conservative estimate should be adequate to produce monolayer coverage.

Electron microscopy

Large liposomes containing 70 mol% dipalmitoyl phosphatidylcholine, 23 mol% dioleoylphosphatidylcholine, and 7 mol% beef-brain gangliosides, were prepared by hydrating dried lipid films in 10 mM sodium phosphate buffer. Tiny sonicated vesicles (25-40 nm diameter) were made by sonicating an aliquot of the large liposome mixture above for several 1 minute bursts on a Branson Model W-350 microtip probe sonicator with ice bath cooling.

Samples were prepared for electron microscopy by incubating large liposomes, sonicated vesicles, and WGA at 4°C, followed by pelleting the large liposomes at 8000 x g. The pellet was resuspended in a drop of buffer and placed on gold discs. Samples were quenched from 4°C in Freon 22 cooled in liquid nitrogen. Freeze-etching was performed for 2 minutes at -100°C using a Balzers BA501 vacuum coating device. Replicas were cleaned with commercial bleach followed by acetone prior to examination using a Philips EM 200.

RESULTS AND DISCUSSION

Ganglioside distribution by electron microscopy

Freeze-etching is a technique of sample preparation for electron microscopy whereby a quick-frozen droplet of membrane suspension is first "cleaved", then "etched" in vacuum at -100°C, and finally shadowed with platinum. "Cleaving" exposes extensive regions of membrane hydrophobic interior, while "etching" sublimes away buffer to expose outer surface features. Simple lipid bilayers appear smooth and featureless when prepared in this way - although, under appropriate conditions, rigid (gel phase) membrane regions can be distinguished by a ripple pattern (the P_β' phase of Tardieu et al.[23]).

Although glycolipid headgroups protrude above the bilayer surface (Fig. 1), they are too insubstantial to cast a platinum

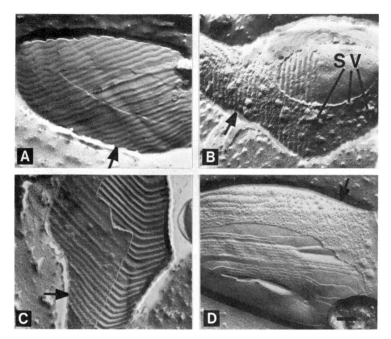

Fig. 2. Freeze-etch electron micrographs of lipid bilayer structures bearing 7 mol% gangliosides and exposed to wheat germ agglutinin (WGA) plus small sonicated vesicles (SV). See text for details. Arrowheads point from ice towards the membrane etch face. (Bar = 100 nm)
  A. Control lipid containing no gangliosides; exposed to 1 mg/ml WGA and SV.
  B. Ganglioside bearing lipid liposomes; exposed to 1 mg/ml WGA and SV.
  C,D. Same as B but using 0.1 mg/ml WGA.

shadow that would permit their direct identification and localization. Freeze-etched outer surfaces of cell membranes often display an unremarkable, finely granular appearance - presumably resulting from the presence of surface proteins. The standard approach to visualization of the distribution of specific cell membrane proteins has been to mark them with appropriate antibodies or lectins tagged with 600,000 M_r ferritin (which casts a distinctive shadow). We have found that, on the otherwise smooth surface of lipid bilayer membranes, lectins alone cast platinum shadows just sufficient to permit localization of their glycolipid attachment sites (Fig. 2B-D). Each micrograph in Fig. 2 shows a convex liposome membrane with etch face (marked by arrow) and fracture face. In Fig. 2A the liposome bore no ganglioside receptor, although it was exposed to WGA as per samples in 2B-D.

The ripple texture to large areas of the various liposomes demarcate bilayer regions in the rigid $P_{\beta'}$ phase alluded to above. The distinctly granular appearance of the etch faces in Figs. 2B–D reflects the presence of lectin bound to ganglioside receptor. At lower lectin concentrations (Figs. 2C, 2D) the density of surface granularity is distinctly reduced and it is possible to resolve individual "bumps" of lectin/receptor complex. The latter are probably larger than would be expected for a single 36,000 M_r molecule of WGA – suggesting that they may well represent small clusters of glycolipid and/or lectin.

In addition, we have been experimenting with lipid vesicles as objects large enough to mark membrane receptors. Sonicated vesicles have a size range of 25 to 40 nm and show up clearly when attached to the surface of large liposomes. They become useful tags for receptor sites when they themselves bear receptors with bound lectin. This is clearly illustrated in Fig. 2B where the high lectin concentration has led to a heavy coverage of the liposome etch face with bound vesicles of some 30 nm dimensions.

We have been encouraged by these results that it is quite feasible to monitor glycolipid distribution in model membranes to a resolution approaching 4 nm. The use of sonicated vesicles has the added feature that it inherently marks sites of strong receptor-mediated membrane-membrane attachment (which might be expected to be sensitive to membrane properties).

In an initial attempt to test the effect of host matrix on glycolipid distribution, we have experimented with a two-component phospholipid matrix (containing 7 mol% gangliosides). The lipid matrix chosen was dipalmitoyl/dioleoyl phosphatidylcholine. Micrographs of this system have been used for Fig. 2. At 4°C rippled, rigid domains enriched in dipalmitoyl phosphatidylcholine exist in equilibrium with smooth, fluid domains enriched in dioleoyl phosphatidylcholine. Note that the granular etch face features related to bound lectin are essentially uniformly distributed between smooth and rippled membrane regions. It would appear therefore, that the ganglioside gross distribution is roughly uniform. Similarly the points of lipid vesicle attachment show no obvious predilection for receptors in fluid or rigid domains.

Surface macromolecules do not induce glycolipid crypticity

As described in the Materials and Methods section, Dextrans and serum albumin with covalently attached fatty acids not only adsorb to bilayer membranes but also remain firmly attached in the face of extensive washing procedures. The question we have asked was whether adsorbed layers of such material would

Table 1. Effect of Tightly Bound Surface Macromolecules on the Enzymatic Release of N-acetyl Neuraminic Acid from Gangliosides in Bilayer Membranes.

		% Reduction in NANA Release Compared to Uncoated Bilayers	
Experiment[a]		Dextran Surface Coat	Serum Albumin Surface Coat
Fluid Bilayer	1	14	-35
	2	14	15
	3	8	-14
	4	-12	-28
	5	10	3
	6	-58	-76
	7	-11	3
Average of 1-7 ($\pm$SD)		-2($\pm$26)	-22($\pm$30)
Rigid Bilayer	8	-3	-
Standard Substrate	9[b]	6	7
	10[b]	0	9

[a] This table lists the results of experiments carried out on different occasions. Each separate experiment was carried out in quintuplicate; and each sample was done with a paired blank, which was identical except for having no enzyme added. Technical details are described in Materials and Methods. Percent reduction in NANA release when membranes bore a surface coat is listed (a minus sign indicates that the measured release was greater in the presence of surface macromolecules). Note that according to a pair-wise student's t-test there was no significant difference in release of NANA in the presence of surface macromolecules.

[b] Neuraminidase activity on the soluble substrate, neuraminyl lactose in the presence and absence of surface-coat material.

physically shield ganglioside from specific binding to macromolecules (neuraminidase in this case), and hence significantly reduce neuraminidase-induced NANA release from lipid bilayers bearing small amounts of ganglioside. The results are summarized in Table 1. As there is a certain variability inherent to the assay procedure (depending presumably upon factors such as variability in the lipid preparations), each experiment was done in quintuplicate. Our expectation was that, when release of NANA from membranes densely coated with macromolecules was compared with that from uncoated bilayers, there would be a measurable reduction in the former case. This "reduction" has been listed in Table 1 as a % of the uncoated membrane value.

The basic message from Table 1 seems to be that a substantial surface layer of tightly bound serum albumin or high molecular weight Dextran leads to no significant reduction in headgroup hydrolysis (experiments 1-7). This observation does not seem to depend upon receptor mobility, since the same was true in a rigid lipid matrix of distearoyl phosphatidylcholine (experiment 8). Furthermore, glutaraldehyde fixation of the serum albumin-coated membranes did not alter the result (data not included). Experimental variability was such that we could miss a difference of anywhere up to a factor of 0.3. However, previous reports of glycolipid crypticity in cell membranes have recorded factors of 10 to 1000 difference in "accessibility" between membranes.[18] Note that there was no measurable influence of surface coat material on the enzyme itself as monitored with the standard substrate, neuraminyl lactose (experiments 9, 10).

Our conclusion is that simple, non-specific shielding by surface macromolecules may play a relatively minor role in previously reported cases of glycolipid crypticity in cell membranes. We would be more inclined to consider the alternative possibilities - that binding events and enzymatic headgroup hydrolysis are sensitive to one or more of: glycolipid dynamics, interaction with other lipids, distribution, orientation, and perhaps specific associations. In this regard it is noteworthy that lipid composition of model membranes has been demonstrated to dramatically influence complement fixation and antibody binding to lipids,[24] as well as enzymatic reactivity of gangliosides.[25]

SUMMARY

A 3-dimensional scale drawing has been produced which purports to illustrate key features of glycolipid environment in the eukaryote plasma membrane. An attempt has been made to isolate two such features for separate study using bilayer model membranes. In particular, i) physical shielding of glycolipid headgroups by membrane-associated macromolecules has been investigated as a possible source of glycolipid crypticity; and ii) freeze-etch EM has been employed in an attempt to understand factors that control glycolipid receptor function and distribution.

The first of these problems was approached by measuring sialic acid released (by the enzyme, neuraminidase) from gangliosides in lipid bilayers. The bilayers could be coated with a firmly attached layer of protein or high molecular weight polysaccharide in order to mimic the presence of macromolecules at a cell surface. In fact, surface coat material did not measurably reduce sialic acid release from gangliosides in fluid or rigid phosphatidylcholine membranes. The implication drawn is that non-specific shielding by surface glycoproteins is an unlikely source of glycolipid crypticity in cell membrames.

The second problem was attacked by direct visualization of lectins and bilayer vesicles on the platinum-shadowed outer surface of large liposomes. To a first approximation bound lectins (and hence gangliosides) were equally distributed between coexisting fluid and rigid phospholipid domains. However the appearance was suggestive of a tendency for gangliosides to exist in small clusters. Lipid vesicles bearing lectin receptors were used to mark points of lectin-mediated membrane-membrane attachment on large liposomes. There was no obvious difference in the extent of such attachment to fluid and rigid domains.

ACKNOWLEDGEMENTS

This research was supported by grants from the Medical Research Council and the National Cancer Institute of Canada. MWP is the holder of a Medical Research Council studentship.

REFERENCES

1. C. W. M. Grant, Lateral phase separations and the cell membrane, in: "Membrane Fluidity," R. C. Aloia, ed., Academic Press, New York, in press.
2. M. L. Brash, and D. J. Lyman, Adsorption of proteins and lipids to nonbiological surfaces, in: "Chemistry of Biosurfaces Vol. 1," M. L. Hair, ed., Marcel Dekker, New York (1971).
3. M. W. Peters, K. R. Barber, and C. W. M. Grant, Headgroup behaviour of an uncharged complex glycolipid, Biochim. Biophys. Acta 693:417 (1982).
4. F. J. Sharom, and C. W. M. Grant, Glycosphingolipids in membrane architecture, J. Supramol. Struct. 6:249 (1977).
5. W. L. Hubbell, and H. M. McConnell, Molecular motion in spin-labelled phospholipids and membranes,

J. Am. Chem. Soc. 93:314 (1971).

6. F. J. Sharom, and C. W. M. Grant, A model for ganglioside behaviour in cell membranes, Biochim. Biophys. Acta 507:280 (1978).

7. P. L. Harris, and E. R. Thornton, Carbon-13 and proton nuclear magnetic resonance studies of gangliosides, J. Am. Chem. Soc. 100:6738 (1978).

8. L. O. Sillerud, H. H. Prestegard, R. K. Yu, D. E. Schafer, and W. H. Konigsberg, Assignment of the ^{13}C nuclear magnetic resonance spectrum of aqueous ganglioside GM1 micelles, Biochemistry 17:2619 (1978).

9. P. M. Lee, N. V. Ketis, K. R. Barber and C. W. M. Grant, Ganglioside headgroup dynamics, Biochim. Biophys. Acta 601:302 (1980).

10. L. O. Sillerud and R. K. Yu, Comparison of the ^{13}C-NMR spectra of ganglioside GM1 with those of GM1-oligosaccharide and asialo-GM1, Carbohydrate Res. 113:173 (1983).

11. A. G. Lee, Lipid phase transitions and phase diagrams II. Mixtures involving lipids, Biochim. Biophys. Acta 472:285 (1977).

12. V. W. Hu, and B. J. Wisnieski, Photoreactive labelling of M13 coat protein in model membranes by use of a glycolipid probe, Proc. Natl. Acad. Sci. (USA) 76:5460 (1979).

13. C. W. Lingwood, S. Hakomori, T. H. Ji, A glycolipid and its associated proteins: Evidence by cross-linking of human erythrocyte surface components, FEBS Lett. 112:265 (1980).

14. K. Watanabe, S. Hakomori, M. E. Powell, and M. Yokota, The amphipathic membrane proteins associated with gangliosides: The Paul-Bunnell antigen is one of the gangliophilic proteins, Biochem. Biophys. Res. Commun. 92:638 (1980).

15. J. A. Skrivanek, R. W. Ledeen, R. U. Margolis and R. K. Margolis, Gangliosides associated with microsomal subfractions of brain: Comparison with synaptic plasma membranes, J. Neurobiol. 13:95 (1982).

16. M. Delmelle, S. P. Dufrane, R. Brasseur, and J. M. Ruysschaert, Clustering of gangliosides in phospholipid bilayers, FEBS Lett. 121:11 (1980).

17. M. R. Bunow, and B. Bunow, Phase behaviour of ganglioside-lecithin mixtures, Biophys. J. 27:325 (1979).

18. S. Hakomori, Glycosphingolipids in cellular interaction, differentiation, and oncogenesis, Ann. Rev. Biochem. 50:733 (1981).

19. D. E. Wolf, J. Schlessinger, E. L. Elson, W. W. Webb, R. Blumenthal, and P. Henkart, Diffusion and patching of macromolecules on planar lipid bilayer membranes, Biochemistry 16:3476 (1977).

20. Y. Lapidot, S. Rappoport, and Y. Wolman, Use of esters of N-hydroxysuccinimide in the synthesis of N-acylamino

acids, J. Lipid Res. 8:142 (1967).
21. J. Warren, The thiobarbituric acid assay of sialic acids, J. Biol. Chem. 234:1971 (1959).
22. M. W. Peters, C. J. Singleton, K. R. Barber, and C. W. M. Grant, Glycolipid crypticity in membranes - not a simple shielding effect of macromolecules, Biochim. Biophys. Acta 731:475 (1983).
23. A. Tardieu, V. Luzzati, and F. C. Reman, Structure and polymorphism of the hydrocarbon chains of lipids: A study of lecithin-water phases, J. Mol. Biol. 75:711 (1973).
24. C. R. Alving and R. L. Richards, Immunological aspects of liposomes, in: "The Liposomes," M. Ostro, ed., Marcel Dekker, New York (1981).
25. M. Masserini, S. Sonnino, R. Ghidoni, V. Chigorno and G. Tettamanti, Galactose oxidase action on GM1 ganglioside in micellar and vesicular dispersions, Biochim. Biophys. Acta 688:333 (1982).

Distribution and Transport

GANGLIOSIDE DISTRIBUTION AT DIFFERENT LEVELS OF ORGANIZATION AND

ITS BIOLOGICAL IMPLICATIONS

Yoshitaka Nagai and Masao Iwamori

Department of Biochemistry, Faculty of Medicine
University of Tokyo, Bunkyo-ku, Tokyo 113, Japan

INTRODUCTION

More than 100 molecular species of glycosphingolipids (GSL) having different carbohydrate chains have so far been isolated and characterized from various mammalian tissues and cells. They are classified into seven general types according to their asialo-carbohydrate structures (Table 1), which are biosynthesized via their own defined pathways. Although the systematic analysis of GSL on various cell surfaces, especially on erythrocyte membranes of various animal species, indicates that the biosynthetic pathways including the termination or modification of carbohydrate chain is under the regulation of genetic control, as shown in the case of blood group antigens, recent observations strongly indicate that the environmental or nongenomic information-dependent modification of carbohydrate chains including the induction of new biosynthetic pathways occurs at various cellular levels outside of the nuclei. Such dual nature of GSL imposes some difficulty when we attempt to identify the exact role of GSL in cell function. The knowledge of

Table 1. Basic carbohydrate chains of glycosphingolipids

Basic carbohydrate	Structure
ganglio-series	Gal(β1-3)GalNAc(β1-4)Gal(β1-4)Glc-ceramide
lacto-series	Gal(β1-3)GlcNAc(β1-3)Gal(β1-4)Glc-ceramide
lactoneo-series	Gal(β1-4)GlcNAc(β1-3)Gal(β1-4)Glc-ceramide
globo-series	GalNAc(β1-3)Gal(α1-4)Gal(β1-4)Glc-ceramide
globoiso-series	GalNAc(β1-3)Gal(α1-3)Gal(β1-4)Glc-ceramide
muco-series	Gal(β1-3)Gal(β1-4)Gal(β1-4)Glc-ceramide
gala-series	Gal(α1-4)Gal-ceramide

GSL distribution is helpful in recognizing the basic structure and the type of modification of the carbohydrate chain in morphologically or functionally specified cells in relation to cell function, differentiation, transformation and other cellular activities.

GANGLIOSIDE DISTRIBUTION IN NERVOUS TISSUES

Brain is a unique organ characterized by an exceedingly high concentration of gangliosides, compared to the extraneural organs or tissues, and also by the higher molecular diversity and complexity of gangliosides. Our recognition of this character, beginning with mapping analysis of brain gangliosides,[1,2] has been deepened by the discovery of increasing numbers of newly found molecular species of gangliosides not only in brain but also in extraneural tissues. The number of such new compounds now exceeds 60 molecular species. Most of these new compounds appear to be minor components of tissue GSL (Fig. 1). The functional importance of gangliosides is not yet clear, but the unique distribution of ganglioside molecules in subcellular and cellular levels allows us to presume their special functional roles in the nervous system (see review articles 3-6). The general profile characteristic of ganglioside distribution in the nervous system, for example, is summarized as follows:

1. GM4 and GM1 are contained in human myelin in a concentration higher than in other brain regions, reflecting the fact that white matter contains more GM4 and GM1 than gray matter.

2. The concentration of gangliosides in gray matter is significantly higher than other regions.

3. Spinal cord contains GD3 in a relatively higher concentration than other regions.

4. Astroglia usually contains gangliosides at twice the concentration of neurons.

5. Human oligodendroglia gives a different ganglioside pattern from neuron and astroglia in containing high concentrations of GM4, GM3 and GD3.

Typical data for ganglioside distribution at different levels of organization in the central nervous system are listed in Table 2.

MOLECULAR DIVERSITY AND ITS CELL BIOLOGICAL BACKGROUND

The diverse nature of the molecular species of GSL is particularly evident in the nervous, haematopoietic and digestive systems. Since Yamakawa first directed our attention to the species- and

Table 2. Ganglioside Distribution in Nervous Tissue.[22-24]

	Brain					Spinal cord		Isolated cells			
	human			chicken	bovine	human		human		rat	
	gray matter	white matter	myelin	white matter	white matter	whole s.cord	myelin	neurons	oligodendroglia	neurons	astroglia
Lipid-bound sialic acid	875	275	632	(µg / g wet weight) 620	326	87	322	(µg / mg protein) 1.3	0.3	(µg / g dry weight) 1900	4300
Sialic acid distribution (%)											
GM4	1.5	8.6	26.6	10.8	2.7	12.8	25.3	0.9	5.9	-	-
GM3	2.7	4.8	1.2	6.9	2.6	14.0	3.5	4.1	8.1	4.8	5.2
GM2	4.1	2.5	3.7	2.3	2.3	3.7	6.6	4.5	5.7	6.0	5.1
GM1	14.9	21.6	34.7	18.9	31.0	16.8	38.2	22.6	20.1	13.8	14.9
GD3	5.4	8.8	2.7	13.1	6.0	16.4	5.0	7.2	11.9	-	-
GD1a	21.7	16.6	6.9	16.4	24.1	4.2	0.7	21.5	16.4	33.5	35.8
GD1a-GalNAc	0.4	1.1	-	-	5.1	-	-	-	-	-	-
GT1a	1.8	2.2	0.5	5.1	2.6	0.5	0.1	1.3	2.7	-	-
GD2	8.0	3.1	0.3	1.3	1.4	0.9	0.4	3.9	3.3	5.8	5.8
GD1b	18.2	16.9	16.1	9.1	10.5	18.7	11.1	20.7	15.0	11.1	11.0
GT1b	16.3	11.1	6.3	9.8	9.1	9.1	7.1	11.3	9.3	15.0	13.1
GQ1b	5.0	2.7	1.0	4.9	1.2	2.9	2.2	2.2	1.9	10.0	9.2

More detailed information is obtained from R. W. Ledeen's recent review,[6] and "Structure and distribution of gangliosides" in Complex Carbohydrates of Nervous Tissue, R. U. Margolis and R. K. Margolis, eds., Plenum Press, New York, pp. 1-23 (1979).

tissue-specific distribution of GSL with his report on erythrocyte membranes (Mosbach colloquium, 1965),[7] the concept of specific distribution has subsequently been extended to subspecies, strain and individual levels and finally to the cellular level. It is of utmost importance to know whether such molecular diversity can be related to cellular function, particularly the special type of cell function. As in the case of primate brain GM4 ganglioside, most cases in which the distribution of GSL was at first regarded to be narrowly or characteristically limited were later found to be not so. This poses a serious problem when we try to attribute specific physiological functions to individual molecular species. Compared to extraneural tissues, in which we frequently encounter with the sporadic occurrence of GSL molecular species, the brain appears to have relatively limited ganglioside and neutral GSL compositions, especially in mammals. It should be added that there are exceptional cases even for mammalian brain as later shown in Table 4.

As already mentioned, three systems, nervous, haematopoietic and digestive (especially epithelial cells of the small intestine, as shown by Karlsson et al. and others), are characterized by the wide molecular diversity of GSL, that is, the molecular diversification of gangliosides in brain, gangliosides and neutral GSL in the haematopoietic system and neutral GSL in intestinal tissues. Although at present we do not have enough convincing experimental evidence to explain such diversification, the general concept schematized in Fig. 2 of the possible cell biological background generating such diversity is possible. In this scheme the brain is assigned a particular position among the three systems based on the postulation of its high biopotential to generate cell individuality.

Table 3. Distribution of Lipid-Bound Sialic Acid in Individual Gangliosides of Various Rat Tissues. (%)

Tissue	GM4	GM3	GM2	GM1	FucGM1	GD3	GD2	GD1a	GD1b	GT1b	GQ1b	Others
Cerebrum		tr	tr	9.4		1.5	tr	40.2	16.4	24.3	8.2	
Cerebellum		tr	tr	4.5		3.5	7.9	31.2	8.8	30.0	13.6	
Spinal cord		tr	tr	9.3		4.3	tr	14.3	22.1	31.1	18.8	
Thymus		15.2				7.1						77.7
Lung		72.9				2.2		16.8				8.4
Heart		93.1	tr			2.4		2.5				2.0
Liver		52.3	tr			0.7		32.0	5.2			7.3
Stomach		62.9		tr	tr	14.8		10.5				11.8
Spleen		55.7	tr	tr		4.0		30.7	2.4			7.2
Intestine		74.9	tr			20.7		1.2				3.2
Kidney	1.2	64.9	tr		tr	28.8		0.8				4.3
Testis		16.4			tr			66.8				16.8
Bone marrow		27.6	7.1	1.6	tr			59.7				4.0
Buffy coat								100.0				
Erythrocytes				36.5	22.3			41.2				

Table 4. Extraneural Gangliosides Having The Same Carbohydrate Structure as That of Major Brain Gangliosides.

Tissue	GM4	GM2	GM1	GD1a	GD1b	GT1a	GT1b	GQ1b
Thyroid			B P	B P				
Thymus			M R					
Lung				R				
Liver			B R	B R	R		R	R
Spleen			B H	B R H				
Kidney	R	B	B	B H				
Testis				B R				
Bone marrow		R		R				
Adrenal gland	G		B	B				
Adipose tissue			P	P			P	
Mammary gland			B		B			
Nasal cartilage			B	B	B	B	B	B
Erythrocyte	M	M	R	R				
Lymphocyte			H					
Egg yolk	C							

B, cow; P, pig; M, mouse; R, rat; H, human; G, guinea pig; C, chicken.

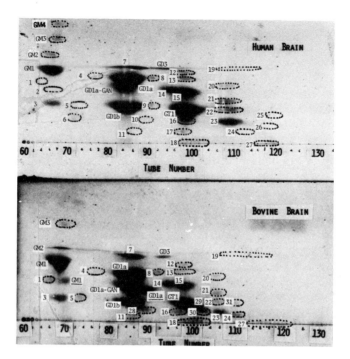

Fig. 1. Molecular profiles of brain gangliosides demonstrated by ganglioside mapping.

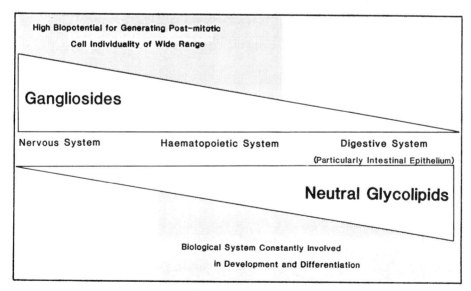

Fig. 2. Possible cell biological background for generation of molecular diversity of glycolipids in three representative mammalian tissues.

At present there is not sufficient evidence either for or against this hypothesis, thus the word "potential" is employed.

DISTRIBUTION OF GANGLIOSIDES IN VARIOUS RAT TISSUES

In the early stages of ganglioside research, ganglio-N-tetraose-containing gangliosides were thought to be exclusively and specifically contained in nervous tissues. However, recent studies showed that ganglio-N-tetraose-containing gangliosides were widely distributed in various extraneural tissues of mammals, expecially of rodents. For example, GM1, fucosyl-GM1 and GD1a were characterized as major gangliosides of rat erythrocytes,[8] and GD1b, GT1b and GQ1b were shown to be present in high concentration in rat hepatic parenchymal cells.[9] In addition, various ganglio-N-tetraose-containing gangliosides appeared in rat tissues and cells after malignant transformation by chemical carcinogens.[10,11] Therefore, the present experiment attempted to clarify the distribution of gangliosides of the ganglio-series in rat tissues.

The results of thin-layer chromatography (TLC) of gangliosides from various rat tissues and the distribution of lipid-bound sialic acid in individual gangliosides are shown in Fig. 3 and Table 1, respectively. In contrast with the finding that non-thymic rabbit extraneural tissues contained predominantly lactose-containing gangliosides such as GM3 and GD3,[12] the tissues containing GM3 and GD3 at concentrations of more than 92% of lipid-bound sialic acid, were only heart, intestine and kidney. Other extraneural tissues contained polar gangliosides with ganglio-N-tetraose in significant

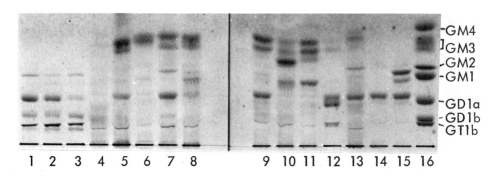

Fig. 3. TLC of gangliosides obtained from various tissues of rat (Wistar strain). 1, cerebrum; 2, cerebellum; 3, spinal cord; 4, thymus; 5, lung; 6, heart; 7, liver; 8, stomach; 9, spleen; 10, intestine; 11, kidney; 12, testis; 13, bone marrow; 14, buffy coat; 15, erythrocytes; 16, reference standard. Developing solvent: $CHCl_3$-MeOH-0.5% $CaCl_2$ (55:45:10), Detection: resorcinol- HCl.

concentrations. In particular, ganglio-N-tetraose was the sole asialo carbohydrate of the gangliosides in buffy coat and erythrocytes. Buffy coat contained a single ganglioside, GD1a and erythrocytes contained GM1, fucosyl-GM1 and GD1a. Also in testis and bone marrow the concentration of ganglio-N-tetraose-containing gangliosides was higher than that of lactose-containing gangliosides. In addition, liver, spleen, lung and stomach contained high concentrations of GD1a and GD1b. Other characteristic observations of gangliosides in rat tissues were GM4 in kidney, GM2 in bone marrow and GM3 with α-hydroxy fatty acid, phytosphingosine and N-glycolyl neuraminic acid in intestine. These findings indicate that rat tissues have the potential to biosynthesize various GSL with ganglio-N-tetraose, including GSL characterized as tumor-specific molecules.[10,11] The distribution of gangliosides with the same structure of major brain gangliosides in extraneural tissues is summarized in Table 3. It is clear that ganglio-series gangliosides are widely distributed in extraneural tissues of various species. In addition, GM4 which was originally characterized as a human myelin-specific ganglioside,[13] has been recognized in several extraneural tissues. Thus it is important in future studies of glycolipid chemistry to discover the biological necessity of the above wide-ranging molecular diversity in various tissues, and particularly the genetic as well as evolutionary mechanisms for prescribing and regulating molecular diversity.

AN APPROACH TO THE ANALYSIS OF CELL SURFACE TOPOGRAPHY OF GANGLIOSIDES

Although various pieces of evidence indicated that GSL are involved in the receptor function on the cell surface, previous experiments have been mainly based on an indirect procedure including binding inhibition, which occasionally led to misunderstanding of the results. However, the procedure reported by Ginsburg et al.,[14] which directly analyzed the binding components to glycolipid receptor developed on TLC plates, largely overcame the shortcomings of the indirect methods. We also attempted direct analysis of the binding ability of biologically active compounds to glycolipid on a native cell surface. Using sheep red blood cells (SRBC) as the target cell, the binding of anti-Forssman antibody was quantitatively determined with a fluorescence-activated cell sorter (FACS). As shown in Figs. 4 and 5, the fluorescence intensity of SRBC stained by indirect membrane immunofluorescence technique with anti-Forssman antibody and fluorescein-conjugated goat anti-rabbit IgM and IgG antibodies was found to be completely linear as a function of serum dilution, i.e., protein concentration. The standard curve for quantitative measurement of IgG anti-Forssman antibody was made using IgG antibody eluted from a Forssman antigen-conjugated affinity column. This procedure could detect 0.25 ng or more of IgG protein.[15] Also, the linearity and sensitivity of the

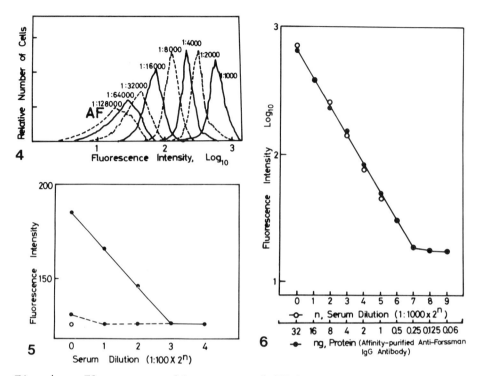

Fig. 4. Fluorescence histograms of SRBC stained with various dilutions of rabbit anti-Forssman antiserum by the indirect membrane immunofluorescence procedure. AF, autofluorescence.

Fig. 5. Titration curves of anti-Forssman antiserum and the affinity-purified IgG antibody analyzed on a FACS. ○, rabbit anti-Forssman antiserum; ●, affinity-purified IgG anti-Forssman antibody.

Fig. 6. Fluorescence intensity of rat erythrocytes stained with various dilutions of anti-fucosyl-GM1 (●) and anti-GM1 (○) antisera. Second antibodies were fluorescein-conjugated goat anti-rabbit IgM (———) and IgG (----) antibodies.

FACS procedure were superior to those of hemolysis and enzyme-linked immunosorbent analysis (ELISA). The same procedure was applied to analyses of reactivity of glycolipids on rat erythrocytes.[8] As described above, rat erythrocytes contained high concentrations of the ganglio-series gangliosides, GM1, fucosyl-GM1 and GD1a. The reactivities of fucosyl-GM1 and GM1 on rat erythrocyte with rabbit anti-fucosyl-GM1 and anti-GM1 antisera were measured by FACS analysis. As shown in Fig. 6, anti-fucosyl-GM1 antibody clearly bound to rat erythrocytes and the fluorescence intensity of rat erythrocytes stained with anti-fucosyl-GM1 antiserum was reduced by a dilution of the antiserum. In contrast, no binding of anti-GM1 antibody on rat erythrocytes was found using both fluorescein-conjugated anti-rabbit IgG and IgM antisera as second antibodies. Consistent with this, immunohemolysis of rat erythrocytes with anti-fucosyl-GM1 antiserum in the presence of guinea pig complement was demonstrated, but no hemolytic activity was observed with anti-GM1 antiserum. Since anti-GM1 antibody bound to GM1 coated on ELISA plate and GM1-containing liposome in a complement fixation assay, the above observation seemed to indicate the cryptic epitope of GM1 on rat erythrocyte membrane. The topographical configuration of glycolipid may be regulated by various factors subjected to unknown conditions of cell exterior as well as interior. Although the significance of crypticity and/or organization of cell surface glycolipids has been emphasized by Hakomori,[16] its underlying molecular mechanism remains unclear. The regulation of glycolipid expression on cell surface might be closely related to the receptor function. Rat erythrocytes should provide a useful model for analyzing the functional significance of gangliosides.

SUBCELLULAR DISTRIBUTION OF GLYCOSPHINGOLIPIDS

Previously we showed by immunocytochemical studies that galactocerebrosides (GC) were present not only in plasma membrane but in cytoplasm in close association with colchicine-sensitive, freeze-fragile, microtubule-like cytoskeltal structures in cultured epithelial cell lines, such as JTC-12, HeLa and MDCK, in which GC was detected chemically (Fig. 7).[17,18] Other cell lines, Vero and

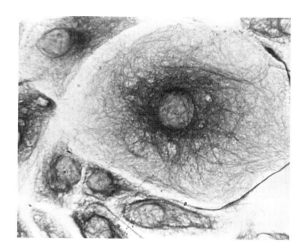

Fig. 7. Colchicine-sensitive, freeze-fragile cytoplasmic fibers of JTC-12 cultured monkey kidney cells. The fibers were visualized by immunoperoxidase labeling after treatment of the cells with affinity-purified anti-galactocerebroside antibody.

8999c, which did not contain GC were not positive for this immunocytochemical reaction. Affinity-purified anti-GC antibody also labelled the cytoplasm of epithelial cells in particular areas of hamster kidney, liver and lung, i.e., the distal tubular epithelium of the cortex of kidney, the epithelial cells of the periportal hepatic bile ducts and bronchiolar epithelium and type II alveolar cells of lung.[19] Similar observations were made on cholera toxin receptor ganglioside GM1 by Streuli et al.[20] Using the immunocytochemical technique they showed that the toxin receptor of monolayer cultured mouse fibroblasts, BALB/c-3T3 cells, was resistant to solubilization with neutral detergents and remained in association with Triton X-100 cytoskeletons. Immunofluorescence studies by Marcus and Janis[21] showed that globoside was present in both plasma membrane and cytoplasm of epithelial cells of proximal convoluted tubuli of human kidney. However, they suggested an association of globoside with mitochondria.

We called this type of distribution of GSL cytoskeleton-associated glycolipid (CAG) because the CAG phenomenon may be true of other GSL and ubiquitous.[18] The cell biological meaning of CAG is not known at present. We consider that CAG is a fundamental component of the cytoskeleton and that CAG is presumably present in a form of GSL-protein (carrier) complex which is most probably loosely bound to the cytoskeleton.[18] In this regard we suggested an important relation of CAG to so-called cytosolic GSL, the existence of which has recently been reported by many workers (i.e., sulfatides: Hershkowitz, et al., 1968; Pleasure and Prockop, 1972; Jungalwala, 1974; Kishimoto et al., 1980; GC: Radin et al., 1977, 1980, 1982; Kishimoto et al., 1980; gangliosides: Tettamanti et al., 1979, 1981; GSL: Bloj and Zilversmit, 1981, Yamada and Sasaki, 1982).[18] We also suggested that CAG may possibly play roles in several important cell biological phenomena, for example, with regard to the transport mechanism of GSL from their site of synthesis to the plasma membrane, the internalization mechanism, receptor-mediated endocytosis, the mechanism of transmitting biological signals received at the plasma membrane to the nucleus, the regulatory mechanism of the assembly and dissociation of the tubulin-related cytoskeleton per se, and other cytoskeleton-related cellular activities. Thus CAG still poses many challenging problems.

SUMMARY

The molecular diversity and complexity of gangliosides are surveyed in nervous and various extraneural tissues with particular emphasis on the different levels of organization such as tissues, cell types, subcellular structures, some cell organelles and topographical molecular membrane architecture. The characteristics revealed with such survey are discussed in an attempt to elucidate their cell biological background.

ACKNOWLEDGEMENTS

These studies were supported in part by a grant from Torey Science Foundation, a grant from the Ministry of Education, Science and Culture of Japan and also a research grant for Intractable Diseases from the Ministry of Health and Welfare of Japan.

REFERENCES

1. Y. Nagai and M. Iwamori, Brain and thymus gangliosides: Their molecular diversity and its biological implications and a dynamic annular model for their function in cell surface membranes, Mol. Cell. Biochem. 29:81 (1980).
2. M. Iwamori and Y. Nagai, A new chromatographic approach to the resolution of individual gangliosides: ganglioside mapping, Biochim. Biophys. Acta 528:257 (1978).
3. C. C. Sweeley and B. Siddiqui, Chemistry of mammalian glycolipids, in: "The Glycoconjugates" Vol. II, M. L. Horowitz and W. Pigman, eds., Academic Press (1977).
4. G. Morgan, G. Gombos, and G. Tettamanti, Glycoproteins and glycolipids of the nervous system, in: "The Glycoconjugates" Vol. I, M. L. Horowitz and W. Pigman, eds., Academic Press (1977).
5. H. Wiegandt, The gangliosides, in: "Advances in Neurochemistry" Vol 4, B. W. Agranoff and M. H. Aprison, eds., Plenum Press (1982).
6. R. W. Ledeen, Gangliosides, in: "Handbook of Neurochemistry", Vol 3 (2nd Ed.), A. Lajtha, ed., Plenum Press (1983).
7. T. Yamakawa, Glycolipids of mammalian red blood cells, in: "Lipoide", E. Schutte, eds., Springer-Verlag, Berlin-Heidelberg-New York (1966).
8. M. Iwamori, J. Shimomura, S. Tsuyuhara, M. Mogi, S. Ishizaki, and Y. Nagai, Differential reactivities of fucosyl-GM1 and GM1 gangliosides on rat erythrocyte membrane revealed by the analysis with anti-fucosyl GM1 and GM1 antisera, J. Biochem. 94:1 (1983).
9. K. Ueno, Y. Kushi, C. Rokukawa, and S. Handa, Distribution of gangliosides in parenchymal and nonparenchymal cells of rat liver, Biochem. Biophys. Res. Commun. 105:681 (1982).
10. E. H. Holms and S. Hakomori, Isolation and characterization of a new fucoganglioside accumulated in precancerous rat liver and in rat hepatoma induced by 2-N-acetyl aminofluorene, J. Biol. Chem. 257:7698 (1982).
11. T. Hirabayshi, T. Taki, M. Matsumoto, and K. Kojima, Comparative study on glycolipid composition between two cell types of rat ascite hepatoma cells, Biochim. Biophys. Acta 529:96 (1978).
12. M. Iwamori and Y. Nagai, Comparative study on ganglioside compositions of various rabbit tissues. Tissue-specificity

in ganglioside molecular species of rabbit thymus, <u>Biochim. Biophys. Acta</u> 665:214 (1981).
13. R. W. Ledeen, R. K. Yu, and L. F. Eng, Gangliosides of human myelin: sialosylgalactosylceramide (G7) as a major component, <u>J. Neurochem.</u> 21:829 (1973).
14. J. L. Magnani, D. F. Smith, and V. Ginsburg, Detection of gangliosides that bind cholera toxin: direct binding of ^{125}I-labeled toxin to thin-layer chromatograms, <u>Anal. Biochem.</u> 109:399 (1980).
15. M. Iwamori, M. Mogi, Y. Hirano, M. Nishio, H. Nakauchi, K. Okumura, and Y. Nagai, A quantitative analysis of cell surface glycosphingolipid with a fluorescence activated cell sorter, <u>J. Immunol. Methods</u> 57:381 (1983).
16. S. Hakomori, Glycosphingolipids in cellular interaction, differentiation and oncogenesis, <u>Ann. Rev. Biochem.</u> 50:733 (1981).
17. K. Sakakibara, T. Momoi, T. Uchida, and Y. Nagai, Evidence for association of glycosphingolipid with a colchicine-sensitive microtubule-like cytoskeletal structure of cultured cells, <u>Nature</u> 293:76 (1981).
18. Y. Nagai and K. Sakakibara, Cytoskeleton-associated glycolipid (CAG) and its cell biological implications, <u>in</u>: "New Vistas in Glycolipid Research", A. Makita, S. Handa, T. Taketomi, and Y. Nagai, eds., Plenum Press, New York (1982).
19. K. Sakakibara, M. Iwamori, T. Uchida, and Y. Nagai, Immunohistochemical localization of galactocerebroside in kidney, liver and lung of golden hamster, <u>Experientia</u> 37:712 (1981).
20. C. H. Streuli, B. Patel, and D. R. Critchley, The cholera toxin receptor ganglioside GM1 remains associated with Triton X-100 cytoskeletons of BALB/c 3T3 cells, <u>Exp. Cell Res.</u> 136:247 (1981).
21. D. M. Marcus and R. Janis, Localization of glycosphingolipids in human tissues by immunofluorescence, <u>J. Immunol.</u> 104:1530 (1970).
22. S. Ando, N.-C. Chang, and R. K. Yu, High-performance thin-layer chromatography and densitometric determination of brain ganglioside compositions of several species, <u>Anal. Biochem.</u> 89:437 (1978).
23. K. Ueno, S. Ando, and R. K. Yu, Gangliosides of human, cat and rabbit spinal cords and cord myelin, <u>J. Lipid Res.</u> 19:863 (1978).
24. R. K. Yu and K. Iqbal, Sialosylgalactosyl ceramide as a specific marker for human myelin and oligodendroglial perikarya: gangliosides of human myelin, oligodendroglia and neurons, <u>J. Neurochem.</u> 32:293 (1979).

THE LABELING OF THE RETINA AND OPTIC TECTUM GANGLIOSIDES AND

GLYCOPROTEINS OF CHICKENS IN DARKNESS OR EXPOSED TO LIGHT

Ranwel Caputto

Departamento de Quimica Biologica. Fac. de Ciencias
Quimicas. Universidad Nacional de Cordoba and CIQUIBIC-
CONICET.
Casilla de Coreo 61. 5016 Cordoba, Argentina

INTRODUCTION

Different authors have reported observations showing that modifications of gangliosides occur in the process of acquiring memory in an avoidance test[1-3] or that the injection of GM_1 ganglioside antibody interfers with memory acquisition.[4,5]

It is a matter of direct observation that the primary functional activity of the nervous system is the reception of stimuli. From the protopathic perception of touch in the most primitive animals to the dicriminative perceptions of the evolved mammals, perception is the initial act of alertness and memory and learning. Under this notion - probably correct, albeit somewhat simplistic - we undertook to study whether gangliosides are plausible candidates to be a type of substance that leaves a permanent trace in the CNS after the perception of the light stimulus.

The optic tract of the chicken has special advantages to study the problem. These animals have relatively large eyes and can receive 2 or 3 succesive intraocular injections of labeled precursor without showing any permanent damage. The second advantage is the observation that practically all of the axons in the optic nerve are decussated in the optic chiasm, thus favoring the determination of the material arriving at the optic tectum from the retina through axonal transport.

BEHAVIOR OF GANGLIOSIDES

It was previously found that chickens receiving an intraocular injection of [^{3}H]N-acetylmannosamine (^{3}H-ManNAc) after 48 h in darkness that were then exposed to light had more labeling in their optic tecta than chickens receiving a similar injection that were then maintained in the dark.[6,7] The increase was variable from experiment to experiment suggesting either that the effect of light can be indirect through some unknown behavioral factor or else that other factor(s) may influence the effect of light. Following the studies in the tectum, the retina was studied. Results showed that the labeling in total retina did not change by exposure to light but it changed in the isolated ganglion cell layer. The increase in the ganglion cell layer was usually less than in the tectum when the comparison was made in percent with respect to the labeling in darkness.

The increase of ganglioside labeling was not due to turning the light on or off, but to a continuous effect of light that lasts for at least 5 h. This was shown in experiments in which lights were turned on and off every 10 or 20 minutes for 5 h. The differences shown by these animals were substantially less than the differences between the animals in continuous light and darkness. In animals maintained for 24 h in continuous light or continuous darkness the differences disappeared.

In the experiments related up to this point, the chickens were in darkness for 48 h and the comparison was made between chickens that, after the injection of labeled precursor, were passed to light or maintained in darkness. We also did experiments in which chickens that had been for 48 h in light were given the injection of precursor and one half of them maintained in light while the other half were put in darkness. In this case the labeling of the animals that were passed to darkness decreased with respect to the labeling of those that remained in light. The labelings of the animals maintained for the previous 48 h in darkness or light were not significantly different: those that from darkness passed to light increased whereas those that from light passed to darkness decreased (Fig. 1).

BEHAVIOR OF PROTEINS AND GLYCOPROTEINS

No extensive studies on the specificity of these changes brought about by light have been carried out so far. Because the proteins are importantly involved in the phenomena of learning and memory we thought to investigate their behavior in the simpler phenomenon of stimulus perception. The results were a surprise. Whereas the labeling of the sialyl fraction of the protein preparation increased in light at about the same proportion as the gang-

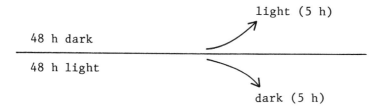

Fig. 1. Labeling of optic tectum gangliosides

liosides, the incorporation of proline into the peptide chains did not show significant differences (Table 1). The interpretation of this apparent discrepancy is very complex and perhaps it is not worthwhile to attempt it without further information. The conclusion that can stand for now is that under the action of light the synthesis or the renewal of sialyl groups of the glycoproteins is accelerated more than the synthesis or the renewal of peptide chains containing proline groups. This concept includes the possibility that under the effect of light the glycosidic chains of the glycoproteins become either more numerous or simply richer in sialyl groups.

ON THE MECHANISM THAT CAUSES THE INCREASE

The determinant of the increase of the sialyl-labeling of gangliosides may be in the retina, i.e. a direct effect of light on the ganglion cells, or, more probably, an effect of the nerve impulses that arise by stimulation of the photoreceptors and reach the ganglion cells through the bipolar or other cells that make synaptic contact with the ganglion cells. Another possibility is that the signal for the increased labeling comes back in a retrograde fashion from the nerve terminal of the optic nerve in the optic tectum to the cell body where the synthesis of gangliosides occurs. Since the separation of the eye cup from its connection with the optic tectum has in every case nullified the effect of light we are considering two possible retrograde mechanisms. One possibility is that the signal comes back from the nerve terminal

Table 1. Labeling of The Optic Tectum Gangliosides, The Glycosidic Moiety of Glycoproteins And of Proteins, After An Intraocular Injection of a Mixture Containing [^{3}H]N-Acetylmannosamine And [^{3}H]Proline.

	Chickens in		Δ%	P
	Light	Darkness		
Gangliosides	1232 (10)	867 (10)	42	<0.01
Glycoproteins (glycosidic moiety)	704 (10)	587 (9)	20	<0.025
Proteins	2762 (10)	2595 (9)	6	N.S.

The animals had been for 48 h in darkness when they received the intraocular injection. Immediately after the injection they were put in light or back in darkness and sacrificed 6 h later.

or optic tectum through the same or other axons. The other possibility is that through the optic nerve a signal is sent to some secretory cell or organ to increase or to reduce the production of a regulator of the synthesis of glycosidic chains of gangliosides and glycoproteins. These two possibilities are compatible with the results obtained with chickens in light or in darkness after the optic nerve was severed, but with no evidence of blood flow impairment. In experiments in which the optic nerve was cut the light ceased to increase the labeling of gangliosides and glycoproteins but rather the labeling in darkness increased to reach the level of labeling in light. It was also shown that the effect was not due simply to increase of entrance into the ganglion cells of the labeled precursor. At least another mechanism should be working because it was verified that in the optic tectum of chickens under light the labeling was not only increased but a higher proportion of labeled ganglioside was in position to be degraded by neuraminidase.[6] Since this means that at the moment of the experiment a part of the gangliosides in both groups of animals are in different positions, the difference in labeling cannot be due only to difference of the specific radioactivity of the precursors. Since most of the gangliosides are produced in the Golgi apparatus and from there they move to their neuraminidase accesible position in the plasma membrane through vesicles where they are in neuraminidase-unaccesible position,[6] we infer from this experiment that the gangliosides and glycoproteins under illumination arrive in less time to their position in the plasma membrane than the gangliosides and glycoproteins of animals in darkness.

REGULATION OF ENZYMATIC STEPS

There is at present a lack of knowledge on the ways the biosynthesis of gangliosides may be regulated. We described years ago[8] the presence of an inhibitor of the sialyl transferase present in the brain tissue. Progress on this subject has provided an electrophoretic homogeneous preparation that is stained by coomasie blue and has a MW of about 5000 D. Judging by the capability of pronase to destroy the inhibitory property of this compound we assume that it is a peptide. Besides this probable peptide another inhibitor(s) present in the brain homogenate was found to be a nucleotide or a mixture of as yet unidentified nucleotides (Irene Albarracin, unpublished).

Inhibition by gangliosides

Probably more important towards the subject of the regulation of the labeling of gangliosides by exposure to light is the observation that the presence of gangliosides inhibits the retinal UDP-galNAc-GM_3 GalNAc transferase (G. A. Nores and R. Caputto, submitted for publication). The polysialogangliosides GD_{1a} and GT_1 are more active as inhibitors than GM_1 and asialogangliosides (Table 2). Apparently most of the inhibitory activity of the gangliosides is exerted by incorporation of the inhibitory ganglioside into the membranes that carry the transferase (Table 3).

Table 2. Inhibitory Capability of Asialoganglioside, GM_1, GD_{1a} And GT_1 Gangliosides on The Activity of UDP-GalNAc-GM_3 GalNAc Transferase.

Ganglioside added to the incubation mixture (μM)		per cent inhibition
Asialoganglioside	(140)	12
Asialoganglioside	(440)	22
GM_1 ganglioside	(150)	37
GM_1 ganglioside	(430)	68
GD_{1a} ganglioside	(150)	59
GD_{1a} ganglioside	(430)	81
GT_1 ganglioside	(150)	58
GT_1 ganglioside	(450)	82

Table 3. Effect on The Enzyme Activity of Preincubating UDP-GalNAc-GM_3 GalNAc Transferase Active Membranes With an Inhibitor (GT_1) or a Substrate (GM_3), Followed by The Separation of The Membranes by Centrifugation.

Preincubation Additions to membranes	Incubation*	pmol/h/mg prot.	Inhibition (%)
--	GM_3	2.70	--
--	$GM_3 + GT_1$	1.40	48
GT_1	GM_3	1.41	48
GM_3	--	2.74	--
GM_3	GT_1	1.83	32

*The incubation system contained, besides the substrates or inhibitor stated, all the factors required for the reaction.

Inhibitory activities in blood

There is present in blood an inhibitor of the retinal UDP-GalNAc-GM_3 GalNAc transferase. An interesting characteristic of this inhibition is that the inhibitory capability of the blood serum of animals that have been for 48 h in the dark is higher than controls that were in light.

A MODEL FOR THE MECHANISM OF THE LIGHT EFFECT

Fig. 2 gives a graphical account of the results so far obtained with added hypothesis that advances ideas on the possible mechanisms that produce the effect. Light impinging on the photoreceptors initiates a nervous wave that reaches the ganglion cells and from these, through the optic nerve, the optic tectum. The wave arriving to the nerve terminal prods the elimination of neurotransmitter that stimulates the next neurone. The elimination of neurotransmitter is assumed to accelerate the deposition of gangliosides in the plasma membrane or to displace forward the ganglioside containing vesicles which are presumed to make a gradient from the Golgi apparatus in the cell body to the nerve terminal. Since the presence of gangliosides close to the membranes that synthesize them inhibits the synthesis, the displacement forward of vesicles should disinhibit the membranes and so, in the presence of labeled precursors, more labeled gangliosides are produced.

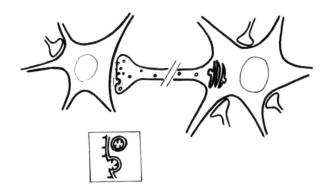

Fig. 2. Graphic representation of the hypothetical mechanism of the increased labeling of gangliosides and glycoproteins under the influence of light. Some of the ganglioside or glycoprotein containing vesicles in the nerve terminal accelerate their incorporation into a neuraminidase accesible position in the terminal. This increases the requirement of gangliosides and glycoproteins from the Golgi apparatus. Cutting of the optic nerve interrupts the whole process.

Another mechanism may be related to the presence of inhibitory substance(s) of UDP-GalNAc-GM_3 galactosaminyltransferase in the blood. This inhibitor increases in the animals maintained in darkness with respect to those that are exposed to light. In this case the disappearance of the differences in dark or light of the animals with severed optic nerve but with unimpaired blood circulation could be explained by analogy with some hormones which are secreted more in either light or darkness only under the influence of the light perceived by the retina.

The increased production of gangliosides or glycoproteins in darkness when the optic nerve had been cut is not explained by either of the two mechanisms, unless the cutting of the nerve causes a loss of axonal material into the meningeal space. In that case the hypothesis of the inhibitory activity of gangliosides can be applied.

SUMMARY

Chickens that received an intraocular injection of ^{3}H-ManNAc and were exposed to light had more labeled gangliosides in the retina ganglion cell layer and in the contralateral optic tectum than similarly treated animals that remained in darkness. The effect is not due to the turning on or off of the light. The sialyl groups of sialoglycoproteins showed similar effect but the labeling of proteins in chickens that received ^{3}H-proline did not show significant differences. So far the effect has been obtained only with retina linked to the optic tectum through the optic nerve. If the nerve is severed the effect disappears.

The gangliosides GD_{1a} and GT_1 are powerful inhibitors of the GM_3-N-acetylgalactosaminyl transferase. The main effect of those gangliosides is expressed when they are linked to the membranes containing the enzyme in such a form that they are not released by washing with water. The hypothesis is advanced that the utilization of gangliosides in the nerve ending during the interneuronal transmission produces a small decrease in their concentration that in turn is transmitted backwards to the neuronal perikarya where it accelerates the synthesis of new gangliosides.

REFERENCES

1. L. N. Irwin and F. E. Samson, Content and turnover of gangliosides in rat brain following behavioral stimulation, J. Neurochem. 18:203 (1971).
2. A. Dunn and E. L. Hogan, Brain gangliosides: Increased incorporation of 1-[^{3}H]-glucosamine during training, Pharmacol. Biochem. Behav. 3:605 (1975).
3. H. E. Savaki and G. M. Levis, Changes in rat brain gangliosides following active avoidance conditioning, Pharmacol. Biochem. Behav. 7:7 (1977).
4. S. E. Karpiak, L. Graf, and M. M. Rapport, Antiserum to brain gangliosides produces recurrent epileptiform activity, Science 194:735 (1976).
5. S. E. Karpiak, L. Graf, and M. M. Rapport, Antibodies to GM_1 gangliosides inhibit a learning avoidance response, Brain Res. 131:637 (1978).
6. B. L. Caputto, G. A. Nores, B. N. Cemborain, and R. Caputto, The effect of light exposure following a intraocular injection of [^{3}H]N-acetylmannosamine on the labeling of gangliosides and glycoproteins of retina ganglion cells and optic tectum of singly caged chickens, Brain Res. 245:231 (1982).
7. R. Caputto, B. L. Caputto, G. A. Nores, and B. N. Cemborain, Effect of light on the labeling of gangliosides and glycoproteins of individually caged chickens, Neural transmission, learning and memory, IBRO Monograph Series Vol. 10, R. Caputto and C. Ajmone Marsan, eds., 179 (1983).
8. R. Duffard and R. Caputto, A natural inhibitor of sialyl transferase and its possible influence on the enzyme activity during brain development, Biochemistry 11:1396 (1972).

BIOSYNTHESIS AND TRANSPORT OF GANGLIOSIDES IN PERIPHERAL NERVE

Allan J. Yates, Ulka R. Tipnis, John H. Hofteig, and
Jean K. Warner

Division of Neuropathology, College of Medicine
Ohio State University
Columbus, Ohio 43210

INTRODUCTION

For several years after the discovery of gangliosides it was believed that they were restricted to the neuronal cell body and not present in peripheral nerve. This idea was reinforced by the results of relatively insensitive early analytical procedures which were unable to demonstrate gangliosides in peripheral nerve. MacMillan and Wherrett[1] first demonstrated a complex pattern of gangliosides in human sciatic nerve and this has been confirmed for several different species.[2-4] The physiological roles of gangliosides are still unknown. However, the finding that exogenously administered gangliosides promote axonal sprouting in regenerating nerve makes it important to determine the sites of synthesis and characteristics of ganglioside transport within nervous tissues.[5] We have studied the metabolism and transport of gangliosides in rabbit sciatic nerve using [^{3}H]-glucosamine, a radiolabelled precursor of ganglioside synthesis. In such studies it is imperative to remove acid soluble contaminants from the ganglioside fractions. Therefore, we have also studied the behavior of several such potential contaminants in procedures commonly employed to isolate gangliosides. The results from both of these studies are discussed herein.

STUDIES ON NORMAL NERVE

Methods

Adult New Zealand white rabbits were given Ketamine (60mg/kg) intramuscularly. When they were sufficiently anesthetized they were intubated and maintained on nitrous oxide, oxygen and halothane. Both L-7 dorsal root ganglia (DRG) were exposed by perforating the laminae with a hand drill, and injected with 50 Ci of [^{3}H]-glucosamine in 10 μl saline. This method has been published in detail.[6] The animals were allowed to survive for several different time intervals (1, 2, 7, 14, 21 days) at which time they were killed and the DRG, lumbosacral trunks (LST) and sciatic nerves (SN) were removed bilaterally. Similar tissues from both sides were pooled and stored at −40°C.

Pairs of tissues from both sides were pulverized in liquid nitrogen and gangliosides extracted and partitioned by the method of Suzuki.[7] The upper phases were treated first with snake venom phosphodiesterase and E. coli alkaline phosphatase[8] and then alkaline methanol.[9] Cold carrier gangliosides were added and the sample dialyzed overnight with several changes of water. The contents of the dialysis bag were loaded onto a Unisil column; neutral lipids were removed with chloroform-methanol (C/M) (4:1) and the final purified ganglioside preparation with C/M (1:1) and methanol. Radioactivity was determined on aliquots of initial homogenate, trichloroacetic acid (TCA) soluble and TCA precipitable fractions, washed lipid extract and final purified ganglioside preparation. Efficiency of radioactivity counting was determined using an external standard. After conversion from cpm to the natural logarithms of dpm, data were analysed by a multiple analysis of variance and Tukey's post-hoc tests.

Results

TCA soluble radioactive material significantly ($p < .05$) decreased with time in DRG, LST and SN from the highest values which occurred one day after injection. TCA precipitable radioactivity continuously increased in SN ($p < .05$) over the 21 days studied. It also increased in LST up to 14 days after which there was no significant change. Radioactivity in washed lipids of SN increased with time to where the values at days 14 and 21 were significantly higher ($p < .05$) than at the earlier times. In both LST and SN the amounts of ganglioside radioactivity 14 and 21 days after injection were significantly higher ($p < .05$) than at days 1 and 2.

The amounts of radioactivity in different anatomical sites are shown as percentages of the combined radioactivity in homogenates of DRG, LST and SN (Fig. 1a-c). The proportion of total nerve homogenate radioactivity present in the acid precipitable fraction of DRG decreased over the 3 week period studied. The proportion of TCA precipitable radioactivity in both LST and SN increased between 1 and 14 days but decreased between 14 and 21 days. The proportion of acid soluble radioactivity decreased markedly between 1 and 7 days in all 3 anatomical sites with only a slight further decrease through to 21 days.

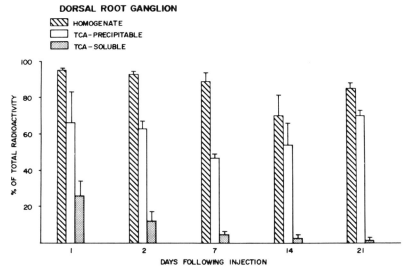

Fig. 1a. Distribution of radioactivity in homogenate, TCA soluble and TCA-precipitable fractions of dorsal root ganglion at several times after injecting DRG with [^{3}H]-glucosamine. Each bar represents the percent of total radioactivity in combined homogenates of DRG, LST and SN for that time point. Every result is the mean and standard deviation from 3 separate experiments.

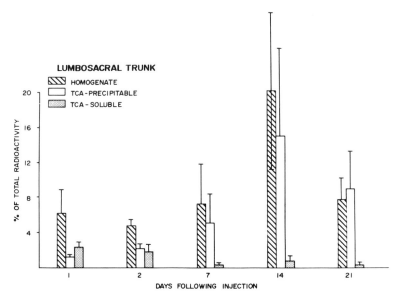

Fig. 1b. Results for lumbosacral trunk expressed in the same way as for Fig. 1a.

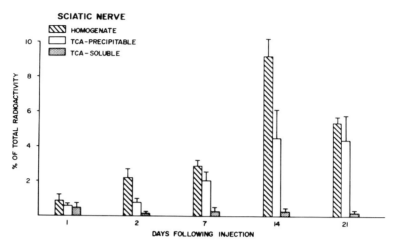

Fig. 1c. Results for sciatic nerve expressed in same way as for Fig. 1a.

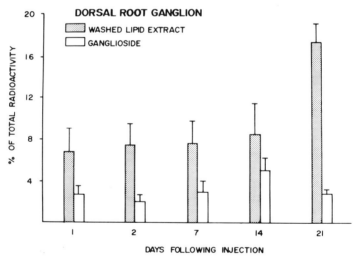

Fig. 2a. Distribution of radioactivity in washed lipid extract and ganglioside fractions of dorsal root ganglion. Results are expressed the same as in Fig. 1a.

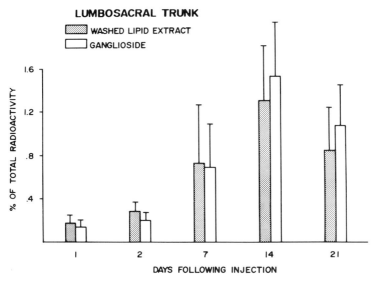

Fig. 2b. Results for lumbosacral trunk expressed in the same way as Fig. 2a.

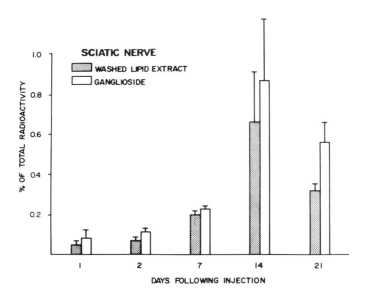

Fig. 2c. Results for sciatic nerve expressed in the same way as Fig. 2a.

Figures 2a-c show the percent of total nerve radioactivity present in the washed lipid and ganglioside fractions of each anatomical site. In DRG, the percent of total radioactivity present in washed lipid was constant between 1 and 14 days, but increased between 14 and 21 days. There was little change in the proportion of radiolabelled ganglioside in DRG with time, but the highest value occurred at 14 days. In both LST and SN the proportion of radioactivity in washed lipid and gangliosides continually increased between 1 and 14 days but fell between 14 and 21 days.

Discussion

The subcellular site of neuronal ganglioside biosynthesis is still not definitely established. Results of some subcellular fractionation studies on brain indicate that synaptosomes contain the appropriate enzymes for ganglioside synthesis.[10,11] However, a more recent study by Ng and Dain[12] indicated that most of the ganglioside sialyltransferase activity in rat brain is not associated with synaptosomes, but rather is in smooth microsomal membranes and Golgi complex derivatives. This is consistent with the findings of Keenan et al.[13] who found that the glycosyltransferase activities involved in ganglioside biosynthesis in rat liver are highest in Golgi apparatus membranes. Maccioni et al.,[14] determined the activities of the enzymes which transfer either sialic acid or galactose to ganglioside acceptors in synaptosomes and neuronal perikarya of rat brain. Both enzyme activities were much higher in perikarya than synaptosomes. Furthermore, the incorporation of radiolabelled N-acetyl-mannosamine into gangliosides occurred much sooner in the perikarya than synaptosomes. From this they concluded that gangliosides are synthesized in neuronal cell bodies as "precursor pools" and are translocated to synaptosomes where they exist as "end product pools". In a subsequent study by Landa et al.,[15] radiolabelled N-acetyl-mannosamine was injected into the eyes of chicks and radioactivity measured in both gangliosides and some of their precursors in retina and optic tectum. Radiolabelled gangliosides appeared in synaptosomes and synaptosomal membranes of optic tecta before tectal perikarya. Furthermore, the labelled ganglioside pattern was similar to that of retina, and colchicine inhibited the appearance of labelled tectal gangliosides. However, when the radiolabelled precursor was injected intracranially, the labelled gangliosides appeared in perikarya before synaptosomes and the pattern resembled that of optic tectum gangliosides. The activtiies of both ganglioside sialytransferase and galactosyltransferase were higher in tectal perikarya than synaptosomes. This evidence gives considerable support to the possibility that gangliosides are synthesized in the neuronal cell body and transported distally where they reside in end product pools.

Our results show that radioactivity from [^{3}H]-glucosamine injected into DRG is incorporated into several different classes of compounds and transported into the axon. During the first 24 hours most of the radiolabelled material in both LST and SN is in the non-lipid fractions, much of which is acid soluble. Over the ensuing 3 weeks there is a considerable decrease in acid soluble radioactivity in all three sites to 2.9% (DRG), 11.6% (LST) and 14.8% (SN) of the day 1 values. During this same time radioactivity in the ganglioside fractions increase in LST (540%) and SN (560%). In both of these sites at 21 days following injection the amounts of radioactivity in the ganglioside fractions were greater than those in the acid soluble fractions. This makes it highly unlikely that the radiolabelled gangliosides in either LST or SN were locally synthesized.

Further evidence that gangliosides were synthesized in DRG and transported into LST and SN was obtained by injecting colchicine (80 nanomoles) into one DRG 2 hours before injecting [^{3}H]-glucosamine into both L-7 DRG. Specimens were removed 24 hours later and lipids extracted. The amounts of radioactivity in the total lipid extract of LST and SN on the side which received no colchicine were 23% and 9% of that in the lipid of the ipsilateral DRG. However, the radioactivity in lipids of both LST and SN of the colchicine-treated side was less than 1% of that in the lipids of their DRG.

There are several reports of studies on the visual system similar to ours on peripheral nerve. Forman et al.,[16] measured the distribution of radioactivity in optic nerves of goldfish after injecting into eyes with [^{3}H]-glucosamine. They found almost all of the radioactivity to migrate with fast axonal flow. Sixty-five percent of the label was in the particulate fraction, 1/3 of which was lipid bound. These results suggest that gangliosides are transported along the axon.

Holm[17] investigated this possibility in the visual pathway of rabbits by injecting [^{3}H]-acetate or [^{14}C]-glucosamine into the eye and measuring the amounts of radiolabelled ganglioside in retina, optic nerve, optic tract and lateral geniculate body at several time intervals. He interpreted his results as indicating that either the injected precursors were rapidly distributed throughout the visual pathway, or a small pool of gangliosides was rapidly transported. Forman and Ledeen[18] employed the visual system of goldfish for similar studies and concluded that their results and those of Holm could be interpreted as evidence for rapid axonal transport of gangliosides. Subsequent studies by Ledeen et al.,[19] indicated that gangliosides in transit to nerve terminals are transported by the rapid flow system, but those gangliosides intrinsic to the axon may either move with the slow system or be part of a slowly metabolizing pool. On the basis of similar double labelling studies, these same authors were able to determine that the entire ganglioside molecule is metabolized as a unit and does not undergo partial degradation, a finding consistent with the concept of both a precursor and end-product pool.

Rösner et al.[20] also studied ganglioside axoplasmic transport in the goldfish visual system. They found that the TCA soluble radioactivity was transported by the fast axonal transport system when either [^{3}H]-N-acetyl-mannosamine or [^{3}H]-N-acetyl-glucosamine was injected intraocularly. They also found that the amount of radiolabelled ganglioside distant from the eye increased with time. This was compatible with rapid transport of gangliosides, but they could not absolutely rule out only the transport of radiolabelled precursor with its being peripherally incorporated into gangliosides. Rösner[21] obtained similar results from studies on chicks and also concluded that the data were compatible with both of these possibilities. However, Tirri and Ledeen[22] found that the specific activity of neuraminidase-labile ganglioside sialic acid to be 3 times as high as neuraminidase-resistance sialic acid of gangliosides isolated from the lateral geniculate body of rabbits injected intraocularly with [^{3}H]-N-acetyl-mannosamine. They interpreted this as evidence for synthesis of gangliosides in the cell body prior to axonal transport.

STUDIES ON TRANSECTED NERVE

Following transection of an axon there is a complex sequence of morphological and biochemical events in the neuronal cell body referred to as "central chromatolysis", "axon reaction", "retrograde reaction", and "cell body response".[23] The degree and timing of this response varies among different types of neruones, and is influenced by a variety of conditions. Nevertheless, there are some common features of the response: swelling of the cell body, increased synthesis of RNA, lipid and "structural proteins", but decreased synthesis of proteins involved with synaptic transmission. In general, these changes are directed towards reconstituting the axon and its connections distal to the site of injury.[24] These events occur as a consequence of either crush or transection injuries. However, if the proximal stump of a transected nerve is not closely apposed to the distal end, the attempt at axonal regeneration is usually abortive. If gangliosides are normally involved in sprouting of damaged axons then it is reasonable to suspect that ganglioside synthesis may increase following nerve trauma. Therefore, we studied the radiolabelling of gangliosides in normal and transected rabbit sciatic nerves.

Adult New Zealand white rabbits were anesthetized with intraperitoneal fetanyl and droperidol as previously described.[25] The left sciatic nerve was transected and reflected distally to prevent reinnervation. The animals were allowed to survive for 1, 2, 3 or 4 weeks before injecting both L-7 DRG as described above. To control for the shorter proximal stump on the transected side, at the time of the second surgery the right sciatic nerve was ligated at the same level as the previous transection. Two days later the animals were killed, both DRG, and LST were removed, weighed and frozen. Gangliosides were extracted and radioactivity estimated as in the above experiment.

The results were expressed as the amount of radioactivity in the ganglioside fraction of DRG or LST as a per-cent of total radioactivity in DRG and LST combined.[26] These data were studied by an analysis of variance, and no significant differences were found comparing values for the transected with the control side.

There are several possible interpretations of these results. First, ganglioside metabolism may not change following axonal damage of any type. If this is the case then it makes it less likely that gangliosides normally play a role in nerve regeneration. Second, ganglioside metabolism may change in response to axonal trauma only if the two damaged ends are closely apposed. Alberghina[27] found that both crushed and transected, surgically reconstructed nerve incorporated more transported phospholipid than control nerves. However, this increase persisted for a longer period of time in the cut, reanastomosed nerves. What role if any the distal stump may play in such metabolic responses is unknown. However, an influence of the distal stump on ganglioside metabolism in the proximal nerve segment could be tested by repeating this type of experiment but employing a crush rather than transection injury. Third, there may have been an alteration in ganglioside metabolism but it was not detected because it either occurred transiently in the first week, or it was too small for our methods to detect. Fourth, there may have been a change in both the cut and control (ligated) nerve. This seems least likely because the magnitude of the change would have to have been the same 2 days after ligation as 30 days after transection.

The question of whether or not gangliosides play a critical role in initiating axonal sprouting is extremely important. Therefore, future experiments should be designed to determine which of these interpretations is correct.

BEHAVIOR OF RADIOLABELLED CONTAMINANTS IN GANGLIOSIDE ISOLATION PROCEDURES

In studies such as those just discussed it is essential that all non-ganglioside constituents which could become radiolabelled must be separated from the final ganglioside preparation. One commonly used method for isolating gangliosides involves a Folch partition and dialysis before purification with silicic acid column chromatography. Another method employs column chromatography both for lipid class separation and desalting. The latter was developed for 2 reasons: (a) because some gangliosides do not partition quantitatively into the upper phase; (b) to obviate the hazard of ganglioside losses during dialysis. We have studied the behavior of some of the likely radiolabelled contaminants in several stages of these two different procedures for ganglioside isolation.

Extraction, Partition and Dialysis

Known amounts of radiolabelled glucosamine hydrocholoride (GlcN), uridine diphosphate galactose (UDP-gal), uridine diphosphate N-acetyl-D-galactosamine (UDP-galNAc) and cytidine 5'-monophosphate sialic acid (CMP-NeuAc) were added separately to one gram portions of human cerebral cortex homogenized in 20 volumes of chloroform-methanol 2:1 (C/M-2:1). Gangliosides were extracted and partitioned by the method of Suzuki.[7] Significant amounts of GlcN (77%) and nucleotide sugars (39-75%) were dissolved in the initial lipid extract and almost all of that partitioned into the upper phase. Therefore, these fractions could contain significant amounts of acid soluble precursors and cannot alone be used as estimates of lipid-bound radioactivity.

Usually an attempt is made to remove these contaminants from the upper phase by dialysis. We found that after 12 hours of dialysis and 6 water changes over 97% of free glucosamine dialysed out of the bag both in the presence and absence of gangliosides. However, this is not the case for 3 nucleotide sugars (UDP-gal, UDP-galNAc, CMP-NeuAc) we studied. Even after 48 hours of dialysis and 8 water changes significant amounts of radioactivity remained in the bag when the nucleotide sugars were dialysed alone (6-22%). Even more (35-99%) was retained when dialysed with gangliosides. It appears that some of the nucleotide sugar is trapped within the ganglioside micelles. This problem was recognized several years ago by Kanfer[8] who suggested that prior to dialysis the upper phase be treated with both snake venom phosphodiesterase and E. coli alkaline phosphatase. We treated the same sugar nucleotides plus guanosine diphosphate fucose (GDP-fuc), uridine disphosphate glucose (UDP-glc), and uridine diphosphate N-acetylglucosamine (UDP-glcNAc), all in the presence of gangliosides, with this double enzyme digestion prior to dialysis. This treatment removed 97.8 - 99.8% of radioactivity from diphosphonucleotide sugars, but only 53.5% of CMP-NeuAc radioactivity. Compared with this enzyme treatment, alkaline methanolysis[9] prior to dialysis was less effective in removing diphosphonucleotide sugars but slightly more effective in removing CMP-NeuAc (14% remained). Therefore, although most free glucosamine and diphosphonucleotide sugars can be removed by enzyme treatment and dialysis, some CMP-NeuAc could still remain in the sample.

Silicic Acid Column Chromatography

Unisil column chromatography is frequently used to remove contaminating lipids less polar than gangliosides which elute from this column in C/M (1:1) and methanol. All six nucleotide sugars which we studied also eluted in these fractions. It is of interest that their mobility shifted more from the methanol to C/M (1:1) fractions when chromatographed in the presence of gangliosides.

More recently Iatrobeads frequently have been used instead of Unisil. We studied the behavior of the same six nucleotide sugars on Iatrobead columns. Almost no radioactivity appeared in the less polar C/M (85:15) fraction, but large amounts eluted as soon as the solvent was changed to C/M (1:2) in which gangliosides elute. Therefore, neither Unisil nor Iatrobeads is useful in removing nucleotide sugars from gangliosides with these solvent schemes.

Sephadex Column Chromatography

DEAE-Sephadex A-25 columns are commonly used to separate neutral from acidic lipids as one of the initial steps in ganglioside purification. We studied the same 6 radiolabelled nucleotide sugars and glucosamine as above. These were separately added to total lipid extracts from 100 mg normal human cerebral cortex and loaded onto a DEAE-Sephadex A-25 column according to the method of Ledeen and Yu.[28] Almost all of the GlcN radioactivity eluted in the first fraction - C/M/water (30:60:8) where the neutral lipids elute. Most of the CMP-sialic acid eluted in the second fraction - C/M/Na acetate (30:60:8), where gangliosides and other acidic lipids elute. Very low recoveries of the other 5 nucleotide sugars were obtained. Even with the more harsh solvent C/M/0.8M KCl (30:60:2) only very little additional UDP-GlcNAc radioactivity was eluted from the column. Therefore, large amounts of diphosphonucleotide sugars are retained by these columns, but significant contamination of both neutral and acidic lipid fractions with radiolabelled acid soluble contaminants could still occur.

Sephadex LH-20 column chromatography is useful in desalting ganglioside preparations following alkaline methanolysis to degrade phospholipids (R. Ledeen - personal communication). When chromatographed alone in an isocratic system with C/M (1:1) as the solvent all 6 nucleotide sugars studied eluted in the same fractions that gangliosides normally appear. A somewhat surprising finding was that when the diphosphonucleotide sugars were chromatographed in the presence of ganglioside, their mobilities were retarded so that most of the radioactivity eluted later than ganglioside. Although most of the radioactivity from CMP-NeuAc eluted later than ganglioside, a small amount did co-elute with ganglioside. Alkaline methanolysis also increased the rate of mobility of diphosphonucleotide sugars. Sephadex LH-20 interacts with aromatic compounds such as nucleic acid bases. Cleaving the base from the sugar, as occurs with alkaline methanolysis, would eliminate this retarding factor and probably accounts for this accelerated mobility of the sugar phosphate. However, treatment of diphosphonucleotide sugars with phosphodiesterase and alkaline phosphatase results in an elution pattern similar to a free amino sugar, the bulk of which separates from the ganglioside peak. The most likely explanation is that removal of the phosphate group by the alkaline phosphatase is removing ionic interactions and it is now being retarded mainly on the basis of gel filtration.

The one nucleotide sugar which we studied that was not separated from ganglioside even after enzyme treatment was CMP-sialic acid. While the reason for this is not entirely clear, it may be related to the fact that carbon 3 (adjacent to the anomeric carbon bearing the phosphate group) has no hydroxyl group. Therefore, a 2,3-cyclic phosphate essential for the hydrolysis reaction to proceed could not be formed.

From these studies we conclude that free GlcN and diphosphonucleotide sugars can be removed by dialysis following enzyme treatment, but silicic acid column chromatography is of no value. A considerable amount of these substances are also removed by DEAE-Sephadex and most can be removed on Sephadex-LH-20 following enzyme treatment. However, CMP-sialic acid is not completely removed by any of these procedures and some may be contaminating ganglioside preparations even after extensive purification.

SUMMARY

Radiolabelled glucosamine was injected into L-7 dorsal root ganglion (DRG) of rabbits. At several different times after injection DRG, lumbosacral trunks (LST) and sciatic nerves (SN) were removed and gangliosides extracted. Two and 3 weeks after injection the amounts of radioactivity in the ganglioside fractions of LST and SN were significantly higher than at days 1 and 2. The TCA soluble radioactivity decreased dramatically over the same time period. Colchicine prevented the appearance of radiolabelled lipid in LST and SN. From these experiments we conclude that some ganglioside is synthesized in the neuronal cell bodies of DRG and transported into the axons of the sciatic nerve. In another experiment the sciatic nerve was transected and ends separated to prevent regeneration. Ganglioside synthesis and transport were studied in these animals in the same way as the previous experiment. There was no difference in the amount of radiolabelled ganglioside that was isolated from DRG or LST of transected compared with control nerves. The behavior of several potential acid soluble contaminants was studied in several steps used to isolate gangliosides. Of those studied only CMP-NeuAc could cause significant contamination of the final ganglioside preparation.

ACKNOWLEDGEMENTS

The authors would like to thank Pamela Horn for typing this manuscript. This work was supported by N.I.H. grant NS-10165. U.R.T. and J.H.H. were supported by N.I.H. 5T32 N507091.

REFERENCES

1. V. H. MacMillan and J. R. Wherrett, A modified procedure for the analysis of mixtures of tissue gangliosides, J. Neurochem. 16:1621 (1969).
2. A. J. Yates and J. R. Wherrett, Changes in the sciatic nerve of the rabbit and its tissue constituents during development, J. Neurochem. 23:993 (1974).
3. J. H. Hofteig, J. R. Mendell, and A. J. Yates, Chemical and morphological studies on Garfish peripheral nerves, J. Comp. Neurol. 198:265 (1981).
4. K. H. Chou, C. E. Nolan, and F. B. Jungalwala, Composition and metabolism of gangliosides in rat peripheral nervous system during development, J. Neurochem. 39:1547 (1982).
5. A. Gorio, G. Carmignoto, L. Facci, and M. Finessa, Motor nerve sprouting by ganglioside treatment. Possible implications for gangliosides on nerve growth, Brain Res. 197:236 (1980).
6. H. W. Palay, U. R. Tipnis, A. J. Yates, and R. R. Yates, A method for injecting dorsal root ganglia in rabbit, J. Neurol. Methods 6:173 (1982).
7. K. Suzuki, The pattern of mammalian brain gangliosides. II. Evaluation of the extraction procedures, postmortem changes and the effect of formalin preservation, J. Neurochem. 12:629 (1965).
8. J. N. Kanfer, Preparation of gangliosides, Methods in Enzymology 14:62 (1969).
9. G. Dawson, Glycosphingolipid levels in an unusual nerovisceral storage disease characterized by lactosylceramide hydrolase deficiency: Lactosylceramidosis, J. Lipid Res. 13:207 (1970).
10. H. Den, B. Kaufman, and S. Roseman, The sialic acids XIII. Subcellular distribution of several glycosyltransferases in embryonic chicken brain, J. Biol. Chem. 250:739 (1975).
11. J. DiCesare and J. Dain, Localization, solubilization and properties of N-acetylgalactosaminyl and galactosyl ganglioside transferases in rat brain, J. Neurochem. 19:403 (1972).
12. S. S. Ng and J. A. Dain, Sialyltransferases in rat brain: intracellular localization and some membrane properties, J. Neurochem. 29:1085 (1977).
13. T. Keenan, D. Morre, and S. Basu, Ganglioside biosynthesis: concentration of glycosphingolipid glycosyltransferases in Golgi apparatus from rat liver, J. Biol. Chem. 249:310 (1974).
14. H. J. F. Maccioni, S. S. Defilpo, C. A. Landa, and R. Caputto, The biosynthesis of brain gangliosides. Ganglioside-glycosylating activity in rat brain neuronal perikarya fraction, Biochem. J. 174:673 (1978).
15. C. A. Landa, H. J. F. Maccioni, and R. Caputto, The site of synthesis of gangliosides in the chick optic system, J. Neurochem. 33:825 (1979).

16. D. S. Forman, B. S. McEwen, and B. Grafstein, Rapid transport of radioactivity in goldfish optic nerve following injection of labelled glucosamine, Brain Res. 28:119 (1971).
17. M. Holm, Gangliosides of the optic pathway: biosynthesis and biodegradation studied in vivo, J. Neurochem. 19:623 (1972).
18. D. S. Forman and R. W. Ledeen, Axonal transport of gangliosides in the goldfish optic nerve, Science 177:630 (1972).
19. R. W. Ledeen, J. A. Skrivanek, L. J. Tirri, R. K. Margolis, and R. U. Margolis, Gangliosides of the neuron: localization and origin, Advances Exp. Med. Biol. 17:83 (1976).
20. H. Rösner, H. Wiegandt, and H. Rahmann, Sialic acid incorporation into gangliosides and glycoprotein of the fish brain, J. Neurochem. 21:655 (1973).
21. H. Rösner, Incorporation of sialic acid into ganglioside and glycoproteins of the optic pathway following an intraocular injection of N-[^{3}H]-acetyl-mannosamine in the chicken, Brain Res. 97:107 (1975).
22. L. Tirri and R. Ledeen, Rapid axonal transport of serine and mannosamine labelled lipids, Abst. Fifth Mtg. Am. Soc. Neurochem. p. 177 (1974).
23. B. Grafstein, The nerve cell body response to axotomy, Exp. Neurol. 48:32 (1975).
24. B. Grafstein and D. S. Forman, Intracellular transport in neurons, Physiol. Rev. 60:1167 (1980).
25. A. J. Yates and D. K. Thompson, Ganglioside composition of nerve underdoing Wallerian degeneration, J. Neurochem. 30:1649 (1978).
26. A. J. Yates, H. H. Hofteig, and U. R. Tipnis, Changes in the lipid composition of peripheral nerve following trauma, in: "Recent Progress in Neural Trauma," H. R. Winn, ed., Raven Press, New York, in press.
27. M. Alberghina, M. Viola, and A. M. Giuffrida, Rapid axonal transport of glycerophospholipids in regenerating hypoglossal nerve of the rabbit, J. Neurochem. 40:25 (1983).
28. R. W. Ledeen and R. K. Yu, Gangliosides: structure isolation and analysis, Method. Enzymol. 83:139 (1982).

CELLULAR LOCALIZATION OF GANGLIOSIDES IN THE MOUSE CEREBELLUM:

ANALYSIS USING NEUROLOGICAL MUTANTS

Thomas N. Seyfried and Robert K. Yu

Department of Neurology
Yale University School of Medicine
333 Cedar Street
New Haven, CT 06510

Understanding the function of gangliosides in the central nervous system (CNS) requires knowledge of their cellular distribution. It would be important to know if certain gangliosides are more or less concentrated in specific neural cell types and to know how gangliosides are distributed over the surface of neurons. In other words, are all gangliosides randomly distributed over the entire cell surface or are there domains on the neuronal surface (synapse, dendrite, perikaryon, axon hillock, axon) where the concentration of one ganglioside predominates over that of another. The distribution of gangliosides between cells and within cells may also vary with age. Hence, a series of neurological mouse mutants that lose specific populations of cerebellar neurons at various stages of development provides an excellent system for studying the cellular distribution of gangliosides in the CNS.

The weaver (wv) mutation destroys granule cells,[1,2] whereas the Purkinje cell degeneration (pcd) mutation destroys Purkinje cells.[3,4] The staggerer (sg) and lurcher (Lc) mutations destroy both granule and Purkinje cells.[5-9] Also, reactive gliosis accompanies neuronal cell loss in the sg/sg and pcd/pcd mutants,[3,10-12] but not in the weaver (wv/wv) mutant.[11]

In our first study,[13] cerebellar gangliosides were analyzed in the wv/wv, pcd/pcd, sg/sg, and Lc/+ mutants at adult ages, i.e., after the granule cells or Purkinje cells had degenerated. By comparing the distribution of cerebellar gangliosides in these mutants, we inferred that certain gangliosides were more or less concentrated in granule cells, Purkinje cells and reactive glia.

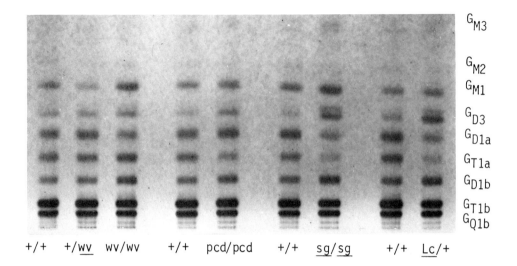

Fig. 1. Thin-layer chromatogram of cerebellar gangliosides in adult normal (+/+) and mutant mice. The ages of the mice and thin-layer conditions were as presented previously,[13] where this illustration appeared originally.

Fig. 1 shows the thin-layer chromatographic distribution of gangliosides in the various mutants and their wild type (+/+) controls, and Table 1 summarizes the influence of these mutations on both the cellularity and concentration of individual gangliosides in the cerebellum.

We found that the loss of certain cell types in these mutants caused significant changes in the distribution of certain gangliosides. The concentration of ganglioside GD1a was significantly reduced in those mutants that lost granule cells (wv/wv, sg/sg and Lc/+, but was not reduced in the pcd/pcd mutant that lost Purkinje cells (Fig. 1 and Table 1). GT1a, on the other hand, was significantly reduced in those mutants that lost Purkinje cells (pcd/pcd, sg/sg and Lc/+, but was not reduced in the wv/wv mutant that lost granule cells (Fig. 1 and Table 1). We concluded that GD1a was enriched in granule cells and that GT1a was enriched in Purkinje cells. These gangliosides may, therefore, serve as useful cell surface markers for changes occurring in granule cell and Purkinje cell membranes during cerebellar development.

The two polysialogangliosides, GT1b and GQ1b, were significantly reduced in all of the mutants, whereas GD1b was not re-

Table 1. Influence of Neurological Mutations on the Cellularity and Ganglioside Concentration of the Adult Mouse Cerebellum.

Genotype	Major Cytological Changes[a]			Ganglioside Concentration[b]						
	Granule Cells	Purkinje Cells	Gliosis	GM1	GD3	GD1a	GT1a	GD1b	GT1b	GQ1b
+/+	+	+	-	+	+	+	+	+	+	+
wv/wv	-	+	-	↑	+	-	+	+	-	-
pcd/pcd	+	-	↑	+	↑	+	-	+	-	-
sg/sg	-	-	↑	+	↑	-	-	+	-	-
Lc/+	-	-	↑	+	↑	-	-	+	-	-

[a]The (+) indicates a normal cell number or ganglioside concentration. The (-) indicates an absence or highly significant reduction in cell number or concentration. The ↑ indicates an abnormal abundance or accumlation.
[b]On dry weight basis.

duced seriously in any of the mutants (Table 1). These findings suggest that GT1b and GQ1b are concentrated in both granule and Purkinje cells and that GD1b may not be selectively enriched in either cell type. Since myelin is not seriously affected by any of the mutations,[13] the failure to find marked reductions in GD1b concentration may be due to its partial enrichment in myelin.[14]

Although the percentage distribution of GM1 was elevated in all of the mutants, its concentration was elevated only in the wv/wv mutant[13] (Fig. 1). Because GM1 is a major myelin ganglioside,[14-17] it is possible that the content or composition of myelin is abnormal in the wv/wv cerebellum.

Ganglioside GD3 was significantly elevated in the pcd/pcd, sg/sg and Lc/+ mutants, but was not affected significantly in the wv/wv mutant (Fig. 1 and Table 1). Because astrocytic proliferation has been documented in the pcd/pcd and sg/sg mutants,[3,10-12] but not in the wv/wv mutant,[11] our findings suggest that GD3 is associated with astrocytic proliferation or serves as a marker for reactive gliosis. Because the Lc/+ and sg/sg mutants share similar histological abnormalities and because both express elevated GD3, we predict that reactive gliosis should also occur in the Lc/+ mutant.

If GD3 is a reliable marker for reactive gliosis, then elevated amounts of GD3 should be found in all neurological disorders where reactive gliosis occurs. A review of the literature on neurological disease strongly supports this notion (Table 2). Regardless of disease etiology, reactive gliosis is associated with increased amounts of GD3. Since no exceptions to this general phenomenon have yet been found, GD3 may be considered a good marker for reactive gliosis in neurological disease.

A differential cellular enrichment of gangliosides in the mouse cerebellum is the simplest interpretation of our findings. Certain gangliosides appear to be more concentrated in some cell types than in others. It is unlikely that the ganglioside alterations found result from inherited defects in ganglioside metabolism because no marked ganglioside alterations were detected in the cerebrums of these mutants. An inherited defect in ganglioside metabolism should be expressed to some degree in all CNS regions. We do not, however, rule out the possibility that the ganglioside alterations arise through an epigenetic phenomenon, where the loss of one cell type may alter ganglioside metabolism in a remaining cell type. With the exception of the association between GD3 and reactive gliosis, there is no compelling evidence to support the epigenetic hypothesis. Although several interpretations for the ganglioside changes may be valid, we feel that the data support the differential cellular enrichment hypothesis.

Table 2. Neurological Disorders Where Increased Amounts of Ganglioside GD3 Have Been Found in Association with Reactive Gliosis.

Multiple Sclerosis[18,19]

Subacute Sclerosing Leukoencephalitis[20,21]

Kuru[22]

Creutzfeld – Jakob Disease[23-25]

Alzheimer's Disease[25]

Huntington's Disease[26]

Metachromatic Leukodystrophy (white matter)[27]

Congenital Amaurotic Idiocy[28]

Krabbe's Disease[29]

Adrenoleukodystrophy[30,31]

The second phase of our study involves the analysis of gangliosides in these mutants during early cerebellar development, i.e., during periods of active neuronal degeneration. It would be important to determine if the loss of granule cells and Purkinje cells can be correlated developmentally with alterations in specific gangliosides. This study was recently completed for the weaver mutant,[32] and the other mutants are now being studied.

The wv mutation affects cerebellar development by destroying the majority of granule cells in the external granule cell layer prior to their translocation across the molecular layer.[1,2,33] This destruction causes a drastic reduction in cerebellar size and an almost complete absence of granule cell synaptic membranes. Although the Purkinje cells are reduced in size and disoriented in the wv/wv cerebellum, these cells are morphologically similar to those seen in the +/+ mice.[2,34-36] Moreover, the Purkinje cells in wv/wv mice form numerous postsynaptic dendritic spines even though the presynaptic granule cell parallel fibers are absent.[2,35]

The wv mutation also shows a dosage effect since granule cell migration and differentiation are slowed or retarded in the +/wv heterozygotes. A large number of granule cells have yet to complete their translocation or establish synaptic contacts by 12 days of age in the +/wv mice.[33,34,37] The period of maximum synaptogenesis in the +/+ mouse cerebellum occurs from 9 to 12 days with synaptogenesis mostly complete by 15 days of age.[38] In the +/wv heterozygotes, however, the lag in granule cell migration prolongs granule cell differentiation and retards the onset of maximum synaptogenesis by approximately 2 days.

The loss of granule cells in the wv/wv mutant caused significant reductions in cerebellar dry weight and ganglioside concentration at all ages studied (Fig. 2). These observations support our previous findings and suggest that gangliosides are abundant in granule cell neurons.[13,32] Cerebellar ganglioside concentration in the +/wv mice was most reduced during the younger ages (9-12 days) which is consistent with the severity of the histological abnormalities found in these mice.[33,34]

The progressive loss of undifferentiated granule cells in the wv/wv mutant was associated with marked changes in the concentration and distribution of certain gangliosides. Especially noteworthy is that some gangliosides were significantly more affected than others. Here we discuss the changes found for GD1a and GT1a. A discussion of the other gangliosides is given in our recent paper.[32] Since granule cells and Purkinje cells comprise the vast majority of neuronal membranes in the cerebellum,[39,40] the ganglioside changes that we found were correlated with the changes occurring in the number and structure of these cell types.

A sharp drop in both the concentration and percentage distribution of GD1a occurred beyond 9 days of age in the wv/wv mice (Fig. 3). Since the sharpest decline in GD1a paralleled the period of most active granule cell synaptogenesis, GD1a may be concentrated in the synaptic and dendritic membranes of these cells. Although some undifferentiated granule cells have degenerated in the 9 day-old wv/wv mice, the percentage distribution of GD1a was actually elevated at this age (Fig. 3). These findings suggest that GD1a is not enriched in undifferentiated granule cells, but becomes enriched in granule cells during their differentiation and synaptogenesis. An enrichment of GD1a in synaptic and dendritic membranes is consistent with the observations of other investigators.[41-46]

The lag in granule cell differentiation and synaptogenesis in the +/wv mice, which occurs as a consequence of retarded granule cell migration, coincided temporarily with a significant lag in GD1a accretion (Fig. 3). In other words, the delayed accretion of GD1a in the +/wv mice was closely associated with the delayed onset of granule cell differentiation and synaptogenesis. Since most granule cells in the +/wv mice will eventually descend

CELLULAR LOCALIZATION OF GANGLIOSIDES 175

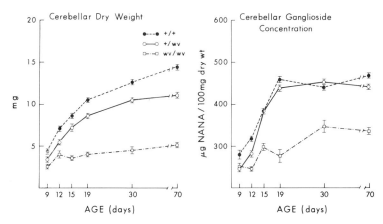

Fig. 2. Developmental profile of cerebellar dry weight and total ganglioside concentration in the +/+, +/wv, and wv/wv mice. The values are expressed as means (± SEM) and the number of separate cerebellar samples analyzed for each genotype at each age were as presented previously,[32] where this illustration appeared originally.

into the granule layer and establish normal synaptic contact, it seems reasonable that the sharp rise in GD1a accretion from 12 to 15 days (Fig. 3) corresponds to the delayed formation of synapses in these mice. Although our findings are consistent with an enrichment of GD1a in granule cell synapses and dendrites, we cannot rule out the possibility that GD1a is evenly distributed over the entire granule cell surface.

A continuous drop occurred in the concentration and distribution of GD1a beyond 15 days of age in the +/+ cerebellum (Fig. 3). This drop became more pronounced from 30 to 70 days in all genotypes. Indeed, the gradual reduction of GD1a beyond the period of active synaptogenesis appears to be a general phenomenon in vertebrate CNS development.[47-50] This reduction in the mouse also occurs earlier in subcortical structures (cerebellum and brain stem) than in the cerebrum.[50] It would be intriguing if the reduction of cerebellar GD1a was associated with the phenomenon of "synaptic adhesion waning", which involves a reduction in both the number and size of initial synaptic adhesions with age.[38]

The loss of granule cells in the wv/wv mice or their retarded migration in the +/wv mice had no significant influence on the concentration of GT1a (Fig. 4). These findings support our previous suggestion that GT1a is not enriched in granule cell membranes.[13]

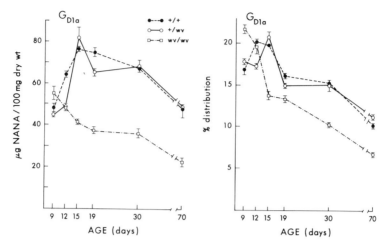

Fig. 3. Developmental profile of the concentration and percentage distribution of ganglioside GD1a in the cerebellums of the +/+, +/wv, and wv/wv mice. The conditions are the same as described in Fig. 2 and in ref. 32, where this illustration appeared originally.

Although the Purkinje cells in the wv/wv mice are somewhat reduced in size and disoriented, they are morphologically similar to those seen in the +/+ mice and can autonomously form post-synaptic dendritic spines in the absence of the presynaptic parallel fibers. The percentage distribution of any ganglioside enriched in the dendritic spines of Purkinje cells should, therefore, appear significantly elevated in the wv/wv cerebellum. In other words, the loss of granule cells should magnify the percentage distribution, but not the concentration of any ganglioside enriched in Purkinje cells. This is precisely what is found for GT1a in the wv/wv cerebellum (Fig. 4), and again suggests that GT1a is enriched in the Purkinje cell dendritic spines. Support for this hypothesis could come from a study of cerebellar gangliosides in the staggerer mutant, which has defects in the development of Purkinje cell dendritic spines.[8]

The noticeable reduction in both the concentration and the percentage distribution of GT1a from 30 to 70 days of age in the wv/wv mice (Fig. 4) may result from a resorption of some of the unattached Purkinje cell dendritic spines. A gradual reduction in the number of Purkinje cell dendritic spines beyond 20 days of age appears to be a natural phenomenon in the normal mouse cerebellum.[51] Possibly, this process is responsible for the slight reduction seen in both the concentration and the percentage distribution of GT1a beyond 19 days of age in the +/+ cerebellum (Fig. 4).

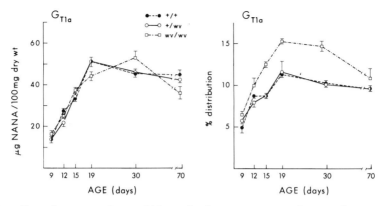

Fig. 4. Developmental profile of the concentration and percentage distribution of ganglioside GTla in the cerebellums of the +/+, +/wv, and wv/wv mice. The conditions are the same as described in Fig. 2 and in ref. 32, where this illustration appeared originally.

Our findings provide strong indirect evidence for a nonrandom cellular distribution of gangliosides in the mouse cerebellum. Ganglioside GDla appears to be more concentrated in granule cells than in Purkinje cells, whereas the opposite occurs for GTla. It also appears that these gangliosides are most actively synthesized during precise developmental epochs, e.g., synaptogenesis. The enrichment of GDla in granule cells, and of GTla in Purkinje cells suggests that these gangliosides may be important for specific granule cell and Purkinje cell functions.

Any hypothesis concerning ganglioside function in the CNS should account for the differential cellular distribution of these molecules. We recently proposed that GD3 may function as an enhancer of membrane permeability.[32,52] This hypothesis emerged from finding an enrichment of GD3 in a variety of neural membranes characterized by having enhanced permeability to ions and metabolites.[32,52] Hence, the continued study of ganglioside distribution during early cerebellar development in the various neurological mutants may provide additional insight into the cellular distribution and function of gangliosides in the CNS.

SUMMARY

We have used genetic dissection to study the cellular localization of gangliosides in the mouse cerebellum. This method employs a series of mouse mutations that destroy specific populations of cerebellar neurons at precise stages of development. By correlating the well documented histological changes occurring in these mutants with changes in ganglioside composition, we have obtained strong evidence for a non-random cellular distribution of gangliosides. Most notably, GD1a is more enriched in granule cells that in Purkinje cells, whereas the opposite is true for GT1a. GD3, on the other hand, is heavily enriched in reactive glia and may serve as a useful biochemical marker for the presence of reactive glia in neurological disease. The continued study of gangliosides in the various mouse mutants will help elucidate their cellular localization in the CNS.

ACKNOWLEDGEMENTS

This work was supported by USPHS grants NS11853, and NS17704, and NSF grant BNS-8305449. T. N. Seyfried was also supported by a Research Career Development Award (NS0059).

REFERENCES

1. P. Rakic and R. Sidman, Weaver mutant mouse cerebellum: Defective neuronal migration secondary to abnormality of Bergmann glia, Proc. Natl. Acad. Sci. (USA) 70:240 (1973).
2. C. Sotelo, Anatomical, physiological, and biochemical studies of the cerebellum from mutant mice. II. Morphological study of cerebellar cortical neurons and circuits in the weaver mouse, Brain Res. 94:19 (1975).
3. R. J. Mullen, E. M. Eicher, and R. L. Sidman, Purkinje cell degeneration, a new neurological mutation in the mouse, Proc. Natl. Acad. Sci. (USA) 73:208 (1976).
4. S. C. Landis and R. J. Mullen, The development and degeneration of Purkinje cells in the pcd mutant mouse, J. Comp. Neurol. 177:125 (1978).
5. R. L. Sidman, P. W. Lane, and M. M. Dickie, Staggerer, a new mutation in the mouse affecting the cerebellum, Science 137:610 (1962).
6. D. B. Wilson, Histological defects in the cerebellum of adult lurcher (Lc) mice, J. Neuropathol. Exp. Neurol. 35:40 (1976).
7. K. W. T. Caddy, The numbers of Purkinje, granule cells and olive neurons in the lurcher mutant mouse, J. Physiol. 277:8 (1978).
8. S. C. Landis and R. L. Sidman, Electron microscopic analysis of postnatal histogenesis in the cerebellar cortex of staggerer mutant mice, J. Comp. Neurol. 179:831 (1978).
9. K. Herrup and R. J. Mullen, Regional variation and absence of large neurons in the cerebellum of the staggerer mouse, Brain Res. 172:1 (1979).
10. D. S. Sax, A. Hirano, and R. J. Shofer, Staggerer, neurolog-

ical murine mutant, Neurology (Minneap.) 18:1093 (1968).
11. A. Hirano and H. M. Dembitzer, The fine structure of astrocytes in the adult staggerer, J. Neuropath. Exp. Neurol. 35:63 (1976).
12. B. Ghetti, L. Truex, B. Sawyer, S. Strada, and M. Schmidt, Exaggerated cyclic AMP accumulation and glial cell reaction in the cerebellum during Purkinje cell degeneration in pcd mutant mice, J. Neurosci. Res. 6:789 (1981).
13. T. N. Seyfried, R. K. Yu, and N. Miyazawa, Differential cellular enrichment of gangliosides in the mouse cerebellum: Analysis using neurological mutants, J. Neurochem. 38:551 (1982).
14. R. K. Yu and K. Iqbal, Sialosylgalactosyl ceramide as a specific marker for human myelin and oligodendroglial perikarya: Gangliosides of human myelin, oligodendroglia and neurons, J. Neurochem. 32:293 (1979).
15. K. Suzuki, S. E. Poduslo, and W. T. Norton, Gangliosides in the myelin fraction of developing rats, Biochim. Biophys. Acta 144:375 (1967).
16. K. Suzuki, S. E. Poduslo, and J. F. Poduslo, Further evidence for a specific ganglioside closely associated with myelin, Biochim. Biophys. Acta 152:576 (1968).
17. K. Ueno, S. Ando, and R. K. Yu, Gangliosides of human, cat and rabbit spinal cords and cord myelin, Lipid Res. 19:863 (1978).
18. R. K. Yu, R. W. Ledeen, and L. F. Eng, Ganglioside abnormalities in multiple sclerosis, J. Neurochem. 23:169 (1974).
19. R. K. Yu, K. Ueno, G. H. Glaser, and W. W. Tourtellotte, Lipid and protein alterations of spinal cord and cord myelin of multiple sclerosis, J. Neurochem. 39:464 (1982).
20. R. Ledeen, K. Salsman, and M. Cabrera, Gangliosides in subacute sclerosing leukoencephalitis: Isolation and fatty acid composition of nine fractions, J. Lipid Res. 9:129 (1968).
21. W. T. Norton, S. E. Poduslo, and K. Suzuki, Subacute sclerosing leukoencephalitis. II. Chemical studies including abnormal myelin and abnormal ganglioside pattern, J. Neuropath Exp. Neurol. 25:582 (1966).
22. R. K. Yu, R. W. Ledeen, D. C. Gajdusek, and C. J. Gibbs, Ganglioside changes in slow virus disease: Analysis of chimpanzee brains infected with kuru and Creutzfeldt-Jakob agents, Brain Res. 70:103 (1974).
23. R. K. Yu, and E. E. Manuelidis, Ganglioside alterations in guinea pig brains at end stages of experimental Creutzfeldt-Jakob disease, J. Neurol. Sci. 35:15 (1978).
24. Y. Tamai, H. Kojima, F. Ikuta, and T. Kumaishi, Alterations in the composition of brain lipids in patients with Creutzfeldt-Jakob disease, J. Neurol. Sci. 35:59 (1968).
25. K. Suzuki and G. Chen, Chemical studies on Creutzfeldt-Jakob

disease, J. Neuropath Exp. Neurol. 25:396 (1966).
26. H. Bernheimer, G. Sperk, K. S. Price, and O. Hornykiewicz, Brain gangliosides in Huntingtons disease, Adv. Neurol. 23:463 (1979).
27. K. Suzuki, in: "Inborn Disorders of Sphingolipid Metabolism," S. M. Aronson and B. W. Volk, eds., Pergamon, New York (1967).
28. B. Hagber, G. Hultquist, R. Ohman, and L. Svennerholm, Congenital amaurotic idiocy, Acta. Paediat. Scandinavica 54:116 (1965).
29. L. Svennerholm, and M. T. Vanier, Brain gangliosides in Krabbe disease, Adv. Exp. Med. Biol. 19:499 (1972).
30. Y. Suzuki, S. H. Tucker, L. B. Rorke, and K. Suzuki, Ultrastructural and biochemical studies of Schilder's disease, J. Neuropath. Exp. Neurol. 29:405 (1970).
31. M. Igarashi, D. Belchis and K. Suzuki, Brain gangliosides in adrenoleukodystrophy, J. Neurochem. 27:327 (1976).
32. T. N. Seyfried, N. Miyazawa, and R. K. Yu, Cellular localization of gangliosides in the developing mouse cerebellum: Analysis using the weaver mutant, J. Neurochem. 41:491 (1983).
33. P. Rakic, and R. Sidman, Sequence of developmental abnormalities leading to granule cell deficit in cerebellar cortex of weaver mutant mice, J. Comp. Neurol. 152:103 (1973).
34. P. Rakic, and R. Sidman, Organization of the cerebellar cortex secondary to deficit of granule cells in weaver mutant mice, J. Comp. Neurol. 152:133 (1973).
35. C. Sotelo, Dendritic abnormalities of Purkinje cells in the cerebellum of neurologic mutant mice (weaver and staggerer), Adv. Neurol. 12:335 (1975).
36. P. Bradley, and M. Berry, The Purkinje cell dendritic tree in mutant mouse cerebellum. A quantitative golgi study of weaver and staggerer mice, Brain Res. 142:135 (1978).
37. Z. Rezai, and C. H. Yoon, Abnormal rate of granule cell migration in the cerebellum of the weaver mouse, Develop. Biol. 29:17 (1972).
38. L. M. H. Larramendi, Analysis of synaptogenesis in the cerebellum of the mouse, in: "Neurobiology of Cerebellar Evolution and Development," R. Llinas, ed., American Medical Association Education and Research Federation, Chicago (1969).
39. S. L. Palay, and V. C. Palay, "Cerebellar Cortex," Springer-Verlag, New York (1974).
40. G. M. Shepherd, "Synaptic Organization of the Brain," Oxford University Press, London (1974).
41. H. K. M. Yusuf, and J. W. T. Dickerson, Disialoganglioside GD1a of rat brain subcellular particles during development, Biochem. J. 174:655 (1978).
42. L. N. Irwin, and C. C. Irwin, Developmental changes in gang-

lioside composition of hippocampus, retina, and optic tectum, Dev. Neurosci. 2:129 (1979).

43. E. R. Engel, J. G. Wood, and F. E. Byrd, Ganglioside patterns and cholera toxin peroxidase labeling of aggregating cells from the chick optic tectum, J. Neurobiol. 10:429 (1979).
44. H. Rosner, Ganglioside changes in the chicken optic lobe as biochemical indicators of brain development and maturation, Brain Res. 236:49 (1982).
45. R. Hilbig, H. Rosner, G. Merz, K. Segler-Stahl, and H. Rahmann, Developmental profiles of gangliosides in mouse and rat cerebral cortex, Wilhelm Rouxs Arch. 191:281 (1982).
46. K. C. Leskawa and A. Rosenberg, The organization of gangliosides and other lipid components in synaptosomal plasma membranes and modifying effects of calcium ion, Cell Molec. Biol. 1:373 (1981).
47. K. Suzuki, The pattern of mammalian brain gangliosides. III. Regional and developmental differences, J. Neurochem. 12:969 (1965).
48. A. Merat, S. Sajjadi, and J. W. T. Dickerson, Effect of development on the gangliosides of rabbit brain, Biol. Neonate 36:25 (1979).
49. L. Svennerholm, and P. Fredman, A procedure for the quantitative isolation of brain gangliosides, Biochim. Biophys. Acta 617:97 (1980).
50. T. N. Seyfried, T. Itoh, G. H. Glaser, N. Miyazawa, and R. K. Yu, Cerebellar gangliosides and phospholipids in mutant mice with ataxia and epilepsy: The tottering/leaner syndrome, Brain Res. 219:429 (1981).
51. G. M. Weiss, and J. J. Pysh, Evidence for loss of Purkinje cell dendrites during late development: A morphometric golgi analysis in the mouse, Brain Res. 154:219 (1978).
52. T. N. Seyfried and R. K. Yu, Cellular localization and function of ganglioside GD3 in the CNS, Trans. Am. Soc. Neurochem. 13:94 (1982).

GANGLIOSIDES AND RELATED COMPOUNDS AS BIOLOGICAL RESPONSE MODIFIERS

Yoshitaka Nagai, Shuichi Tsuji and Yutaka Sanai

Department of Biochemistry, Faculty of Medicine
University of Tokyo, Bunkyo-ku, Tokyo 113, Japan

MODIFICATION OF CELLULAR ACTIVITIES BY VIRTUE OF THE MANIPULATION
OF CELL SURFACE GLYCOSPHINGOLIPID ANTIGEN

In the previous chapter[1] we already discussed the curious fact that a considerably large number of glycosphingolipids (GSL) occur sporadically in nature and that this sporadic character seemingly makes it difficult to presuppose a special physiological function of these GSL except for practical utility in differentiating individual cells from each other. The significance of individual GSL species as cell surface markers or surface differentiation markers has already been recognized, as in the case of blood group GSL, Forssman antigen, globosides and asialo-GM1 in the subpopulation analysis of the haematopoietic system and in the analysis of developmental and differentiation processes of the haematopoietic system. Thus these markers, though they have no definite intrinsic physiological function, provide a useful tool to analyse these important biological processes at a cellular level. Moreover, we can specifically manipulate or control cell activities <u>in vivo</u> of tumor cells[2] and subpopulations of lymphoid cells[3,4] by virtue of such cell marker recognition, for example, using specific antibodies or ligands to these surface markers. We will discuss this problem by taking asialo-GM1 (GA1 or Gg_4Cer) ganglioside as representative.

Asialo-GM1 is one of the common antigens between the nervous and haematopoietic systems. Initially it was demonstrated to be an important cell surface marker of natural killer (NK) cells in mice[5-7] and also to be a useful surface differentiation marker for analyzing fetal thymocytes in differentiation.[9] However, later it was shown to be present also on the surface of mouse suppressor T cells [10] and then on mouse T cell clones including killer T

cells.[11] In rats, asialo-GM1 was first identified to be the rat
T lymphocyte-macrophage-associated antigen (TLMA)[12] and later was
shown to be present also in nylon column-enriched T cells, bone
marrow cells and some subpopulations of other immunocytes.[13,14] As
mentioned in the previous chapter[1] the mere presence of a glycolipid antigen in plasma membranes does not certify that it is
always in a reactive state to the respective antibody or ligand.
The reactivity depends, for example, upon the nature, size and
conformation of the ligand molecule used as a probe, the static as
well as dynamic topographical molecular organization of surface
membranes, cytoplasmic or cytoskeletal structures, and many other
factors. Thus, at present the establishment of asialo-GM1 as
surface marker including its cell surface reactivity requires
further inquiry and precise analysis.

On the other hand, it has been known that the anti-asialo-GM1
antibody or antiserum, when injected into individual animals, has a
potency to exert several interesting biological effects most
probably mediated by the modulation of particular cell activities
of the hematopoietic system. Thus, subcutaneous injection of 0.5
ml of anti-asialo-GM1 antiserum twice a week into golden hamsters
which had previously been transplanted with Yoshida ascitic
hepatoma cells in their cheek pouch resulted in this heterotransplanted tumor growing in unlimited fashion until the animal's death
without rejection of the heterotransplant[3] (Fig. 1). Instead the
treatment of the animals with control preimmune normal rabbit serum
(NRS) or with anti-GM1-antiserum always resulted in regression of
the transplanted tumor (Fig. 1). The same observation was also

anti asialo GM1-treated

NRS-treated (control)

Fig. 1. Effects of anti-asialo-GM1 serum on the growth of
heterotransplanted tumor in hamster. An arrow indicates
the site where tumor cells were transplanted.

made on nude mice transplanted with YAC-1, Moloney virus-induced lymphoma cells[4] and on mice transplanted with NK-sensitive and NK-resistant subline of the murine lymphoma L51178Y.[15] The similar type of experiment was performed with a strain of mouse which spontaneously develops mammary tumors. Subcutaneous injection of 50 μl of anti-asialo-GM1 antiserum twice a week caused earlier appearance of the tumor and the tumor incidence was also enhanced (Fig. 2). Mice infected with mouse hepatitis virus (JHM) showed a high rate of death and higher virus titer in liver, brain and spleen, when the animals were subjected to the same treatment with anti-asialo-GM1 antiserum[16] (Table 1). In this case, we measured NK cell activity and the level of interferon production. In the antiserum-treated mice, NK cell activity decreased rapidly, while the production of interferon maintained almost the same level.[16] It is likely that most of these phenomena in mice may be ascribed to the specific damage of NK cells or some particular subpopulation of lymphoid cells by exogenously administered anti-asialo-GM1 antibody. And, therefore, it is not unreasonable for us to hope that the particular cell activity might be controlled or modulated by the in vivo manipulation of the surface marker glycolipid molecules with exogenous application of specific antibodies or ligands to them. Hakomori attempted a similar type of experiment on tumor cells, targeting asialo GM2 (GA2) of the cell surfaces and aiming at the tumor therapy.[2]

Table 1. Effects of Anti-Asialo GM1 Serum on The Infection of Mice With Mouse Hepatitis Virus, JHM.[1]

	anti-GA1 serum	NRS
cumulative mortality (%)	80	25
virus titer ($\log_{10}$PFU / 0.2g tissue)		
liver	5	2
spleen	6	2
brain	6	2
NK cell activity (specific ^{51}Cr release;%)	5	22.5
interferon titer[2]	160	20

(1) These data were obtained day 7 after inoculation of virus except for the cumulative mortality which was determined from the observation for two weeks. Anti-asialo GM1 serum (GA1) and normal rabbit serum (NRM) were administered intraperitonially at day 0.

(2) Interferon titer was expressed by the reciprocal of the dilution of a sample producing 50% reduction in plaque number as compared with control vesicular stomatitis virus infected culture.

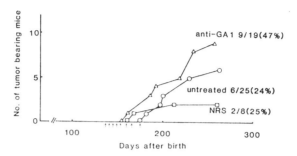

Fig. 2. Effect of anti-asialo-GM1 serum on spontaneous incidence of mammary tumor in mice (C3HxB/c) F1. Arrows indicate the day when antisera were injected and the number in parentheses indicates the percent incidence of the tumor.

Our successful experience in modulating a particular type of cell activity of the haematopoietic system suggests the possibility that under certain pathological conditions the nervous and haematopoietic systems may interact with each other by means of the common antigen or derived antibody as seen in the case of asialo-GM1 which is shared by the two systems. In fact, our finding of the interrelationship between asialo-GM1 and NK cells was originally derived from the fact that anti-brain-associated T cell antiserum (BAT) has a potency to kill mouse NK cells in the presence of complement-like factor of serum.

Our pursuit after the responsible antigen led to the discovery of the fact that rabbit anti-mouse brain antisera contain a high titer of anti-asialo-GM1 antibody and that anti-asialo-GM1 antibody raised in rabbit also showed similar NK-cell killing activity as the anti-brain antisera.

In humans, elevation of anti-asialo-GM1 titer has recently been reported in the case of systemic lupus erythematosus accompanied by neurological disorders,[17,18] neurobehcet disease,[19] multiple sclerosis,[18] and patients who have experienced ischemic injury or trauma to the central nervous system.[18] In some cases of systemic lupus erythematosus accompanied by neurological disorders the antibody level fluctuated down and then up in parallel with exacerbation and remission.[17] It is interesting to note that the suppressor T cell level was reduced in peripheral blood of patients with multiple sclerosis.[20,21] It should be added that anti-asialo-GM1 antibody was found to be cytotoxic not to the human NK cells but presumably to the mitogen-induced human suppressor T cells (data not shown). Although there is available only circumstantial evidence at present, it is not unreasonable to consider that the above-mentioned interplay between the nervous and hematopoietic systems may be operative also

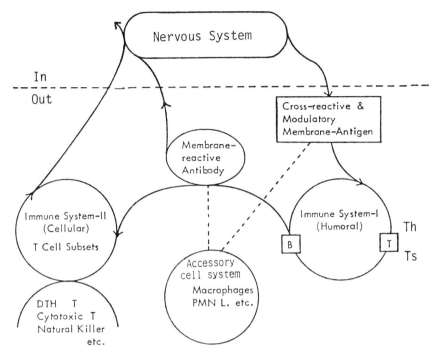

Fig. 3. Self-regulatory interaction between brain and immune system.

in human diseases and that the parameters related to the interactions may provide useful information for understanding some aspects of the pathogenesis of certain neurological diseases and also for diagnosis as well as prognosis of the diseases (Fig. 3). It is interesting that E. S. Golub[22] has recently suggested the possible connections among the nervous, haematopoietic and, in addition, germ-cell systems, though viewed from different aspect from ours.

SEARCH FOR BIOACTIVE GANGLIOSIDES: TETRASIALOGANGLIOSIDE GQ1b AS A NOVEL NERVE GROWTH PROMOTER

Several recent reports describe a special physiological function conferred by exogenous gangliosides, e.g., promotion of neurite outgrowth,[23] enhancement of the interaction of fibronectin with ganglioside-deficient cells,[24] prolongation of cell cycle in a gemistocytic astrocytoma cell line,[25] inhibition of phorbol ester promotion of cell transformation,[26] return of TSH receptor function in TSH receptor-deficient rat thyroid tumor cells,[27] etc. These works were mostly performed using total gangliosides or their subfractions.

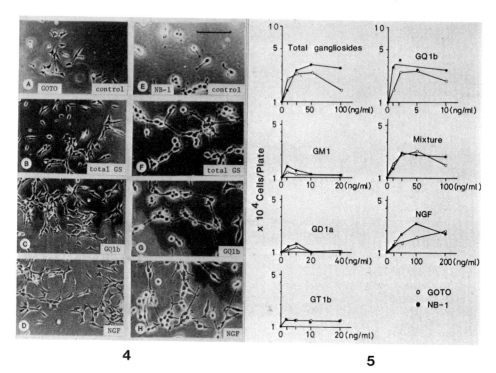

Fig. 4. Photomicrographs of cultures of neuroblastoma cells grown for 24 hr with or without exogenous total gangliosides, purified GQ1b and NGF. Each bar in the photos indicates 100 μm. A-D, GOTO cells; E-H, NB-1 cells; total GS, bovine brain total ganglioside fraction; NGF, 7S-nerve growth factor.

Fig. 5. Dose dependence of various gangliosides and NGF on the cell number of neuroblastoma cells.

We recently found that a single molecular species of tetrasialoganglioside, GQ1b, greatly enhanced neurite outgrowth and cell proliferation in in vitro cell cultures of neuroblastoma cell lines at a nanogram per ml level of concentration.[28] Biological activity was found only in this particular minor component of brain gangliosides but not in others. We used two neuroblastoma cell lines (GOTO and NB-1) established by Sekiguchi et al.[29] and Miyake et al.,[30] respectively. Gangliosides were extracted from brain and each molecular species used was purified according to the method of Nagai and Iwamori.[31]

For assessment of the biological effect of gangliosides, cells were first grown exponentially to $2-6 \times 10^4$/35 mm polystyrene dish,

Structure of GQ1b:

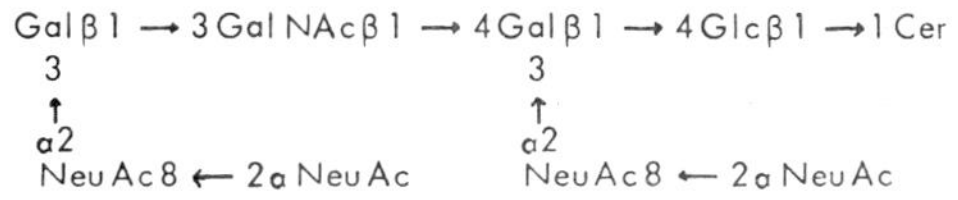

$$\text{Gal}\beta 1 \rightarrow 3\text{GalNAc}\beta 1 \rightarrow 4\text{Gal}\beta 1 \rightarrow 4\text{Glc}\beta 1 \rightarrow 1\text{Cer}$$
$$\begin{array}{cc} 3 & 3 \\ \uparrow & \uparrow \\ \alpha 2 & \alpha 2 \\ \text{NeuAc}8 \leftarrow 2\alpha\text{NeuAc} & \text{NeuAc}8 \leftarrow 2\alpha\text{NeuAc} \end{array}$$

Fig. 6. Structure of GQ1b.

and after removal of the medium containing serum (45% RPMI, 45% MEM and 10% fetal calf serum), fresh serum free medium (50% RPMI and 50% MEM) with or without various amounts of bovine brain total gangliosides (GS) was added. Twenty-four hours later, we measured the cell number, the number of neurites and the total length of neurites. In the presence of 50 ng/ml of GS, the number of living cells was 2.2 (GOTO) or 3.1 (NB-1) times as much as that of the control without GS ($p < 0.001$). The number and total length of neurites in GS treated cells were significantly larger than those of controls (Table 3). Remarkably, GS had an influence on the cell proliferation and neurite outgrowth (Fig. 4 and Table 3). These two effects were dependent on the GS concentration (Fig. 5). The dose response in the promotion of neurite outgrowth was almost the same as that in cell proliferation, except that at concentrations of more than 50 ng/ml they were maintained at the same levels within the range of experimental error. One may particularly consider the possibility that certain neuronal stimulating factors, like nerve growth factor (NGF), might contaminate the GS fraction, because the GS fraction used in this study represented Folch's upper phase fraction. 7S-NGF also had an effect on our two cell lines in terms of the cell number, the neurite number and the total neurite length (Table 3). If proteins possibly present in GS really were responsible for the above functions, it is expected that these activities would be reduced to the control level after protease treatment. However, Pronase E exerted absolutely no effect on the proliferation or neurite outgrowth promotion of GS under any conditions examined (Table 3). On the other hand, treatment with neuraminidase from Arthrobacter ureafaciens as well as from Vibrio cholerae was found to reduce this stimulation activity nearly to control levels. By neuraminidase treatment, all gangliosides used were completely converted to GM1. Thus, the above stimulating effects were not due to possibly contaminating proteins but to a certain ganglioside(s) itself, other than GM1.

The next question was what sort of ganglioside(s) plays a part in the above effects. We examined purified homogeneous GM1, GD1a, GT1b and GQ1b in the same manner. It was clearly shown that only GQ1b showed the same effect as that of total gangliosides. Though other gangliosides used had a small effect on the cell number, no effect could be observed on neurite growth promotion other than levels within experimental error. The optimum concentration of

Table 3. Effects of Exogeneous Gangliosides on Cell Number And Neurite Outgrowth of Neuroblastoma Cell Lines.

	GOTO			NB-1		
	Cell number ($\times 10^4$/plate)	Neurite number /cell	Total neurite length (µm/cell)	Cell number ($\times 10^4$/plate)	Neurite number /cell	Total neurite length (µm/cell)
Control	3.89±0.47	2.12±0.32	48.3±3.24	1.92±0.69	2.26±0.36	82.8±3.36
Total gangliosides (50 ng/ml)						
Native	8.70±0.54*	2.62±0.36**	63.8±2.97*	5.92±0.71*	3.16±0.46**	121.4±4.14*
Pronase E treated	8.44±0.45*	2.70±0.24**	63.1±4.78*	4.53±0.45*	3.03±0.26**	114.6±3.62*
Neuraminidase (from *Arthrobacter ureafaciens*) treated	4.36±0.70	2.25±0.41	46.2±2.46	2.16±0.49	2.28±0.49	92.8±7.52
GQ1b (5 ng/ml)	9.26±0.78*	2.56±0.40**	62.6±3.88*	5.09±0.62*	3.10±0.27**	124.5±2.45*
GQ1b (5 ng/ml) from nasal septum	8.88±0.42*	2.52±0.28**	60.8±2.18*	4.09±0.79*	3.06±0.48**	109.5±4.26*

*: $P<0.001$, **: $P<0.05$.
Gangliosides were from human brain unless otherwise noted.

GQ1b necessary for cell proliferation was 2-5 ng/ml. This concentration was comparable to that of GQ1b present in GS. Again, the activity observed in GQ1b was also abolished by treatment with two kinds of neuraminidases but not with protease. In addition, GQ1b isolated from different sources, for example, bovine nasal septum, was also effective for the above stimulation (Table 3). All these facts support our conclusion that the activity is possessed by the GQ1b molecular species. Thus, we could show that GQ1b has two effects, like those of NGF, on the two neuroblastoma cell lines: (a) it increases the cell number, and (b) increases the neurite number and the total length of neurites. Moreover, from the comparison of active or inactive chemical structures it may be presumed that the terminal disialosyl residues linked to the terminal galactose of ganglio-N-tetraose may play an important role in these activities. It is of great importance to understand the molecular mechanism of action of this autacoid (local hormone)-like ganglioside. In the future it is expected that other biologically active gangliosides may be discovered among such minor species.

SUMMARY

Several possibilities are discussed with regard to the biological potency of gangliosides and related compounds as biological response modifier. Evidence is presented that a ganglioside, GQ1b, but not other gangliosides examined exhibits a strong biological activity at a few nanomolar concentration for the promotion of neurite outgrowth as well as cell division in two cell lines of neuroblastoma.

ACKNOWLEDGMENTS

This study was supported in part by a grant from Toray Foundation, a grant from the Ministry of Education Science and Culture of Japan, and also a research grant for intractable diseases from the Ministry of Health and Welfare of Japan.

REFERENCES

1. Y. Nagai and M. Iwamori, Ganglioside distribution at different levels of organization and its biological implications; this volume.
2. D. L. Urdel and S. Hakomori, Tumor-associated ganglio-N-triosylceramide: target for antibody-dependent, avidin-mediated drug killing of tumor cells, J. Biol. Chem. 255:10509 (1980).
3. K. Sakakibara, T. Uchida, and Y. Nagai, Immunosuppressive effect of anti-asialo GM1 antiserum on hamsters heterotransplanted with Yoshida ascitic hepatoma cells, Proc. Japan. Cancer Assoc. The 29th Ann. Meeting, Tokyo, p. 101 (1980).
4. S. Habu, H. Fukui, K. Shimamura, M. Kasai, Y. Nagai, K. Okumura, and S. Tamaoki, In vivo effect of anti-asialo-GM1. I. Reduction of NK activity and enhancement of transplantation tumor growth in nude mice, J. Immunol. 127:34 (1981).
5. M. Kasai, M. Iwamori, Y. Nagai, K. Okumura, and T. Tada, A glycolipid on the surface of mouse natural killer cells, Eur. J. Immunol. 10:175 (1980).
6. W. W. Young, Jr., S. Hakomori, J. M. Durdik, and C. S. Henney, Identification of ganglio-N-tetraosylceramide as a new surface marker for murine natural killer (NK) cells, J. Immunol. 124:199 (1980).
7. G. A. Schwarting, A. Summers, R. D. Stout, D. R. Parkinson, and S. D. Waksal, Association of asialo GM1 with natural killer activity in mice, Fed. Proc. 39:931 (1980).
8. G. A. Schwarting and A. Summers, Gangliotetraosylceramide is a T cell differentiation antigen associated with natural cell-mediated cytotoxicity, J. Immunol. 124:1691 (1980).
9. S. Habu, M. Kasai, Y. Nagai, N. Tamaoki, T. Tada, L. A. Herzenberg, and K. Ikumura, The glycolipid asialo GM1 as a new differentiation antigen of fetal thymocytes, J. Immunol. 125:2284 (1980).
10. T. Nakano, Y. Imai, M. Naiki, and T. Osawa, Characterization of mouse helper and suppressor T cell subsets separated by lectins, J. Immunol. 125:1928 (1980).
11. B. Kniep, T. R. Hunig, F. W. Fitch, J. Heuer, E. Kolsch, and P. F. Muhlradt, Neutral glycosphingolipids of murine myeloma cells and helper, cytolytic, and suppressor T lymphocytes, Biochemistry 22:251 (1981).

12. T. Momoi, H. Wiegandt, R. Arndt, and H. Thiele, Gangliotetraosylceramide, the rat T lymphocyte-macrophage-associated antigen: chemical detection and cellular distribution, J. Immunol. 125:2496 (1980).
13. T. Taki, K. Takagi, R. Kamada, M. Matsumoto, and K. Kojima, Study of asialo gangliosides on surface membranes of rat bone marrow cells and macrophages, J. Biochem. 90:1653 (1981).
14. T. Momoi, K. Nakajima, K. Sakakibara, and Y. Nagai, Localization of a glycosphingolipid, asialo GM1, in rat immunocytes, J. Biochem. 91:301 (1982).
15. I. Kawase, D. L. Urdal, C. G. Brooks, and C. S. Henney, Selective depletion of NK cell activity in vivo and its effect on the growth of NK-sensitive and NK-resistant tumor cell variants, Int. J. Cancer 29:567 (1982).
16. F. Taguchi, Y. Sanai, K. Fujiwara, and Y. Nagai, Role of natural killer cells in the infection of mice with mouse hepatitis virus, JHM, as demonstrated by the use of anti-asialo GM1 serum, submitted to Infect. Immun.
17. T. Hirano, H. Hashimoto, Y. Shiokawa, M. Iwamori, Y. Nagai, M. Kasai, Y. Ochiai, and K. Okumura, Antiglycolipid autoantibody detected in the sera from systemic lupus erythematosus patients, J. Clin. Invest. 66:1437 (1980).
18. T. Endo, D. D. Scott, S. Stewart, S. K. Kundu, and D. M. Marcus, Antibodies to glycosphingolipids in patients with multiple sclerosis and SLE, in: "Glycoconjugates", Proc. 7th Internatl. Symp. on Glycoconjugates, Lund-Ronneby, Sweden, p. 244 (1983).
19. G. Inaba and J. Aoyama, Anti-glycolipid antibodies in neuro-behcet's syndrome, in: "Behcet's Disease", Pathogenetic Mechanism and Clinical Features, Proc. Internatl. Conf. on Behcet's Disease, G. Inaba, ed., p. 145, Univ. Tokyo Press, Tokyo (1982).
20. M. A. Bach, F. Phan-Dinh-Tuy, E. Tourier, L. Chatenoud, J.-F. Bach, C. Martin, and J. D. Degos, Deficit of suppressor T cells in active multiple sclerosis, Lancet 1221 (1980).
21. E. L. Reinherz, H. L. Weiner, S. L. Hauser, J. A. Cohen, J. A. Distaso, and S. F. Schlossman, Loss of suppressor T cells in active multiple sclerosis. Analysis with monoclonal antibodies, New Engl. J. Med. 303:125 (1980).
22. E. S. Golub, Connections between the nervous, haematopoietic and germ-cell systems, Nature 299:483 (1982).
23. M. M. Rapport and A. Gorio (eds.), "Gangliosides in Neurological and Neuromuscular Function, Development, and Repair", Raven Press, New York (1981).
24. K. M. Yamada, D. R. Critchley, P. H. Fishman, and J. Moss, Exogenous gangliosides enhance the interaction of fibronectin with ganglioside-deficient cells, Exp. Cell Res. 143:295 (1983).

25. C. Icard-Liepkalns, V. A. Liepkalns, A. J. Yates, and R. E. Stephens, Cell cycle phases of a novel human cell line and the effect of exogenous gangliosides, Biochem. Biophys. Res. Commun. 105:225 (1982).
26. L. Srinivas, T. D. Gindhart, and N. H. Colburn, Tumor-promoter-resistant cells lack trisialoganglioside response, Proc. Natl. Acad. Sci. USA 79:4988 (1982).
27. P. Laccetti, E. F. Grollman, S. M. Aloj, and L. D. Kohn, Ganglioside dependent return of TSH receptor function in a rat thyroid tumor with a TSH receptor defect, Biochem. Biophys. Res. Commun. 110:772 (1983).
28. S. Tsuji, M. Arita, and Y. Nagai, GQ1b, a bioactive ganglioside that exhibits novel nerve growth factor (NGF)-like activities in the two cell lines of neuroblastoma, J. Biochem. 94:303 (1983).
29. M. Sekiguchi, T. Oota, K. Sakakibara, N. Inui, and G. Fujii, Establishment and characterization of a human neuroblastoma cell line in tissue culture, Japan. J. Exp. Med. 49:67 (1979).
30. S. Miyake, Y. Shimo, T. Kitanuma, Y. Nojyo, T. Nakamura, S. Imashuku, and T. Abe, Characteristic of continuous and functional cell line NB-1, derived from a human neuroblastoma, Autonomic Nerv. System 10:115 (1973).
31. Y. Nagai and M. Iwamori, Brain and thymus gangliosides: Their molecular diversity and its biological implications and a dynamic annular model for their function in cell surface membranes, Mol. Cell. Biochem. 29:81 (1980).

Metabolism

AN OUTLINE OF GANGLIOSIDE METABOLISM

G. Tettamanti

Department of Biological Chemistry
The Medical School
University of Milan
Via Saldini 50, 20133
Milan, Italy

INTRODUCTION

Gangliosides are mainly located in the cell plasma membrane and are asymmetrically disposed on the outer membrane leaflet. Their hydrophobic portion (the ceramide) is inserted into the membrane layer and the oligosaccharide moiety protrudes on the membrane surface. Very small amounts of ganglioside are also present in intracellular structures and compartments. These gangliosides are likely the expression of the transient forms moving from the site of biosynthesis to the plasma membrane or migrating from the plasma membrane to the site of degradation. They constitute a small "metabolic" pool as compared to the large "final residence" pool.

Gangliosides display a highly differentiated carbohydrate composition. The molecular species belonging to the "ganglio" series, which are the most abundant in the nervous system, differ from each other in terms of the number and position of sialic acid residues, and/or the presence of additional saccharide units, like fucose. Conversely, the main difference in the species of the "neo-lacto" series resides in the number of repeating units (Galβ1→4GlcNAcβ1→3) present in the molecule.

Most of the information regarding ganglioside metabolism pertains to gangliosides of the ganglio series and refers to brain gangliosides. However, important contributions were also provided by studies carried out on extraneural tissues such as liver, spleen, kidney, mammary gland and thyroid; these involve gangliosides of both the ganglio and neo-lacto series. Moreover the use of cultured cells, such as transformed and tumoral cells, or of primary cultures of brain cells - neurons and glial cells - opened new and simpler approaches to the elucidation of some aspects of ganglioside metabolism.

Ganglioside metabolism has been extensively reviewed in recent years.[1-6] Details on the in vivo studies of ganglioside metabolism and on the enzymatic aspects of ganglioside biosynthesis and degradation can be found in these reviews. The same reviews are also the source for reference to original contributions.

COMPREHENSIVE PICTURE OF GANGLIOSIDE METABOLISM

The first approach to ganglioside metabolism consisted in injecting into animals, generally rats, various radiolabelled precursors (glucose, galactose, hexosamine, serine, fatty acid, sphingosine, acetate), and then determining the amount and rate of radioactivity incorporation into gangliosides, generally of the brain. These studies showed half life values for gangliosides ranging from 10 to 60 days depending on the choice of precursor and animal age.[7-9] Under the same condition the turnover rate of gangliosides was faster than that of other glycolipids,[10] but slower than that of glycoproteins. Injection of labelled fucose was also followed by formation of a number of brain gangliosides radiolabelled in their fucose moiety.[11]

In vivo[8] as well as in vitro[12] studies also demonstrated that the rates of labelling and turnover were similar for the different gangliosides. This indicated that none of the gangliosides residing in the plasma membrane acted as precursor or product of the others and that, in consequence, each of them should have been biosynthesized and degraded separately.

BIOSYNTHETIC PATHWAYS

In principle, the biosynthesis of a ganglioside is the result of sequential additions of saccharide units starting from ceramide. Each reaction is catalyzed by a glycosyltransferase, the saccharide unit being transferred from the corresponding sugar-nucleotide to the acceptor. The sequence of glycosylations follows a certain order which has been established for most gangliosides.[3,4,13] This was the result of studies where individual purified enzymes were used,[6] and of investigations with subcellular preparations containing assembled sets of enzymes.[14] The biosynthetic pathway for the gangliosides of the ganglio series is illustrated in Fig. 1. Lactosylceramide (GA3) appears as the common precursor for all the gangliosides of this series. One biosynthetic route is based on early insertion of sialic acid and leads to the gangliosides carrying a single sialic acid residue linked to inner galactose. A second route, based on early insertion of a disialosyl residue, leads to the gangliosides carrying a disialosyl residue on inner galactose. A third route, based on early insertion of a trisialosyl residue, is presumed to lead to the gangliosides carrying a trisialosyl residue on inner galactose. The fourth route, in which the sialic acid residue is attached later (the "aminoglycolipid route"), leads to the monosialoganglioside of the ganglio series carrying sialic acid on the terminal galactose (GM1b). The aminoglycolipid route was originally suggested[15] as an alternative pathway for the biosynthesis of GM1 and higher gangliosides. It is now clear[16] that the sialyltransferase that converts GgOse$_4$Cer to a ganglioside produces GM1b rather than GM1.

The known and presumed steps for the biosynthesis of gangliosides of the neo-lacto series are shown in Fig. 2. It is not known whether a unique, or several, galactosyl transferases and N-acetyl-glucosaminyltransferases (the pivotal enzymes of this pathway) are involved in the process. The synthesis of ganglioside GM4 is the result of galactosylation of ceramide, followed by sialosylation of the galactose residue.[17]

DEGRADATION PATHWAY

Ganglioside degradation occurs in a sequential manner with liberation of the individual sugar moieties starting from the non reducing terminus of the oligosaccharide.[1,3,5] The process is catalyzed by glycohydrolases which lead to formation of ceramide. This is split by ceramidase into sphingosine and fatty acid. Most of the enzymes which are involved in ganglioside catabolism have the features of lysosomal enzymes.[18] Some of these enzymes (β-glucosidase, β-galactosidase, βN-acetylgalactosaminidase) display full activity only in the presence of protein "cofactors" or "activators", which seem to occur in the lysosomes too.[19,20]

The degradation pathway of the gangliosides of the ganglio series is illustrated in Fig. 3. The degradation is initiated by the action of sialidase which removes sialic acid. One feature of brain and other tissues sialidases is their sensitivity to some steric characteristics of the region where the α-ketosidic linkage of sialic acid resides.[3] For instance, GT1b is preferentially degraded to GD1b rather than GD1a, indicating that the sialosyl-galactose linkage is more accessible to the enzyme than the sialosyl-sialic acid linkage. A strong steric hindrance, likely resulting from an oxygen cage surrounding the α-ketosidic linkage, is responsible for the almost complete resistance to tissue sialidase of GM1 and GM2.[3,5] The site(s) of subcellular location of the sialidase(s) involved in ganglioside catabolism will be discussed later.

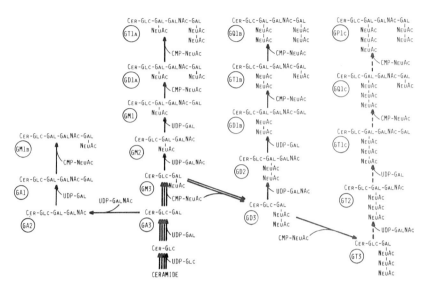

Fig. 1. Gangliosides of the ganglio series: pathways for biosynthesis. Dotted lines indicate proposed reactions.

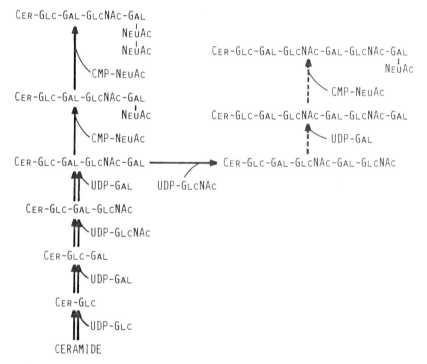

Fig. 2. Gangliosides of the neo-lacto series: pathways for biosynthesis. Dotted lines indicate proposed reactions.

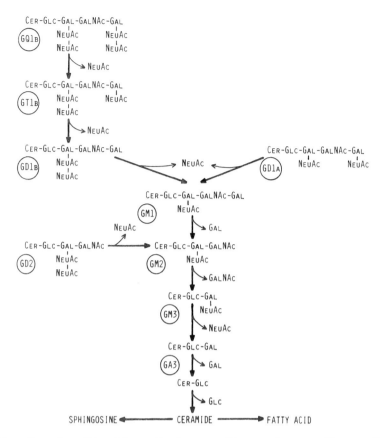

Fig. 3. Gangliosides of the ganglio series: pathways for biodegradation.

MECHANISM AND SITE OF BIOSYNTHESIS AND DEGRADATION

The sequential order of glycosylations governing ganglioside biosynthesis is assumed to be guaranteed by the assembly of the involved glycosyltransferases in a "multienzyme complex" (see Fig. 4).[21] In this complex the glycolipid product of one glycosyltransferase serves as substrate for the next enzyme. Transient intermediates are tightly bound to the complex which releases only the final product.[14] The gangliosides liberated by the biosynthetic machinery can no longer be acceptors of glycosyl units until their insertion in the plasma membrane.[2,14] The "multienzyme complex" hypothesis has received substantial, although indirect, experimental support.[14,22,23] According to this view the similar turnover rates observed for the different gangliosides in in vivo studies are explained assuming that each ganglioside is synthesized by a separate multienzyme complex.[14]

The multienzyme complexes responsible for ganglioside biosynthesis appear to be part of the Golgi apparatus (Fig. 4).[4,5] This statement is directly supported by observations made on extraneural tissues but can be extended to nervous tissues too.[24-26] The starting precursor for ganglioside biosynthesis as well as the mechanism of insertion into the multienzyme complex are still matters of hypotheses. A recent study[27] carried out on cultured neurotumor cells suggests that the initial glucosylation of ceramide might take place in the cis-Golgi cisternae, or in the endoplasmic reticulum, and further glycosylations in the trans-Golgi cisternae. This would imply that glucosylceramide, rather than ceramide, is the precursor of gangliosides in the multienzyme complex and poses the problem of how the precursor is selected for the glycosylations leading to gangliosides. In this regard the ceramide composition, which is known to be different in the different gangliosides, may play an important role in both the biosynthetic projections of the molecule and the regulation of the metabolic correlations among the different glycolipid families.[28] It is clear from the above scheme that each of the gangliosides found in tissues can be the end product of one multienzyme complex, which has been incorporated into the plasma membrane or an intermediate of another multienzyme complex producing ganglioside species with a higher degree of glycosylation.

The degradation of gangliosides seems to occur at two sites.[1,3,5] This particularly applies to the brain where the process has been studied in more detail. Because of the presence of an active sialidase in neuronal plasma membranes[29,30] it is suggested that the degradation of major multisialogangliosides is initiated at this very site, further breakdown taking place in the lysosomes.[2] However it is becoming evident that brain lysosomes contain a sialidase activity too.[31,32] Therefore the degradation of gangliosides might complete its course in the lysosomes (Fig. 4), the plasma membrane bound sialidase having a complementary function in ganglioside catabolism or a quite different metabolic role. It should be emphasized that the lysosomal involvement in ganglioside degradation has received direct demonstration by some inborn lysosomal diseases where the absence of a lysosomal enzyme, like β-galactosidase and β-hexosaminidase, is followed by intralysosomal accumulation of particular gangliosides (GM1 or GM2).[33] Furthermore, intoxication with chloroquine - a drug which accumulates in the lysosomes and reduces the activity of several lysosomal enzymes - leads to intralysosomal storage of gangliosides in the nervous system.[34] The observation that in the brain of chloroquine intoxicated animals the increase of gangliosides affects also the more complex multi-sialosylated species may support the hypothesis that the whole process of ganglioside degradation takes place in the lysosomes.

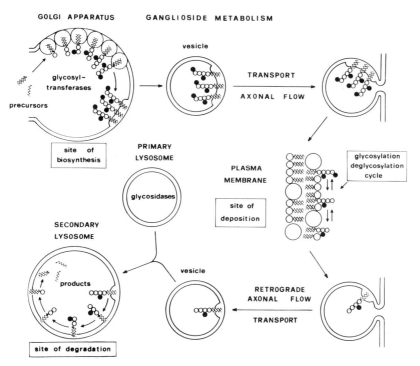

Fig. 4. General picture of intracellular sites of ganglioside biosynthesis and biodegradation, and of ganglioside transport

TRANSPORT FROM THE SITE OF BIOSYNTHESIS TO THE PLASMA MEMBRANES AND FROM THESE TO THE SITE OF DEGRADATION

The mechanism by which gangliosides are transported to and from the plasma membrane is largely unknown. It was recently demonstrated[35] that in cultured neurotumor cells newly synthesized gangliosides remain inside the cell for about 20 minutes before they appear at the cell surface. These intracellular gangliosides appear to be associated with membranes.[27] Studies in rat brain showed that newly synthesized gangliosides are membrane bound[36] in a manner unavailable to sialidase action. This may suggest, in analogy with sulfatide transport,[37] that at special sites of the Golgi apparatus a vesiculation process takes place with formation of ganglioside-carrying vesicles (Fig. 4). In these vesicles gangliosides are associated with the membrane and asymmetrically located on the inner surface. A mechanism could be postulated by which gangliosides, possibly by lateral diffusion, leave the biosynthetic machinery and reach the site of Golgi membrane available for vesiculation. Fusion of the transport-vesicle with the plasma membrane would automatically expose gangliosides on the outer surface of the same membrane (Fig. 4).

Following the general trend of viewing the plasma membrane-lysosome relationships (Fig. 4), it can be suggested that some surface events would lead a patch of the plasma membrane, carrying gangliosides, to internalize giving rise to a vesicle.[1] This would fuse with a primary lysosome and form a secondary lysosome where the degradation of gangliosides takes place.

METABOLISM IN THE BRAIN AND AXONAL FLOW

Both neuronal and glial cells contain gangliosides and are capable to biosynthesize their own ganglioside.[5] Gangliosides are present in all the plasma membrane surrounding neurons, that is at the level of cell body, axon, dendrites and nerve terminals.[5] It is still controversial whether the membranes of the synaptic junctions are more enriched and have a ganglioside pattern different from that of other portions of the neuron. A debate arose concerning the site of biosynthesis of neuronal gangliosides, in the cell body and/or in the nerve terminals.[1] It is now clear that neuronal gangliosides are biosynthesized at the level of the Golgi apparatus located in the cell body (no evidence has been provided so far for the presence of Golgi cisternae in the nerve endings).[5] The gangliosides which reside in the plasma membrane surrounding the axon and nerve endings reach their final location through rapid axonal flow (Fig. 4).[5,38,39] This was the result of accurate studies performed with rabbit, chick and goldfish optic systems after intraocular injection of a proper labelled precursor (for instance ^{3}H-N-acetylmannosamine).[1,5] The different gangliosides located in the nerve terminals appear to incorporate radioactivity at similar rates and to undergo transport simultaneously.[38,39] Labelling patterns of the small amounts of gangliosides in the soluble fraction of the nerve endings led to the suggestion[40] that some gangliosides in the nerve ending may receive part of their sialic acid locally, after their arrival by axonal transport. The mechanism by which the gangliosides, located in the nerve endings, reach the lysosomal apparatus is unknown. The observation that in GM1- and GM2-gangliosidoses gangliosides are stored in the nerve cell perikaryon or in the axon close to the perikaryon, but not in the nerve endings, led to the postulate[2] of a retrograde axonal transport of gangliosides from the nerve endings to the lysosomal apparatus present in the cell body.

METABOLIC REACTIONS AT THE PLASMA MEMBRANE LEVEL

The presence of glycosyltransferases and glycohydrolases has been described in the plasma membranes of various cells.[29,41,42] On this basis the hypothesis can be accepted that plasma membrane-bound glycoconjugates, gangliosides included, undergo local chemical modifications. The concept arises that, besides the metabolic operations which define the biosynthesis and the degradation of the whole molecule, reactions occurring at the plasma membrane level may be responsible for local fluctuations of ganglioside composition. The problem is open and difficult to solve owing to the difficulty of preparing plasma membranes devoid of Golgi apparatus contamination. However there is at least one example of a fairly homogeneous preparation of neuronal plasma membranes (obtained from isolated and purified nerve endings), which contains two enzymes - sialidase and sialyltransferase - capable of affecting gangliosides.[30,43] These enzymes can promote a "sialylation-desialylation" cycle on gangliosides which can be viewed as a regulatory mechanism for the charges and binding sites supplied by gangliosides to the same membranes. Support for this hypothesis can be found in the recent report[44] that the various forms of sialic acid - free, CMP bound, lipid bound, protein bound - had similar specific radioactivities after intraventricular injection of labelled N-acetylmannosamine in the rat, indicating active recycling of sialic acid molecules among different sialosyl derivatives. The local availability of CMP-sialic acid for sialylation constitutes a crucial point for the "sialylation-desialylation" hypothesis, especially if the only site for CMP-sialic acid formation has to be considered the nuclei.[45] However recent investigations[46] yet to be extended to brain tissue, recognized an alternative subcellular site for the biosynthesis of CMP-sialic acid.

FACTORS INFLUENCING AND REGULATING METABOLISM

The enzymes involved in ganglioside metabolism are gene products. Genetic defects of enzymes affecting ganglioside metabolism (β-galactosidase in GM1 gangliosidosis; β-N-acetylhexosaminidase in GM2 gangliosidosis) are known to cause intralysosomal accumulation of gangliosides followed by a gradually fatal impairment of cell and tissue function.[33] Some glycosyltransferases are directly affected in primary neoplastic transformation, being thus responsible for the compositional changes of cell surface glycoconjugates (including gangliosides) which are characteristic of transformed cells.[1] In general malignant transformation is accompanied by decreased ganglioside synthesis, especially of the more complex species.[47]

A possible but largely unknown mechanism for regulating ganglioside metabolism would be the control of the involved enzymes at the level of their biosynthesis or turnover. Puromycin, injected into brains of young animals, was shown to decrease the incorporation of labelled sugar precursors into gangliosides.[48]

The same drug was also demonstrated[49] to reduce the activity of the enzyme UDP-glucose: ceramide glucosyltransferase, which initiates ganglioside synthesis. It has also been established[50] that exposure of HeLa cells to butyrate was followed by increased synthesis of gangliosides GM3 and GM1, with concomitant striking enhancement of a sialosyltransferase active on glycolipids. The increase of enzyme activity was blocked by addition of actinomycin D and cycloheximide, while it was not affected by cyclic AMP or drugs known to increase the intracellular level of cyclic AMP. Thus it was concluded[50] that butyrate is able to induce sialosyltransferase activity.

The mode of assembly of the different glycosyltransferases in the ordered systems of enzymes which synthesize gangliosides could also be an important point of regulation. It is conceivable that both availability and stoichiometry among the different enzymes is crucial for normal ganglioside production. Nothing is known about this matter.

The production and accessibility at the luminal side of Golgi cisternae of the sugar-nucleotides required for glycosylation are also processes of potential interest for regulation. In fact, 2,4-dintrophenol, which impairs the energy supply of the cell, depresses also the rate of incorporation of labelled sugars into gangliosides (reported in Reference 1). Moreover, tunicamycin was recently shown[51] to inhibit ganglioside biosynthesis in vitro by blocking the transport of the sugar-nucleotides into Golgi-derived vesicles.

Investigations carried out on neurotumor cells tend to suggest that some glycosyltransferases involved in ganglioside metabolism can undergo reversible transition between active and inactive forms through a process of phosphorylation and dephosphorylation.[52] The protein kinase promoting phosphorylation and activating the glycosyltransferase would be dependent on the intracellular level of cyclic AMP. This suggestion derived from the observation that opiates and opioid peptides, which are known to block adenylate cyclase through the GTP regulatory protein, were able to reversibly inhibit the biosynthesis of gangliosides in mouse neuroblastoma cells. In particular the N-acetylgalactosaminyltransferase which converts GM3 to GM2 appeared to be sensitive to the above effect.[53] Since this enzyme is responsible for the biosynthesis of the more complex gangliosides, its regulation may result in an overall control of ganglioside biosynthesis. The concept of metabolic regulation based on phosphorylation-dephosphorylation of glycosyltransferases was postulated[52,53] assuming that the involved glycosyltransferases reside in the plasma membrane, where the opiate receptor is located. Since cyclic AMP is the molecule which mediates the activation of protein kinase, it is possible to conceive that the regulatory mechanism may apply also to the glycosyltransferases linked to the Golgi apparatus.

Most of the enzymes affecting gangliosides are membrane bound. Therefore a control of their activity can be accomplished by specific changes of the lipid environment around the enzyme, or by modifications of the membrane physico-chemical properties. Some phospholipids, namely phosphatidylglycerol, were found[51] to markedly stimulate the biosynthesis of gangliosides in isolated rat liver Golgi vesicles, probably by facilitating the entrance of sugar-nucleotides. In addition the activity of brain membrane bound sialidase was shown[54] to be dependent on the fluidity of the membrane bilayer, the activity of the enzyme increasing by enhancing fluidity.

PROBLEMS AND PERSPECTIVES

Many are the open questions concerning ganglioside metabolism. The mechanism of intracellular transport of gangliosides, the control of precursor availability for their biosynthesis, the assembling mechanism and internal dynamics of the multiglycosyltransferase systems, the possible synchronization of biosynthetic and degradative processes in response to the functional demand of the cell, the role of plasma membrane-bound glycosyltransferases and glycohydrolases are only some of them.

In recent years enormous progress has been made in the field of glycolipid structural and conformational analysis, of chemical derivation and labelling of gangliosides, of subcellular fractionation and tissue culture methodology. All this is expected to provide the experimental support and rationale for substantial improvement in the studies and knowledge of ganglioside metabolism.

Abbreviations used

Symbols for gangliosides and saccharides follow the Svennerholm[55] system and the recommendations of the IUPAC-IUB Commission on Biochemical Nomenclature.[56]

REFERENCES

1. E. Brunngraber, Biosynthesis of gangliosides; Catabolism of gangliosides, glycosaminoglycans and glycoproteins, in: "Neurochemistry of Aminosugars," C. C. Thomas, Springfield, Ill., pp. 332-356 and 410-437 (1979).
2. L. Svennerholm, Structure and biology of cell membrane gangliosides, in: "Cholera and Related Diarrheas," O.

Ouchterlony and J. Holmgren, eds., S. Karger, Basel, pp. 80-87 (1980).
3. A. Rosenberg, Biosynthesis and metabolism of gangliosides, in: "Complex Carbohydrates of Nervous Tissue," R. U. Margolis and R. K. Margolis, eds., Plenum Press, New York, pp. 25-43 (1980).
4. H. Wiegandt, The gangliosides,, in: "Advances in Neurochemistry," B. W. Agranoff and M. H. Aprison, eds., Plenum Publ. Corp., New York, Vol. 4, pp. 149-223 (1982).
5. R. W. Ledeen, Gangliosides, in: "Handbook of Neurochemistry," A. Lajtha, ed., Vol. 3 (2nd Ed.), Plenum Publ. Corp., New York, pp. 41-90 (1983).
6. S. Basu and M. Basu, Expression of glycosphingolipid glycosyltransferases in development and transformation, in: "The Glycoconjugates," M. I. Horowitz and W. Pigman, eds., Academic Press, New York, Vol. III, pp. 265-286 (1982).
7. R. M. Burton, L. Garcia-Bunuel, M. Golden, Incorporation of radioactivity of D-glucosamine-1-^{14}C, D-glucose-1-^{14}C, D-galactose-1-^{14}C and DL-serine-3-^{14}C, into rat brain glycolipids, Biochem. 2:580 (1963).
8. K. Suzuki, Formation and turnover of the major brain gangliosides during development, J. Neurochem. 14:917 (1967).
9. M. Holm and L. Svennerholm, Biosynthesis and biodegradation of rat brain gangliosides studied in vivo, J. Neurochem. 19:609 (1972).
10. J. A. Skrivanek, R. W. Ledeen, R. U. Margolis and R. K. Margolis, Gangliosides associated with microsomal subfractions of brain: comparison with synaptic plasma membranes, J. Neurobiol. 13:95 (1982).
11. R. W. Ledeen, J. E. Haley, and J. A. Skrivanek, Study of ganglioside patterns with two-dimensional thin layer chromatography and radioautography: detection of new fucogangliosides and other minor species in rabbit brain, Analyt. Biochem. 112:135 (1981).
12. A. Arce, H. J. Maccioni, and R. Caputto, The biosynthesis of gangliosides. The incorporation of galactose, N-acetylgalactosamine and N-acetylneuraminic acid into endogenous acceptors of subcellular particles from rat brain in vitro, Biochem. J. 121:483 (1971).
13. P.H. Fishman and R. O. Brady, Biosynthesis and function of gangliosides, Science 194:906 (1976).
14. R. Caputto, H. J. Maccioni, A. Arce, and R. F. A. Cumar, Biosynthesis of brain gangliosides, Adv. Exptl. Med. Biol. 71:27 (1976).
15. M. C. M. Yip and J. A. Dain, The enzymic synthesis of ganglioside: I. Brain uridine diphosphate D-galactose: N-acetylgalactosaminyl-galactosyl-glucosylceramide galactosyl transferase, Lipids 4:270 (1969).

16. A. Stoffyn, P. Stoffyn and M. C. M. Yip, Chemical structure of monosialoganglioside GM1b biosynthesized in vitro, Biochim. Biophys. Acta 409:97 (1975).
17. R. K. Yu and S. H. Lee, In vitro biosynthesis of sialosylgalactosyl-ceramide (G_7) by mouse brain microsomes, J. Biol. Chem. 251:198 (1976).
18. S. Gatt, Enzymatic aspects of sphingolipid degradation, Chem. Phys. Lipids 5:235 (1979).
19. K. Sandhoff and E. Conzelmann, Activation of lysosomal hydrolysis of complex glycolipids by non-enzymic proteins Trends in Biochem. Sci. 4:231 (1979).
20. S. C. Li and Y. T. Li, Protein activators for the hydrolysis of GM1 and GM2 gangliosides, in: "Methods in Enzymology," V. Ginsburg ed., Vol. 83, pp. 588-595 (1982).
21. S. Roseman, The synthesis of complex carbohydrates by multiglycosyl-transferase systems and their potential function in intracellular adhesion, Chem. Phys. Lipids 5:270 (1970).
22. C. A. Landa, H. J. Maccioni, A. Arce, and R. Caputto, The biosynthesis of gangliosides. Separation of membranes with different ratios of gangliosides-sialylating activity to gangliosides, Biochem. J. 168:325 (1977).
23. H. J. Maccioni, S. S. De Filpo, C. A. Landa, and R. Caputto, The biosynthesis of gangliosides, ganglioside glycosylating activity in rat brain neuronal perikarya fraction, Biochem. J. 174:673 (1978).
24. T. W. Keenan, Membranes of mammary gland. IX. Concentration of glycosphingolipid galactosyl and sialyltransferases in Golgi apparatus from bovine mammary gland, J. Dairy Sci. 57:189 (1974).
25. T. W. Keenan, D. J. Morre, and S. Basu, Concentration of glycosphingolipid glycosyltransferase in Golgi apparatus from rat liver, J. Biol. Chem. 249:310 (1974).
26. C. A. Landa, H. J. Maccioni, and R. Caputto, The site of synthesis of gangliosides in the chick optic system, J. Neurochem. 33:825 (1979).
27. H. Miller-Podraza and P. H. Fishman, Soluble gangliosides in cultured neurotumor cells, J. Neurochem. 41:860 (1983).
28. R. Kannagi, E. Nudelman, and S. I. Hakomori, Possible role of ceramide in defining structure and function of membrane glycolipids, Proc. Natl. Acad. Sci. USA 79:3470 (1982).
29. C.-L. Schengrund and A. Rosenberg, Intracellular location and properties of bovine brain sialidase, J. Biol. Chem. 245:6196 (1970).
30. G. Tettamanti, I. G. Morgan, A. Gombos, G. Vincendon, and P. Mandel, Sub-synaptosomal localization of brain particulate neuraminidase, Brain Res. 47:515 (1972).
31. L. Caimi, T. Burkart, U. N. Wiesmann, and G. Tettamanti, Subcellular localization of sialidase in myelinating

mouse brain, Proceedings of the International Symposium "Sialidase and sialidoses," Genoa, September (1980).
32. R. W. Veh and M. Sander, Differentation between ganglioside and sialyllactose sialidases in human tissues, in: "Sialidases and Sialidoses," G. Tettamanti, P. Durand, and S. Di Donato, eds., Edi Ermes, Milan, Italy, pp. 71-109 (1981).
33. K. Sandhoff, The biochemistry of sphingolipid storage diseases, Angew. Chem. Int. Ed. Engl. 16:273 (1977).
34. G. W. Klinghardt, P. Fredman, and L. Svennerholm, Chloroquine intoxication induces ganglioside storage in nervous tissue: a chemical and histopathological study of brain, spinal cord, dorsal root ganglia and retina in the miniature pig, J. Neurochem. 37:897 (1981).
35. H. Miller-Podraza and P. H. Fishman, Translocation of newly synthesized gangliosides to the cell surface, Biochem. 21:3265 (1982).
36. C. A. Landa, S. S. Defildo, H. J. Maccioni, and R. Caputto, Disposition of gangliosides and sialosyl glycoproteins in neuronal membranes, J. Neurochem. 37:813 (1981).
37. T. Burkart, L. Caimi, H. P. Siegrist, N. N. Herschkowitz, and U. N. Wiesmann, Vesicular transport of sulfatide in the myelinating mouse brain, functional association with lysosomes?, J. Biol. Chem. 257:3151 (1982).
38. R. W. Ledeen, J. A. Skrivanek, J. Nunez, J. R. Sclafani, W. T. Norton and M. Farooq, Implication of the distribution and transport of gangliosides in the nervous system, in: "Gangliosides in Neurological and Neuromuscular Function, Development and Repair," M. M. Rapport and A. Gorio, eds., Raven Press, New York, pp. 211-223 (1981).
39. H. Rosner and G. Merz, Uniform distribution and similar turnover rates of individual gangliosides along axons of retinal ganglion-cells in the chicken, Brain Res. 236:63 (1982).
40. R. W. Ledeen, J. A. Skrivanek, L. J. Tirri, R. K. Margolis, and R. U. Margolis, Gangliosides of the neuron: localization and origin, Adv. Exptl. Med. Biol. 71:83 (1976).
41. L. M. Patt and W. J. Grimes, Cell surface glycolipid and glycoprotein glycosyltransferases of normal and transformed cells, J. Biol. Chem. 249:4157 (1974).
42. H. B. Bosman, Cell surface glycosyltransferases and acceptors in normal and RNA- and DNA virus transformed fibroblasts, Biochem. Biophys. Res. Commun. 43:1118 (1972).
43. A. Preti, A. Fiorilli, A. Lombardo, L. Caimi, and G. Tettamanti, Occurrence of sialyltransferase activity in the synaptosomal membranes prepared from calf brain cortex, J. Neurochem. 35:281 (1980).
44. W. Ferwerda C. M. Blok, and J. Heijlman, Turnover of free sialic acid, CMP-sialic acid, and bound sialic acid in rat brain, J. Neurochem. 36:1492 (1981).

45. D. Van Den Eijnden, The subcellular location of cytidine 5´-monophospho-N-acetylneuraminic acid synthetase in calf brain, J. Neurochem. 21:949 (1973).
46. W. Ferwerda, C. M. Blok, and J. Van Rinsum, CMP-N-acetylneuraminic acid; is it synthesized in the nucleus?, in: "Glycoconjugates: Proceedings of the 7th Int. Symposium on Glycoconjugates," M. A. Chester, D. Heinegard, A. Lundblad, and S. Svensson, eds., pp. 733-734 (1983).
47. S. I. Hakomori, Structures and organization of cell surface glycolipids. Dependency on cell growth and malignant transformation, Biochim. Biophys. Acta 417:55 (1975).
48. J. Kanfer and R. L. Richards, Effect of puromycin on the incorporation of radioactive sugars into gangliosides in vivo, J. Neurochem. 14:513 (1967).
49. S. N. Shah and N. A. Peterson, Effects of cycloheximide and puromycin on glycosyltransferases of microsomal fractions from rat brain, Biochim. Biophys. Acta 239:126 (1971).
50. P. H. Fishman and R. C. Henneberry, Induction of ganglioside biosynthesis in cultured cells by butyric acid, in: "Cell Surface Glycolipids," C. C. Sweeley, ed., ACS Symposium Series 128, American Chemical Society, Washington, D.C., pp. 223-239 (1980).
51. H. K. M. Yusuf, G. Pohlentz, G. Schwarzmann and K. Sandhoff, Ganglioside biosynthesis in rat liver Golgi apparatus: stimulation by phosphatidylglycerol and inhibition by tunicamycin, in: "Glycoconjugates: Proceedings of the 7th Int. Symposium on Glycoconjugates," M. A. Chester, D. Heinegard, A. Lundblad, and S. Svensson, eds., pp. 773-774 (1983).
52. G. Dawson, R. W. McLawhon, G. Schoon, and R. J. Miller, Modulation of ganglioside synthesis by enkephalins, opiates and prostaglandins. Role of cyclic AMP in glycosylation, in: "Cell Surface Glycolipids," C. C. Sweeley, ed., ACS Symposium Series 128, American Chemical Society, Washington, D.C., pp. 359-372 (1980).
53. R. W. McLawhon, G. S. Schoon, and G. Dawson, Possible role of cyclic AMP in the receptor-mediated regulation of glycosyltransferase activities in neurotumor cell lines, J. Neurochem. 37:132 (1981).
54. K. Sandhoff, G. Scheel, and H. Nehrkorn, Membrane bound sialidase. Regulation of ganglioside GD1a degradation by the physical state of the membrane, in: "Sialidases and Sialidoses," G. Tettamanti, P. Durand, and S. Di Donato, eds., Edi Ermes, Milan, Italy, pp. 125-143 (1981).
55. L. Svennerholm, Ganglioside designation, Adv. Exptl. Med. Biol. 125:11 (1980).
56. IUPAC-IUB Commission on Biochemical nomenclature. The nomenclature of lipids, Lipids 12:455 (1977), Abbreviated terminology of oligosaccharide chains, J. Biol. Chem. 257:3347 (1982).

ACTIVATOR PROTEINS FOR THE CATABOLISM OF GLYCOSPHINGOLIPIDS

Y.-T. Li and S.-C. Li

Department of Biochemistry and Delta Regional Primate
Research Center, Tulane University Medical Center
New Orleans, Louisiana 70112, U.S.A.

INTRODUCTION

After Buchner's discovery in 1897[1] that the alcoholic fermentation of glucose could be carried out by yeast-juice (cell free extract of yeast), Harden and Young[2] reported in 1904 that the fermentation of glucose by yeast-juice was greatly increased by the addition of boiled filtered yeast-juice, although the boiled yeast-juice itself was incapable of setting up fermentation. This was the first demonstration of the requirement of an "activator" (co-enzyme) for an enzymic reaction. The biological importance of co-enzymes has been well appreciated, since most water soluble vitamins are components of co-enzymes. Until recently, it had not occurred to biochemists that hydrolysis of glycosphingolipids may require a co-factor. Through the work of many laboratories, it has been well recognized that the hydrolysis of sugar chains in glycosphingolipids cannot proceed by the action of enzymes alone; the hydrolysis requires the presence of an activator protein. In this chapter we would like to give a brief overview on activator proteins related to the catabolism of glycosphingolipids and to use our studies on chemical pathology of Type-AB GM2-gangliosidosis as an example to illustrate the biological importance of the activator proteins

ACTIVATOR PROTEINS RELATED TO THE CATABOLISM OF GLYCOSPHINGOLIPIDS

Although lysosomal glycosidases have been postulated to be responsible for the catabolism of glycosphingolipids in vivo, it has been difficult to demonstrate the hydrolysis of glycosphingolipids by lysosomal glycosidases in vitro. Due to their lipophilic nature, the enzymic hydrolysis of glycosphingolipids requires the addition of a bile salt such as sodium taurodeoxycholate (TDC) to the reaction mixture. Since bile salts do not exist in tissues other than liver, the in vitro hydrolysis of glycosphingolipids in the presence of a bile salt may not reflect the reactions taking place in vivo. In the past decade, several activator proteins have been isolated from various tissues. In the presence of these activator proteins, several glycosphingolipids have been found to be hydrolyzed by glycosidases without the aid of a detergent. Three key observations led to the revelation of an activator requirement for the enzymic hydrolysis of glycosphingolipids: (a) Mehl and Jatzkewitz[3] observed the separation of porcine kidney arylsulfatase A into heat-stable and heat-labile protein fractions and demonstrated that the heat-stable fraction stimulated the heat-labile arylsulfatase A to carry out the hydrolysis of cerebroside sulfate; (b) Ho and O'Brien[4] reported that the heat-stable fraction of Gaucher spleen stimulated the heat-labile, membrane-bound, glucosidase prepared from normal spleen; and (c) Li et al.[5] found that the crude human hepatic β-hexosaminidase (Hex) preparation could convert GM2 into GM3 in the absence of detergent, but further purification of the enzyme resulted in a loss of this activity. They further showed that the crude enzyme contained an activating factor which was removed during the purification.

The Activator Protein for the Enzymic Hydrolysis of Cerebroside Sulfate

While studying the catabolism of cerebroside sulfate, Mehl and Jatzkewitz[3] separated the arylsulfatase A isolated from porcine kidney into heat-labile and heat-stable protein fractions by a carrier-free high voltage electrophoresis. They found that the heat-labile fraction was identical to arylsulfatase A. This fraction alone could hydrolyze synthetic arylsulfates but not cerebroside sulfate. The heat-stable fraction, however, was enzymatically inactive, but able to stimulate the heat-labile arylsulfatase A to carry out the hydrolysis of cerebroside sulfate. This was the first observation alluding to the necessity of an activator protein for the enzymic hydrolysis of a glycosphingolipid. This activator protein has been isolated from human liver by Fisher and Jatzkewitz.[6] The purified activator protein migrated as a single protein band when subjected to polyacrylamide gel electrophoresis. Its apparent molecular weight was determined to be 21,500 ± 1,500. The isoelectric point of the activator protein was found to be between pH 3.8 - 5.0 by isoelectrofocusing. The activating effect was abolished by Pronase E. Mraz et al. further reported that human hepatic activator protein which was specific for the stimulation of the hydrolysis of cerebroside sulfate by human enzyme could also stimulate the same reaction catalyzed by acidic sulfatases isolated from various invertebrates. Thus, the effect of the human activator protein was not restricted to the stimulation of human arylsulfatase A.

The Activator Protein for the Enzymic Hydrolysis of Glucosylceramide

Ho and O'Brien,[4] using 4-methylumbelliferyl-β-D-Glc as substrate, detected an unexpected increase in β-glucosidase activity when spleen homogenates from a control and a Gaucher's patient were mixed. They found that two factors were responsible for the reconstitution of β-glucosidase activity. One was called factor C from control spleen, and the other, factor P, from the patient's spleen. Factor P was identified as an acidic glycoprotein which was very soluble, heat-stable and devoid of β-glucosidase activity. Factor C, a crude particulate preparation from normal human spleen was heat-labile, insoluble, and was associated with minimal amounts of β-glucosidase activity. Ho et al.[8] further demonstrated that GlcCer cleaving activity from normal spleen could also be reconstituted from factors P and C. This activator has been isolated from normal and Gaucher's spleen[9] and from bovine spleen.[10] Wenger et al.[11] reported that a crude activator preparation from Gaucher spleen stimulated not only human GlcCer β-glucosidase but also GalCer β-galactosidase and sphingomyelinase. Christomanou[12] also reported that an activator preparation isolated from Gaucher spleen stimulated the degradation of both GlcCer and sphingomyelin. This activator protein was found to be different from that described by Ho and O'Brien,[4] since it could not stimulate the enzymic hydrolysis of 4-methylumbelliferyl-β-Glc.

The Activator Protein (GM1-activator) for the Enzymic Hydrolysis of GM1 Ganglioside

The GM1-activator for the enzymic hydrolysis of GM1 ganglioside was discovered by Li et al.[5] while studying the degradation of GM2 ganglioside by human β-hexosaminidases. They found that a crude β-hexosaminidase fraction prepared by ammonium sulfate fractionation of human liver extract or urine could efficiently convert GM2 into GM3 in the absence of a detergent. However, after the separation of Hex A from Hex B by DEAE-cellulose chromatography, both isozymes became inactive toward the hydrolysis of GM2. They subsequently found that a heat-stable, nondialyzable preparation obtained from the crude human hepatic hexosaminidase fraction could stimulate the hydrolysis of GM2 by Hex A but not Hex B. This crude, heat-stable preparation was also found to stimulate the conversion of GM1 into GM2, and $GbOse_3Cer$ into LacCer by human hepatic β-galactosidase and α-galactosidase, respectively. Therefore, at that time, it was assumed that the same activator protein stimulated the enzymic hydrolysis of several glycosphingolipids. Since GM2 was not readily available, Li and Li[13] used the stimulatory effect on the hydrolysis of GM1 carried out by human hepatic β-galactosidase to follow the purification of the activator protein from human liver. This activator has been purified to an electrophoretically homogeneous form. Its molecular weight was estimated to be about 22,000. It was heat-stable and withstood heating in a bath of boiling water for 30 min at pH 7.0. The pI of this activator was estimated to be about pH 4.1 by disc gel isoelectrofocusing. The activator protein was antigenic and sensitive to Pronase. Inui and Wenger[14] using the same purification method of Li and Li[13] found that the yield of this activator from the liver of a patient with Type 1 GM1-gangliosidosis was about 35 times that from the control. This activator could stimulate the hydrolysis of GM1 and asialo-GM1 by GM1-β-galactosidase. However, this activator could not stimulate the hydrolysis of GalCer and LacCer by GM1-β-galactosidase. Li and Li[15] observed that this activator stimulated the hydrolysis of the following reactions: GM1 and GM2 by Clostridial sialidase; $GbOse_3Cer$ by human placental α-galactosidase; $GbOse_4Cer$ by jack bean β-hexosaminidase or by human hepatic Hex A and Hex B; asialo-GM2 by human hepatic Hex A and Hex B; and, GalCer sulfate catalyzed by arylsulfatase A. Thus, GM1-activator may function as an agent which interacts with the lipophilic glycosphingolipid substrate to facilitate the reaction catalyzed by various enzymes.

The Activator Protein (GM2-activator) for the Enzymic Hydrolysis of GM2 Ganglioside

The presence of a GM2-activator which stimulates the hydrolysis of $GM2_5$ catalyzed by Hex A was first detected in human liver by Li et al.[5] and subsequently by Hechtman.[16] Conzelmann and Sandhoff, on the other hand,[17] detected the presence of a similar activator in human kidney.[17] This activator has been isolated in electrophoretically homogeneous form from human kidney,[17] liver[18] and brain.[19] By Sephadex G-75 filtration, the molecular weight of GM2-activator isolated from human liver[18] and brain[19] was estimated to be 23,500. The pI of this activator was found to be about pH 4.8 by isoelectrofocusing.[19] The GM2-activator isolated from human kidney[17] exhibited a pI of pH 4.8 and a molecular weight of 25,000 as estimated by Sephadex G-100 filtration. Rabbit antisera against GM2-activator isolated from brain, kidney and liver have been prepared. It has been found that GM2-activators isolated from liver, kidney and brain were immunologically identical (unpublished observations). The hydrolysis of GM2 by Hex A in the presence of the activator protein was severely inhibited by a buffer of high ionic strength (Fig. 1). No such inhibition was observed in the hydrolysis of GM1 by β-galactosidase. The GM2-activator stimulated the hydrolysis of GM2 much better than that of asialo-GM2 in a low ionic strength buffer ($\Gamma/2 = 0.01$), whereas in a high ionic strength buffer ($\Gamma/2 = 0.2$), the hydrolysis of asialo-GM2 was greater than that of GM2. Compared with the above two substrates, the activator exerted the least stimulation for the hydrolysis of $GbOse_4Cer$. It is, therefore, very important to standardize the assay condition before comparing the specificity of activator proteins isolated from different tissues.

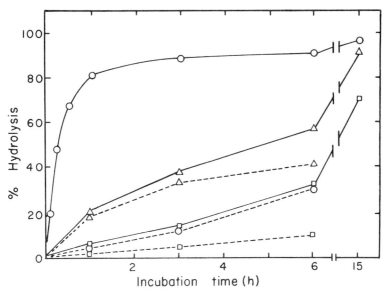

Fig. 1. Effect of ionic strength on the hydrolysis of GM2, asialo-GM2, and GbOse$_4$Cer by human Hex A in the presence of GM2-activator. Solid lines indicate that the reaction was carried out in 0.01 M acetate buffer, pH 4.6 ($\Gamma/2 = 0.01$). Dotted lines indicate that the same reaction was carried out in 0.1 M citrate buffer, pH 4.6 ($\Gamma/2 = 0.2$). (o), hydrolysis of GM2; (△), hydrolysis of asialo-GM2; (□), hydrolysis of GbOse$_4$Cer.

GM2-ACTIVATOR AND BIOCHEMICAL PATHOLOGY OF TYPE-AB GM2-GANGLIOSIDOSIS

GM2-gangliosidosis has been classified into three biochemically distinct types.[20] Type-B (classical Tay-Sachs disease) has been established to be caused by the deficiency of Hex A and Type-O (Sandhoff's disease), by the deficiency of both Hex A and Hex B. Type-AB (late infantile or juvenile GM2-gangliosidosis) is characterized by the presence of a normal or elevated level of both Hex A and Hex B. Since there is no apparent enzyme deficiency, the accumulation of GM2 in the tissues of Type-AB GM2-gangliosidosis cannot be easily explained. Concerning the etiology of Type-AB GM2-gangliosidosis, Conzelmann and Sandhoff[21] reported that the heated, crude extract derived from the kidney of a patient with Type-AB GM2-gangliosidosis was deficient in a factor which stimulates the hydrolysis of GM2 catalyzed by human hepatic Hex A. Hechtman et al.[22] also reported the deficiency of a similar activator in the crude liver extract of a patient with Type-AB GM2-gangliosidosis. By using immunological methods we also found that the brain of one of the two cases of Type-AB GM2-gangliosidosis reported by Goldman et al.[23] was devoid of GM2-activator,[19] however, the other case was not due to the activator deficiency. We found that this case was due to the defective Hex A.[24]

Levels of GM2-activator and β-Hexosaminidase in the Brains of Three Types of GM2-gangliosidoses

Table 1 summarizes the levels of GM2-activator and β-hexosaminidases in the brains of one case of Type-B, two cases of Type-O,[23] and two cases (case 1 and case 2) of Type-AB GM2-gangliosidosis together with two normal brains. With the exception of case 2 Type-AB GM2-gangliosidosis, all pathological brains contained higher levels of GM2-activator than the two controls. Compared with the two normals, the level of GM2-activator in the brain of case 2 Type-AB GM2-gangliosidosis was found to be extremely low. This result[21] was similar to those reported by Conzelmann and Sandhoff[21] and Hechtman et al.[22] Case 1, on the other hand, represents a new variant.

Properties of β-Hexosaminidases and GM2-Activator Isolated from the New Variant of Type-AB GM2-Gangliosidosis

Since the level of GM2-activator was markedly elevated in the new variant (case 1), and no deficiency of β-hexosaminidases was detected using synthetic substrates, we then examined the hydrolysis of GM2 by β-hexosaminidases prepared from the brain of this patient in the presence of pure GM2-activator isolated from normal human brain. As shown in Fig. 2, the β-hexosaminidases (containing isoenzyme A and B) isolated from the brain of this patient could not hydrolyze GM2 in the presence of the pure GM2-activator. Under the same conditions, the β-hexosaminidases obtained from a normal control could hydrolyze GM2. The cross examination of the enzymes and the activators isolated from a normal control (Fig. 2) revealed that the β-hexosaminidases from this patient could not hydrolyze GM2 in the presence of the partially purified activator isolated either from the brain of this patient or from the control brain. On the other hand, the β-hexosaminidases isolated from the control brain could hydrolyze GM2 in the presence of the partially purified activator isolated from the patient's brain or from the control brain. In order to study the properties of Hex A and Hex B in the brain of case 1 Type-AB GM2-gangliosidosis, the crude β-hexosaminidase fraction obtained from the brain of this patient was separated into three peaks, designated as A_1, A_2 and B, by DEAE-Sephadex A-50 chromatography as shown in Fig. 3. The distribution of these activities based on p-nitrophenyl-β-GlcNAc as substrate were A_1, 17.6%; A_2, 25.3% and B, 57.3%. The electrophoresis on Cellogel (Fig. 3b) revealed that the enzymes A_2 and B had mobilities similar to those of the pure Hex A and Hex B isolated from normal human liver. The enzyme A_1 migrated between A_2 and B. The ability of these enzymes to hydrolyze GM2 was examined in the

Table 1. The Levels of GM2-activator and Total β-hexosaminidases in the Brains of Three Types of GM2-gangliosidosis

The GM2-activator and β-hexosaminidases from the following samples were partially purified according to the method described previously.[18] For the normal brains, duplicates of 10 g tissue were analyzed to check the reproducibility of the method.

	$N_{2.7}$	$N_{2.5}$	Type-AB Case 1	Type-AB Case 2	Type-0 Case 1	Type-0 Case 2	Type-B
Tissue (g)	10 10	10 10	10.3	10.4	10	10	10
Total β-hexosaminidase (unit)	5.4 5.2	5.0 4.8	7.9	14.2	0	0	7.2
GM2-activator ($\times 10^{-5}$ unit)[a]	0.89 0.65	0.82 0.54	2.94	0.04	2.26	3.20	3.80
Protein after octyl-sepharose (mg)	1.96 2.18	2.18 2.52	3.34	3.24	5.67	6.46	3.57

[a] These values were determined by using C_{18} beads as described in reference 19. They are slightly lower than that determined by dialysis method.[24]

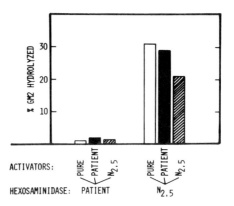

Fig. 2. Hydrolysis of GM2 using β-hexosaminidases and activators derived from the pathological brain and a normal brain of 2.5 years of age ($N_{2.5}$). The incubation mixture contained [^{3}H]-labeled GM2 (1.7×10^4 cpm), 10 nmole; acetate buffer (10 mM), pH 4.6; β-hexosaminidases, 0.4 unit; and approximately 2×10^3 units of GM2-activator. The sources for β-hexosaminidases and GM2-activators are indicated. The incubation was carried out at 37°C for 16 hours. The pure activator was isolated from normal human brain using the method described previously.[19]

presence of either the GM2-activator isolated from normal brain or TDC. As shown in Fig. 3c, neither isozyme A_1 nor A_2 isolated from the pathological brain could hydrolyze GM2 in the presence of GM2-activator. Hex B from brains of normal and case 1 appeared similar in that neither could hydrolyze GM2 in the presence of GM2-activator but both could hydrolyze GM2 slowly in the presence of TDC. On the other hand, the hydrolysis of GM2 by A_1 and A_2 in the presence of TDC was much less than that of isozyme B. It is important to note that TDC can stimulate the hydrolysis of GM2 carried out by Hex A and Hex B. The GM2-activator, on the other hand, only stimulates Hex A, but not Hex B, in carrying out the hydrolysis of GM2. If only TDC was used to examine the ability of the β-hexosaminidases from this pathological brain to hydrolyze GM2, the abnormality of this β-hexosaminidase would not have been detected. These results indicate that this patient had a defective Hex A, but not the activator. This β-hexosaminidase was found to cross-react with the antibody against the α-chain of normal Hex A.[19]

Differentiation of Two Variants of Type-AB GM2-Gangliosidosis Using p-Nitrophenyl-β-GlcNAc-sulfate as Substrate

The two variants of Type-AB GM2-gangliosidosis cannot be distinguished by the analysis of β-hexosaminidases using PNP-GlcNAc,

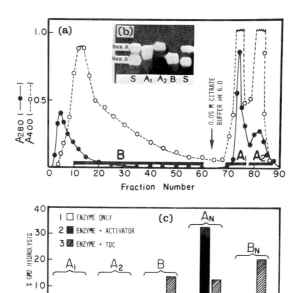

Fig. 3. (a), DEAE-Sephadex A-50 chromatography of β-hexosaminidases. From 69 g of the brain, 38 units (in 5.5 ml) of β-Hex was obtained by $(NH_4)_2SO_4$ precipitation and Sephadex G-200 filtration.[5] This preparation was dialyzed against 50 mM Na-phosphate buffer, pH 7.0 and applied to a DEAE-Sephadex A-50 column (1 x 25 cm). After washing off the unadsorbed Hex B with the same buffer, the column was eluted with 50 mM Na-citrate buffer, pH 6.0. Hex A was split into A_1 and A_2. ●, absorbance at 280 nm. ○, absorbance at 400 nm for enzyme activity using PNP-β-GlcNAc as substrate. (b), Cellogel electrophoresis of the fractions in (a). (c), Hydrolysis of GM2 using the fractions from (a) in the presence of GM2-activator (4×10^3 units) or TDC (200 μg). For each assay 0.3 units of the enzyme were used. A_N and B_N are Hex A and Hex B from normal brain, respectively.

4MU-GlcNAc, or by the detection of the accumulation of GM2. Furthermore, the analysis of the levels of GM2-activator and the determination of the hydrolysis of GM2 are very laborious. We found that these two variants could be distinguished by using p-nitrophenyl-6-sulfo-2-acetamido-2-deoxy-β-D-glucopyranoside (PNP-GlcNAc-6-SO_4) as substrate.[25] As shown in Table 2, the Hex A from the variant with activator deficiency (case 2 of Goldman et al.[23]) had a normal activity toward both PNP-GlcNAc and PNP-GlcNAc-6-SO_4. The variant caused by a defective Hex A (case 1 of

Goldman et al.[23]), on the other hand, exhibited severely attenuated activity toward PNP-GlcNAc-6-SO$_4$ but normal activity toward PNP-GlcNAc. The results presented in Table 2 suggest that by using PNP-GlcNAc-6-SO$_4$, Hex A from the new variant behaves like Hex B.

<u>Immunological Detection of GM2-Activator in the Two Variants of Type-AB GM2-Gangliosidosis</u>

Since the <u>in vitro</u> assay cannot distinguish the activator "deficiency" (CRM-negative) from the activator "defect" (CRM-positive but functionally impaired), we used immunoelectrophoresis to examine the presence or absence of the GM2-activator in the two variants of Type-AB GM2-gangliosidosis. Since normal tissues contain both GM1- and GM2-activator,[26] these two activators were examined simultaneously. As shown in Fig. 4, all brain samples with the exception of case 2 Type-AB GM2-gangliosidosis contained GM1- and GM2-activator. Case 2, on the other hand, contained only GM1-activator, but was devoid of GM2-activator. This is the first description of the CRM-negative GM2-activator deficiency in Type-AB GM2-gangliosidosis. Theoretically GM2-gangliosidosis can be also caused by the presence of CRM-positive but functionally negative GM2-activator (see Table 3). The <u>in vitro</u> assay reported by Conzelmann and Sandhoff[21] and Hechtman et al.[22] cannot distinguish the "activator deficiency" from "activator defect". Table 3 summarizes the possible causes for Type-AB GM2-gangliosidosis.

Table 2. Hydrolysis of PNP-GlcNAc and PNP-GlcNAc-6-SO$_4$ by Hex A and Hex B[25] Isolated from the Brains of Two Variants of Type-AB GM2-Gangliosidosis

Cases	Hydrolysis of		PNP-GlcNAc / PNP-GlcNAc-6-SO$_4$
	PNP-GlcNAc	PNP-GlcNAc-6-SO$_4$	
	μmol/min/10 g tissue		Ratio
Case 1 (4.7 yrs):			
Hex A	2.65	0.11	24.1
Hex B	3.57	0.11	32.5
Case 2 (3.5 yrs):			
Hex A	4.88	1.13	4.3
Hex B	4.26	0.17	25.1
Normal (2.7 yrs):			
Hex A	3.99	0.96	4.2
Hex B	1.32	0.04	33.0

Table 3. Three Possible Variants of Type-AB GM2-Gangliosidosis

Variants	Hex PNP-GlcNAc A B	Hex A on GM2-Hydr with GM2-Act	GM2-Act Immuno Assay	GM2-Act In vitro Assay
I	+ +	+	−	−
II	+ +	+	+	−
III	+ +	−	+	+

I = GM2-activator deficiency: "true AB variant".
II = GM2-activator defect: "iso AB variant" - to be discovered.
III = Hex A defect: "pseudo AB or new AB variant".

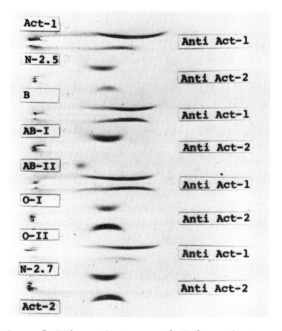

Fig. 4. Detection of GM1-activator and GM2-activator in pathological brains by immunoelectrophoresis. The samples in the wells were: Act-1, pure GM1-activator; Act-2, pure GM2-activator; N-2.5 and N-2.7, activator fractions derived from the two normal controls aged 2.5 and 2.7 years. The activator fractions obtained from three types of GM2-gangliosidosis were: B, Type-B; AB-I, case 1 of Type-AB; AB-II, case 2 of Type-AB; O-I, case 1 of Type-O; O-II, case 2 of Type-O. Anti Act-1, rabbit antiserum against GM1-activator; Anti Act-2, guinea pig antiserum against GM2-activator.

Presence of Activator Proteins in Normal Human Urine

The diagnosis of sphingolipidoses caused by the deficiency of activator proteins depends on the analysis of the presence or absence of activator proteins. So far, the analyses have been made only in postmortem tissues such as brain, kidney, liver and spleen. Recently, we detected the presence of activator proteins which stimulate the enzymic hydrolysis of GM1, GM2 and cerebroside sulfate in normal human urine.[27] The discovery of the presence of activator proteins in normal human urine may have potential diagnostic significance, since it could lead to the development of a method for the detection of lipid storage diseases due to activator deficiencies in living patients.

CONCLUSION

Our studies on the catabolism of GM2 led to the discovery of the requirement of two separate protein activators for the enzymic hydrolysis of GM1 and GM2. The analysis of the levels of GM2-activator and the study on β-hexosaminidases of two cases of Type-AB GM2-gangliosidosis enabled us to reveal a new variant of Tay-Sachs disease. This new variant contained a defective β-hexosaminidase A and an elevated level of GM2-activator. Our recent finding of activator proteins in normal human urine may facilitate the routine analysis of activator in Type-AB GM2-gangliosidosis patients and also in other sphingolipidoses due to their respective activator deficiencies. Our work on chemical pathology of Type-AB GM2-gangliosidosis clearly showed the physiological importance of activator proteins. The discovery of activator proteins for the enzymic hydrolysis of glycosphingolipids may represent one of the most exciting developments in the field of glycosphingolipid catabolism.

ACKNOWLEDGEMENTS

We are grateful to Dr. K. Suzuki of Albert Einstein College of Medicine and Dr. Y. Suzuki of Tokyo University for supplying us with pathological samples. This investigation was supported by Grants NS 09626 and RR 00164 from the National Institutes of Health.

REFERENCES

1. E. Buchner, Alkoholische gahrung ohne hefezellen, Berichte 30:117 (1897).
2. A. Harden and J. W. Young, The alcoholic ferment of yeast-juice, J. Physiol. 32:a, Proc. of November 12, 1904.

3. E. Mehl and H. Jatzkewitz, Eine cerebrosidsulfatase aus schweineniere, Hoppe-Seyler's Z. Physiol. Chem. 339:260 (1964).
4. M. W. Ho and J. S. O'Brien, Gaucher's disease: Deficiency of 'acid' β-glucosidase and reconstitution of enzyme activity in vitro, Proc. Natl. Acad. Sci. USA 68:2810 (1971).
5. Y.-T. Li, M. Y. Mazzotta, C.-C. Wan, R. Orth, and S.-C. Li, Hydrolysis of Tay-Sachs ganglioside by β-hexosaminidase A of human liver and urine, J. Biol. Chem. 248:7512 (1973).
6. G. Fischer and H. Jatzkewitz, The activator of cerebroside sulphatase. Purification from human liver and identification as a protein, Hoppe-Seyler's Z. Physiol. Chem. 356:605 (1975).
7. W. Mraz, G. Fischer, and H. Jatzkewitz, The activator of human cerebroside sulphatase. Activating effect on the acidic forms of the sulphatases from invertebrates, Hoppe-Seyler's Z. Physiol. Chem. 357:201 (1976).
8. M. W. Ho, J. S. O'Brien, N. S. Radin, and J. S. Erickson, Glucocerebrosidase: Reconstitution of activity from macromolecular components, Biochem. J. 131:173 (1973).
9. S. P. Peters, P. Coyle, C. J. Coffee, R. H. Glew, M. S. Kuhlenschmidt, L. Rosenfeld, and Y. C. Lee, Purification and properties of a heat-stable glucocerebrosidase activating factor from control and Gaucher spleen, J. Biol. Chem. 252:563 (1977).
10. S. L. Berent and N. S. Radin, β-Glucosidase activator protein from bovine spleen ("Coglucosidase"), Arch. Biochem. Biophys. 208:248 (1981).
11. D. A. Wenger and S. Roth, Isolation of an activator protein for glucocerebrosidase in control and Gaucher disease brain, Biochem. Intl. 5:705 (1982).
12. H. Christomanou, Niemann-Pick disease, Type-C: Evidence for the deficiency of an activating factor stimulating sphingomyelin and glucocerebroside degradation, Hoppe-Seyler's Z. Physiol. Chem. 361:1489 (1980).
13. S.-C. Li and Y.-T. Li, An activator stimulating the enzymic hydrolysis of sphingoglycolipids, J. Biol. Chem. 251:1159 (1976).
14. K. Inui and D. A. Wenger, Properties of a protein activator of glycosphingolipid hydrolysis isolated from the liver of a patient with GM1-gangliosidosis, type 1, Biochem. Biophys. Res. Commun. 105:745 (1982).
15. Y.-T. Li and S.-C. Li, Activator proteins related to the hydrolysis of glycosphingolipids catalyzed by lysosomal glycosidases, in: "Lysosomes in Biology and Pathology," Vol.7, J. Dingle and R. T. Dean, eds., Elsevier Biomedical Press, Amsterdam, The Netherlands (in press).
16. P. Hechtman, Characterization of an activating factor required for hydrolysis of GM2 ganglioside catalyzed by hexosaminidase A, Can. J. Biochem. 55:315 (1977).

17. E. Conzelmann and K. Sandhoff, Purification and characterization of an activator protein for the degradation of glycolipids GM2 and GA2 by hexosaminidase A, Hoppe-Seyler's Z. Physiol. Chem. 360:1837 (1979).
18. S.-C. Li, Y. Hirabayashi and Y.-T. Li, A protein activator for the enzymic hydrolysis of GM2 ganglioside, J. Biol. Chem. 256:6234 (1981).
19. Y. Hirabayashi, Y.-T. Li, and S.-C. Li, The protein activator specific for the enzymic hydrolysis of GM2 ganglioside in normal human brain and brains of three types of GM2-gangliosidosis, J. Neurochem. 40:168 (1983).
20. K. Sandhoff, K. Harzer, W. Wassle, and H. Jatzkewitz, Enzyme alterations and lipid storage in three variants of Tay-Sachs disease, J. Neurochem. 18:2469 (1971).
21. E. Conzelmann and K. Sandhoff, AB variant of infantile GM2 gangliosidosis: Deficiency of a factor necessary for stimulation of hexosaminidase A-catalyzed degradation of ganglioside GM2 and glycolipid GA2, Proc. Natl. Acad. Sci. USA 75:3979 (1978).
22. P. Hechtman, B. A. Gordon, and N. M. K. Ng Ying Kin, Deficiency of the hexosaminidase A activator protein in a case of GM2 gangliosidosis; variant AB, Pediatr. Res. 16:217 (1982).
23. J. E. Goldman, T. Yamanaka, I. Rapin, M. Adachi, K. Suzuki, and K. Suzuki, The AB-variant of GM2-gangliosidosis: Clinical, biochemical, and pathological studies of two patients, Acta Neuropathol. (Berl) 52:189 (1980).
24. S.-C. Li, Y. Hirabayashi, and Y.-T. Li, A new variant of Type-AB GM2-gangliosidosis, Biochem. Biophys. Res. Commun. 101:479 (1981).
25. Y.-T. Li, Y. Hirabayashi, and S.-C. Li, Differentiation of two variants of Type-AB GM2-gangliosidosis using chromogenic substrates. Am. J. Hum. Genet. 35:520 (1983)
26. S.-C. Li, T. Nakamura, A. Ogamo, and Y.-T. Li, Evidence for the presence of two separate protein activators for the enzymic hydrolysis of GM1 and GM2 gangliosides, J. Biol. Chem. 254: 10592 (1979).
27. Y.-T. Li, I. A. Muhiudeen, R. DeGasperi, Y. Hirabayashi, and S.-C. Li, Presence of activator proteins for the enzymic hydrolysis of GM1 and GM2 gangliosides in normal human urine, Am. J. Hum. Genet. 35:629 (1983)

GANGLIOSIDE BIOSYNTHESIS IN RAT LIVER GOLGI APPARATUS: STIMULATION BY PHOSPHATIDYLGLYCEROL AND INHIBITION BY TUNICAMYCIN

Harun K. M. Yusuf*, Gottfried Pohlentz, Günter Schwarzmann, and Konrad Sandhoff

Institut für Organische Chemie und Biochemie der Universität Bonn, Gerhard-Domagk-Str. 1, D-5300 Bonn 1 FRG. *Present address: Dept. Biochemistry, Univ. of Dacca, Dacca-2, Bangladesh

INTRODUCTION

Gangliosides are synthesized mainly in membranes in a stepwise manner by the sequential addition of individual sugars to the growing glycolipid (for review see ref. 1). A prerequisite for glycoconjugate biosynthesis in the inner compartment of the Golgi[2-5] is the entrance of sugar nucleotides from the cytosol where they are synthesized.[6-7] Recently Hirschberg and associates[4] described experimental evidence for a carrier mediated transport of CMP-NeuAc and GDP-Fuc.

All the in vitro studies reported so far on ganglioside biosynthesis[8-15] used detergents for stimulation of the overall process and for solubilization of the membranes carrying the sugar transferases. Hence any carrier mediated transport of the activated sugars is easily overlooked.

Abbreviations: CMP-NeuAc, cytidine 5'monophospho-N-acetylneuraminic acid; GDP-Fuc, guanosine 5'-diphosphofucose; PG, phosphatidylglycerol; G_{M1}, II^{3}Neu5AcGgOse$_4$Cer; G_{M2}, II^{3}Neu5AcGgOse$_3$Cer; G_{M3}, II^{3}Neu5AcLacCer; G_{D3}, II^3(Neu5Ac)$_2$LacCer; G_{D1a}, IV^3, II^3 (Neu5Ac)$_2$GgOse$_4$Cer; UDP-GalNAc, uridine 5'-diphospho-2-deoxy-2-acetamido-galactopyranose; TLC, thin layer chromatography; Dol-P, dolichylphosphate.

We tried to develop a system which allowed us to study ganglioside biosynthesis in vitro in the absence of detergents.

METHODS AND RESULTS

Preparation of Golgi vesicles

Golgi derived vesicles were isolated from rat liver essentially as described by Sandberg et al.[16] With respect to G_{M2}- and G_{D3}-synthase activities the Golgi membrane fraction was enriched 50-60 fold over the total homogenate or microsomes.[17,18] Other Golgi marker enzymes (e.g. G_{M1}- and G_{D1a}-syntheses) were also similarly enriched in the Golgi fraction. Determinations of the specific activities of glucose-6-phosphatase (marker enzyme for endoplasmic reticulum), acid phosphatase (lysosomes) and 5'-nucleotidase (plasma membrane) indicated a maximun of 3-5% contamination of the Golgi fraction with plasma membranes and less than 1% with either microsomal or lysosomal membranes.

Stimulation of ganglioside synthesis by phosphatidylglycerol (PG)

Figures 1-4 show the phosphatidylglycerol-induced stimulation of G_{M2}-, G_{M1}-, G_{D3}- and G_{D1a}-synthase activities.

Optimal stimulation (6-fold) of G_{M2}-synthase was found in the absence of detergents at a PG concentration of 0.04% (20 μg in 50 μl) with 150 M G_{M3} as lipid acceptor, 20 μM UDP-GalNAc and 50 μg Golgi protein (Fig. 1). On the other hand, optimal stimulation by the detergent octylglucoside was less than that observed with PG.[17] The stimulation by either PG or octylglucoside was almost linear up to 15 min and up to 100 μg Golgi protein. Dependence of ganglioside G_{M2}-synthesis on the UDP-GalNAc concentration and on the acceptor ganglioside G_{M3} concentration have been assayed.[17]

Other phospholipids tested were also stimulatory for G_{M2}-synthesis, but less active compared to PG: phosphatidylserine (at the optimal concentration of 0.04%), phosphatidylethanolamine (0.08%) and phosphatidylinositol (0.06%) were 47-58% as active. Phosphatidylcholine (0.02%), phosphatidic acid (0.05%) and cardiolipin (0.01%) were less stimulatory, around 30%, while sphingomyelin had no effect and lysophosphatidylcholines (both palmitoyl and stearoyl) were rather inhibitory.

Maximum stimulation (about 21 fold) of G_{M1} synthesis was found with 30 μg PG in a 50 l assay or with Triton X-100 up to 0.2% (w/v) (Fig. 2). In this assay phosphatidylethanolamine (at optimal concentration of 0.08%) was about 90% as active as PG. Phosphatidylcholine (at optimal concentration of 0.02%), on the other hand, was only 30% as active.

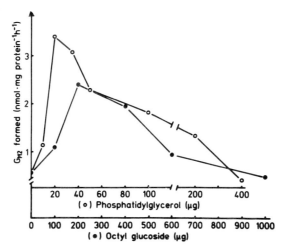

Fig. 1. Phosphatidylglycerol stimulates G_{M2} synthesis in Golgi membranes isolated from rat liver.[17] The enzyme assays were done with 150 μM G_{M3}, 64 mM cacodylate-HCl, pH 7.35, 5 mM CDP-choline, 20 mM $MnCl_2$, 0.3 M sucrose, 20 μM UDP-[^{14}C]GalNAc (47 mCi/mmol), 50 μg freshly prepared Golgi-rich membrane protein, and the indicated amounts of phosphatidylglycerol or octylglucoside, in a total volume of 50 μl. Chloroform:methanol (2:1, v/v) solutions of lipid acceptor, and of phospholipid or of the detergent were evaporated to complete dryness under N_2; buffer, salt and sucrose solution were added and the mixtures were ultrasonicated for 30 sec in a cup-horn sonicator at 100 W. Golgi membranes and the labelled sugar nucleotides were then added.

In contrast to G_{M2}- and G_{M1}-synthesis, the synthesis of ganglioside G_{D3} was virtually independent of PG (Fig. 3). For G_{D1a}-synthesis PG at an optimal concentration of 0.15% (w/v) was only one-third as active as Triton CF-54 (Fig. 4).

Product identification

The reaction products of synthase systems (G_{M2}, G_{M1}, G_{D1a}, G_{D3}) were determined to be identical to the respective authentic glycosphingolipids in TLC.[17,18]

Tunicamycin inhibition of ganglioside biosynthesis

Tunicamycin is known to be a specific inhibitor of glycoprotein biosynthesis by the Dol-P pathway. Therefore we were surprised to see that tunicamycin inhibits ganglioside G_{M2} (Fig. 5) and G_{M1}-synthesis (Fig. 6), too.[17,18]

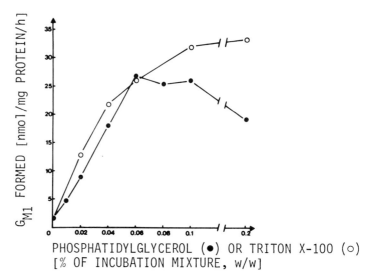

Fig. 2. Stimulation by phosphatidylglycerol of ganglioside G_{M1} synthesis in intact Golgi vesicles isolated from rat liver.[18]

The inhibition was observed only when the enzymes were measured in the presence or absence of PG but not when they were assayed in the presence of detergents (Triton X-100 or octyl-β-D-glucoside). This observation already indicates that sugar transferases themselves were not inhibited by this antibiotic. On the other hand freezing and thawing or ultrasonication of freshly prepared Golgi vesicles indicated that stimulation of G_{M1}- and G_{M2}-synthesis by PG and its inhibition by tunicamycin was highly dependent on the vesicular integrity of the Golgi membranes (Table 1). With freshly prepared unpelleted Golgi vesicles, the stimulation of G_{M2}-synthase by PG was maximal (about 5.5 fold) and the inhibition by tunicamycin was also strong (86% with 10 μg of this antibiotic). If the membranes were pelleted or frozen and thawed several times the basal enzyme activity (without phospholipid) increased, while the activity in presence of PG remained almost unchanged, so that the net stimulatory effect of this phospholipid decreased. The degree of inhibition by tunicamycin decreased concomitantly. When Golgi vesicles were disrupted by ultrasonication (4 x 1 min) prior to the incubation, phosphatidylglycerol stimulation disappeared completely (there was rather a slight inhibition) and there was virtually no inhibition by tunicamycin any more. Ultrasonication, however, resulted in about 30% loss of maximal enzyme activity.

The above results indicate that tunicamycin does not inhibit the glycosyltransferases themselves. This lends support to the

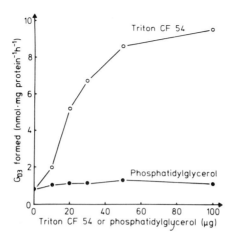

Fig. 3. Phosphatidylglycerol does not stimulate G_{D3} synthesis in rat liver Golgi membranes.[17] The enzyme assays were done with 150 μM G_{M3}, 150 mM cacodylate-HCl, pH 6.6, 10 mM $MgCl_2$, 10 mM 2-mercaptoethanol, 0.3 M sucrose, 0.5 mM CMP-[^{14}C]NeuAc (3500 cpm/nmol), 50 μg Golgi-rich membrane protein, and the indicated amounts of Triton CF-54 or phosphatidylglycerol, in a total volume of 50 μl. Further details are as in Fig. 1.

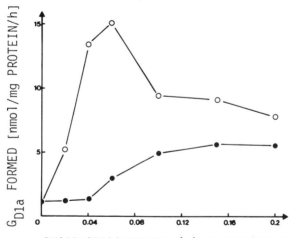

Fig. 4. Phosphatidylglycerol stimulates G_{D1a} synthesis in rat liver Golgi membranes up to one third of the Triton CF-54 stimulation. The enzyme assays were done with 150 μM G_{M1}, 150 mM cacodylate-HCl, pH 6.6, 10 mM $MgCl_2$, 10 mM 2-mercaptoethanol, 0.3 M sucrose, 0.5 mM CMP-[^{14}C]NeuAc (3500 cpm/nmol), 50 μg Golgi-rich membrane protein, and the indicated amounts of Triton CF-54 or phosphatidylglycerol, in a total volume of 50 μl. Further details are as in Fig. 1.

idea that tunicamycin might inhibit the transport of sugar nucleotides across intact Golgi membranes. Kuhn and White[19,20] have already proposed carrier systems for the transport of UDP-Gal across rat mammary gland Golgi membranes. Sommers and Hirschberg[4] have also recently reported on the existence of carrier systems mediating the transport of CMP-NeuAc and GDP-Fuc in rat liver Golgi vesicles.

Based on results obtained after pretreatment of Golgi vesicles with pronase, these authors[4] concluded that the carriers face, at least partially, the cytoplasmic side of the vesicles. Therefore we investigated the effect of pretreatment of Golgi vesicles with pronase on the subsequent effect of tunicamycin on ganglioside G_{M1} synthesis. When freshly prepared Golgi vesicles were pretreated with pronase (50 µg Golgi protein with 1 µg pronase) (Table 2), the resulting membranes lost their dependence on Triton X-100 or phosphatidylglycerol for optimal enzyme activity, and tunicamycin did not inhibit the synthetic activity any more, even in the presence of the phospholipid. The control membranes (treated exactly similarly, but without pronase), on the other hand, showed both effects, although they were now relatively small (only 3.4-3.5-fold stimulation by Triton or phospholipid and up to 72% inhibition by 5 µg tunicamycin) as compared to those found with Golgi vesicles used directly (see Figs. 2 and 6). The data of Table 2 also show that

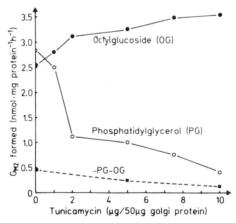

Fig. 5. Tunicamycin inhibits G_{M2} synthesis in rat liver Golgi vesicles.[17] Golgi membranes (50 µg protein) were incubated with phosphatidylglycerol (20 µg) or octylglucoside (200 µg) in the presence of varying amounts of tunicamycin. Assay conditions are as described for Fig. 1, except that the final volume was 60 µl. Reactions (15 min, 37°C) were stopped with 1 ml chloroform:methanol (2:1, v/v) and analyzed as described.[17]

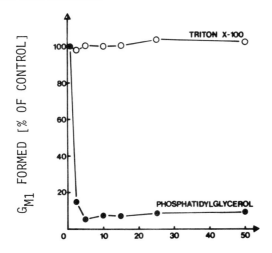

Fig. 6. Inhibition by tunicamycin of ganglioside G_{M1} synthesis in Golgi-derived vesicles isolated from rat liver.[18] 100% values in the presence of Triton X-100 (50 μg per 60 μl assay) and phosphatidylglycerol (30 μg per 60 μl assay) were 29.1 and 20.3 nmoles/mg/h, respectively.

the pronase-treated membranes retained full enzyme activity: the more than 2-fold specific activity in the pronase-treated membranes is presumably due to the fact that pronase treatment resulted in about 65% loss of vesicle protein, compared to 15% loss in controls, and that equal amounts (50 μg) of membrane protein from each group were used for subsequent enzyme assays and thus twice the amount of Golgi vesicles in case of pronase treatment.

The above results strongly suggest the existence of carriers for UDP-Gal across rat liver Golgi vesicles, which are partly or fully cleaved off by pronase, so that the resulting membranes do not show any more susceptibility towards tunicamycin. After pronase treatment the vesicles obviously became leaky for UDP-Gal but not for pronase. This is suggested by the fact that ganglioside G_{M1} synthase activity was fully retained after pronase treatment. On the other hand G_{M1}-synthase activity was reduced to 50% by pronase treatment of ultrasonicated Golgi vesicles.

To obtain further evidence that tunicamycin inhibits the transport of UDP-Gal across Golgi membranes, Golgi vesicles were incubated with UDP [^{3}H]Gal (125 μM) for 10 min at 30°C in the presence or absence of tunicamycin. After incubation (also at 0°C) the membranes were recovered by centrifugation, washed, and the soluble radioactivity associated with the pellet was extracted in water. The results on the rate of transport, after correction for

Table 1. Phosphatidylglycerol (PG) Dependence And Tunicamycin Inhibition of Ganglioside G_{M2} Synthesis In Different Golgi Membrane Preparations.

Golgi preparation	G_{M2} formed nmol/mg protein/h			Stimulation by PG (-fold)	Inhibition by tunicamycin (%)
	without PG	with PG (20 µg)	with PG (20 g) + tunicamycin (10 µg)		
Unpelleted, fresh	0.518	2.826	0.395	5.46	86.0
Pelleted, frozen and thawed	1.039	2.734	1.079	2.63	60.5
Pelleted, two times frozen and thawed	1.755	2.740	1.539	1.56	43.8
Unpelleted, 4 x 1 min ultrasonicated	2.001	1.656	1.651	none	0.3

unspecific adsorption (0°C), are presented in Fig. 7. The results show an uptake rate for UDP-Gal of about 85 pmoles/mg protein/min at 30°C. This rate was monitored from both the radioactivity that was firmly bound to the membranes (about 92% of total pellet activity) and from the water extractable radioactivity. Tunicamycin inhibited, in a concentration-dependent manner, the rate of uptake, the inhibition reaching about 71% at the antibiotic concentration of 0.28 µg per 50 µg Golgi protein. The radioactivity extractable with water represented, as indicated above, only 7-8% of total radioactivity. Tunicamycin also decreased this radioactivity to about 20% (Fig. 7).

Incubation of Golgi vesicles with UDP-Gal in the presence of phosphatidylglycerol (10 µg per 50 µg protein) did not alter either water-soluble or -insoluble radioactivity associated with the pellet, while tunicamycin (0.14 µg/50 µg protein) in this case, too decreased (by about 25%) the membrane bound radioactivity (Fig. 7).

Incubation of Golgi vesicles in the presence of phosphatidylglycerol for 10 min at 30°C did not alter (decrease or increase) either the soluble or insoluble radioactivity of the pellets, indicating either that the phospholipid does not facilitate the transport of the nucleotide sugar or that the pellets are already satur-

Table 2. G_{M1} Synthesis In Rat Liver Golgi Vesicles Pre-treated With Pronase.[18]

	G_{M1} formed (nmol/mg/h)					
Tunicamycin (μg)	0	2.5	5	0	2.5	5
substance added (per assay)	control			pronase treated		
None	2.69 (100)	1.99 (73.9)	1.53 (56.8)	24.89 (100)	23.50 (94.4)	20.46 (82.2)
Triton X-100 (40 μg/assay)	9.61 (100)	9.82 (102.2)	10.29 (107.1)	19.29 (100)	21.02 (108.8)	21.32 (110.5)
Phosphatidyl-glycerol (40 μg/assay)	9.20 (100)	5.39 (58.6)	2.55 (27.7)	18.61 (100)	17.76 (95.4)	18.20 (97.8)

Freshly prepared, unpelleted Golgi vesicles were treated with or without pronase (50 μg Golgi protein to 1 μg pronase). The membranes were then collected by centrifuging at 192,000 x g for 1 hr. The pellet surfaces were gently washed twice with ice-cold water, resuspended in 1 M sucrose, and aliquots (50 μg protein) used for enzyme assay. Recovery of protein after pronase treatment was 37.4% compared to 84.1% in the control. Figures in parentheses are the percent of values obtained in absence of tunicamycin. Average values of two different determinations are shown. UDP-[^{14}C] Gal was used in the assays.

ated with radioactivity in the absence of phospholipid after a 10 min exposure. In the latter case a stimulation of uptake of UDP-Gal by the phospholipid would not show up in the experiment. Therefore, a time-dependent analysis (0-10 min) of the transport in the presence of phospholipid should be done before any firm conclusion can be drawn.

DISCUSSION

The results presented depend critically on the integrity of the Golgi vesicles used. Several lines of evidence support the 'right-side-out' orientation of the vesicles employed.[18] Ganglioside G_{D1a}-, G_{D3}- and G_{M1}-synthases showed 12 to 21 fold higher

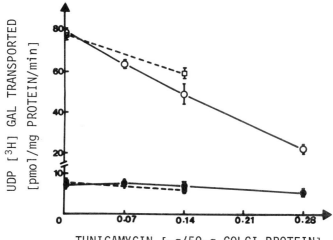

Fig. 7. Inhibition by tunicamycin of UDP-Gal transport across rat liver Golgi membranes.[18] Freshly prepared unpelleted Golgi vesicles (260 μg protein) were incubated along with UDP-[^{3}H]Gal, in the presence of the indicated amounts of tunicamycin which were added as follows: Tunicamycin (1 mg) was suspended in 192 μl water, thoroughly vortexed, and then ultrasonicated for 10 min (100 W). After centrifugation (5 min, 10,000 x g) aliquots of the clear supernatant were added to the incubation mixtures. (Tunicamycin concentration in the supernatant was determined by measuring adsorption at 260 nm). Symbols: Continued lines, without phosphatidylglycerol; o———o, membrane-bound radioactivity; •———•, soluble radioactivity. Broken lines, with phosphatidylglycerol; □---□ , membrane-bound radioactivity, ■---■ , soluble radioactivity.

activities in the presence of detergents. The basal enzyme activities (in the absence of stimulatory agents) increased whereas the activity in the presence of additives remained almost unaltered when the vesicles were pelleted, frozen and thawed several times or disrupted by ultrasonication. Furthermore freshly prepared Golgi vesicles were unable to glycosylate a macromolecular substrate like ovalbumin even in the presence of phosphatidylglycerol at the concentrations used. Disintegration of the vesicles by ultrasonication or by addition of Triton X-100 led to full activity towards ovalbumin which was not inhibited by tunicamycin.[18]

Based on the results presented in this paper, we would like to propose the following: Rat liver Golgi vesicles contain carriers that mediate the transport of UDP-Gal and UDP-GalNAc. The carriers, like those for CMP-NeuAc and GDP-Fuc,[4] face the cytoplasmic side of the vesicles. Tunicamycin inhibits ganglioside biosynthesis in Golgi membranes by blocking the carriers and not by inhibiting the transferases.

Tunicamycin should be more efficient in inhibiting the in vivo biosynthetic process, because the vesicles in this situation are by far more intact than when isolated by repeated ultracentrifugation. Recently tunicamycin was, indeed, found to inhibit strongly the in vivo biosynthesis of gangliosides in neuronal cells.[21] Although the exact mechanism of phospholipid-induced stimulation of in vitro ganglioside biosynthesis remains unclear, the system provides an excellent tool for the study of tunicamycin action on this process. Further work along this line is expected to lead to a better understanding of the mechanism in situ of ganglioside biosynthesis.

SUMMARY

Golgi vesicles were isolated and purified from rat liver, in which the specific activities of glycosyltransferases (e.g. G_{M3}: CMP-NeuAc sialytransferase, G_{D3}-synthase; G_{M3}: UDP-GalNAc galactosaminyltransferase, G_{M2}-synthase) were 50-60 times enriched relative to microsomes or total homogenate. Synthesis of gangliosides G_{M2} and G_{M1} in such Golgi vesicles is, in the absence of any detergents, stimulated 6- and 20-fold, respectively, by phosphatidylglycerol. Other phospholipids like dolichyl phosphate, phosphatidylethanolamine and phosphatidylserine are also significantly stimulatory. Tunicamycin inhibits the synthesis of gangliosides G_{M2} and G_{M1} in isolated Golgi vesicles, but only in the absence of detergents. The dependence on phosphatidylglycerol and the degree of inhibition by tunicamycin of the synthetic activities are strictly dependent on the intactness of the Golgi vesicles: both phenomena become increasingly less evident when the vesicles are pelleted, and frozen and thawed several times, and completely disappear when the vesicles are solubilized by the detergents or disrupted by ultrasonication. Furthermore, tunicamycin inhibition is reversible by increased concentration of phosphatidylglycerol.

In pronase-treated Golgi vesicles, which retain full enzyme activitiy, both phospholipid-dependence and tunicamycin inhibition of the synthetic activity disappear completely. When freshly prepared Golgi vesicles are incubated with 125 µM UDP [^{3}H]Gal for 10 min at 30°C, the nucleotide sugar is found to be transported into the vesicles at the rate of about 85 pmoles/mg protein/min, 92% of radiolabel remaining firmly bound with membrane. Tunicamycin inhibits this transport in a concentration-dependent manner. The results show that, while the mechanism of phosphatidylglycerol induced stimulation of the synthetic activity remains unclear, tunicamycin inhibits ganglioside biosynthesis by blocking the transport of the nucleotide sugar across Golgi vesicles and not by inhibiting the transferase enzyme directly.

ACKNOWLEDGEMENTS

This work was supported by a research grant (Sa 257/11-3) from the Deutsche Forschungsgemeinschaft. H. K. M. Yusuf was supported by a research fellowship of the Alexander von Humboldt Foundation in the Federal Republic of Germany.

REFERENCES

1. H. Schacter and S. Roseman, Mammalian glycosyltransferases: their role in the synthesis and function of complex carbohydrates and glycolipids, in: "The Biochemistry of Glycoproteins and Proteoglycans" W. J. Lennarz, ed., Plenum Press, New York pp. 85-160 (1980).
2. B. Fleischer, Orientation of glycoprotein galactosyltransferase enzymes in vesicles derived from rat liver Golgi apparatus, J. Cell Biol. 89:246 (1981).
3. D. J. Carey and C. B. Hirschberg, Topography of sialoglycoproteins and sialyltransferases in mouse and rat liver Golgi, J. Biol. Chem. 256:989 (1981).
4. L. W. Sommers and C. B. Hirschberg, Transport of sugar nucleotides into rat liver Golgi, J. Biol. Chem. 257:10811 (1982).
5. K. E. Creek and D. J. Morre, Translocation of cytidine 5'-monophosphosialic acid across Golgi apparatus membranes, Biochim. Biophys. Acta 643:292 (1981).
6. D. J. Carey, L. W. Sommers, and C. B. Hirschberg, CMP-N-acetylneuraminic acid: Isolation from and penetration into mouse liver microsomes, Cell 19:597 (1980).
7. S. W. Coates, T. Gurney Jr., L. W. Sommers, M. Yeh, and C. B. Hirschberg, Subcellular localization of sugar nucleotide synthetases, J. Biol. Chem. 255:9225 (1980).
8. B. Kaufman, S. Basu, and S. Roseman, Enzymatic synthesis of disialogangliosides from monosialogangliosides by sialyltransferases from embryonic chicken brain, J. Biol. Chem. 243:5803 (1968).
9. S. Basu, B. Kaufman, and S. Roseman, Enzymatic synthesis of glucocerebroside by a glycosyltransferase from embryonic chicken brain, J. Biol. Chem. 248:1388 (1973).
10. T. W. Keenan, D. J. Morre, and S. Basu, Ganglioside biosynthesis, J. Biol. Chem. 249:310 (1974).
11. J. C. Steigerwald, S. Basu, B. Kaufman, and S. Roseman, Sialic acids, J. Biol. Chem. 250:6727 (1975).
12. F. E. Wilkinson, D. J. Morre, and T. W. Keenan, Ganglioside biosynthesis. Characterization of uridine diphosphate

galactose: G_{M2} galactosyltransferase in Golgi apparatus from rat liver, J. Lipid Res. 17:146 (1976).
13. C. L. Richarson, T. W. Keenan, and D. J. Morré, Characterization of CMP-N-acetylneuraminid acid: Lactosylceramide sialyltransferase in Golgi apparatus from rat liver, Biochim. Biophys. Acta 488:88 (1977).
14. C. M. Eppler, D. J. Morré, and T. W. Keenan, Ganglioside Biosynthesis in rat liver, Biochim. Biophys. Acta 619:318 (1980).
15. H.-J. Senn, C. Cooper, P. C. Warnke, M. Wagner, and K. Decker, Ganglioside biosynthesis in rat liver, Eur. J. Biochem. 120:59 (1981).
16. P.-O. Sandberg, L. Marzella, and H. Glauman, A method for rapid isolation of rough and smooth microsomes and Golgi apparatus from rat liver in the same sucrose gradient, Exptl. Cell Res. 130:393 (1980).
17. H. K. M. Yusuf, G. Pohlentz, G. Schwarzmann, and K. Sandhoff, Ganglioside biosynthesis in rat liver Golgi apparatus: Stimulation by phosphatidylglycerol and inhibition by tunicamycin, Eur. J. Biochem. 134L47 (1983).
18. H. K. M. Yusuf, G. Pohlentz, and K. Sandhoff, Tunicamycin inhibits ganglioside biosynthesis in rat liver Golgi apparatus by blocking sugar nucleotide transport across membrane vesicles Proc. Natl. Acad. Sci. USA in press (1983).
19. N. J. Kuhn and A. White, The topography of lactose synthesis, Biochem. J. 148:77 (1975).
20. N. J. Kuhn and A. White, Evidence for specific transport of uridine diphosphate galactose across the Golgi membrane of rat mammary gland, Biochem. J. 154:243 (1976).
21. S. P. Guarnaccia, J. H. Shaper, and R. L. Schnaar, Tunicamycin inhibits ganglioside biosynthesis in neuronal cells, Proc. Natl. Acad. Sci. USA 80:1551 (1983).

INCORPORATION RATE OF GM1 GANGLIOSIDE INTO MOUSE BRAIN MYELIN:

EFFECT OF AGING AND MODIFICATION BY HORMONES AND OTHER COMPOUNDS

Susumu Ando, Yasukazu Tanaka, Yuriko Ono, and Kazuo Kon

Department of Biochemistry, Tokyo Metropolitan
Institute of Gerontology, Itabashi-ku, Tokyo

INTRODUCTION

Myelin formation in the rat brain is activated at a particular postnatal age of between two to five weeks, and the rate is remarkably reduced to very low level in adulthood.[1] Hence, myelin once formed had been considered to be metabolically inactive or stable. However, it has been found that considerable incorporation of newly synthesized material into myelin still occurs after the active myelinogenesis period.[2] Structural features of myelin membranes may limit the replacement of the constituents or the remodeling of the membranes, and indeed, the turnover of myelin components was reported to be much slower than those of the corresponding molecules in microsomes.[2,3] The metabolic activity of GM1 ganglioside in myelin of young rat brain was compared with the activity in the whole brain by Suzuki,[4] who demonstrated that the labeled GM1 accumulated at a slower rate in myelin for the first one month after the injection of radioactive glucosamine.

Turnover rates reported to date vary over a wide range, depending on the types of radioactive precursors used and the experimental design (e.g., intervals at which the decay of radioactivities are determined).[2] One of the most serious problems is the recycling of radioactive labels, leading to apparently longer turnover rates. Although 2-tritiated glycerol is considered to be a good pulse label for brain phospholipids because of reduced recycling,[5] ideal precursors for metabolic studies of sphingolipids have not yet been found. We have developed a method by which newly synthesized molecules are almost universally labeled and the molecular species formed at the beginning of the experiment can be discriminated by calculation from the species formed later due to recycling of the label.[6] This method, involving deuterium labeling by deuterium oxide and determination of deutered molecules by mass fragmentography, has the following advantages as compared to ordinary radioisotope labeling experiments: 1, deuterium oxide may not seriously disturb precursor pools; 2, different compounds are universally labeled; 3, absolute amounts of newly synthesized molecules are determined without data on the specific activities of the particular precursor pool; 4, the proportion of newly formed to preexisting molecules, that is, the replacement rate, can be calculated; 5, very accurate determination of labeled molecules is possible by mass fragmentography. In the present study we have examined the metabolic activity of GM1 in myelin by this method in order to determine how it is affected by aging or by various chemical agents.

MATERIALS AND METHODS

Different age groups of C57BL/6 mice were supplied from the aging colony of this institute. Each group was given 30% deuterium oxide as drinking water for four to six weeks, and maintained with a laboratory chow diet. Cerebella were taken from mice at intervals of one or two weeks during the administration of deuterium oxide. Myelin was obtained by the method of Norton and Poduslo.[7] Total lipids extracted were separated into neutral and acidic fractions by DEAE-Sephadex chromatography.[8] The acidic fraction was subjected to a mild base treatment and applied to a Sephadex LH-20 column (Pharmacia Fine Chem., Sweden). Gangliosides were eluted with methanol from the column, and were freed from most of the sulfatide and salts (recovery of gangliosides, 76%; unpublished method). The total ganglioside sample was injected into a high performance liquid chromatograph equipped with a silica gel column (250 x 8 mm i.d.; theoretical plates, 20,000), and gangliosides were eluted with a mixed solvent of isopropanol/n-hexane/ethanol/aq 10 mM NaCl (55:17:10:18) (unpublished, basically according to the method of Watanabe and Arao[9]). Purified GM1 (about 10 μg as sialic acid) was permethylated with methyliodide/NaH in dimethylformamide.[10] A half of the product was methanolyzed at 100° with 3% HCl in methanol to produce permethylated neuraminic acid. The other half was hydrolyzed for 20 hours at 80° with 0.8N sulfuric acid in 80% acetic acid, and derivatized to alditol acetates.[11] These monosaccharide samples were introduced into a gas chromatograph-mass spectrometer equipped with a 1% OV-17 column (Shimadzu-LKB 900B, Shimadzu Scientific Co., Kyoto) which was operated in the chemical ionization mode with ammonia as a reactant gas. Quasi-molecular ions corresponding to deuterated and non-deuterated sugars were monitored, and their abundance was determined by mass fragmentography. Molar distribution patterns of deuterated molecules were analyzed by curve fitting using the Bernoulli distribution equation.[6]

INCORPORATION RATE OF GM1 GANGLIOSIDE

RESULTS AND DISCUSSION

The method used here was basically established in the previous study.[6] In this study a chemical ionization (CI) technique was employed to obtain ions (quasi-molecular ions) representing the intact molecular sizes, because electron impact (EI) ionization causes intense fragmentation. Fig. 1 compares the mass spectra of 2,6-O-methyl-1,3,4,5-O-acetylgalactitol and permethylated neuraminic acid taken in the EI and CI modes. Sensitivity for detection of

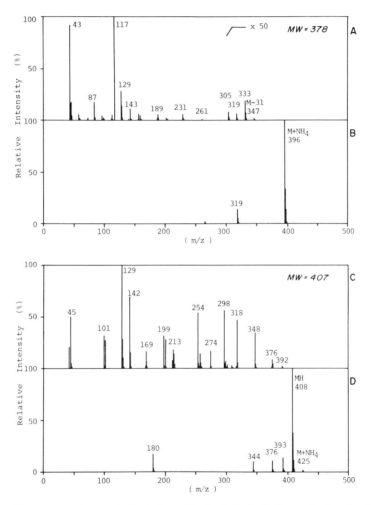

Fig. 1 Mass spectra of 2,6-O-methyl-1,3,4,5-O-acetylgalactitol (A,B) and permethylated N-methyl,acetylneuraminic acid (C,D). A,C were taken in the EI mode, and B,D in the CI mode using ammonia gas.

these sugar derivatives appears to be remarkably enhanced by employing the ammonia CI method. For routine measurement in mass fragmentography, 10-20 ng of sample was injected per peak. This method provides excellent reproducibility in measurement, that is, less than 0.2% variation.

Myelin was isolated from mice given deuterium oxide, and each myelin ganglioside was isolated by high performance liquid chromatography. Purified GM1 was permethylated, followed by degradation to provide monomeric sugar derivatives, permethylated neuraminic acid and alditol acetates. This method enables one to distinguish internal and external (non-reducing end) galactoses from each other as well as glucose, galactosamine and sialic acid. The molar distribution patterns of the deuterated species of these sugars were delineated by the procedure reported previously.[6] Calculation by curve fitting with the Bernoulli distribution equation revealed that portions of non-deuterated species were included in the newly synthesized species. Therefore, the total percentage of newly formed species including the above non-deuterated fraction was taken as the percent replacement rate of the corresponding pool in myelin with new molecules. The fractional replacement rate may indicate actual incorporation or turnover rate of a membrane component in terms of absolute rate. It is difficult to obtain such an absolute rate from usually designed radioisotope experiments.[2]

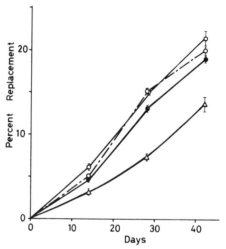

Fig. 2. Percent replacement rates of newly synthesized sugar moieties in GM1. Percentages of deuterated molecules were determined by mass fragmentography. Symbols used are as follows: o———o , internal galactose; o—--—o , external galactose; •———• , glucose; and -△————△ , neuraminic acid.

Table 1. Changes in Replacement Rate by Newly Synthesized Molecules of the Sugar Components in GM1

Age (day)	49	230	490
Glc	11.3	4.1	3.2
int-Gal	12.1	4.9	4.0
ext-Gal	14.2	4.6	3.5
NeuAc	4.2	2.3	2.0

The rate of incorporation of GM1 into myelin membranes can be deduced from the data on the replacement rates of the sugar components. Fig. 2 shows the time courses of the replacement of each sugar component in myelin GM1 isolated from 230 day-old mice. The decreasing trends in replacement rates of the sugar components of GM1 with advancing age are shown in Table 1. Neutral sugar moieties appeared to be replaced more rapidly than sialic acid. The apparently slow replacement with newly formed sialic acid may imply an effect of dilution by non-labeled molecules. If one assumes that galactose moieties were little reutilized and that the replacement rate of extGal is similar to the incorporation rate of GM1 (Barkai[12] suggested that carbohydrate half-lives were closely related to the turnover rates of glycolipids), it can be suppposed that about 50% of GM1 incorporated into myelin carried newly synthesized sialic acid and that the remaining 50% reutilized previously existing sialic acid released from somewhere else. Ferwerda et al.[13] also suggested the reutilization of sialic acid based on the finding that the specific radioactivities of CMP-sialic acid and bound sialic acid remained at the same level in rat whole brain after injection of radioactive N-acetylmannosamine. Slower replacement rates for glucose than for the two galactoses (Table 1) may indicate that utilization of preexisting glucocerebroside occurs to some extent (about 20%).

In order to investigate the possibility that metabolic activities of myelin membranes could be affected by exogenous factors, the replacement rates of GM1 were determined by the present method in animals treated with chemical agents. Fig. 3 shows the data obtained with cerebral white matter GM1 of mice treated as indicated. The mice treated with thyroxine showed increased incorporation of GM1, whereas decreased tendency in the incorporation was observed in the PTU treated mice. It is well known that myelin synthesis in the hyperthyroid state is initiated earlier but is terminated earlier too.[14] Our results suggest that the turnover of adult myelin is also enhanced by thyroid hormone. Estradiol was reported to enhance the cerebroside sulfotransferase activity which is related to myelination.[15] S-Adenosylmethionine is supposed to increase the membrane fluidity[16] which might accelerate the mem-

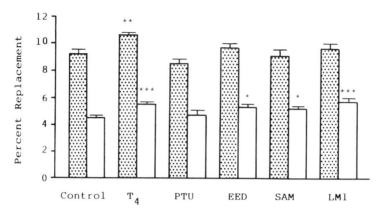

Fig. 3 Effects of hormones and various compounds on the incorporation rates of GM1 (▨▨▨, for external galactose moiety; ▭, for neuraminic acid). Adult C57BL mice (30 week-old, female) were used. The following chemicals were given for four weeks: T_4 (L-thyroxine), 0.001% in drinking water; PTU (6-n-propyl-2-thiouracil), 0.05% in drinking water; EED (17α-ethynylestradiol), 3 μg/day; SAM (S-adenosylmethionine, 1 mg/day; and ganglioside LM1 (sialosylparagloboside), 30 μg/day. Statistical significance vs control: *, $p<0.05$; **, $p<0.01$; and ***, $p<0.005$.

brane turnover. Gangliosides were recently tested for their activities in neurite formation.[17] In this study the latter three compounds seemed to show only minor effects on the GM1 incorporation. Thus, it is suggested that the metabolic rates of myelin membranes, and probably those of other CNS membranes too, may be modified by exogenous chemical factors, and this raises the possibility that some alterations that occur with aging[18] and the impairment of membrane function due to pathological events may be reversed by administration of suitable agents.

SUMMARY

The turnover rate of GM1 ganglioside in myelin was examined to reveal age-related alterations in the metabolic activity. Three different age groups of mice were given deuterium oxide, and myelin was prepared from cerebella at intervals of two weeks. GM1 was isolated from the total myelin gangliosides by high performance liquid chromatography. Deuterated sugar moieties of GM1 were determined by chemical ionization-mass spectrometry which provided prominent quasimolecular ions. This method made it possible to determine separately the incorporation of deuterium into internal and external galactoses as well as other components. The incorporation rate of GM1 into myelin was clearly shown to be decreased with advancing age. Lower incorporation of newly synthesized sialic acid into GM1 than that of other sugars may indicate reutilization of sialic acid at about 50%.

The possibility of modification of the myelin metabolism by exogenous factors was examined by monitoring the incorporation rate of GM1 in animals treated with chemical agents. It was revealed that thyroxine enhanced the incorporation of GM1 into adult brain myelin, whereas propylthiouracil reduced the incorporation. Other chemicals, estradiol, S-adenosylmethionine and LM1 ganglioside, showed only minor effects on the myelin turnover.

REFERENCES

1. W. T. Norton and S. E. Poduslo, Myelination in rat brain: Changes in myelin composition during brain maturation, J. Neurochem. 21:759 (1979).
2. J. A. Benjamins and M. E. Smith, Metabolism of myelin, in: "Myelin," P. Morell, ed., Plenum Press, N.Y. (1977).
3. S. L. Miller, J. A. Benjamins, and P. Morell, Metabolism of glycerophospholipids of myelin and microsomes in rat brain: Reutilization of precursors, J. Biol. Chem. 252:4025 (1977).
4. K. Suzuki, Formation and turnover of myelin ganglioside, J. Neurochem. 17:209 (1970).
5. J. A. Benjamins and G. M. McKhann, [2-^{3}H]Glycerol as a precursor of phospholipids in rat brain: Evidence for lack of recycling, J. Neurochem. 20:1111 (1973).
6. Y. Tanaka and S. Ando, Increased stearate formation by chain elongation in streptozotocin diabetic mice: Compensation for decreased de novo fatty acid synthesis, Biomed. Res. 2:404 (1981).
7. W. T. Norton and S. E. Poduslo, Myelination in rat brain: Method of myelin isolation, J. Neurochem. 21:749 (1973).
8. R. W. Ledeen, R. K. Yu, and L. F. Eng, Gangliosides of human myelin: Sialosylgalactosylceramide (G7) as a major component, J. Neurochem. 21:829 (1973).
9. K. Watanabe and Y. Arao, A new solvent system for the separation of neutral glycosphingolipids, J. Lipid Res. 22:1020 (1981).
10. T. Imanari and Z. Tamura, Gas chromatography of glucuronides, Chem. Pharm. Bull. 15:1677 (1967).
11. W. Stoffel and P. Hanfland, Analysis of amino sugar-containing glycosphingolipids by combined gas liquid chromatography and mass spectrometry, Z. Physiol. Chem. 354:21 (1973).
12. A. I. Barkai, Turnover in vivo of lactosylceramide and other glycosphingolipids in the adult rat brain, J. Neurochem. 36:317 (1981).

13. W. Ferwerda, C. M. Blok, and J. Heijlman, Turnover of free sialic acid, CMP-sialic acid, and bound sialic acid in rat brain, J. Neurochem. 36:1492 (1981).
14. S. N. Walters and P. Morell, Effects of altered thyroid states on myelinogenesis, J. Neurochem. 36:1792 (1981).
15. H. P. Siegrist, T. Burkart, K. Hoffmann, U. Wiesmann, and N. Hershkowitz, Influence of reduced cholesterol synthesis on the activity of cerebroside sulfotransferase in cultured glioblastoma cells treated with estradiol, Biochim. Biophys. Acta 572:160 (1979).
16. F. Hirata and J. Axelrod, Enzymatic methylation of phosphatidylethanolamine increases erythrocyte membrane fluidity, Nature 21:219 (1978).
17. J. I. Morgan and W. Seifert, Growth factors and gangliosides: A possible new perspective in neuronal growth control, J. Supramol. Struct. 10:111 (1979).
18. Y. Ono, S. Ando, Y. Tanaka, and R. W. Ledeen, Myelin metabolism and aging, Proc. Jap. Conf. Biochem. Lipids 24:353 (1982).

BIOSYNTHESIS IN VITRO OF GANGLIOSIDES CONTAINING Gg- AND Lc-CORES

Subhash Basu, Manju Basu, John W. Kyle and
Hung-Che Chon

Department of Chemistry
Biochemistry, Biophysics and Molecular Biology Program
University of Notre Dame, Notre Dame, Indiana 46556

INTRODUCTION

The structures of at least 40 different gangliosides containing N-acetylgalactosamine[1-7] and N-acetylglucosamine[8-21] have been reported by several laboratories, in addition to four hexosamine-free gangliosides (GM4, GM3, GD3 and GT3).[22-26] All of these gangliosides are widely distributed in animal tissues, particularly in synaptic membranes and on erythrocyte surfaces. In general, the gangliosides may be reclassified as acidic glycosphingolipids (acidic GSLs) containing N-glycolyl-(NeuGc) or N-acetylneuraminic (NeuAc) moieties attached to two distinct oligosaccharide core structures. The gangliosides containing the gangliotetraosylceramide (GgOse$_4$Cer; Galβ1-3GalNAcβ1-4Galβ1-4Glc-Cer) core appears to be ubiquitous in neuronal cells. However, extraneuronal tissues and erythrocytes of various species contain gangliosides with neo-lactotetraosylceramide (Galβ1-4GlcNAcβ1-3Galβ1-4Glc-Cer; nLcOse$_4$Cer) as the core structure (Fig. 1). At least three different galactosyl linkages (Galβ1-4Glc-R, Galβ1-3GalNAc-R' and Galβ1-4GlcNAc-R'') (Fig. 2) and three different sialyl linkages (NeuAcα2-3Galβ1-4Glc-R', NeuAcα2-8NeuAcα2-3Gal-R'' and NeuAcα2-3Galβ1-4GlcNAc-R''') (Fig. 3) are present in gangliosides of the Gg- and Lc-series.

On the basis of our previous studies with an embryonic chicken brain membrane system, we have proposed schemes for the stepwise biosynthesis of GD1a[27-30] and LM1[31-34] gangliosides starting from ceramide (Fig. 4). Based on substrate competition, kinetic and heat inactivation studies, it was already apparent that these act-

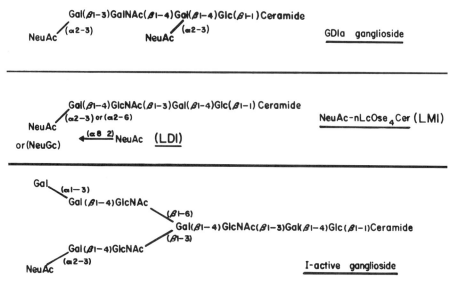

Fig. 1. Gangliosides of animal cell surfaces.

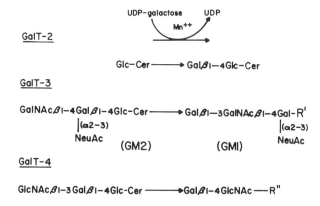

Fig. 2. Three galactosyltransferase activities (embryonic chicken brain).

ivities probably are catalyzed by different proteins. Solubilization of these synthetic enzyme activities from membranes, using various detergents, and their separation by different chromatographic methods and isoelectric focusing are under way in our laboratory. These membrane-bound glycosyltransferases seem to contain both lipophilic and lipophobic centers. One particular detergent may be capable of maintaining the structure-function integrity of the enzyme, even though many detergents can solubilize individual specific glycosyltransferases. We focused on the purification of a functionally active individual glycosyltransferase (e.g., GalT-4 or SAT-3) that was free of other activities (e.g., GalT-2^{35} or GalT-3^{36}) to prove that different glycosyltransferases catalyze different linkages.

MATERIALS AND METHODS

Glycolipid:glycosyltransferases from embryonic chicken brain membrane fraction

Fresh frozen 9- to 13-day-old embryonic chicken brains were homogenized in 3 to 6 volumes of 0.32 M sucrose containing 0.1% 2-mercaptoethanol and 1 mM EDTA, pH 7.0 (SME). The homogenate was centrifuged at 12,000 x g for 30 min, and the GSL:glycosyltransferase activities were located in a light pink membrane layer (the buffy coat) overlying a packed, reddish, heavy lower pellet. The buffy coat (containing 57% GalT-3, 64% GalT-4, and 86% SAT-3 activities) was treated with 0.2 to 0.5% detergent (Nonidet P-40 for GalT-4; Triton CF-54 for SAT-3) for 1 hour at 4°C (Fig. 5). Centrifugation (106,000 x g; 2 hr) of the detergent-treated buffy coat effectively removed particulate matter from the detergent-soluble supernatant (DSS). This solubilization procedure resulted in 40 to 50% recovery of GalT-3, GalT-4 and SAT-3 activities and yielded a homogeneous solution for further micro-isoelectric focusing and column chromatographic separation.

Micro-isoelectric focusing of GSL:galactosyl- and sialyltransferase activities

Isoelectric focusing conveniently purifies and separates enzymes present in our detergent soluble supernatant (DSS). Glass tubes (0.5 x 18 cm) were sealed^{37} at one end with 0.5 ml of polyacrylamide plugs containing 2% ampholine (pH 3 to 10 or 4 to 8) in a solution of 15% polyacrylamide, 0.5% (w/v) N,N'-dimethylene bisacrylamide and 1 µl TEMED (per ml). A discontinuous gradient of 10%, 20% and 30% sucrose (each containing 2% ampholine, pH 3 to 9) was set up in each tube with 20% or 30% sucrose containing the DSS fraction (Fig. 6). The tubes were then placed in disc gel electrophoresis apparatus. Electrofocusing with 0.1 N acetic acid as the anolyte and 0.1 N NaOH as the catholyte was carried out first at

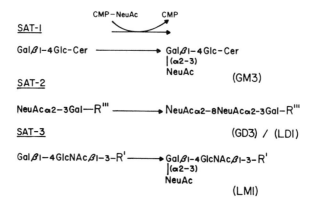

Fig. 3. Three sialyltransferase activities (embryonic chicken brain).

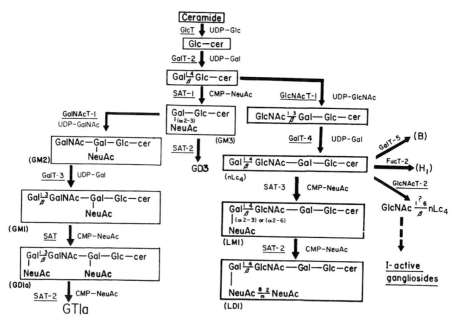

Fig. 4. Proposed pathways for biosynthesis of gangliosides.

200 volts for 2 hrs, then raised to 300 volts for 2 hrs and finally continued at 400 volts for 4 to 12 hours with approximately 5 milliamps per gradient. Each gradient was then fractionated from the bottom, using a peristaltic pump. Approximately 12 to 14 fractions of 0.2 ml each were collected by inserting a needle through the polyacrylamide plug. Each fraction was assayed for GalT-4 and SAT-3 activities (Fig. 7).

Separation of GSL:galactosyltransferase activities by DEAE-Sepharose CL-6B column chromatography

The detergent solubilized supernatant (DSS), containing 50 to 60 mg of protein, was applied to a DEAE-Sepharose CL-6B column (1 x 9 cm) that had been equilibriated with 10 mM Tris-HCl buffer, pH 7.6, containing 3 mM 2-mercaptoethanol and 0.5% Nonidet P-40 (Fig. 8). Nonadsorbed protein was washed from the column with approximately 30 ml of the same buffer which was then used in 40-ml linear KCl gradient from 0 to 400 mM at pH 7.6 to elute the GalT-4 and GalT-3 activities. A flow rate of 0.5 ml/min was maintained throughout development of the column and approximately 1.2 ml fractions were collected. Each fraction was assayed for GalT-4[32,38] and GalT-3[32,36] activities as described before. The active fractions were then stored at -18°C in 50% glycerol to prevent freezing.

RESULTS AND DISCUSSION

Solubilization and characterization of specific GSL:galactosyltransferase activities

At least three different galactosyltransferase catalyzed reactions (Fig. 2) have been characterized in a membrane preparation isolated from embryonic chicken brain. All could be partially solubilized (GalT-2, 68%; GalT-3, 63%; GalT-4, 80%) from the buffy coat layer in the presence of 0.5% Nonidet P-40. An enzyme fraction obtained after gel filtration on Biogel A-1.5 M[39] with Tris-HCl (pH 7.6) buffer containing 0.5% Nonidet P-40, or from a DEAE-Sepharose CL-6B (Fig. 8) was free of both GalT-2[31,35] and GalT-3[31,36] activities. Our GalT-4 activity has been tested with LcOse$_3$Cer (GlcNAc β 1-3Gal β 1-4Glc-Cer) (K_m = 0.1 mM) and p-NO$_2$-Phenyl- β -GlcNAc (K_m = 0.14 mM) as acceptors[39] (J. Kyle, M. Basu, and S. Basu, manuscript in preparation). The latter substrate had a 3- to 5-fold higher V_{max} under similar conditions in the presence of 10.0 mM Mn^{2+} at pH 7.1. However, removal of detergent with Biobead SM-2 resulted in a soluble enzyme preparation that did not require detergent for optimum activity with p-NO$_2$-phenyl-β -GlcNAc as substrate.

GalT-4 activity is not inhibited by the addition of various concentrations of Glc-Cer (Table 1), indicating that the two GSL:

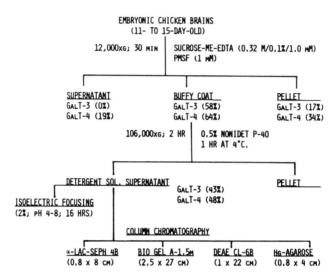

Fig. 5. Procedure for isolation of ECB GSL:Gal transferases.

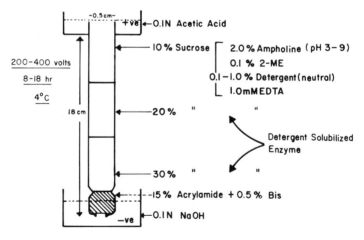

Fig. 6. Micro-isoelectric focusing.

galactosyltransferase activities GalT-2 and GalT-3 are different. The absence of GalT-2 activity from this partially purified soluble GalT-4 fraction was established using different detergents (Cutscum, Triton X-100 and Nonidet P-40) at various concentrations. We have also separated[40] (J. Kyle and S. Basu, manuscript in preparation) GalT-3 and GalT-4 activities from the detergent-soluble supernatant fraction (DSS), using an α-lactalbumin-Sepharose 4B affinity column. Recently Kaplan and Hechtman[41] separated multiple forms of GalT-3 activity from a detergent solubilized supernatant fraction from rat liver Golgi membrane. They did not try to characterize GalT-4 activity in these GalT-3 fractions. Multiple forms of GalT-3 have not yet been detected in embryonic chicken brain systems. In all of our preparation we have used 1.0 mM PMSF. In the absence of PMSF or other proteolytic inhibitors, the rat liver system[41] may have yielded multiple forms of GalT-3 artificially. In order to establish the exact linkage between the terminal galactose and the rest of the chain in GM1 or nLcOse$_4$Cer synthesized enzymatically, we first focused our attention on the separation of GalT-3 and GalT-4 activities. We identified GalT-4, galactosyltransferase activity with LcOse$_3$Cer as acceptor in bovine,[42] rabbit[43] and human[44] sera. We were able to elute the human serum galactosyltransferase that catalyzes synthesis of lactosamine and nLcOse$_4$Cer from a Sepharose 4B-p-aminophenyl-N-acetylglucosamine column. Purified human serum GalT-4 catalyzed the synthesis only of 0.2% lactosylceramide[44] but not of nLcOse$_4$Cer. Whether an activator protein like α-lactalbumin is needed for GalT-4 activity or whether a separate protein (GalT-2) catalyzes the synthesis of lactosylceramide is not yet known. Inhibition of GalT-2 activity in the presence of α-lactalbumin has also been reported in a serum system.[45]

Table 1. Mixed Substrate Competition with Purified GalT-4; DEAE CL-6B Fraction Free of GalT-3

Substrate	^{14}C-Galactose Incorporated		
	'v'	Expected Values For Two Enzymes	% Inhibition
	(nmol/ml/2 hr)		
GlcNAc β 1-3Gal β 1-4Glc-Cer (Lc$_3$)	25.6		
p-NO$_2$-∅-β -GlcNAc (PNPG)	27.2		
Glc β 1-1Cer	0.6		
Lc$_3$ (1.6 mM) + Glc-Cer (0.5 mM)	23.8	26.2	9.0
PNPG (0.2 mM) + Glc-Cer (0.5 mM)	29.4	29.8	1.0

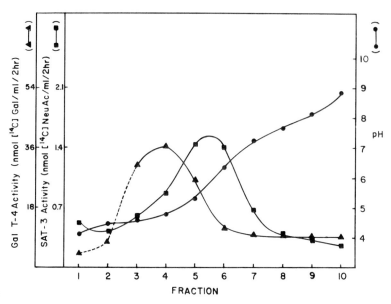

Fig. 7. Isoelectric focusing of GSL:glycosyltransferases (11 day embryonic chicken brain).

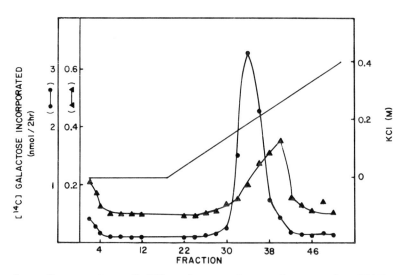

Fig. 8. Separation of GSL:galactosyltransferases by DEAE anion exchange chromatography (embryonic chicken brain).

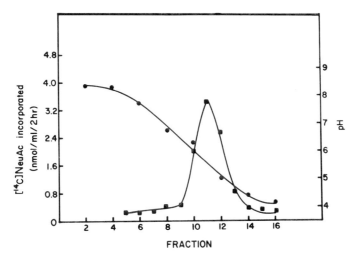

Fig. 9. Isoelectric focusing of GSL:sialyltransferase (SAT-3) (11 day-old embryonic chicken brain).

Solubilization and characterization of specific GSL:sialyltransferases

SAT-3 activity has been solubilized (86%) together with SAT-1 (71%) and SAT-2 (85%) activities from the buffy coat layer described above, using the nonionic detergent Triton CF-54 (0.2 to 0.5%). Our recent competitive studies[34] suggest that the same SAT-3 catalyzes the transfer of NeuAc from CMP-[^{14}C]NeuAc to nLcOse$_4$Cer Galβ 1-4GlcNAcβ 1-3Galβ 1-4Glc-Cer and GgOse$_4$Cer (Galβ 1-3GalNAcβ 1-4Galβ 1-4Glc-Cer). However, unambiguous proof requires prior separation of SAT-3 activity from SAT-2[31,33] and SAT-1[27,31] activities. Using micro-isoelectric focusing (Fig. 9) we have isolated an enzyme fraction that contains mostly SAT-3 activity (pI= 5.5) and little or no SAT-1 or SAT-2 activities. Further resolution of SAT-3 in two peaks has not yet been achieved. The terminal linkage of sialic acid in the enzymatic product, LM1 and GM1b[46] has been proved to be 2-3 in embryonic chicken brain. It will be of interest to determine whether the SAT-3 activity isolated from embryonic chicken brain recognizes Galβ 1-4GlcNAc- or Galβ 1-3GalNAc- as the terminal domain of an oligosaccharide. On the other hand, SAT-3 isolated from embryonic chicken brain does not act in the presence of an oligosaccharide containing a terminal Galβ 1-4Glc- as present in lactosylceramide.

Two GSL:N-acetylglucosaminyltransferases [GlcNAcT-2(β 1-3) and GlcNAcT-3(β 1-6)] which are involved in the biosynthesis of Ii-core gangliosides (Fig. 4) have been solubilized from mouse T-lymphoma (P-1798) with Triton CF-54 (0.1%). The GalT-4 activity has also been separated from the GlcNAcT activities using DEAE-ion exchange column chromatography.[47]

SUMMARY

On the basis of our previous and present studies with embryonic chicken brain system, we have proposed stepwise biosynthesis of GD1a (Gg-series) and LD1 (Lc-series) gangliosides, starting from ceramide (Fig. 4). At least three different galactosyltransferases GalT-2 (UDP-Gal:Glc-Cer), GalT-3(UDP-Gal:GM2) and GalT-4(UDP-Gal:LcOse$_3$-Cer) and three different sialyltransferases SAT-1(CMP-NeuAc:Lac-Cer), SAT-2(CMP-NeuAc:GM3) and SAT-3(CMP-NeuAc:nLcOse$_4$ Cer) are involved in the biosynthesis in vitro of these gangliosides. All six of these glycosyltransferases have been solubilized using nonionic detergents. Two forms of glycolipid:galactosyltransferases (GalT-3 and GalT-4) have been separated by DEAE sepharose CL-6B chromatography from solubilized supernatant of 11- to 13-day-old embryonic chicken brain. Using microisoelectric focusing (pH gradient 3 to 8) the galactosyltransferases (GalT-3 and GalT-4) have been separated from SAT-3. Two β-N-acetylglucosaminyltransferases (GlcNAcT-2(UDP-GlcNAc:nLcOse$_4$Cer(β 1-3)) and GlcNAcT-3(UDP-GlcNAc:nLcOse$_4$Cer(β 1-6)) have also been solubilized from mouse T-lymphoma, P-1798, using Triton CF-54. These enzymes are involved in the synthesis of Ii-core gangliosides and ^{3}H-products have been characterized by methylation studies. Further separation of these two GlcNAcT's are in progress.

ACKNOWLEDGMENTS

This work was supported in part by NIH Grants NS-18005 and CA-14764 to S. Basu and R-23-CA-33751 to M. Basu. We wish to thank Mr. Thomas A. Brown for help in the isolation and purification of glycosidases. The authors are grateful to Dr. Michael Potter of the National Cancer Institute for his gift samples of P-1798 mouse lymphomas.

REFERENCES

1. L. Svennerholm, The chemical structure of normal human brain and Tay Sachs gangliosides, Biochem. Biophys. Res. Commun. 9:436 (1962).
2. R. Kuhn and H. Wiegandt, Die konstitution der gangliotetraose und des gangliosidе G1, Chem. Ber. 96:866 (1963).
3. R. Ledeen and K. Salsman, Structure of the Tay Sachs ganglioside. I., Biochemistry 4:2225 (1965).
4. L. Svennerholm, J. E. Mansson, and Y. T. Li, Isolation and structural determination of a novel ganglioside, a disialo-

pentahexosylceradmide from human brain, J. Biol. Chem. 248: 740 (1973).
5. S. Ando and R. K. Yu, Isolation and characterization of a novel trisialoganglioside, GT1a, from human brain, J. Biol. Chem. 252:6247 (1977).
6. M. Iwamori and Y. Nagai, Isolation and characterization of a novel ganglioside monosialosylpentahexosylceramide from human brain, J. Biochem. 84:1601 (1978).
7. P. Fredman, J. E. Mansson, L. Svennerholm, K. A. Karlsson, I. Pascher, and B. E. Samuelson, The structure of the tetrasialoganglioside from human brain, FEBS Lett. 110:80 (1980).
8. H. Wiegandt and B. Schultz, Spleen gangliosides: The structure of ganglioside $G_{LNnT}1$ (NGNA), Z. Naturforsch Teil B 24b:945 (1969).
9. H. Wiegandt and H. W. Bucking, Carbohydrate components of extraneuronal gangliosides from bovine and human spleen and bovine kidney, Eur. J. Biochem. 15:287 (1970).
10. H. Wiegandt, Gangliosides of extraneural organs, Hoppe Seyler's Z. Physiol. Chem. Bd. 354:1049 (1973).
11. Y. T. Li, J. E. Mansson, M. T. Vanier, and L. Svennerholm, Structure of the major glucosamine containing ganglioside of human tissue, J. Biol. Chem, 248:2634 (1973).
12. J. L. Chien, S. C. Li, R. A. Laine, and Y. T. Li, Characterization of gangliosides from bovine erythrocyte membranes, J. Biol. Chem. 253:4031 (1978).
13. K. Uemura, M. Yuzawa, and T. Taketomi, Characterization of major glycolipids in bovine erythrocyte membrane, J. Biochem. 83:463 (1978).
14. H. Rauvala, T. Krusius, and J. Finne, Disialosylparagloboside, a novel ganglioside isolated from human kidney, Biochim. Biophys. Acta 531:266 (1978).
15. A. Suzuki, I. Ishizuka, and T. Yamakawa, Isolation and characterization of a ganglioside containing fucose from boar testis, J. Biochem. 78:947 (1975).
16. M. Ohashi and T. Yamakawa, Isolation and characterization of glycosphingolipids in pig adipose tissue, J. Biochem. 81:1675 (1977).
17. R. Ghidoni, S. Sonnino, G. Tettamanti, H. Wiegandt, and V. Zambotti, On the structure of two new gangliosides from beef brain, J. Neurochem. 27:511 (1976).
18. J. L. Magnani, M. Brockhaus, D. F. Smith, V. Ginsburg, M. Blaszcyk, K. F. Mitchell, Z. Steplewski, and H. Kaprowski, A monosialoganglioside is a monoclonal antibody defined antigen of colon carcinoma, Science 212:55 (1981).
19. K. Watanabe, S. Hakomori, R. A. Childs, and T. Feizi, Characterization of a blood group I active ganglioside, J. Biol. Chem. 254:3221 (1979).
20. K. Watanabe and S. Hakomori, Gangliosides of human erythrocyte, Biochemistry 24:5502 (1979).

21. K. Watanabe, M. E. Powell, and S. Hakomori, Isolation and characterization of gangliosides with a new sialosyl linkage and core structures, J. Biol. Chem. 254:8223 (1979).
22. R. Kuhn and H. Wiegandt, Weitere gangliosides aus menschenhirn, Z. Naturforsch Teil B 19b:256 (1979).
23. R. W. Ledeen, R. K. Yu, and L. F. Eng, Gangliosides of human myelin, J. Neurochem. 21:829 (1973).
24. T. Yamakawa and S. Suzuki, The chemistry of the lipids of posthemolytic residue or stroma of erythrocytes. J. Biochem. 38:199 (1951).
25. E. Klenk and L. Georgias, Uber zwei weitere komponenten des gemisches der gehirnganglioside, Hoppe Seylers Z. Physiol. Chem. 348:1261 (1967).
26. H. Wiegandt, The gangliosides, in: "Adv. Neurochem." B. W. Agranoff and M. H. Aprison, eds., Plenum Publishing Co., New York (1982).
27. S. Basu, Studies on the biosynthesis of gangliosides. Ph.D. thesis, University of Michigan, Ann Arbor (1966).
28. S. Basu, B. Kaufman, and S. Roseman, Enzymatic synthesis of glucocerebroside by a glucosyltransferase from embryonic chicken brain, J. Biol. Chem. 248:1388 (1973).
29. J. C. Steigerwald, S. Basu, B. Kaufman, and S. Roseman, Sialic acids: Enzymatic synthesis of Tay Sachs ganglioside, J. Biol. Chem. 250:6727 (1975).
30. S. Basu, J. L. Chien, and K. A. Presper, Biosynthesis of gangliosides in tissues, in: "Structure and Function of Gangliosides" L. Svennerholm, H. Dreyfus, P. Urban, and P. Mandel, eds., p 213, Plenum Publishing Co., New York (1980).
31. S. Basu and M. Basu, Expression of glycosphingolipid glycosyltransferases in development and transformation, in: "The Glycoconjugates, Vol. III" M. Horowitz, ed., p 265, Academic Press, New York (1982).
32. M. Basu, K. A. Presper, S. Basu, L. M. Hoffman, and S. E. Brooks, Differential activities of glycolipid glycosyltransferases in Tay-Sachs Disease: Studies in cultured cells from cerebrum, Proc. Natl. Acad. Sci. USA 76:4270 (1979).
33. S. Basu, M. Basu, and H. Higashi, Biosynthesis in vitro of sialoglycosphingolipids, Proc. 6th Internatl. Glycoconjugates p. 41 (1981).
34. M. Basu, S. Basu, A. Stoffyn, and P. Stoffyn, Biosynthesis in vitro of sialyl(α2-3)neolactotetraosylceramide by a sialyltransferase from embryonic chicken brain, J. Biol. Chem. 257:12765 (1982).
35. S. Basu, B. Kaufman, and S. Roseman, Enzymatic synthesis of ceramide glucose and ceramide lactose by glycosyltransferases from embryonic chicken brain, J. Biol. Chem. 243: 5802 (1968).

36. S. Basu, B. Kaufman, and S. Roseman, Conversion of Tay-Sachs ganglioside to monosialoganglioside by brain uridine diphosphate D galactose: Glycolipid galactosyltransferase, J. Biol. Chem. 240:4115 (1965).
37. J. N. Behnke, S. M. Dagher, T. H. Massey, and W. C. Deal, Rapid, multisample isoelectric focusing in sucrose density gradients using conventional polyacrylamide electrophoresis equipment, Anal. Biochem. 69:1 (1975).
38. M. Basu and S. Basu, Enzymatic synthesis of a tetraglycosylceramide by a glycosyltransferase from rabbit bone marrow, J. Biol. Chem. 247:1480 (1972).
39. J. Kyle, M. Basu, and S. Basu, Characterization of glycolipid: galactosyltransferase activities from embryonic chicken brain, Federation Proc. 42:2020 (1983).
40. J. Kyle, M. Basu, and S. Basu, Solubilization and separation of two glycolipid:galactosyltransferase activities from embryonic chicken brain, Federation Proc. 41:1169 (1982).
41. F. Kaplan and P. Hechtman, Purification and properties of two enzymes catalyzing galactose transfer to GM2 ganglioside from rat liver, J. Biol. Chem. 258:770 (1983).
42. J. R. Moskal, J. L. Chien, M. Basu, and S. Basu, Structure and biosynthesis in vitro of bovine erythrocyte pentaglycosylceramide, Federation Proc. 34:645 (1975).
43. J. R. Moskal, Structure of a pentaglycosylceramide from bovine erythrocyte and its biosynthesis in vitro in cultured cells and serum. Ph.D. thesis, University of Notre Dame, Notre Dame, Indiana (1977).
44. J. Zielensk and J. Koscielak, Enzymatic synthesis of neolactotetraosylceramide by the N acetyllactosamine synthesis of human serum, Eur. J. Biochem. 125:323 (1982).
45. K. Yamato and A. Yoshida, Biosynthesis of lactosylceramide and paragloboside by human lactose synthase A protein, J. Biochem. 92:1123 (1982).
46. P. Stoffyn and A. Stoffyn, Biosynthesis in vitro of mono and disialo gangliosides from gangliotetraosylceramide by cultured cell lines and young rat brain. Structure of the products, and activity and specificity of sialosyltransferase, Carbohydr. Res. 78:327 (1980).
47. M. Basu, H. C. Chon, T. Brown, and S. Basu, Biosynthesis in vitro of N acetylglucosamine containing core of Ii ganglioside, Proc. 7th Internatl. Glycoconjugate, Sweden p. 772 (1983).

GENETIC REGULATION OF GM2(NeuGc) EXPRESSION IN LIVER OF MOUSE

Akemi Suzuki, Yasuhiro Hashimoto, Mikiko Abe,
Yoshihiro Kiuchi* and Tamio Yamakawa

Metabolism Section and *Laboratory Animal Science
Section, The Tokyo Metropolitan Institute of Medical
Science, Honkomagome, Bunkyo-ku, Tokyo

INTRODUCTION

Remarkable diversity of the carbohydrate structure of glycolipids has been well described and the change in the expression of glycolipids during embryogenesis, differentiation, organogenesis, maturation and malignant transformation is also recognized.[1] During the course of our study, we found that the expression of sulfated glycolipid, Seminolipid, was closely related to spermatogenesis or the differentiation of germinal cells.[2,3] We also noticed that the Forssman antigen was detected only in the mesenchymal tissues of small intestine in the adult mouse,[4,5] while it was expressed in every cell of the embryo at the stage of the late morulae.[6] These phenomena may indicate that the expression of glycolipids is strictly regulated at the level of the gene expression. At present, however, we do not know whether the regulated expression of glycolipids will further trigger various kinds of physiological as well as morphological changes in the tissue, or if these are simply produced by the change in the expression of other inevitable functions.

Recently, we found that WHT/Ht, an inbred strain of mouse, lacked GM2(NeuGc) in liver, whereas other strains such as DBA/2, BALB/c, C3H/He and C57BL/10 contained it. Thus, we carried out genetic analyses on the expression of GM2(NeuGc) in mice and the following three observations are reported here. (i) The expression of GM2(NeuGc) was regulated as an autosomal dominant trait[7] and its expression was directly controlled by the activity of UDP-GalNAc: GM3(NeuGc) N-acetylgalactosaminyltransferase;[8] (ii) mouse hepatocytes were stained with anti-GM2(NeuGc) antibody by an immunofluorescence technique; (iii) the brain gangliosides obtained from WHT/Ht, which lacked GM2(NeuGc) in its liver, were indistinguishable from those obtained from BALB/c which had GM2(NeuGc) in its liver.

RESULTS

GM2(NeuGc) expression in mouse liver is regulated as an autosomal dominant trait

As shown in Fig. 1, WHT/Ht, an inbred strain of mouse, established by H. B. Hewitt,[9] lacks GM2(NeuGc) in the liver, while DBA/2 contains GM2(NeuGc) as a major ganglioside together with GM3(NeuGc) and GM3(NeuAc) as minor components. Thus, we performed a mating experiment between DBA/2 and WHT/Ht based on a working hypothesis that the GM2(NeuGc) expression would be a dominant trait. A thin layer chromatogram of liver gangliosides from F_1 mice produced by the (DBA/2 x WHT/Ht) combination is shown in Fig. 1. The result indicates that the expression of GM2(NeuGc) is dominant and is not linked to the sex chromosome. This conclusion was also confirmed by the ganglioside analysis of F_1 produced by (WHT/Ht x DBA/2), which is the reversed combination of the parents. By the backcross experiment of F_1 to WHT/Ht, GM2(NeuGc) expression was segregated into two groups, one with GM2(NeuGc) and the other with GM3(NeuGc). As shown in Fig. 2, it is demonstrated by the analysis of liver gangliosides of 7 individual mice of a litter that 4 mice did not have GM2(NeuGc), like WHT/Ht, but 3 mice had GM2(NeuGc) like DBA/2. The ratio of the number of GM2(NeuGc) positive mice to that of negative mice came close to the theoretical value of 1:1 when a larger population was analyzed. These results clearly indicate that the expression of GM2(NeuGc) is regulated as an autosomal dominant trait by a single gene.

Since these results remind us of the genetic regulation of ABO blood group antigens, we examined the possibility that the expression of GM2(NeuGc) might be regulated by the presence of an enzyme, N-acetylgalactosaminyltransferase, which catalyzes the reaction to synthesize GM2(NeuGc) from GM3(NeuGc). The activity was measured with the liver microsomal fractions obtained from BALB/c, WHT/Ht and their F_1, using GM3(NeuGc) as a glycolipid acceptor and UDP-[^{14}C]GalNAc as a sugar nucleotide donor. As shown in Fig. 3, WHT/Ht strain did not exhibit any enzyme activity, and on the other hand BALB/c gave a value of 1.0 - 1.5 nmoles/mg protein/h. F_1 also showed an activity whose level was half of that in BALB/c. As the amount of protein in the microsomal fraction was similar among these three groups, the total enzyme activities in liver should be parallel to the specific activities.

The activity in the backcross of F_1 to WHT/Ht was also

GENETIC REGULATION OF GM2(NeuGc) EXPRESSION

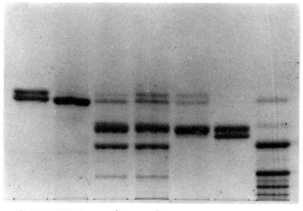

Fig. 1. Thin layer chromatogram of the gangliosides in the liver of WHT/Ht, DBA/2 and (DBA/2 x WHT/Ht)F_1. The gangliosides were prepared by the method of Ledeen et al.[10] which consisted of DEAE-Sephadex column chromatography, alkaline treatment and Unisil column chromatography. Lane GM3, a standard mixture of GM3(NeuAc) and GM3(NeuGc) purified from dog and horse erythrocytes; lane GM2, GM2(NeuGc) prepared from the erythrocytes of C3H/He as a reference and lane DBG, a ganglioside mixture of dog brain. The solvent system was chloroform-methanol-0.2% $CaCl_2$ in water (55:45:10) and spots were visualized by heating the plate at 95°C with resorcinol reagent.

measured and found to be segregated into two levels. One level was similar to that of F_1 and the activity of the other level was almost zero. Furthermore, every mouse having activity similar to that of F_1 contained GM2(NeuGc) in its liver and all the mice having no activity did not contain GM2(NeuGc).

By these two observations, we can conclude that GM2(NeuGc) expression in mouse liver is genetically regulated as an autosomal dominant trait through the activity of UDP-GalNAc:GM3(NeuGc) N-acetylgalactosaminyltransferase.

It is of interest to note that an enhanced elongation of sugar chains of the liver gangliosides from GM2(NeuGc) to GM1(NeuGc) and then to GD1 was observed in F_1 between DBA/2 and WHT/Ht (Fig. 1).

Immunohistochemical localization of GM2(NeuGc) in the liver of BALB/c mouse

Anti-GM2(NeuGc) antibodies produced in a rabbit were purified

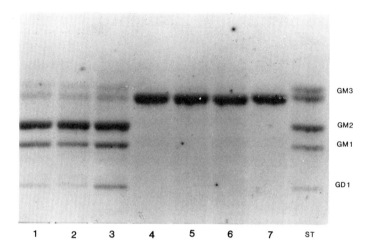

Fig. 2. Thin layer chromatogram of the liver gangliosides in the mice of the backcross of (DBA/2 x WHT/Ht)F_1 to WHT/Ht. Lanes 1 to 7 exhibit the gangliosides of 7 individual mice of a litter. Note that lanes 1 to 3 show the similar pattern to F_1 and lanes 4 to 7 show the similar pattern to WHT/Ht. The conditions for TLC were same as those of Fig. 1.

by chromatography on a column of aminopropyl silica gel conjugated with the purified GM2(NeuGc)[11] and used for fluorescence staining in combination with fluorescinated goat anti-rabbit IgG as a secondary antibody. The purified anti-GM2(NeuGc) was weakly cross-reactive to GM3(NeuGc) but was not to GM3(NeuAc), GM2(NeuAc), or GM1(NeuGc) and (NeuAc) when examined by an enzyme linked immunosorbent assay.[5] As shown in Fig. 4, cytoplasm in the hepatocytes was positively stained.

Brain gangliosides in WHT/Ht

Since WHT/Ht grows quite normally and does not show any neurological symptoms we expected that gangliosides in the brain of WHT/Ht would be similar to those of other strains. We analyzed brain gangliosides by the mapping technique of Iwamori and Nagai[12] with a minor modification. For the first dimension, the gangliosides were separated by high performance liquid chromatography with a DEAE-silica gel column[13] and programmed elution with a linear gradient of ammonium acetate from 0 to 0.12 M in methanol. The gangliosides in the fractionated effluents were analyzed on a silica gel TLC plate (Fig. 5). As expected, the mapping pattern of the brain gangliosides in WHT/Ht was quite similar to that in BALB/c.

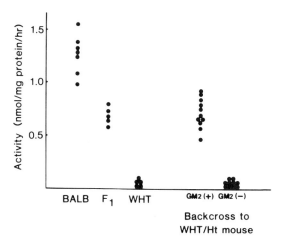

Fig. 3. The activity of UDP-GalNAc:GM3(NeuGc) N-acetylgalactosaminyltransferase in the liver microsomal fractions of WHT/Ht, BALB/c and their progeny. Each dot represents the transferase activity of individual mice. In the backcross of F_1 to WHT/Ht, GM2(+) represents the enzyme activity in mouse which expresses GM2(NeuGc) in liver and GM2(-) indicates the activity in mouse which lacks GM2(NeuGc). Incubation mixture was composed of GM3(NeuAc), 50 μg; UDP-[^{14}C]GalNAc, 10 nmol; microsomal protein, 100 μg; Triton X-100, 300 μg; sodium cacodylate, 100 μmol at pH 7.3; $MnCl_2$, 10 μmol in a total volume of 100 μl. The gangliosides were separated by a Sep-Pak C18 cartridge from water soluble radioactive compounds and radioactivities of gangliosides were counted. The product was confirmed to be GM2(NeuGc) by autoradiography of TLC.[8]

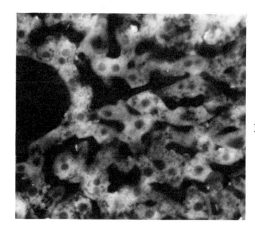

Fig. 4. A fluorescence micrograph of BALB/c liver stained with the purified rabbit anti-GM2 (NeuGc) antibody and fluorescinated goat anti-rabbit IgG.

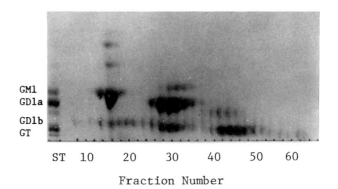

Fig. 5. Ganglioside mapping of WHT/Ht brain. The brain gangliosides of WHT/Ht were obtained in the upper phase of Folch's partition. Lyophilized materials obtained after dialysis were dissolved in methanol-water (1:1, by volume) and injected onto a DEAE-silica gel column (0.3 x 25 cm) connected to a high performance liquid chromatograph at the flow rate of 0.5 ml methanol/min. After the neutral fraction was eluted during 20 min, gangliosides were separated by a linear gradient elution from 0 to 0.18 M ammonium acetate in methanol for 60 min at an increased flow rate of 1 ml/min and effluents were collected in 1 ml fractions. Aliquots were analyzed on a TLC plate developed with chloroform-methanol-5N ammonium hydroxide-0.04% $CaCl_2$ in water (60:40:4:5, by vol.). The spots were visualized by heating the plate at 95°C with resorcinol reagent.

We could not find any detectable amount of GM1(NeuGc), GD1a(NeuGc) and GD1b(NeuGc) even in the brain of BALB/c (data are not shown).

DISCUSSION

The expression of GM2(NeuGc) in mouse liver was directly regulated as an autosomal dominant trait through the activity of UDP-GalNAc:GM3(NeuGc) N-acetylgalactosaminyltransferase. However, we do not know what kind of defect is in the enzyme molecule as well as the genetic information on WHT/Ht and what is the function of GM2(NeuGc) ganglioside in the liver; further investigations are necessary to answer such questions.

Since the composition of brain gangliosides of WHT/Ht was similar to that of BALB/c which could synthesize GM2(NeuGc) in liver, there are two possibilities on the nature of N-acetylgalactosaminyltransferase. First, if one assumes that there exists only one kind of transferase which has a similar affinity to GM3(NeuAc) as to GM3(NeuGc), the defect in the enzyme activity should be limited to liver and erythrocytes in WHT/Ht,[14] which means that the expression of the enzyme activity is closely related to organ differentiation. Second, if there are two kinds of N-acetylgalactosaminyltransferase in mouse [one acts on GM3(NeuAc) but not on GM3(NeuGc), which participates in the biosynthesis of brain gangliosides containing only NeuAc, and the other acts on GM3(NeuGc) but not on GM3(NeuAc), which participates in the biosynthesis of liver gangliosides containing NeuGc], WHT/Ht should lack the latter transferase. This question has to be answered by the ganglioside analysis and the determination of substrate specificity of the transferase in various organs of mice including WHT/Ht and other strains expressing GM2(NeuGc).

As mentioned in the results, the enhanced elongation of the sugar chains of gangliosides from GM2(NeuGc) to GM1(NeuGc) and then to GD1(NeuGc) was observed in the F_1 mice. One possible explanation for this observation is that F_1 mice may acquire the abilities to convert GM3(NeuGc) to GM2(NeuGc) from DBA/2 and GM2(NeuGc) to GM1(NeuGc) and then GD1(NeuGc) from WHT/Ht as shown below. This is an interesting observation which has to be explained in biochemical terms as in the case of the P antigen expression.[15-17]

```
WHT/Ht   GM3 ─┼─ GM2 ─── GM1( ─── GD )
F1              ·········
DBA/2    GM3 ─── GM2 ─┼─ GM1( ─── GD )
```

Willison et al. suggested the possibility that expression of glycolipids may be genetically regulated in mouse embryos and linked to H-2 complex on chromosome 17.[6] However, we observed that the activity is not linked to either H-2 complex nor other chromosomal markers so far as we tested, such as Idh-1 (chromosome 1), a (chromosome 2), Car-3 (chromosome 3), b (chromosome 4), Pgm-1 (chromosome 5), Hbb or Gpi-1 (chromosome 7) and Es-3 (chromosome 11). Mapping the N-acetylgalactosaminyltransferase gene on a mouse chromosome is in progress.

We feel that basic knowledge on the genetics of the regulatory mechanism for the expression of the carbohydrate structure in glycolipids is extremely limited. As mouse is considered to be one of the most preferred experimental animals for genetic analysis, we will be able to develop a field of glycolipid genetics based not only on glycolipid biochemistry but also on enzymology and molecular biology by use of various mutants.

SUMMARY

GM2 containing NeuGc was a major ganglioside in mouse liver of inbred strains such as DBA/2, BALB/c, C57BL/10 and C3H/He, which are commonly used for biochemical and immunological studies. On the other hand, the liver of WHT/Ht, an inbred strain, contained GM3(NeuGc) as a major ganglioside and lacked GM2(NeuGc). We report here that the GM2(NeuGc) expression was analyzed in the liver of the progeny between WHT/Ht and DBA/2 and the positive expression of GM2(NeuGc) was proved to be a dominant trait regulated by an autosomal single gene. Moreover, the N-acetylgalactosaminyltransferase activity to convert GM3(NeuGc) to GM2(NeuGc) was measured in the liver microsomal fraction of WHT/Ht, BALB/c and their F_1. F_1 expressed almost half of the activity in BALB/c and WHT/Ht did not express a detectable amount of activity. The backcross of F_1 to WHT/Ht segregated into two groups. One expressed both GM2(NeuGc) and the transferase activity and the other expressed neither of them. There was no exceptional individual which was not grouped into either of these two groups. These results indicate that GM2(NeuGc) expression is directly regulated by the N-acetylgalactosaminyltransferase activity, the expression of the enzyme activity is regulated by an autosomal single gene and WHT/Ht is a mutant of the recessive homozygote which cannot express the enzyme activity in its liver.

WHT/Ht does not develop any neurological symptoms but grows and breeds well. The brain ganglioside composition was proved to be identical to those in BALB/c brain. The result suggests that WHT/Ht has N-acetylgalactosaminyltransferase to convert GM3(NeuAc) to GM2(NeuAc) in its brain. It is a subject for further study to elucidate what kind of defect is involved in the GM2(NeuGc) biosynthesis of WHT/Ht liver.

ACKNOWLEDGEMENTS

We would like to thank Dr. K. Suzuki, Dr. K. Sudo, Prof. K. Odaka, (University of Tokyo, Institute of Medical Science), Dr. K. Moriwaki (National Institute of Genetics), and Dr. K. Iwasaki (Tokyo Metropolitan Institute of Medical Science) for valuable discussions, and Dr. S. K. Kundu (Baylor College of Medicine) for providing us with aminopropyl silica gel and DEAE-silica gel. This study was supported in part by grants from the Ministry of Education, Science and Culture of Japan, the Technology Agency of Japan and the Foundation for the Promotion of Research on Medical Resources.

REFERENCES

1. S. Hakomori, Glycosphingolipids in cellular interaction, differentiation, and oncogenesis, Ann. Rev. Biochem. 501: 733 (1981).
2. S. Handa, K. Yamato, I. Ishizuka, A. Suzuki, and T. Yamakawa, Biosynthesis of seminolipid: Sulfation in vivo and in vitro, J. Biochem. 75:77 (1974).
3. A. Suzuki, M. Sato, S. Handa, Y. Muto, and T. Yamakawa, Decrease of seminolipid content in the testes of rats with vitamin A deficiency determined by high performance liquid chromatography. J. Biochem. 82:461 (1977).
4. A. Suzuki and T. Yamakawa, The different distribution of asialo GM1 and Forssman antigen in the small intestine of mouse demonstrated by immunofluorescence staining, J. Biochem. 90:1551 (1981).
5. A. Suzuki, Y. Umesaki, and T. Yamakawa, Localization of asialo GM1 and Forssman antigen in the small intestine of mouse, Adv. Exp. Med. Biol. 152:415 (1982).
6. K. R. Willison, R. A. Karol, A. Suzuki, S. K. Kundu, and D. M. Marcus, Neutral glycolipid antigens as developmental markers of mouse teratocarcinoma and early embryos: An immunologic and chemical analysis, J. Immunol. 129:603 (1982).
7. Y. Hashimoto, H. Otsuka, K. Sudo, K. Suzuki, A. Suzuki, and T. Yamakawa, Genetic regulation of GM2 expression in the liver of mouse, J. Biochem. 93:895 (1983).
8. Y. Hashimoto, Y. Kiuchi, A. Suzuki, and T. Yamakawa, Genetically regulated expression of UDP-GalNAc:GM3(NeuGc) N-acetylgalactosaminyltransferase activity in mouse liver, J. Biochem. submitted.
9. J. Staats, Standardized nomenclature for inbred strains of mice: Seventh listing, Cancer Res. 40:2087 (1980).
10. R. W. Ledeen, R. K. Yu, and L. F. Eng, Gangliosides of human myelin: Sialogalactosylceramide (G7) as a major component, J. Neurochem. 21:829 (1973).
11. S. K. Kundu and S. K. Roy, Aminosilica gel as a solid support for preparation of glycolipid immunoadsorbent and purification of antibodies, J. Lipid Res. 20:825 (1978).
12. M. Iwamori and Y. Nagai, A new chromatographic approach to the resolution of individual gangliosides. Ganglioside mapping, Biochim. Biophys. Acta 528:257 (1978).
13. S. K. Kundu and S. K. Roy, A rapid and quantitative method for the isolation of gangliosides and neutral glycosphingolipids by DEAE-silica gel chromatography, J. Lipid Res. 19:390 (1979).
14. S. Hamanaka (nee Yasue), S. Handa, and T. Yamakawa, Ganglioside composition of erythrocytes from various strains of inbred mice, J. Biochem. 86:1623 (1979).
15. M. Fellous, A. Garbal, G. Nobillot, and J. Weils, Studies on

the biosynthetic pathway of human P erythrocyte antigen using genetic complementation tests between fibroblasts of P^k and p blood group erythrocytes, Vox Sang. 32:262 (1977).
16. S. Kijimoto-Ochiai, M. Naiki, and A. Makita, Defects of glycosyltransferase activities in human fibroblasts of P^k and p blood group phenotypes, Proc. Natl. Acad. Sci. 74: 5407 (1977).
17. D. M. Marcus, S. K. Kundu, and A. Suzuki, The P blood group system: Recent progress in immunochemistry and genetics, Semin. Hematol. 18:63 (1981).

NEW APPROACHES IN THE STUDY OF GANGLIOSIDE METABOLISM

G. Tettamanti, R. Ghidoni, S. Sonnino, V. Chigorno,
B. Venerando, A. Giuliani and A. Fiorilli

Dept. of Biological Chemistry, the Medical School
University of Milan, Milan, Italy

INTRODUCTION

It is generally accepted that gangliosides are biosynthesized in the Golgi apparatus and then move to the plasma membrane which is the main site of their cellular location.[1-3] The process of ganglioside biosynthesis consists of sequential additions of monosaccharide units to a starting precursor (ceramide or psychosine) catalyzed by specific glycosyltransferases.[1,2,4,5] Gangliosides are degraded in the lysosomes, where several glycohydrolases remove sequentially the individual saccharide units and the formed ceramide is split by ceramidase into sphingosine and fatty acid.[6-8] In the case of nerve cells the process of both biosynthesis and degradation of gangliosides is assumed to occur primarily in the perikaryon.[7-9] Transport of gangliosides to and from the plasma membranes is mediated by fast axonal flow.[9,10]

Many are the open questions concerning ganglioside metabolism. For instance little is known[11] on the mechanism of ganglioside transport - as vesicles, or free molecules, or lipoproteic complexes - from the Golgi apparatus to plasma membranes and from these to lysosomes. Moreover the glycosyltransferases and glycosidases in the plasma membranes may produce changes in the chemical composition of gangliosides, possibly in relation to functional performance. Finally the intimate relation between the Golgi apparatus and the lysosomes suggests a possible synchronization of biosynthetic and degradative processes in response to the functional demand of the cell.

In the present paper we attempt to provide an experimental basis for answering some of these questions.

EXPERIMENTAL APPROACH AND BASIC EXPERIMENTAL CONDITIONS

So far ganglioside metabolism was studied by two main approaches: (a) injection into animals, or treatment of subcellular particles or cells in culture, with radioactive precursors (glucose, galactose, hexosamine, fatty acid, acetate), followed by detection of the radioactivity incorporated into individual gangliosides; (b) recognition and characterization of the different enzymes involved in ganglioside metabolism. The approach we propose here consists of injecting animals with highly radioactive and isotopically labeled ganglioside (GM1), and recognizing its degradation pathway and possible involvement in the neosynthesis of complex lipids. Previous studies[12-14] already showed that gangliosides injected into mice or rats are distributed in the various organs and tissues, taken up by the cells and metabolized. Since liver played a major role in metabolism of exogenously administered gangliosides we used a liver for perfecting our experimental design with the aim to extend it to other tissues.

<u>Animals</u>: inbred strains of C3H/HcNCr 1BR mice (Charles River), 25 g average body weight.

<u>Labeled ganglioside</u>: ganglioside GM1, prepared in pure form (over 99%) from calf brain, according to Ghidoni et al.;[15] isotopically ^{3}H-labeled at the level of terminal galactose by a perfected galactose-oxidase method[16] ([Gal-^{3}H]GM1: specific radioactivity, 1.5 Ci/mmole), or at the C-3 of long chain base by the DDQ/^{3}H-NaBH$_4$ method,[17] with removal of the formed threo stereoisomers by HPLC[18] ([Sph-^{3}H]GM1: specific radioactivity, 1.3 Ci/mmole) (in both cases radiochemical purity better than 99%).

<u>Injection</u>: 10-50 µCi of [Gal-^{3}H]GM1 or [Sph-^{3}H]GM1 intraperitoneally.

<u>Animal and liver treatment</u>: animals killed by decapitation; liver removed, washed and submitted to subcellular fractionation,[19-21] the following purified subcellular fractions being prepared: "plasma membrane fraction", "lysosome fraction", Golgi apparatus fraction", and "soluble fraction"; liver perfused and treated with collagenase and submitted to separation of parenchymal and non-parenchymal cells.[22]

<u>Lipid and ganglioside analysis</u>: tissue specimens submitted to lipid extraction[23] and fractionation into individual entities.[14,15,24]

<u>Determination of radioactivity</u>: by liquid scintillation counting; radiochromatoscanning; radio-GLC; and fluorography.[14,17]

INCORPORATION AND DISTRIBUTION OF RADIOACTIVITY IN THE INDIVIDUAL LIVER GANGLIOSIDES

After injection of 10-50 µCi of ^{3}H-labeled GM1 liver retained a substantial portion of radioactivity. The maximum of incorporation was reached at 4 hr from injection with [Gal-^{3}H]GM1, and at 5-6 hr with [Sph-^{3}H]GM1; at 24 hr liver incorporated radioactivity was less than half obtained at the maximum. Incorporated radioactivity was proportionately higher with [Sph-^{3}H]GM1 than with [Gal-^{3}H]GM1; at the corresponding maximum it was 7-10% of total injected radioactivity with [Gal-^{3}H]GM1 and 18-23% with [Sph-^{3}H]GM1. The volatile portion of incorporated radioactivity was up to 40% with [Gal-^{3}H]GM1, and never exceeded 5% with [Sph-^{3}H]GM1. In either case the remainder was non-volatile radioactivity. This difference, together with the finding that liver incorporated less radioactivity with [Gal-^{3}H]GM1 than with [Sph-^{3}H]GM1, likely means that the removal of galactose from injected GM1 occurs earlier than that of sphingosine and reflects the much more rapid metabolic fate of galactose than of sphingosine.

Almost all the radioactivity incorporated into liver was present in the total lipid extract. Using [Gal-^{3}H]GM1 the radioactivity present in the lipid extract distributed mainly in the aqueous phase, up to 60% being dialysable, and almost no radioactivity was present in the dried organic phase. Using [Sph-^{3}H]GM1 the proportion of radioactivity carried by the dialysed aqueous phase (total ganglioside fraction) diminished with time with concurrent increase of the portion carried by the dried organic phase (Fig. 1). This indicates that injected GM1 is metabolized also to other lipids to an extent which increases with time after injection. The radioactivity present in the total ganglioside fraction was carried by gangliosides GM1 and GD1a-(NeuAc,NeuGl) after administration of [Gal-^{3}H]GM1, and by gangliosides GM3, GM2, GM1 and GD1a-(NeuAc,NeuGl), after administration of [Sph-^{3}H]GM1. At 24 hr from injection (Fig. 2 and Table 1) the distribution of radioactivity in the different gangliosides was as follows: (a) with [Gal-^{3}H]GM1, 97% in GM1, 1.6% in GD1a-(NeuAc,NeuGl), the remainder in minor components; (b) with [Sph-^{3}H]GM1, 26% in GM1, 64% in GM2, 6% in GM3 and 2% in GD1a-(NeuAc,NeuGl), the remainder in minor components.

Abbreviations used: ganglioside nomenclature according to Svennerholm;[24] DDQ, 2,3-dichloro-5,6-dicyanobenzoquinone; TLC, thin layer chromatography; HPTLC, high performance thin layer chromatography; GLC, gas-liquid chromatography; HPLC, high performance liquid chromatography.

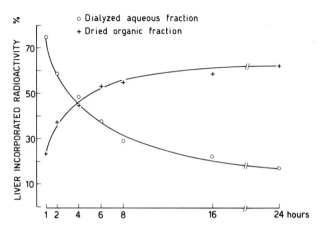

Fig. 1. Distribution of liver incorporated radioactivity in the dialyzed aqueous fraction and dried organic fraction obtained from total lipid extract at different times after injection of 10 µCi of [Sph-^{3}H]GM1.

The individual radioactive gangliosides present in the liver after 24 hr from injection of 50 µCi of ^{3}H-labelled GM1 were isolated and the specific radioactivity and radioactivity distribution in the molecule were then determined. The radioactivity of GM1 and GD1a-(NeuAc,NeuGl) obtained after [Gal-^{3}H]GM1 injection was entirely carried by the sugar moiety of the ganglioside, and was retained by the tetrahexosyl-ceramide produced by partial acid hydrolysis. This indicates that in both gangliosides the radioactivity was carried by the galactose residue located in terminal position. The radioactivity of GM3, GM2, GM1, and GD1a-(NeuAc, NeuGl) obtained after [Sph-^{3}H]GM1 injection was carried exclusively by the long chain base moiety and was maintained in the neutral glycosphingolipids produced by partial acid hydrolysis, indicating that the radioactivity was carried in the long chain base. Radioactive GD1a-(NeuAc,NeuGl) obtained after injection of [Gal-^{3}H]GM1 and [Sph-^{3}H]GM1 gave rise upon exhaustive sialidase treatment to GM1 having the same specific radioactivity as the starting ganglioside. The specific radioactivity values of all the isolated gangliosides are given in Table 1.

DISTRIBUTION OF RADIOACTIVE GANGLIOSIDES IN THE DIFFERENT LIVER SUBCELLULAR FRACTIONS

Figure 3 shows the pattern of radioactive gangliosides in the different subcellular fractions prepared from liver after 24 hr injection of 50 µCi of [Sph-^{3}H]GM1. The proportions among the different gangliosides were different in the different fractions and had a different trend with time (Fig. 4). In the "lysosome frac-

Table 1. Distribution of Radioactivity and Specific Radioactivity of the Different Radioactive Gangliosides Isolated From Mouse Liver After 24 hr From Injection of 50 µCi of [Gal-^{3}H]GM1 or of [Sph-^{3}H]GM1.

	[Gal-^{3}H]GM1		
	Distribution of Radioactivity		
	10^{-3} x dpm/g of fresh tissue	%	Specific radioactivity (mCi/mole)
GM3	0.6	0.1	5
GM2	1.9	0.3	7
GM1	618.8	97.2	2880
GD1a-(NeuAc,NeuGl)	12.0	1.6	86
Minor components	5.1	0.8	--
	[Sph-^{3}H]GM1		
	Distribution of Radioactivity		
	10^{-3} x dpm/g of fresh tissue	%	Specific radioactivity (mCi/mole)
GM3	77.1	6.2	652
GM2	783.7	63.0	3190
GM1	325.9	26.2	1301
GD1a-(NeuAc,NeuGl)	24.9	2.0	161
Minor components	32.4	2.6	--

tion" GM1 sharply diminished with time, with concurrent increase of GM2 and, to a lesser extent, of GM3, while GD1a-(NeuAc,NeuGl) was virtually absent. In the "Golgi apparatus fraction" GM1 was at all times the most abundant ganglioside; all the other radioactive gangliosides progressively increased with time and, among them, GD1a-(NeuAc,NeuGl) reached in this fraction and after 24 hr the highest percentage content (7%). The "plasma membrane fraction"

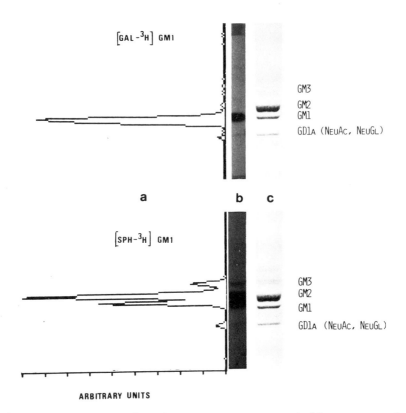

Fig. 2. Distribution of radioactivity in the different gangliosides of mice liver after 24 hr from injection of 20 µCi of [Gal-^{3}H]GM1 and [Sph-^{3}H]GM1. (a) Radiochromatoscanning; (b) TLC, autoradiography (fluorography); (c) TLC, colorimetric visualization.

had a behavior between that of lysosomes and the Golgi apparatus. The percentage of GD1a-(NeuAc,NeuGl) at 24 hr (3.4%) almost doubled that of the starting homogenate.

These results indicate that exogenously administered GM1 enters the routes of ganglioside metabolism in the liver. In lysosomes the major events appear to be its degradation to GM2 and GM3. The pattern of radioactive gangliosides in the Golgi apparatus may be interpreted as due to involvement of GM1 in the biosynthesis of new gangliosides. This interpretation particularly fits to GD1a-(NeuAc,NeuGl) which can result from direct sialylation of taken up GM1, or by utilization of a by-product of GM1 degradation (ceramide?, sphingosine?) into the multiglycosyltransferase complex producing this ganglioside. The same interpretation applies also to radioactive GM2 and GM3, although some contamination with lysosomal

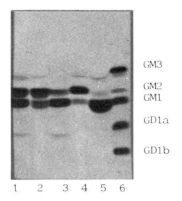

1) STARTING HOMOGENATE 4) LYSOSOMAL FRACTION
2) PLASMA MEMBRANE FRACTION 5) SOLUBLE FRACTION
3) GOLGI APPARATUS FRACTION 6) STANDARD RADIOACTIVE GANGLIOSIDES

Fig. 3. TLC autoradiography (fluorography) of gangliosides in the different liver subcellular fractions after 24 hr from injection of [Sph-^{3}H]GM1.

membranes, or functional interactions between the Golgi apparatus and lysosomes, might have introduced some of these gangliosides from the lysosomal pool. The presence of radioactive gangliosides in the plasma membranes may well result from the continuous exchange of membranous materials between plasma membranes and both lysosomes and Golgi derived vesicles. The relatively high content of radioactive GD1a-(NeuAc,NeuGl) in the plasma membrane fraction may in part be due to direct glycosylation of exogenous GM1, inserted into the membrane, by the sialyltransferase which is known to occur in liver plasma membranes.[25]

RELATIONSHIP BETWEEN INJECTED ^{3}H-LABELED GM1 AND OTHER LIPID CONSTITUENTS OF THE LIVER

The lipids present in the dried organic phase were submitted to class fractionation[24] and the radioactivity carried by the individual fractions (after 24 hr from injection of [Sph-^{3}H]GM1) was counted. Four main peaks of radioactivity were obtained. The first eluted peak of radioactivity corresponded to the fraction containing ceramide and neutral lipids; the second peak to that containing sulphatides, cerebrosides, and lower neutral glycolipids; the third peak to that containing mainly sphingomyelin; the fourth, minor, peak to that containing higher neutral glycolipids and phospholipids. The third peak, which carried about 10% of the starting organic phase radioactivity, was composed of three components, the most abundant of which cochromatographed with an

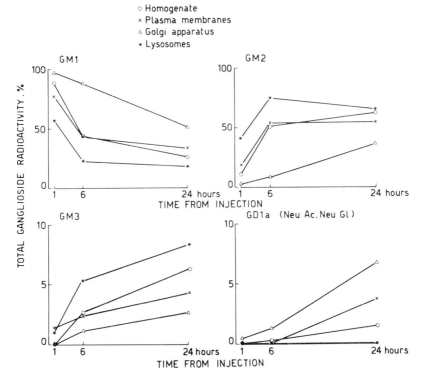

Fig. 4. Radioactive ganglioside composition of liver subcellular fractions after different times from injection of 50 μCi of [Sph-^{3}H]GM1.

authentic sample of sphingomyelin (Fig. 5). The sphingomyelin nature of this component was proved by verification of the presence of phosphorus and long chain bases, and of sensitivity to sphingomyelinase. All the radioactivity of the third peak was carried by the sphingomyelin component, as shown by TLC radiochromatoscanning (see Fig. 5) and by radio GLC of the long chain bases herein contained. This evidence is consistent with the hypothesis that a by-product of injected GM1, possibly sphingosine or ceramide, is involved in biosynthetic pathways concerning a class of sphingolipids different from gangliosides.

CELLULAR (PARENCHYMAL OR NON-PARENCHYMAL) LOCATION OF INCORPORATED RADIOACTIVITY IN THE LIVER

Cellular fractionation of liver provided a "parenchymal fraction" which was more than 98% homogeneous in hepatocytes, with over 95% of viability. As shown in Table 2, the recovery of the typical marker of parenchymal cells, glucose-6-phosphate-phosphatase, in the "parenchymal fraction" was 63%. Under the same experimental conditions 58% of total liver incorporated radioactivity was recovered in the "parenchymal fraction". This means that the greatest portion (at least 90%) of liver incorporated ^{3}H-labeled GM1 was taken up by hepatocytes. This finding agrees with Lang's[13] observation that administration of ^{3}H-labeled gangliosides in the rat was followed by elimination of non-volatile radioactivity through the bile tract.

CONCLUSION

The present study has demonstrated that exogenously administered GM1 is rapidly metabolized in mouse liver. The pathways of this metabolic fate consist of degradation to smaller products, and re-utilization of the by-products for endogenous biosynthesis of gangliosides and other lipids, namely sphingomyelin. The experimental approach presented here seems to be valuable for studying the routes of cellular ganglioside metabolism, the synchronization between biosynthesis and degradation, and the relationships between the metabolism of gangliosides and that of sphingolipids in general. This approach is expected to be extendable to other organs and tissues.

SUMMARY

Ganglioside GM1, ^{3}H-labeled in the sphingosine or terminal galactose moiety was injected into mice and its metabolic fate in

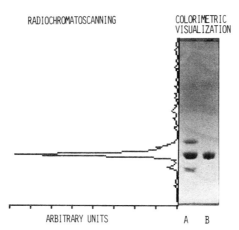

Fig. 5. Incorporation of radioactivity into liver sphingomyelin after injection of 30 µCi of [Sph-^{3}H]GM1.

Table 2. Glucose-6-Phosphate Phosphatase (G-6-Pase) and Incorporated Radioactivity in the Purified Hepatocyte Fraction (Parenchymal Fraction) Prepared From Mice After 6 hr From Injection of 10 µCi of [Sph-^{3}H]GM1. The Data Refer to 1 g Starting Fresh Tissue, and are the Means of Four Experiments. S.D. Values Were Lower Than 10% of Mean Values.

	Incorporated radioactivity	Recovery*	G-6-Pase, total µmoles released	Recovery*
	total dpm	%	phosphate/min	%
Liver	4.86×10^6	--	22.1	--
Hepatocyte fraction	2.81×10^6	58	13.9	63

*Referred to liver

the liver was followed. After administration of sphingosine-labeled GM1 all major liver gangliosides (GM3, GM2, GM1, GD1a-NeuAc, NeuGl) became radioactive, the radioactivity residing in all cases on the sphingosine moiety. The specific radioactivity was highest on GM1, followed by GM2, GM3 and GD1a-NeuAc,NeuGl. Several neutral glycosphingolipids and sphingomyelin were also formed. After administration of galactose-labelled GM1 the only radioactive gangliosides present in the liver were GM1 and GD1a-NeuAc,NeuGl, both carrying the radioactivity on the terminal galactose residue, with no formation of labelled neutral glycosphingolipids. Subcellular studies gave clear evidence that GM1, after being taken up by the liver, was mainly degraded to GM2, GM3 and neutral glycosphingolipids at the level of lysosomes. A part of it was sialylated to more complex gangliosides and some of its metabolic by-products were used for the biosynthesis of other sphingolipid species, likely at the level of the Golgi apparatus. All this suggests that exogenous GM1 is introduced in the metabolic routes of endogenous gangliosides and of other sphingolipids, which are operating in the liver.

ACKNOWLEDGEMENTS

This work was supported in part by grants from the Consiglio Nazionale dele Ricerche (C.N.R.), Rome, Italy.

REFERENCES

1. R. Caputto, H. J. F. Maccioni, and A. Arce, Biosynthesis of brain gangliosides, Mol. Cell Biochem. 4:97 (1974).
2. P. H. Fishman and R. O. Brady, Biosynthesis and function of gangliosides, Science 194:906 (1976).
3. H. Miller-Podraza and P. H. Fishman, Translocation of newly synthesized gangliosides to the cell surface, Biochemistry 21:3265 (1982).
4. S. Roseman, The synthesis of complex carbohydrates by multi-glycosyl-transferase systems and their potential function in intercellular adhesion, Chem. Phys. Lipids 5:270 (1970).
5. H. J. F. Maccioni, S. S. Defilpo, S. A. Landa, and R. Caputto, The biosynthesis of brain gangliosides. Ganglioside-glycosylating activity in rat brain neuronal perikarya fraction, Biochem. J. 174:673 (1978).
6. S. Gatt, Enzymatic aspects of sphingolipid degradation, Chem. Phys. Lipids 5:235 (1970).
7. L. Svennerholm, Structure and biology of cell membrane gangliosides, in: "Cholera and related diarrheas", O. Ouchterlony and J. Holmgred, eds., S. Karger, Basel p. 80 (1980).
8. H. Wiegandt, The gangliosides, in: "Advances in Neurochemistry", B. W. Agranoff and M. H. Aprison, eds., Vol. 4, Plenum Publ. Corp., New York, p. 149 (1982).
9. R. W. Ledeen, Gangliosides, in: "Handbook of Neurochemistry", A. Lajtha, ed., Vol. 3 (2nd Ed.), Plenum Publ. Corp., New York p. 41 (1983).
10. H. Rosner and G. Merz, Uniform distribution and similar turnover rates of individual gangliosides along axons of retinal ganglion cells in the chicken, Brain Research 236: 63 (1982).
11. C. A. Landa, S. S. Defilpo, H. J. Maccioni, and R. Caputto, Disposition of gangliosides and sialosylglycoproteins in neuronal membranes, J. Neurochem. 37:813 (1981).
12. G. Tettamanti, B. Venerando, S. Roberti, V. Chigorno, S. Sonnino, R. Ghidoni, P. Orlando, and P. Massari, The fate of exogenously administered brain gangliosides, in: "Gangliosides in neurological and neuromuscular function, development and repair", M.M. Rapport and A. Gorio, eds., Raven Press, New York p. 225 (1981).
13. W. Lang, Pharmacokinetic studies with ^{3}H-labeled exogenous gangliosides injected intramuscularly into rats, in: "Gangliosides in neurological and neuromuscular function, development and repair", M. M. Rapport and A. Gorio, eds., Raven Press, New York p. 241 (1981).
14. R. Ghidoni, S. Sonnino, V. Chigorno, B. Venerando, and G. Tettamanti, Occurrence of glycosylation and deglycosylation of exogenously administered ganglioside GM1 in mouse liver, Biochem. J. 213:321 (1983).
15. R. Ghidoni, S. Sonnino, M. Masserini, P. Orlando, and G.

Tettamanti, Specific tritium labeling of gangliosides at the 3-position of sphingosines, J. Lipid Res. 22:1286 (1981).
16. P. Orlando, G. Cocciante, G. Ippolito, P. Massari, S. Roberti, and G. Tettamanti, The fate of tritium labeled GM1 ganglioside injected in mice, Pharmacol. Res. Comm. 11:759 (1979).
17. R. Ghidoni, S. Sonnino, G. Tettamanti, N. Baumann, G. Reuter, and R. Schauer, Isolation and characterization of a trisialoganglioside from mouse brain containing 9-0-acetyl,N-acetylneuraminic acid, J. Biol. Chem. 255:6990 (1980).
18. S. Sonnino, R. Ghidoni, G. Kirschner, G. Galli, and G. Tettamanti, Preparation of natural, synthetic and isotopically [^{3}H]-labeled gangliosides, in: "Glycoconjugates", Proceedings of the 7th Int. Symposium, M. L. Chester, D. Heinegard, A. Lundblad, and S. Svensson, eds., Lund, p. 369 (1983).
19. O. Touster, N. N. Aronson, J. T. Dulaney, and H. Hendrickson, Isolation of rat liver plasma membranes. Use of nucleotide pyrophosphatase and phosphodiesterase I as marker enzymes, J. Cell Biol. 47:604 (1970).
20. P. O. Sandberg, L. Marzella, and H. Glausmann, A method for rapid isolation of rough and smooth microsomes and Golgi apparatus from rat liver in the same sucrose gradient, Eptl. Cell Res. 130:393 (1980).
21. P. L. Sawant, S. Shibko, U. S. Kumta, and A. L. Tappel, Isolation of rat liver lysosomes and their general properties, Biochim. Biophys. Acta 85:84 (1964).
22. K. Ueno, Y. Kushi, C. Rokukawa, and S. Handa, Distribution of gangliosides in parenchymal and non-parenchymal cells of rat liver, Biochem. Biophys. Res. Comm. 105:681 (1982).
23. G. Tettamanti, F. Bonali, S. Marchesini, and V. Zanbotti, A new procedure for the extraction, purification and fractionation of brain gangliosides, Biochim. Biophys. Acta 296:160 (1973).
24. L. D. Bergelson, Lipid biochemical preparations, Elsevier North-Holland Biomedical Press (1980).
25. W. E. Pricer and G. Ashwell, The binding of desialylated glycoproteins by plasma membranes of rat liver, J. Biol. Chem. 246:4825 (1971).
26. L. Svennerholm, The gangliosides, J. Lipid Res. 5:145 (1964)

Mutants and Development

GLYCOSPHINGOLIPIDS OF CHICKEN SKELETAL MUSCLE IN EARLY DEVELOPMENT

AND GENETIC DYSTROPHY

Edward L. Hogan, Jaw-Long Chien and Somsankar Dasgupta

Department of Neurology
Medical University of South Carolina
Charleston, SC 29425

INTRODUCTION

Our systematic characterization of the glycosphingolipids (GSL) of chicken skeletal muscle has revealed gangliosides of the lactosyl- (GM3, GD3), lactosylneotetraosyl- or lactosaminyl- (MG-IV*, MG-VI), ganglio- (GD1a, GT1) and globo-(MG-V, DG-V) series. In fact, these latter globo-gangliosides were the first gangliosides described that contained the globoside oligosaccharide sequence.[1,3] The major neutral GSL in chicken muscle of the Leghorn strain is the Forssman hapten pentaglycosylceramide.[4]

---

*Gangliosides are designated according to the nomenclature of Svennerholm, L. (1963) J. Neurochem. Vol. 10:613 and IUPAC-IUB Commission on the Nomenclature of Lipids (1977) Eur. J. Biochem. 79:11-21. We use a simple designation for gangliosides of the nLcOseCer and GbOseCer series on a base formula of two capital letters and a following Roman numeral. The first letter indicates the number of sialyl residues as M (mono), D (di) etc., the second is G (ganglioside) and the following Roman numeral indicates the number of hexosyl moieties in the backbone chain. A lower case letter following the G can be used to indicate the backbone chain (e.g. b=Gb, n=nLc, etc.). Thus, our designations for gangliosides containing glucosamine are MG-IV or MGn-IV = IV^3 NeuAcnLcOse$_4$Cer and MG-VI = VI^3 NeuAcnLcOse$_6$Cer and of the globo-series are MG-V or MGb-V = V^3 NeuAcGalGbOse$_4$Cer and DG-V = V^3 NeuAc$_{(2)}$ GalGbOse$_4$Cer. Abbreviations of the neutral GSL are LacCer = CDH = lactosyl ceramide, CTH = ceramide trihexoside (GbOse$_3$Cer).

Mature skeletal muscle has a multinucleate syncytial nature resulting from a developmental stage of fusion of uninucleate myoblasts to form myotubes. In the embryo, the predominant ganglioside of chicken leg muscle is GD3 with a shift after hatching to GM3 as the major ganglioside.[5] The activities of two galactosyltransferases have been examined in chicken embryo pectoral muscle and found to decline after the 16th day of embryonic life.[6] In rabbit, the sialyltransferase synthesizing GM3 (CMP-sialic acid:LacCer-sialyltransferase) decreased 6 and 18 fold in sarcolemma and sarcoplasmic reticulum respectively in transition from neonate to adult.[7]

In view of the concentration of GSL at the cell surface and their possible involvement in the process of muscle cell plasmalemmal fusion, we have examined the course in embryonic and early postnatal life the acidic and neutral GSL of chicken skeletal (pectoral) muscle as well as the activities of relevant transferases and glycosidases. We have also studied a genetically matched strain of Leghorn chickens bearing a mutant gene manifest as muscular dystrophy. This mutant is useful in seeking antecedent changes since it does not exhibit substantial abnormality of pectoral strength, alteration in serum levels of creatine kinase or pyruvate kinase[8] or muscle histology[9,10] at these ages.

METHODS

Isolation of GSL

Normal chickens and their genetically matched dystrophic strain were obtained from Dr. Louis Pierro at the University of Connecticut. Pectoral muscles from normal and dystrophic chickens were dissected free of extraneous tissues at carefully timed ages of embryonic and the first week of post-hatching life. Approximately 4 g of muscle were homogenized three times with 20 ml of tetrahydrofuran:0.01M KCl(7:1). These supernatants were combined, dried in a rotary evaporator and treated with 6 ml of 0.6N NaOH in methanol. After neutralization they were evaporated to dryness and dialyzed against distilled water at 4°C with three changes of water. The retentate were dried in a Savant evaporator, dissolved in methanol and applied to Sepharose CL-6B (acetate form). Neutral lipids and asialosphingolipids were eluted with three volumes of methanol and the gangliosides were eluted with 0.1 M ammonium acetate in methanol.[11] The ammonium acetate was removed by Sep-Pak according to the method of McCluer.[12]

Determination of sialyltransferase activity

Approximately 0.5 g of muscle was homogenized in 2 ml of homogenization medium containing 0.5 M sucrose, 1% dextran (average

M.W. 225,000), 5 mM mercaptoethanol, 1 mM $MgCl_2$ and 0.04 M Tris-maleate buffer pH 6.4. The homogenates were centrifuged at 3000 x g for 30 minutes and the supernatant used as enzyme source. Three glycolipid substrates were used as acceptors for sialic acid: LacCer, nLcOse$_4$Cer and nLcOse$_6$Cer. They were obtained by neuraminidase treatment of their sialyl counterparts. The assay system includes glycolipid acceptor, 0.025 µmole; Triton CF-54, 80 µg, HEPES buffer pH 6.8, 10 µmole; $MgCl_2$, 0.25 µmole; CMP-[^{14}C]NeuAc, (1.9×10^6 dpm/µmol) 0.016 µmole; and enzyme (15 µl) in a total volume of 40 µl. The glycolipid and detergent were dissolved in chloroform:methanol(2:1, v/v). Solvent was removed in a Savant concentrator, and buffers and CMP-[^{14}C]NeuAc were added. The reaction was initiated by the addition of enzyme, incubated at 37°C for 4 hours and stopped by the addition of 1 ml of methanol. The amount of [^{14}C]NeuAc incorporated was determined as previously described.[13] Activity is expressed as µmoles NeuAc incorporated/mg protein/hr.

Measurement of glycosidases

Tissues were homogenized in the same manner as that used for the assay of sialyltransferase except that dextran was omitted from the medium. An aliquot (15 µl) of homogenate was used for the determination of glycosidases using p-nitro-phenol glycosides (Sigma, St. Louis, MO) as substrate. After incubation for 2 hr, the reaction mixtures were stopped by adding 0.2 M borate buffer pH 9.8, centrifuged and activities measured at 400 µm.

RESULTS

The GSL of pectoral muscle were examined during the last half of embryonic and the first week of post-hatching life. Tissues were obtained at E9 (ninth day *in ovo*) and at 3 day intervals subsequently (E12, E15, E18) and then at the first and seventh days post-hatching (P1, P7). During development the lipid bound sialic acid (LBSA) per gram of wet tissue did not change in concentration to any extent. The major gangliosides found were GM3, GD3, MG-IV (sialylparagloboside) and MG-VI (the hexahexosyl derivative of MG-IV containing an additional lactosamine unit) as we have previously described[11] and has been confirmed.[14] Fig. 1A is a thin-layer chromatogram of pectoral muscle gangliosides during normal development. At P7 the proportion of GD3 decreases abruptly to the low level found in the adult.[2] During development there are no obvious changes in the other gangliosides including MG-VI. The solvent system in this TLC does not resolve GD1a from MG-VI but the GD1a is a minor component. Fig. 1B displays the gangliosides during development in the chickens with muscular dystrophy. We note that GD3 does not diminish at P7, but it is not yet certain whether this

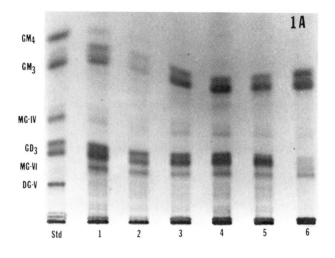
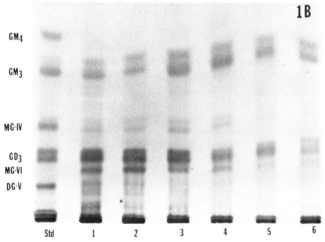

Fig. 1. TLC resolution of pectoral muscle ganglioside at E9, E12, E15, E18, P1 and P7 ages. Fig. 1A; normal muscle. (GD3 is the major ganglioside early but it decreases at P7 and GM3 becomes predominant. Mg-IV increases in the embryo but decreases after hatching. MG-VI remains relatively constant. Chicken embryo muscle contains GM4 which is most prominent at E9 and also GM2 in small concentration at E18.) Fig. 1B; dystrophic pectoral muscle at the same ages. [Ganglioside distribution is similar to normal. At P7, GD3 concentration remains high relative to GM3 and contrasts with the reduction in the normal of GD3 at P7.] Each lane contains glycolipids obtained from 0.6 g of wet tissue with the exception of P1 from dystrophic muscle (0.35g).

is consistent in the mutant. No other differences in the gangliosides of dystrophic, relative to normal, muscle are apparent. It is also noteworthy that at E18, a minor band is observed with an R_f intermediate to GM3 and MG-IV. This is the approximate R_f of GM2 which has been identified in the gangliosides of dystrophic but not normal chicken muscle at age 5 weeks and 2 1/2 years by Kundu et al.[14] At these adult ages there is extensive degeneration of muscle together with increased numbers of satellite, fat and other cells,[9] which hinders interpretation of the appearance of a minor ganglioside band. It now appears that GM2 occurs in embryonic normal as well as dystrophic muscle.

The changes in neutral GSL during development are more conspicuous (Fig. 2). Progressive increases in ceramide trihexoside (GbOse$_3$Cer), globoside and Forssman hapten are seen. The neutral GSL from normal muscle are in the odd-numbered lanes at E9, E12, E15, E18, P1 and P7. The shift to longer oligosaccharide chain lengths in the globo-series is first apparent at E15 and quite obvious at E18. This correlates with the phase in myogenesis characterized by formation of syncytial myotubes and hence corresponds to the stage of in vivo muscle cell fusion. A neutral GSL of longer oligosaccharide chain length is seen through development and diminishes rather dramatically at P7. The structure of this glycolipid is currently being investigated. The even-numbered

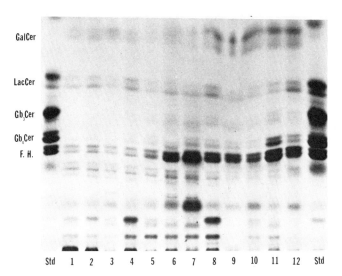

Fig. 2. TLC resolution of pectoral muscle neutral glycosphingolipids at E9, E12, E15, E18, P1 and P7 ages. Odd numbers represent those from normal chickens while even numbers are obtained from dystrophic chickens. Each lane contains glycolipids obtained from 0.6 g of wet tissue with the exception of P1 from dystrophic muscle (0.35 g).

lanes in Fig. 2 contain the glycolipids in the genetic dystrophy. A consistent prominence of galactosylceramide is manifest in the dystrophic tissue beginning at E9. This suggests that a lag in maturation occurs in dystrophy with a higher proportion of the monohexosylceramide.

The activity of sialyltransferase was determined using three lipid acceptors (Table 1). Activity is maximal at E12 and decreases rapidly in later embryonic life and following hatching. These transferase activities are mediated by the same enzyme (Dasgupta, Chien and Hogan, manuscript in preparation). Maximal activity was obtained with the lacto-n-tetraosylceramide ($nLcOse_4Cer$) as substrate. However, in tissue the predominant ganglioside is GM3 suggesting that the sialyltransferase is not the rate-limiting step of muscle ganglioside biosynthesis. No difference in sialyltransferase activity in dystrophic muscle was found at any of these ages of early myogenesis.

The activities of six glycosidases were examined employing artificial substrates. Maximal activities occurred at E12 and declined rapidly thereafter. There was no substantial difference comparing dystrophic and normal tissue though increases ranging from 10 to 20% relative to control were observed for α-galactosidase, α-galactosaminidase and α-mannosidase.

Table 1. Sialyltransferase Activity* of Developing Chick Pectoral Muscles.

AGE (DAYS)	LacCer		$nLcOse_4Cer$		$nLcOse_6Cer$	
	N	D	N	D	N	D
E9	0.157	0.143	0.387	0.356	0.399	0.392
E12	0.670	0.574	1.008	0.899	0.975	0.903
E15	0.160	0.145	0.430	0.371	0.337	0.296
E18	0.084	0.059	0.219	0.204	0.260	0.247
P1	0.030	0.025	0.110	0.080	0.086	0.073
P7	0.005	0.004	0.025	0.026	0.014	0.011

N = Normal   D = Dystrophic
E = Embryonic   P = Post-hatching

*Enzyme activity is µmole SA incorporated/mg protein/hr

Table 2. Glycosidase Activities* of Developing Chick Pectoral Muscles.

AGE (DAYS)	α-Galactosidase		β-Galactosidase		α-Mannosidase	
	N	D	N	D	N	D
E9	0.022	0.027	0.254	0.271	0.236	0.255
E12	0.037	0.043	0.205	0.202	0.171	0.175
E15	0.020	0.020	0.124	0.112	0.127	0.130
E18	0.016	0.020	0.131	0.104	0.125	0.097
P1	0.005	0.012	0.087	0.054	0.020	0.013
P7	0.003	0.002	0.013	0.013	-	-

	α-Galactosaminidase		β-Hexosaminidase		α-Glucosidase	
	N	D	N	D	N	D
E9	0.114	0.158	1.515	1.481	0.027	0.027
E12	0.102	0.109	1.457	1.336	0.016	0.014
E15	0.079	0.096	0.531	0.575	0.010	0.012
E18	0.093	0.114	0.503	0.589	0.006	0.011
P1	0.020	0.040	0.230	0.167	0.007	0.011
P7	0.009	0.011	0.044	0.053	0.008	0.011

*Enzyme Activity in μmole/mg protein/hr

DISCUSSION

The interesting feature of GSL changes in development is the substantial change in GSL occurring at the stage of myogenesis corresponding to the appearance of the myotube. This compositional change involves at least three of the neutral GSL of the globo-series and may involve in addition one or more GSL of longer oligosaccharide length (but whose structure(s) have not yet been determined). The Forssman glycolipid is quantitatively the most prominent of the glycolipids affected. Though much remains to be done in defining the cellular location and possible role of these molecules in muscle cell fusion, it seems likely that they are cell surface constituents and may therefore be involved either directly in cellular recognition and interaction events or in the secondary changes of membrane reorganization. The detection of lectins in embryonic chicken muscle for galactosamine[17] and lactose[18] further strengthens the possibility of the involvement of these glycolipids in this process. Also, we are currently pursuing this finding of fusion-linked change in GSL disposition by examination of GSL in primary muscle cell cultures (K. Leskawa et al., unpublished). In this system, the stages of myogenesis can be synchronized by varying the extracellular concentration of calcium.

There are several lines of evidence suggesting that human muscular dystrophy results from a membrane abnormality affecting the sarcolemma (for review, see 16). In this study, there have not been major changes in gangliosides and/or neutral GSL in dystrophic chicken skeletal muscle during development. Also, there is no evident change in the activity of the sialyltransferase specific for the main gangliosides. The increases found in glycosidase activities are not very great and are not considered likely to be specific for a genetic error in the primary structure of these enzymes. We are inclined to attribute the increases found to an increase in tissue substrate levels, since increased gangliosides and neutral GSL in adult dystrophic muscle have been found.[2,14,15] If the genetic disorder in chicken dystrophy were to affect primarily a membrane constituent, we would have expected a more obvious change in the GSL (which are located mainly in the cell surface membrane) than was found.

SUMMARY

The acidic and neutral GSL of chicken pectoral muscle and the activities of relevant sialyltransferase and glycosidases have been examined during embryonic and early post-hatching development. At this stage of myogenesis, a prominent shift to the neutral GSL of longer oligosaccharide length involving Forssman glycolipid most prominently and also globoside and $GbOse_3Cer$ occurred but the distribution of muscle-type gangliosides was not obviously affected. The glycosidase and sialyltransferase activities decreased dramatically just prior to or at hatching. The fusion-linked change in GSL suggests a role for terminal galactosamine and/or galactose residues in myoblast aggregation. A parallel developmental study of genetic muscular dystrophy revealed similar GSL levels and enzyme activities. A larger proportion of lactosylceramide in dystrophic muscle throughout development suggests a developmental lag in the mutant.

ACKNOWLEDGEMENT

Supported by the Muscular Dystrophy Association and PHS Grant NS-16120. We thank Ms. Dianne Collins for preparing the manuscript.

REFERENCES

1. J.-L. Chien and E. L. Hogan, Novel gangliosides containing gal-globotetraosyl sequence in chicken muscle, Fed. Proc. 39:2183 (1980).
2. E. L. Hogan, R. D. Happel, and J.-L. Chien, Membrane glycosphingolipids in chicken muscular dystrophy, in: "New Vistas in Glycolipid Research," H. Makita, S. Handa, T. Taketomi and Y. Nagai, eds. Plenum Publishing Corporation, New York, (1982).
3. J.-L. Chien and E. L. Hogan, Novel pentahexosyl ganglioside of the globo series purified from chicken muscle, J. Biol. Chem. 258:10727 (1983).
4. J.-L. Chien and E. L. Hogan, Glycosphingolipids of skeletal muscle, in: "Cell Surface Glycolipids," C. C. Sweeley, ed., Am. Chem. Soc., (1980).
5. M. Saito and A. Rosenberg, Glycolipids and their developmental patterns in chick thigh and leg muscles, J. Lipid. Res. 23:3 (1982).
6. S. Ullrich, C. Kent, and P. M. Carlson, Changes in galactosyltransferase activity in chick pectoral muscle during embryonic development, Biochem. J. 196:17 (181).
7. G. F. Clark and P. B. Smith, Studies on glycoconjugate metabolism in developing skeletal muscle membranes, Biochem. Biophys. Acta 755:56 (1983).
8. E. A. Barnard and P. J. Barnard, Use of genetically dystrophic animals in chemotherapy trials and application of serotonin antagonists as antidystrophic drugs, Ann. N. Y. Acad. Sci. 317:374 (1979).
9. J. A. Pizzey and E. A. Barnard, Structural changes in muscles of the dystrophic chicken. I. Quantitative indices, Neuropath. and Applied Neurobiol. 9:21 (1983).
10. J. A. Pizzey and E. A. Barnard, Structural changes in muscles of the dystrophic chicken. II. Progression of the Histopathology in the Pectoralis Muscle, Neuropath. and Applied Neurobiol. 9:149 (1983).
11. J.-L. Chien and E. L. Hogan, Characterization of two gangliosides of the paragloboside series from chicken skeletal muscle, Biochem. Biophys. Acta 620:454 (1980).
12. M. A. Williams and R. H. McCluer, The use of Sep-Pak C_{18} cartridges during the isolation of gangliosides, J. Neurochem. 35:266 (1980).
13. J.-L. Chien, T. Williams, and S. Basu, Biosynthesis of a globoside-type glycosphingolipid by a β-N-acetylgalactosminyl-transferase from embryonic chicken brain, J. Biol. Chem. 248:1778 (1973).
14. S. K. Kundu, L. K. Misra, and G. Luthra, Muscle glycolipids in inherited muscular dystrophy in chickens, FEBS Letters 150:359 (1982).

15. T. Miyatake, K. Nakamura, T. Ariga, and T. Atsumi, Gangliosides in dystrophic chicken muscle, in: "Muscular Dystrophy," Jap. Med. Res. Found. Publication No. 16. Bashi, SE, Ed., Univ. of Tokyo Press (1982).
16. L. P. Rowland, Biochemistry of muscle membranes in Duchenne muscular dystrophy, Muscle and Nerve 3:3 (1980).
17. H. Ceri, D. Kobiler, and S. H. Barondes, Heparin-inhibitable lectin purification from chicken liver and embryonic chicken muscle, J. Biol. Chem. 256:390 (1981).
18. T. P. Nowak, D. Kobiler, L. E. Roel, and S. H. Barondes, Developmentally regulated lectin from embryonic chick pectoral muscle, purification by affinity chromatography, J. Biol. Chem. 252:6026 (1977).

GANGLIOSIDE ALTERATIONS IN THE GENETICALLY-DETERMINED HYPERTROPHIC

NEUROPATHY OF THE MURINE NEUROLOGICAL MUTANT TREMBLER

Marie-Luce Harpin, Jacques Portoukalian*, Bernard Zalc
and Nicole Baumann

Laboratoire de Neurochimie, INSERM U.
134, Hopital de la Salpetriere
75651 Paris Cedex 13
*Laboratoire d'Immunologie
Centre Leon Berard
69008 Lyon, France

INTRODUCTION

Hypertrophic neuropathy is a standard pathological reaction of peripheral nerve that occurs in different diseases. On transverse sections, it is characterized by onion-bulb disposition of proliferated Schwann cells around nerve fibers. These hypertrophic onion-bulb changes were first recognized in the hereditary neuropathies. Until recently, they were considered specific for Dejerine-Sottas disease, but from clinical and experimental studies they are known to be the non-specific consequence of Schwann cell proliferation resulting from repeated segmental demyelination and remyelination.[1]

The Trembler mutant has appeared to be a very useful experimental tool to investigate hypertrophic neuropathies. The mutant has been described by Falconer.[2] It has an autosomal dominant inheritance. Ayers and Anderson[3] by histological examination revealed a peripheral onion-bulb neuropathy. The young mice show retardation of myelin development, generalized myelin degeneration and early onion-bulb development. Adult animals present a picture almost identical to hypertrophic neuropathy in the human with slowing of nerve conduction velocities,[4] hypomyelination, segmental demyelination, onion-bulbs and an increase in endoneural and perineural connective tissues. The mean axonal diameter of the peripheral nerves is always smaller than normal and the ratio of myelin thickness to axon diameter is reduced.

Experiments using transplantation techniques[5] prove that the defect of myelination resides in a primary Schwann cell disorder rather than in an axonal defect. Perkins et al.[6,7] have shown an abnormal persistence of postnatal Schwann cell proliferation in the Trembler mouse nerves which, in unaffected animals, are composed of myelinated fibers. The morphology of the Schwann cells, their number and rate of proliferation appear to be normal in unmyelinated fibers.

The presence of gangliosides in peripheral nerves is now well established.[8-12] They have been localized in myelin[10,13] in unmyelinated nerves,[13] and in those undergoing Wallerian degeneration.[11] Several arguments are in favor of gangliosides being involved in neuronal growth control differentiation[14-19] and neuronal-glial interaction.[19,20] However, it is not yet clear whether they are directly involved in axon-Schwann cell recognition in relation to myelination and Schwann cell proliferation. In this respect, the Trembler mutation seemed interesting to study because it has a defect in PNS myelination related to an impairment in Schwann cell differentiation. This study was performed so as to detect modifications of the ganglioside pattern which could be related to the morphological cellular and subcellular abnormalities. Previous investigations have been performed on 2 month-old animals.[21] Here we report studies at an earlier stage (25-35 day old animals) which allow comparisons at different developmental periods.

MATERIALS AND METHODS

Mice

The animals were raised in our laboratory. The autosomal dominant mutation is on an unknown original background.[22] Males carrying the genetic defect have been mated to B6D2F1 normal females for many generations in our laboratory. The offsprings were used for the experiments and B6D2F1 as controls.

Dissection of the sciatic nerve

Sciatic nerves were carefully dissected from their origin to the knee. They were kept at -30°C until use.

Extraction and purification of gangliosides

Before lipid extraction, the sciatic nerves were lyophilized. Thereafter, they were crushed in a mortar in the presence of Fontainebleau sand which had been previously washed with a mixture of chloroform:methanol:water 70:30:4 (v/v/v) and dried. Lipid extraction was performed on approximately 1-1.5g of wet weight of nerves. Extraction was performed according to Pollet et al.[23] in chloroform:methanol:water 70:30:4 (v/v/v), 1 ml for each 20 mg dry weight using sonication, and further extraction for one hour at +4°C by magnetic stirring. The lipid extract was centrifuged at 900 g, dried under nitrogen and weighed. It was resuspended in chloroform:methanol 2:1 (v/v) and washed as follows according to Folch et al.[24] with modifications: for 10 ml of chloroform:methanol, 2 ml of water were added. After thorough mixing and centrifugation, the upper phase containing the gangliosides was kept. To the lower phase, were added 2 ml of methanol so as to complete the volume to what it was previously, and 2 ml of water. The partition was performed as previously. The same operation was performed once more and the three upper phases were pooled together. The gangliosides that partitioned in the upper phase were further purified using Sep-Pak (Waters Associates) cartridges.[25] The cartridges were washed as described in the procedure using methanol or mixtures of chloroform-methanol, except that the final wash of methanol was followed by an equilibration wash of the cartridge with 10 ml of methanol:saline 1:1 (v/v). To the combined upper phases were added an equal volume of saline; the mixture was then applied to the cartridge with slight pressure, collected and reapplied again to the column. Salts were eluted with 20 ml distilled water. Gangliosides were eluted with 15 ml of chloroform:methanol, 2:1, concentrated under nitrogen and redissolved in the original volume. Before being applied to peripheral nerve, this procedure was checked on brain gangliosides: the yield was satisfactory (over 90%) and the densitometric profile of resorcinol positive spots (see below) was identical.

Ganglioside distribution and characterization

Chromatography of gangliosides was performed on HPTLC thin-layer plates of silica gel G 60 F (Merck Co) and developed in chloroform:methanol:0.25% $CaCl_2$ 60:35:8 (v/v/v). The total material put on each lane was equivalent to 2 µg neuraminic acid. The relative percentage of gangliosides was determined as a function of neuraminic acid concentration by densitometry according to Dreyfus et al.[26] on Cliniscan (Helena) using resorcinol-HCl reagent[27] spray. Gangliosides were characterized according to known standards provided by one of us (J. Portoukalian) and generously given by S. I. Hakomori (University of Washington). The standard ganglioside mixture was provided by Fidia laboratories.

Immunocharacterization of gangliosides

Characterization of gangliosides was also performed using antibodies for specific gangliosides: polyclonal IgG anti GM_1 antibody provided by Dr. M. J. Coulon-Morelec (Pasteur Institute) and a monoclonal human IgM anti GM_3 previously characterized.[28]

The glycolipid antigens were detected on thin-layer chromatograhy by using a modification of the technique of Magnani et al.[29] Briefly, thin-layer chromatography of glycolipid antigens were performed on Schleicher and Schull TLC-ready-foils silica-gel plates F 1500. The technique was performed thereafter as described[29] except that the antibody solution was incubated at 37°C for 3 hours and that an indirect immunoperoxydase procedure was developed instead of using an ^{125}I-labelled second antibody. For immunodetection horseradish peroxydase-conjugated sheep anti rabbit IgG (Pasteur) was used for polyclonal anti GM_1 IgG and horseradish peroxydase conjugated sheep antihuman IgM (Pasteur) for monoclonal anti GM_3 IgM. Development was performed using as reagent 4-chloro-1-naphtol (Merck) and hydrogen peroxide as described by Hawkes et al.[30] Orcinol reagent was used for total glycolipid detection.[31]

RESULTS AND DISCUSSION

Characterization of gangliosides

As shown on Fig. 1, the ganglioside profile of B6D2F1 mice peripheral nerve is different from what is observed for CNS brain gangliosides. In contrast to CNS, peripheral nerve gangliosides are composed of major amounts of GM_3 (Table 1 and Figs. 1 & 2). The disialoganglioside fraction contained mostly gangliosides migrating as GD_{1a} and GD_{1b} was also present. As for rat peripheral nerve,[12] GD_{1a} migrated as a doublet as well as GM_3. As reviewed recently by Chou et al,[12] TLC band splitting can be due to differences in fatty acid chain length, sialic acid, hexosamines, location of sialic acid or any combination of these factors. Sialosylparagloboside (SPG) which is an important component of rat trigeminal nerve, was not evident or barely detectable in our preparations. Densitometry of resorcinol positive spots hardly detected the presence of GM_1; however using a specific IgG antibody against GM_1, this ganglioside could be clearly demonstrated (Fig. 3). The use of an anti GM_3 monoclonal IgM confirmed the presence of GM_3 but it was not possible to distinguish between NeuNAc or NeuNGly GM_3. The data observed did not differ widely from what had been shown previously for 2 month-old animals, confirming that, as for the rat,[12] most changes in development occur before one month of age. An exception was GM_1 which remains more detectable in the adult. Each preparation corresponds to a ganglioside pooled extract from 1 g fresh sciatic nerves. Each determination was repeated twice. Percentage of neuraminic acid were determined on TLC plates by densitometry of resorcinol-HCl positive spots.

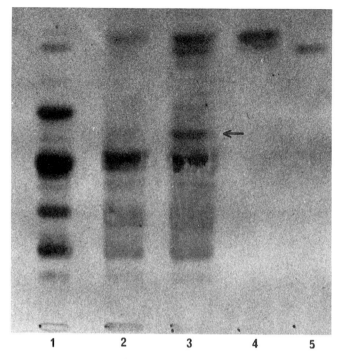

Fig. 1. Thin-layer chromatogram of gangliosies of peripheral nerve from B6D2F1 and Trembler mice. Lane 1: standard mixture of brain gangliosides + GM_3 containing from top to bottom GM_3, GM_1, GD_{1a}, GD_{1b}, and GT_{1b}. Lane 2: gangliosides from sciatic nerves of B6D2F1 mice. Lane 3: gangliosides from Trembler sciatic nerves. Arrow points to GD_3. Lane 4: NeuNAc GM_3 standard. Lane 5: NeuNGly GM_3 standard. 2 µg of ganglioside mixtures and 1 µg of standards were applied to HPTLC plates which were developed with chloroform:methanol:0.25% $CaCl_2$ 60:35:8. Spots were visualized with resorcinol-HCl reagent.

Ganglioside composition of Trembler mice peripheral nerves

As shown on Fig. 1, the ganglioside pattern of peripheral nerve in the Trembler neurological mutant was closely related to what is observed in B6D2F1 controls except for the importance of the content in GM_3 and the presence of GD_3. Although hardly detectable by densitometry, GM_1 was present as shown by immuno-detection (Fig. 3). By densitometry (Fig. 2), GM_3 appeared clearly as 3 bands, the nature of which remained undetermined although the lower band could be related to the presence of NeuNGly GM_3 when compared to the migration of a known standard (Fig. 1). The increase in GD_3 content appeared to be also very

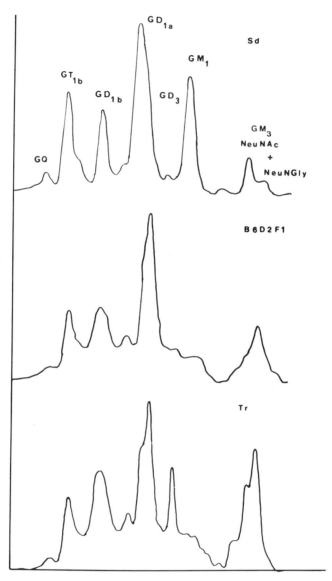

Fig. 2. Densitometric scanning of chromatograms of peripheral nerve gangliosides. Resorcinol-HCl positive spots were measured. Top: standards from pig brain to which were added NeuNAc GM_3 and NeuGly GM_3 (first peak). Center: B6D2F1 control nerves. Bottom: Trembler mutant nerves.

Table 1. Distribution of Gangliosides in Sciatic Nerves of B6D2F1 Mice and Trembler Mutants (Expressed as Percentage of Total Neuraminic Acid).

Ganglioside	B6D2F1	Trembler
GQ	1	1
GT1b	12.7	9
GD1b	13.5	15
?	1.7	2.3
GD1a	39.6	25.2
GD3	0	7
GM1	4	2.5
GM3	27.5	38

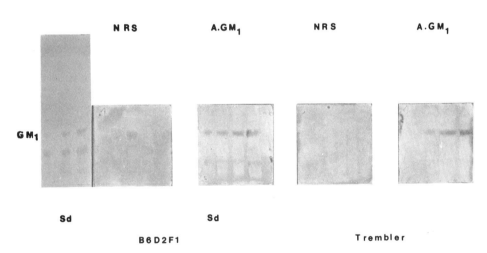

Fig. 3. Immunodetection of peripheral nerve GM_1 on TLC using anti GM_1 polyclonal IgG and an indirect immunoperoxidase procedure. Sd: standard mixture of gangliosides detected by orcinol reagent. NRS: normal rabbit IgG controls. A.GM_1: anti GM_1 IgG. For B6D2F1, 3 lanes were spotted containing from left to right 1.25, 2.5 and 5 µg ganglioside NeuNAc. For Trembler 4 lanes were spotted containing from left to right 0.25, 0.5, 1 and 2 µg ganglioside NeuNAc.

characteristic of the Trembler mutation. The same abnormalities in relation to the normal developmental pattern have been observed in 2 month-old animals. GM_3 and GD_3 are considered to be major constituents of immature and proliferating Schwann cells;[12] as nerve Schwann cells mature, the mole percentage composition of GM_3 and GD_3 declines. The results observed here in the Trembler mutation are consistent with this observation; there is a retardation in peripheral nerve maturation and an abnormal Schwann cell proliferation. The existence of qualitative modifications especially in relation to GM_3 remains to be proven. Further studies will have to be performed to determine whether this abnormal ganglioside pattern is directly related to the Schwann cell ferentiation abnormality causing abnormal proliferation in the mutant.

REFERENCES

1. P. K. Thomas, Hereditary Neuropathies, Trends Neurosci. 4:X (1981).
2. D. Falconer, Two new mutants, Trembler and reeler, with neurological action in the house mouse (Mus Musculus L), J. Genet. 50:192 (1951).
3. M. Ayers, R. Anderson, Onion bulb neuropathy in the Trembler mouse: a model of hypertrophic interstitial neuropathy (Dejerine-Sottas) in man, Acta Neuropath. (Berl.) 25:54 (1973).
4. P. A. Low, J. G. McLeod, Hereditary demyelinating neuropathy in the Trembler mouse, J. Neurol. Sci. 26:565 (1975).
5. A. J. Aguayo, M. Attiwell, J. Trecarten, S. Perkins, G. M. Bray, Abnormal myelination in transplanted Trembler mouse Schwann cells, Nature, Lon. 265:73 (1977).
6. C. S. Perkins, A. J. Aguayo and G. M. Bray, Schwann cell multiplication in Trembler mice, Neuropath. and Applied Neurobiol. 7:115 (1981).
7. C. S. Perkins, A. J. Aguayo and G. M. Bray, Behavior of Schwann cells from Trembler mouse unmyelinated fibers transplanted into myelinated nerves, Exp. Neurobiol. 71:515 (1981).
8. A. J. Yates and J. R. Wherrett, Changes in the sciatic nerve of the rabbit and its tissue constituents during development, J. Neurochem. 23:993 (1974).
9. F. Klein and P. Mandel, Gangliosides of the peripheral nervous system of the rat, Life Sci. 16:751 (1975).
10. J. W. Fong, R. W. Ledeen, S. K. Kundu and S. W. Brostoff, Gangliosides of peripheral nerve myelin, J. Neurochem. 26:157 (1976).
11. A. J. Yates and D. K. Thompson, Ganglioside composition of peripheral nerve undergoing Wallerian degeneration, J. Neurochem. 30:1649 (1978).

12. K. H. Chou, C. E. Nolan and F. B. Jungalwala, Composition and metabolism of gangliosides in rat peripheral nervous system during development, J. Neurochem. 39:1547 (1982).
13. J. H. Hofteig, J. R. Mendell and A. J. Yates, Chemical and morphological studies on garfish peripheral nerves, J. Comp. Neur. 198:265 (1981).
14. J. I. Morgan and W. Seifert, Growth factors and gangliosides: a possible new perspective in neuronal growth control, J. Supramol. Struct. 10:111 (1979).
15. D. P. Purpura and H. J. Baker, Neurite induction in mature cortical neurons in feline GM1-ganglioside storage disease, Nature, 266:553 (1977).
16. M. Willinger and M. Schachner, GM1 gangliosie as a marker for neuronal differentiation in mouse cerebellum, Dev. Biol. 74:101 (1980).
17. J. J. Hauw, S. Fenelon, J. M. Boutry and R. Escourolle, Effet des gangliosides sur la croissance de ganglion spinal de cobaye en culture in vitro. Resultats preliminaires concernant une preparation de gangliosides de cortex cerebral de boeuf, C. R. Acad. Sci. 292:569 (1981).
18. F. J. Roisen, H. Bertfeld, R. Nagele and G. Yorke, Ganglioside stimulation of axonal sprouting in vitro, Science, 214:577 (1981).
19. M. M. Rapport, A. Gorio, eds., Gangliosides in Neurological and Neuromuscular Function, Development and Repair, Raven Press, New York (1981).
20. P. Mandel, P. Dreyfus, H. Yusufi, AN. N. K. Sarlieve, L. Robert, J. Neskovic, N. Harth, S. and G. Rebel, Neuronal and glial cell cultures, a tool for investigation of ganglioside function, Adv. Exp. Med. Biol. 125:515 (1980).
21. M. L. Harpin, J. Portoukalian and N. Baumann, Modifications of ganglioside composition in peripheral nerve of myelin-deficient Trembler mutant mouse, Neurochem. Res. 7:1367 (1982).
22. F. Lachapelle, B. Zalc, N. Baumann and J. L. Guenet, "Production and use of genetically uniform dysmyelinating mutants of the mouse," in: Neurological Mutations Affecting Myelination, N. Baumann, ed., Elsevier/North-Holland, Amsterdam, (1980).
23. S. Pollet, S. Ermidou, F. Le Saux, M. Monge and N. Baumann, Microanalysis of brain lipids: multiple two-dimensional thin-layer chromatography, J. Lipid Res. 19:916 (1978).
24. J. Folch, M. Lees and G. H. Sloane-Stanley, Method for the isolation and purification of total lipids from animal tissues, J. Biol. Chem. 226:497 (1957).
25. M. A. Williams and R. H. McCluer, The use of Sep-Pak C18 cartridges during the isolation of gangliosides, J. Neurochem. 35:266 (1980).
26. H. Dreyfus, P. F. Urban, P. Bosch, S. Edel-Harth, G. Rebel and P. Mandel, Effect of light on gangliosides from calf

retina and photoreceptors, J. Neurochem. 22:1073 (1974).
27. L. Svennerholm, "Ganglioside metabolism," in: Comprehensive Biochemistry, M. Florkin and E. H. Stotz eds., Elsevier/North-Holland, Amsterdam, (1970).
28. Proceedings of the VIth International Symposium on Glycoconjugates, T. Yamakawa, T. Osawa, S. Handa, Japan Society, Soc. Press, Tokoyo, (1981).
29. J. L. Magnani, B. Nilsson, M. Brockhaus, D. Zopf, Z. Steplewski, H. Koprowski and V. Ginsburg, A monoclonal antibody-defined antigen associated with gastrointestinal cancer is a ganglioside containing sialylated lacto-N-fucopentose II. J. Biol. Chem. 257:14365 (1982).
30. R. Hawkes, E. Niday and J. Gordon, A dot-immunobinding assay for monoclonal and other antibodies, Anal. Biochem. 119:142 (1982).
31. V. P. Skipski and M. Barkley, in: Methods in Enzymology, J. M. Lowenstein ed., 14:545 (1969).

COMPARATIVE AND DEVELOPMENTAL BEHAVIOR OF ALKALI LABILE

GANGLIOSIDES IN THE BRAIN

R. Ghidoni, S. Sonnino, V. Chigorno, A. Malesci, and
G. Tettamanti

Dept. of Biological Chemistry, The Medical School
Univ. of Milan, Milan, Italy

INTRODUCTION

In vertebrate tissues gangliosides are more abundant in brain than in other organs. About 50 different molecular species have been isolated and structurally characterized. The number of new gangliosides is rapidly increasing, due to the development of new technologies capable of detecting and characterizing very minor and less accessible components.

Recently a two-dimensional TLC procedure, with densitometric spot quantification, has been devised for the separation of ganglioside mixtures. It has the advantage of recognizing ganglioside species containing ester linkages (alkali labile gangliosides).[1,2] Alkali lability can be attributed to the presence of inner esters (lactones) involving the carboxyl group of sialic acid, or the O-acetylation (or O-acylation, in general) of some of the ganglioside sugar components. The presence in gangliosides of both inner esters and O-acetylated sialic acid has received experimental

---

Abbreviations used: the ganglioside nomenclature of Svennerholm[14] is used; 9-O-Ac-GT1b, GT1b containing 9-O-acetyl,N-acetylneuraminic acid; NeuAc, N-acetylneuraminic acid; 9-O-Ac-GQ1b, GQ1b containing 9-O-acetyl,N-acetylneuraminic acid; TLC, thin layer chromatography; HPTLC, high performance thin layer chromatography; GLC, gas liquid chromatography; MS, mass spectrometry.

support. Lactones of gangliosides, first found by Wiegandt,[3] were observed in rat brain.[5] 9-0-Acetyl,N-acetylneuraminic acid was recognized as one of the sialosyl residues present in ganglioside mixtures extracted from brains of different animal species,[6] and in two individual gangliosides isolated from mouse brain.[7,8] 4-0-Acetyl,N-glycolyneuraminic acid is present in a ganglioside isolated from equine erythrocytes.[9]

The present report presents data that stress the importance of alkali labile gangliosides. Data are presented on the occurrence of alkali labile gangliosides in brains of different animals and on the relationship of this family of gangliosides with some developmental and behavioral phenomena.

RECOGNITION, QUANTIFICATION AND ANALYSIS OF ALKALI LABILE GANGLIOSIDES

The methodological approach to recognition and quantification of alkali labile gangliosides in the brain has been described in detail in previous papers.[1,2] It is based on two-dimensional TLC carried out as follows: (a) use of silica gel HPTLC plates (Merck); (b) first development with the solvent chloroform/methanol/0.2% aqueous $CaCl_2$, 50/42/11 by vol (40 min at 20°C); (c) exposure of the dried plates to ammonia vapor for 5 hours at 20°C; (d) complete removal of ammonia by ventilation; (e) second development with the same solvent as above (40 min at 20°C); (f) location of the spots by treatment with a p-dimethylaminobenzaldehyde spray reagent followed by heating at 110°C for 10 min; (g) quantification of the spots by densitometric scanning with a computer assisted Camag densitometer. The strategy for detection of alkali labile gangliosides consists of submitting one sample of the ganglioside mixture to the above procedure with omission of the alkali treatment step, and a second sample to the complete procedure. Comparison of the two patterns obtained reveals the presence of alkali labile gangliosides. The method employed for compositional analyses, structural determinations and assessment of the chemical nature of ester linkages (0-acyl group; lactone) present in alkali labile gangliosides have already been reported.[2,4,7,8]

CONTENT OF ALKALI LABILE GANGLIOSIDES IN DIFFERENT ANIMAL BRAINS

Ganglioside mixtures were extracted and purified from the whole brain of 24 different animal species by the tetrahydrofuran/potassium phosphate buffer method[10] and were analyzed for alkali labile gangliosides. As shown in Fig. 1, the content of alkali labile gangliosides ranged from a low value of 0.5% of total gang-

Fig. 1. Content of alkali labile gangliosides in the whole brain of different animals. Total gangliosides were determined as bound sialic acid.[15] The content of alkali labile gangliosides is given as percentage of the total ganglioside content.

liosides in ox brain to a high value of 90% in lamprey brain. In some animals (lizard, blind worm, nigerian cichlid, bass fish and lamprey) alkali labile gangliosides were markedly preponderant. Because of the wide variability encountered and the relatively low number of examined animal species, it was not possible to correlate the content of alkali labile gangliosides to the phylogenetic position of the animal. However, lower vertebrates had a content of alkali labile gangliosides which was generally higher than that of birds and mammals. The variability of alkali labile ganglioside content in the 10 fish species examined may suggest that the proportion of alkali labile gangliosides is influenced by environmental factors, such as water temperature and depth, and seasonal variations.

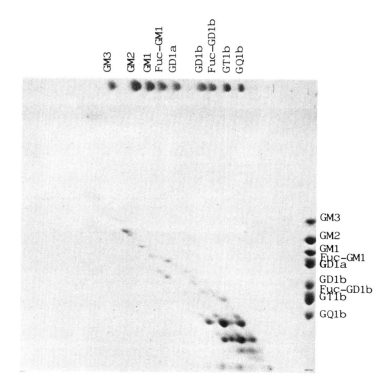

Fig. 2. Two-dimensional TLC with intermediate ammonia treatment of bass fish brain gangliosides. Monodimensional runs of standard gangliosides are shown for comparison.

CHEMICAL STUDIES ON THE ALKALI LABILE GANGLIOSIDES OF THE BRAIN OF SOME ANIMALS

We initiated a systematic study aimed at defining the chemical structure of brain alkali labile gangliosides. For this purpose we chose mouse, bass fish and lamprey since they are readily available and have patterns of alkali labile gangliosides of particular interest.

Mouse brain contains two main alkali labile gangliosides which were isolated in the pure form and were structurally characterized.[7,9] Both of them contain a residue of 9-O-acetyl,N-acetylneuraminic acid. The 9-O-acetyl group is responsible for their alkali lability. They have the basic structure of GT1b and GQ1b gangliosides, respectively. The 9-O-acetylated sialic acid residue is located, in both cases, at the end of the di-sialosyl chain linked to the inner galactose residue.

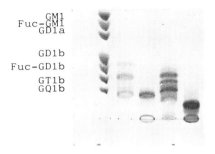

Fig. 3. Monodimensional TLC of the tetrasialo and pentasialoganglioside fractions separated by a DEAE sepharose column, before and after alkaline treatment (0.1M NaOH, overnight at 20°C followed by dialysis). Lane A: standard gangliosides; lane B: tetrasialoganglioside fraction; lane C: same as B after alkaline treatment; lane D: pentasialoganglioside fraction; lane E: same as D after alkaline treatment. Solvent system: chloroform/methanol/0.2% aqueous $CaCl_2$, 50/42/11 by vol.

Bass fish is commonly bred in Italy for alimentary purposes. The total ganglioside content of bass fish brain was 626 g ± 48 g (S.D.) of lipid bound NeuAc/g fresh tissue. The long chain base composition of the total ganglioside mixture was as follows: C18 sphingosine, 86.7%; C18 sphinganine, 12.7%; C20 sphingosine, 1.6%. The most abundant fatty acid was C18:0, which accounted for 49.4% of the total fatty acid content, followed by C18:1 (19.6%), C16:0 (12.7%) and C24:1 (5.6%). The sialic acid/sphingosine ratio was 4.1, indicating that the polysialylated species of gangliosides were particularly abundant. A two-dimensional TLC separation of bass fish brain gangliosides, obtained by the use of intermediate ammonia treatment, is given in Fig. 2. Most of the ganglioside spots appeared to be off the straight line expected[2] for alkali stable gangliosides. This indicates that they underwent a chemical modification during exposure to ammonia; in other words, they were alkali labile. GLS-MS analyses[11] of the sialic acids, released from the total ganglioside mixture by exhaustive sialidase treatment,[7] demonstrated the presence of 9-O-acetyl,N-acetylneuraminic acid. This represented about 30% of the total sialic acid content, the remainder being N-acetylneuraminic acid. Therefore, the alkali lability of bassfish gangliosides is most likely caused by the presence of O-acetylated sialic acid. Moreover, the abundance of polysialylated species of gangliosides suggests that some gangliosides of bass fish brain may contain more than one O-acetylated sialic acid per molecule.

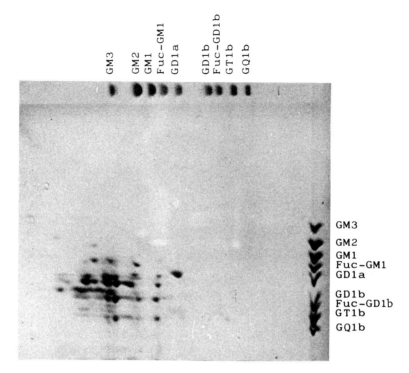

Fig. 4. Two-dimensional TLC with intermediate ammonia treatment of lamprey brain gangliosides. Monodimensional runs of standard gangliosides are shown for comparison.

Initial attempts were made to isolate and characterize the predominant alkali labile gangliosides present in bass fish brain. As the first approach, the total ganglioside mixture was fractionated according to the degree of ganglioside sialylation. For this purpose the DEAD-Sepharose column chromatographic procedure of Iwamori and Nagai[12] was used, with some modifications (an ammonium acetate gradient in methanol was employed). As shown in Fig. 3, both the tetrasialoganglioside and the pentasialoganglioside fractions were composed of several molecular species, displaying different Rf's, when submitted to conventional[7] monodimensional TLC separation. However the same fractions, after treatment with 0.1 M sodium carbonate[2] followed by dialysis, showed a single ganglioside spot. This indicated that each fraction consisted of a "family" of alkali labile gangliosides related to a single alkali stable ganglioside, a tetrasialoganglioside and a pentasialoganglioside, respectively. Structural studies on these fractions are in progress.

Among the examined animals, lamprey brain had the highest content of alkali labile gangliosides (90% of total). Fig. 4 reproduces a two-dimensional TLC separation, with intermediate

ammonia treatment, of lamprey brain gangliosides. The pattern is
extremely complex. Only one alkali stable ganglioside, GD1a, is
clearly visible among a large number of different alkali labile
species. In order to obtain information on the nature of these
alkali labile gangliosides, the total brain gangliosides were
submitted to alkaline treatment and then examined by monodimensional TLC, following the procedure of Sonnino et al.[2] The untreated
ganglioside mixture displayed (Fig. 5) several fast migrating
bands. After overnight treatment with 0.1 M NaOH at 20°C, the TLC
migration pattern dramatically changed with the appearance of two
main slow moving bands. However, after overnight treatment with
17 M ammonia at 20°C, the pattern was intermediate between that of
untreated and NaOH treated gangliosides. It has been shown[2,4] that
the presence of alkali labile linkages decreases the polarity of
gangliosides. Therefore, alkali labile gangliosides have, under
the conditions used for TLC, faster migratory rates than the corresponding alkali stable compounds. Should alkali lability residue
in O-acetylated groups, treatment with NaOH or concentrated ammonia
would give the same results.[2] Conversely, a different behavior is
expected for compounds containing inner esters. In fact, one
lactone group provides, when exposed to NaOH, one derivative containing a free carboxyl group, and when exposed to ammonia vapor,
two derivatives, one with free carboxyl group and the other as the
amide formed by aminolysis.[2,4] This latter compound has a faster
migratory rate than the corresponding compound with a free carboxyl
group. Thus, the different patterns obtained with the two alkaline

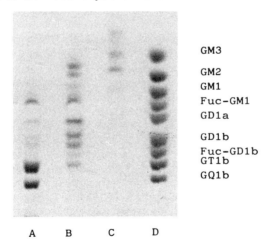

A   B   C   D

Fig. 5. Monodimensional TLC of lamprey brain gangliosides before
and after alkaline treatment. Lane A: brain gangliosides
after overnight treatment at 20°C with 0.1 M NaOH followed
by dialysis; lane B: brain gangliosides after overnight
treatment at 20°C with 17 M NH_4OH followed by dialysis;
lane C: untreated brain gangliosides; lane D: standard
gangliosides. Solvent system as in the legend to Fig. 3.

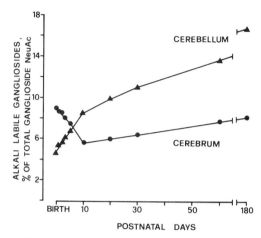

Fig. 6. Changes of the content of alkali labile gangliosides in rabbit cerebrum and cerebellum during postnatal life. The content (as bound NeuAc) of alkali labile gangliosides is expressed as percentage of total ganglioside bound NeuAc.

treatments is highly suggestive of the presence, in lamprey brain, of gangliosides in lactonic or polylactonic form following the suggestions of Gross et al.[5] Verification of this hypothesis is in progress.

DEVELOPMENTAL BEHAVIOR OF ALKALI LABILE GANGLIOSIDES IN RABBIT CEREBRUM AND CEREBELLUM

The developmental behavior of alkali labile gangliosides in rabbit cerebrum and cerebellum from birth to six months of life was studied. The proportion of alkali labile gangliosides of the total labile ganglioside content and the profile of the individual alkali labile gangliosides during postnatal life were different in cerebellum and cerebrum. Fig. 6 shows that: (a) in cerebellum, the proportion of alkali labile gangliosides increased progressively from 8.9% at birth to 16.7% at six months of age, and (b) in cerebrum the proportion decreased from 8.9% at birth to a low value of 5.5% at 10 days of age, followed by a slow increase up to 8.1% at six months of age. The major alkali labile gangliosides were 9-O-Ac-GT1b, 9-O-Ac-GQ1b, and a species which has not yet been identified (symbol I_3 in Fig. 7). All the major alkali labile gangliosides in cerebellum, especially 9-O-Ac-GT1b and 9-O-Ac-GQ1b, tended to accumulate during postnatal life (Fig. 7). In cerebrum only two alkali labile gangliosides, namely 9-O-Ac-GT1b and I_3, increased, starting from the 10th day of life, while the others progressively diminished after birth.

ALKALI LABILE GANGLIOSIDES IN THE BRAIN

Table 1. Content of Alkali Stable and Alkali Labile Gangliosides (as Bound NeuAc, Percentage of Total) in Different Nervous System Structures of Normothermic and Hibernating Dormouse.

	Alkali stable gangliosides		Alkali labile gangliosides	
	Normo-thermic	Hiber-nating	Normo-thermic	Hiber-nating
Cortex	87.0	100.0	13.0	0.0
Cerebellum	81.8	100.0	18.2	0.0
Spinal cord	68.7	100.0	31.3	0.0
Pons	74.1	100.0	25.9	0.0
Bulbus olfactorius	89.0	99.4	11.0	0.6
Regio quadrigemina	77.1	100.0	22.9	0.0
Brain stem	88.2	96.5	11.8	3.5

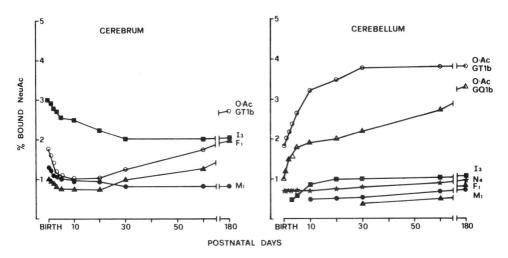

Fig. 7. Quantitative patterns of the individual alkali labile gangliosides of rabbit cerebrum and cerebellum during postnatal life. The lettering system used for the individual ganglioside is referenced.[2]

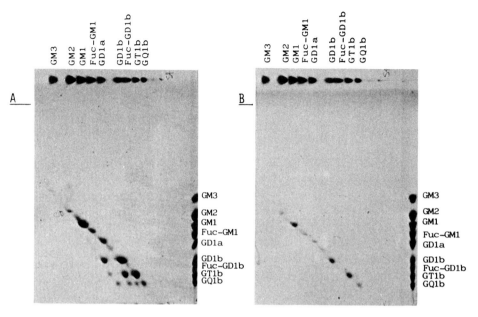

Fig. 8. Two-dimensional TLC, with intermediate ammonia treatment, of spinal cord gangliosides from A) normothermic, and B) hibernating dormouse.

ALKALI LABILE GANGLIOSIDES IN BRAIN OF THE HIBERNATING AND NORMOTHERMIC DORMOUSE

Attention has been recently paid to involvement of gangliosides in thermal adaptation.[13] Following this trend, the content and pattern of alkali labile gangliosides were investigated in the central and peripheral nervous systems of normothermic (22°C) and hibernating (6°C) dormouse. Fig. 8 shows a two-dimensional TLC fractionation, with intermediate ammonia treatment, of the gangliosides from spinal cord of mice under both thermal conditions. It is evident that alkali labile gangliosides, present in the normothermic animals, virtually disappeared in the hibernating animals. Table 1 gives the percentage content of alkali labile gangliosides from the nervous system structures examined in the two thermal conditions. In all cases the percentage of alkali labile gangliosides dramatically diminished during hibernation, and in most brain structures were absent. The hibernating dormouse would therefore constitute an interesting and potentially useful model for investigating the metabolic relationship between alkali labile and alkali stable gangliosides.

CONCLUSION

Alkali labile gangliosides have been found to be present, although in different quantities, in the brain of all animals that were examined. In some species they constitute, from a quantitative point of view, the major portion of total gangliosides. A well established chemical feature of brain alkali labile gangliosides is the presence of 9-O-acetyl,N-acetylneuraminic acid. Two different alkali labile gangliosides containing this residue of sialic acid, namely 9-O-Ac-GT1b and 9-O-Ac-GQ1b, have already been isolated and chemically characterized. The occurrence of gangliosides containing more than one O-acetyl group in their molecule is very likely, as well as that of gangliosides carrying an inner ester. Alkali labile gangliosides represent an important family of gangliosides and their relative abundance seems to be influenced by phylogenetic, ontogenetic as well as environmental adaptive factors. The occurrence of this family of gangliosides poses questions concerning the possible involvement of sialic acid acetylation and/or lactonization in the modulation of physico-chemical and functional properties of gangliosides and in the regulation of ganglioside turnover. It would be of further interest to find out whether O-acylation, or lactonization, occurs during ganglioside biosynthesis or after the gangliosides have reached their final site of location, the cell plasma membranes.

ACKNOWLEDGEMENTS

This work was supported by grants from C.N.R. (Rome). The brains of dormouse, japanese quail, zebra finch, nigerian cichlid, antarctic ice fish and gadus morhua were provided by Drs. H. Rahmann, R. Hilbig, M. Mulheisen, and J. Marx (Stuttgart, FRG); those of gadus callarias, dog fish, rays and lamprey by Dr. N. Avrova (Leningrad, USSR).

REFERENCES

1. V. Chigorno, S. Sonnino, R. Ghidoni, and G. Tettamanti, Densitometric quantification of brain gangliosides separated by two-dimensional thin layer chromatography, Neurochem. Intern. 4:397 (1982).
2. S. Sonnino, R. Ghidoni, V. Chigorno, M. Masserini, and G. Tettamanti, Recognition by two-dimensional thin layer chromatography and densitometric quantification of alkali labile gangliosides from the brain of different animals, Anal. Biochem. 128:104 (1983).
3. H. Wiegandt, Ganglioside, Ergeb. Physiol. Biol. Chem. Exp. Pharmacol. 57:190 (1966).

4. J. E. Evans and R. H. McCluer, Synthesis and characterization of sialosyllactosylceramide inner ester, Fed. Proc. 30:1133 (1971).
5. S. K. Gross, M. A. Williams, and R. H. McCluer, Alkali labile, sodium borohydride reducible ganglioside sialic acid residues in brain, Neurochem. 34:1351 (1980).
6. J. Haverkamp, R. W. Veh, M. Sander, R. Schauer, J. P. Kamerling, and J. F. G. Vliegenthart, Demonstration of 9-O-acetyl,N-acetylneuraminic acid in brain gangliosides from various vertebrates including man, Hoppe-Seyler's Z. Physiol. Chem. 358:1609 (1977).
7. R. Ghidoni, S. Sonnino, G. Tettamanti, N. Baumann, G. Reuter, and R. Schauer, Isolation and characterization of a tri-sialoganglioside from mouse brain, containing 9-O-acetyl,N-acetylneuraminic acid, J. Biol. Chem. 255:6990 (1980).
8. V. Chigorno, S. Sonnino, R. Ghidoni, and G. Tettamanti, Isolation and characterization of a tetrasialoganglioside from mouse brain, containing 9-O-acetyl,N-acetylneuraminic acid, Neurochem. Intern 4:531 (1982).
9. S. Hakomori and T. Saito, Isolation and characterization of a glycosphingolipid having a new sialic acid, Biochemistry 8: 5082 (1969).
10. G. Tettamanti, F. Bonali, S. Marchesini, and V. Zambotti, A new procedure for the extraction, purification and fractionation of brain gangliosides, Biochim. Biophys. Acta 296:160 (1973).
11. J. P. Kamerling, J. Haverkamp, J. F. G. Vliegenthart, C. Versluis, and R. Schauer, Mass spectrometry O-acetylated-N-acetylneuraminic acid, in: Recent development in mass spectrometry in biochemistry and medicine, Vol. 1, Plen. Press, New York, p. 503 (1978).
12. M. Iwamori and Y. Nagai, A new chromatographic approach to the resolution of individual gangliosides, Biochim. Biophys. Acta 528:257 (1978).
13. H. Rahmann, Brain gangliosides and thermal adaption in vertebrates, Zool. Jb. Physiol. 85:209 (1981).
14. L. Svennerholm, The gangliosides, J.Lipid Res. 5:145 (1964).
15. L. Svennerholm, Quantitative estimation of sialic acid. II. A colorimetric resorcinol-hydrochloric acid method, Biochim. Biophys. Acta 24:604 (1957).

PHYLOGENY AND ONTOGENY OF VERTEBRATE BRAIN GANGLIOSIDES

Louis Neal Irwin*

Department of Biology
Simmons College
Boston, MA 02115, U.S.A.

As methodological improvements disclose an increasingly complex array of gangliosides in all brain tissues examined, the quest for meaningful generalizations about the biological functions and biomedical potential of these compounds becomes more formidable. A useful starting point, however, is the observation that ganglioside patterns vary phylogenetically and are developmentally regulated, thus the particular pattern of gangliosides found in the nervous system of any vertebrate is dependent on both the evolutionary history and developmental state of that animal. The aim of this chapter is to briefly summarize what is known about phylogenetic variations and developmental changes in vertebrate brain gangliosides, to explore the relationships between ganglioside phylogeny and ontogeny, and to discuss the implications of this perspective on possible biological roles for gangliosides in neural tissue.

EVOLUTION OF GANGLIOSIDE PATTERNS

Gangliosides are common only in the deuterostomate phyla (echinoderms and chordates),[1] and only for the vertebrates have broad surveys of ganglioside distribution been carried out. From the earliest studies of phylogenetic variation in ganglioside patterns,[2,3] it has been obvious that ganglioside populations extracted from species of different vertebrate Classes differ substantially. Less

---

*Dr. Irwin is an Adjunct Associate Biochemist at the Eunice Kennedy Shriver Center for Mental Retardation, Waltham, U.S.A., where some of the work described here was performed.

clear has been the extent of this variation within the same Class, and the biological factors responsible for the variations that do exist.

Ganglioside patterns in different vertebrate classes

The ancestral pattern for vertebrate gangliosides is not known. Gangliosides from the lamprey, a surviving but specialized member of the jawless fishes, consist largely of two fractions that migrate ahead of D1a*,[3] roughly in the region of D3 to M3; but a number of more polar gangliosides are present as well.[4] The chemical identity of none of the gangliosides is known.

Elasmobranchs. The cartilagenous fishes have a complex set of relatively polar gangliosides, except that the pattern is often dominated by a non-polar ganglioside thought to be D3.[5] Probably a majority of the gangliosides have a shorter sugar backbone that the tetrahexose sequence characteristic of avian and mammalian gangliosides.[5] There is substantial variation in the number of sialic acid residues and degree of saturation in the lipid moiety of gangliosides from different elasmobranchs, with more closely related species having more similar biochemical properties.[6]

Teleosts. The bony fishes have been studied in greater detail than any other group of vertebrates. A large number of polysialated gangliosides, carrying four or more sialic acid residues, is characteristic of this Class, with relatively few components in the fast migrating non-polar region.[7-9] The trisialogangliosides of teleostean neural tissue have been identified as T3, T2, and T1c, and a pathway for their synthesis distinct from that found in tetrapods has been proposed.[10] A significant percentage of sialic acid residues in codfish gangliosides appear to be o-acetylated.[11] Evolutionarily, the teleosts are an extremely diverse group, and this is reflected in a great range of ganglioside specific concentrations and quantitative compositional detail. In general, however, closely related species tend to have similar ganglioside profiles; only at the level of Family or above do qualitative differences in ganglioside patterns become obvious.[12]

Amphibians. A simpler ganglioside pattern is consistently obtained from amphibians than from any vertebrate Class with the possible exception of the jawless fishes.[14,15] Four prominent fractions, identified as a disialo- and three trisialogangliosides, that do not comigrate with any mammalian counterparts[16] have been characterized from the nervous system of Rana catesbeiana. This

---

*Ganglioside designations are based on the nomenclature of Svennerholm.[13]

pattern superficially resembles that of teleosts and has generally been taken as typical of amphibians, but salamanders have a ganglioside pattern distinctly different from that of frogs and toads, showing in some cases a higher proportion of less polar gangliosides.[3,17] A recent analysis of variation among 10 amphibian species reveals that differences within the same Genus and Family are quantitative only, despite great differences in habitat.[18] This is the same correlation between ganglioside variation and taxonomic position observed for teleosts.[12]

Reptiles. An insufficient number of reptiles has been studied to assess the extent of variation within this Class, but reptilian gangliosides have consistently been reported to consist of a complex array, ranging from the most to the least polar gangliosides.[3,14,19] A ganglioside sample from the garter snake (Thamnophis) was recently resolved in my lab into more than 40 components distributed almost evenly throughout the entire two-dimensional range of the chromatogram. Gangliosides that appear to co-migrate with all the major avian and mammalian gangliosides are extractable from reptilian samples, thus the modern metabolic pathways for gangliosides may have originated in the reptilian nervous system.

Birds. Only a few avian species have been examined, but all show a pattern dominated by six major gangliosides -- D3, M1, D1a, D1b, T1b, and Q1 -- with a substantial number of minor components.[3,4,14,20] The avian retina is notable for its high concentration of D3.[21]

Mammals. Ganglioside patterns have been examined from humans and many other, though mostly domesticated and laboratory, species. The concentration of gangliosides in mammalian brain tissue is higher than that found in brains of all the other vertebrate Classes. Qualitatively, mammalian ganglioside patterns are dominated by five major components -- M1, D1a, D1b, T1b, and Q1.[3,4,22,23] Thus the mammalian and avian patterns are quite similar at the level of major components, but two-dimensional thin layer chromatography (2D-TLC) reveals numerous differences among the extensive number of minor gangliosides characteristic of both groups.[23] Significant quantitative differences have been noted among closely related species[24] and inbred strains.[25]

Overview of trends in vertebrate ganglioside evolution

On the basis of very limited information, it appears that the earliest vertebrate gangliosides were relatively non-polar and fewer in number. A proliferation of metabolic pathways resulted in the appearance of a more complex, more polar array in the cartilaginous fishes. The highest concentration of polysialated gangliosides occurs in the teleosts, but this is probably a specialization outside the main sequence of ganglioside evolution. Amphibian gangliosides may represent a simplification of patterns inherited from non-terrestrial ancestors. In contrast, metabolic pathways proliferate at the reptilian level to yield the most complex ganglioside arrays known. This is followed by a relative consolidation of selected reptilian gangliosides into the major fractions, representing a less polar distribution overall, in the birds and mammals. These relationships among the four tetrapod Classes are shown in Fig. 1.

Significance of evolutionary trends

The conservative nature of ganglioside evolution suggests that natural selection has constrained their tendency to vary. Yet the ganglioside variations that do occur are more extensive than would be predicted from the relative uniformity of ultrastructure and function at the subcellular level seen across all vertebrate species. Therefore, fundamental physiological, ecological, or ontogenetic factors common to related species but distinctive across broad taxonomic groups have probably selected for particular ganglioside patterns more than functional mechanisms at the ultrastructural or molecular level.

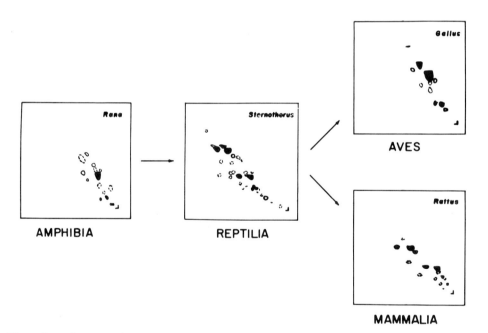

Fig. 1. Presumed evolutionary sequence of ganglioside patterns among the four tetrapod Classes. Patterns were obtained by tracing actual 2D-TLC chromatograms of brain gangliosides from a frog, turtle, chicken, and rat, though a direct phyletic relationship between these particular genera is not intended.

What these biological factors might be is unknown, but the clear differentiation of ganglioside patterns in every vertebrate Class suggests that the evolutionary forces that created each new Class influenced the ganglioside composition of the brain. In particular, the transitions from aquatic (fish) to terrestrial (amphibian) habitats, and from anamniotic (amphibian) to amniotic (reptilian) reproduction seem to have generated substantially new ganglioside patterns. A slightly less dramatic transition correlates with the emergence (from reptiles) of endothermy (in birds and mammals).

Rahmann and his colleagues[3,26,27] have drawn attention to a compelling correlation between the polysialation of gangliosides and cold adaption. Certainly the consolidation of the ganglioside pattern into a smaller number of major components in birds and mammals, which live within narrow internal thermal limits, provides an intriguing contrast to the more complex array of gangliosides found in reptiles, which have to survive at a greater range of temperatures. But related factors could be involved, such as activity level[9] or basal metabolic rate.[28] Osmoregulatory characteristics need to be scrutinized as possible selective factors as well.

ONTOGENY OF BRAIN GANGLIOSIDES

From the earliest years of ganglioside research it has been known that the ganglioside composition of brain tissue changes qualitatively and quantitatively as development proceeds. Until recently, however, the vast majority of developmental studies have concentrated on a small number of avian and mammalian species, and utilized batch preparation methodologies that made analysis of tiny amounts of tissue at early embryonic stages impractical. While the lack of comparative data remains, important new insights into ganglioside ontogeny at early developmental stages have recently been gained.

Ganglioside changes in developing fishes and amphibians

In the rainbow trout, a teleost, ganglioside concentration rises in parallel with acetylcholinesterase activity and synaptogenesis through a larval stage leading to a free-swimming juvenile.[29] Just after hatching, the ganglioside pattern is dominated by components that migrate in the mono- and disialoganglioside region. By the time of metamorphosis, the polysialogangliosides characteristic of mature teleosts have become dominant.

Gangliosides extracted from whole embryos of the frog Rana pipiens migrated in the mono- and disialoganglioside region, in contrast to more polar gangliosides extracted from Rana sylvaticus at a later developmental stage.[30] Gangliosides extracted exclusively from brain tissues of Rana pipiens and Xenopus laevis in my lab have consistently shown the presence of polysialogangliosides at early developmental stages, though less polar gangliosides are present and tend to diminish as development proceeds.[31]

Ganglioside changes during development of the chick

The vast majority of studies on avian ganglioside ontogeny have been conducted on the domestic chicken and duck -- a limitation that should be borne in mind. The availability of extensive morphological information on the development of the chick compensates in part for this weakness. There is clearly a strong coincidence between the accumulation of gangliosides and the period of rapid differentiation, dendritic proliferation, and synaptogenesis in the brain of the chick embryo.[32-34] Prior to differentiation, the ganglioside pattern is dominated by M3 and D3, but as differentiation proceeds, the more polar gangliosides of the mature state come to dominate the pattern.[35-37]

An interesting feature of development first noted by Rösner,[38] then by Irwin et al.,[39] is the transient appearance of polysialogangliosides that either diminish or become overwhelmed by the more dominant components at later stages. These highly polar gangliosides comigrate with gangliosides Tlc, Qlc, Pl, and H thought to be major components in the brains of teleosts.[40] Their appearance correlates with the period of rapid differentiation, while their relative decline occurs as cell arborization and synaptogenesis come to dominate the morphogenetic process. Gangliosides M1 and M4 increase with myelination, but D3 remains a major component of chick brain throughout the life of the animal. In the retina, D3 remains the most abundant ganglioside into adulthood.

Hunter and Dunn[41] have recently completed an exhaustive 2D-TLC analysis of the ganglioside pattern in the developing chick optic tectum, which essentially confirms the work of Rösner[40] and extends it in considerable detail. Over 70 resorcinol-positive spots were distinguished at some point in development, including Tlc, Qlc, P, and H. A number of the gangliosides do not persist into adulthood, or do so without increasing enough to avoid being swamped by the major components. Some gangliosides appear transiently as doublets. And several gangliosides do not appear until late stages of neural development.

Ganglioside changes during development in mammals

Changes in ganglioside pattern during postnatal development have been studied extensively in rodents, but to some extent in a number of other mammals as well. The major trend is rapid accumulation of total gangliosides just prior to and continuing into the period of myelination, marked by a substantial relative increase in M1 and Dla.[42-45] The rise in M1 coincides with the formation of myelin, of which it is the major ganglioside constituent.[46] Trace amounts of M4 also rise with myelination.[47] The marked increase in D1a has been attributed in particular to proliferation of dendritic membranes,[48] though the morphogenetic events that accompany its sharp rise are too complex to rule out other possibilities. There are regional differences in the timing of ganglioside changes, but apparently no ganglioside patterns turn out to be unique to a particular brain region. Peripheral nerves[49] and retinal tissue,[50] however, do give rise to a rather different ganglioside pattern dominated by D3 and other less polar gangliosides, presumably due to the unique cellular components of these tissues.

Recent studies of prenatal ganglioside patterns in rodents have confirmed Tettamanti's[51] previous observation that early fetal gangliosides consist largely of components less polar than M1. There appears to be general agreement now that at very early stages of neurogenesis, M3 and D3 are the dominant gangliosides in the rat and mouse, as they are in the chick.[52-54]

Hunter and her colleagues[23] have amplified these observations by an extensive 2D-TLC analysis of ganglioside changes during fetal development in the mouse neopallium. A total of 66 resorcinol-positive spots were distinguished at some point during development. In addition to the easily recognizable major gangliosides, other spots were tentatively identified as D1a-GalNAc, D1a-NG, D1b-fucose, D2, T1a, T1b-9-OAc, T3, and P. At embryonic day 13 (E13), M3 and D3 clearly dominate the pattern, but then decline progressively as the major gangliosides of adult brain begin to increase rapidly. The pattern thus changes from a relatively simple one at E13 to a much more heterogeneous array by E17. As M1, D1a, D1b, T1b, and Q1b accumulate in large amounts, the minor components become less evident, though careful scrutiny reveals the presence of most of them as late as the time of birth. Some of these trace components increased while others decreased; and some appeared at one stage and disappeared at another. The net effect is a progressive complication of the ganglioside pattern as differentiation proceeds, followed at later stages by a consolidation of the array to a few major components, which results in either a real or apparent simplification of the ganglioside pattern.

ONTOPHYLETICS OF VERTEBRATE BRAIN GANGLIOSIDES

Ontophyletics has been defined[55] as the study of the relationships between ontogeny and phylogeny. Information of sufficient detail on the phylogeny and ontogeny of brain gangliosides may now be available to suggest some of the ways in which evolutionary history has affected the nature and sequence of developmental changes in gangliosides in vertebrate brain tissue.

In the two species -- chick and mouse -- for which detailed information on early developmental stages is now available, the ganglioside pattern passes through three distinct phases. First, the pattern is dominated by one or two non-polar gangliosides (M3 and/or D3). In the second phase, the pattern becomes quite heterogeneous, with gangliosides dispersed over the full range of polarity. In the third phase, the pattern becomes somewhat simplified as the most polar gangliosides diminish and a smaller population of select gangliosides, including the relatively less polar M1 and D1a emerge as dominant components.

Superficially, this ontogenetic sequence mimics the phylogenetic trend from a relatively non-polar pattern in the earliest vertebrates, through the complex patterns of some fishes and reptiles, to the relatively consolidated mixture of both polar and less polar gangliosides in existing birds and mammals. The suggestion[20,38] that ontogeny recapitulates phylogeny in the expression of ganglioside patterns thus appears to have general validity. However, it should be noted that at no point in chick or mouse development does a strictly teleostean or amphibian ganglioside pattern appear. On the other hand, the dispersed pattern that characterizes the middle phase of avian and mammalian brain development resembles the typical reptilian pattern to a striking degree. Thus, it appears that in these species, metabolic pathways have evolved to bypass synthesis of most fish and amphibian gangliosides (except in minor amounts) but not reptilian gangliosides, as development proceeds toward the consolidated pattern seen in the mature brain tissue of birds and mammals.

IMPLICATIONS OF GANGLIOSIDE PHYLOGENY AND ONTOGENY

Phylogenetic variations in ganglioside distribution imply that the role of gangliosides is a relatively recent one in the history of life, since gangliosides do not appear with certainty before the evolution of echinoderms and chordates. Thus, to the extent that gangliosides contribute to ligand binding, ionic permeability, morphogenesis, mitotic regulation, and other fundamental cellular mechanisms found in all animals, their role must be ancillary to or superimposed upon pre-existing mechanisms that did not depend on gangliosides. This reasoning would further predict that functions attributable to M1 and D1a were absent or carried out by other molecules in vertebrates more ancestral than reptiles, since M1 and D1a did not become major components of the pattern until the reptiles evolved.

The ontophyletic evidence explains in part the curious sequence through which ganglioside patterns pass during development in birds and mammals. Essentially, there appears to be a particular sequence of metabolic pathways defined by evolutionary history that must be expressed in arriving at the mature ganglioside pattern. The intermediate stages of that sequence may be evolutionary residues serving no biological function. On the other hand, the opposite may be true: the complex, transient ganglioside pattern of intermediate developmental stages may be essential to a host of morphogenetic, physiological, or metabolic processes uniquely required at a given stage of development.

Finally, the survival into adulthood of five or six gangliosides in sizeable quantities in all birds and mammals suggests that at least these gangliosides play an indispensable role in brain function throughout the life of the animal.

SUMMARY AND CONCLUSIONS

1. Gangliosides evolved relatively recently in the history of life, thus their contribution to fundamental cellular processes must be ancillary to or superimposed on preexisting mechanisms.

2. Brain ganglioside patterns vary along taxonomic lines in a fairly conservative fashion, indicating that general ecophysiological factors have probably provided the major selective constraints.

3. During brain development in birds and mammals, gangliosides pass through a transient stage of pattern complexity that may reflect their reptilian ancestry. While this ganglioside heterogeneity could provide positional information within the developing tissue, it might merely reflect a necessary but incidental transition to the handful of major gangliosides essential to mature brain function.

ACKNOWLEDGEMENT

The author's work was supported by Grant NS 15924 from N.I.H. and a grant from the Simmons College Fund for Research.

REFERENCES

1. V. E. Vaskovsky, E. Y. Kostetsky, V. I. Svetashev, I. G. Zhukova, and G. P. Smirnova, Comp. Biochem. Physiol. 34:163 (1970).
2. E. G. Trams and C. J. Lauter, Biochim. Biophys. Acta 60:350 (1962).
3. N. F. Avrova, J. Neurochem. 18:667 (1971).
4. H. Rahmann and R. Hilbig, J. Comp. Physiol. 151:215 (1983).
5. N. F. Avrova, Y.-T. Li, E. L. Obukhova, J. Neurochem. 32:1807 (1979).
6. E. M. Kreps, Comp. Biochem. Physiol. 68B:363 (1981).
7. I. Ishizuka, M. Kloppenburg, and H. Wiegandt, Biochim. Biophys. Acta 210:299 (1970).

8. R. H. McCluer and B. W. Argranoff, J. Neurochem. 19:2307 (1972).
9. E. M. Kreps et al., Comp. Biochem. Physiol. 52B:283 (1975).
10. R. K. Yu and S. Ando, Adv. Exp. Med. Biol. 125:33 (1980).
11. J. Haverkamp, R. W. Veh, M. Sander, R. Schauer, J. P. Kammerling, and G. F. Vliegenthart, Hoppe-Seyler's Z. Physiol. Chem. 358:1609 (1977).
12. R. Hilbig and H. Rahmann, J. Neurochem. 34:236 (1980).
13. L. Svennerholm, J. Lipid Res. 5:145 (1978).
14. L. N. Irwin and C. C. Irwin, Comp. Biochem. Physiol. 64B:121 (1979).
15. G. D. Hunter, V. M. Wiegant, and A. J. Dunn, J. Neurochem. 37:1025 (1981).
16. M. Ohashi, Adv. Exp. Med. Biol. 152:47 (1982).
17. L. N. Irwin, Biochem. Syst. Ecol. 10:257 (1982).
18. L. N. Irwin and K. Schwartz, Comp. Biochem. Physiol. (in press).
19. N. F. Avrova, Zhur. Evol. Biokh. Fisiol. 4:128 (1968).
20. R. Hilbig, H. Rosner, and H. Rahmann, Comp. Biochem. Physiol. 68B:301 (1981).
21. P.-F. Urban, S. Edel-Harth, and H. L. Dreyfus, Exp. Eye Res. 20:397 (1975).
22. H. C. Price and R. K. Yu, Comp. Biochem. Physiol. 54B:451 (1976).
23. G. D. Hunter, L. N. Irwin, and R. H. McCluer, presented at Satellite Meeting on "Ganglioside Structure, Function and Biomedical Potential", Parksville, B. C., 6 July 1983.
24. A. Reglero, J. Garcia-Alonso, and J. A. Cabezas, J. Neurochem. 34:744 (1980).
25. H. Dreyfus, S. Harth, A. Giulani-Debernardi, M. Roos, G. Mack, and P. Mandel, Neurochem. Res. 7:477 (1982).
26. H. Breer, Life Sci. 16:1459 (1975).
27. H. Rahmann, R. Hilbig, W. Probst, and M. Muhleisen, J. Therm. Biol. 8:107 (1983).
28. L. N. Irwin and K. Loehr, Soc. Neurosci. Abstr. 8:694 (1982).
29. H. Breer and H. Rahmann, Roux's Arch. Dev. Biol. 181:65 (1977).
30. J. A. Yiamouyiannis and J. A. Dain, J. Neurochem. 15:673 (1968).
31. L. N. Irwin and G. D. Hunter, submitted to Roux's Arch. Dev. Biol., Aug. 1983.
32. O. W. Garrigan and E. Chargaff, Biochim. Biophys. Acta 70:452 (1963).
33. C.-L. Schendgrund and A. Rosenberg, Biochem. (Easton) 10:2424 (1971).
34. D. B. Gray and L. N. Irwin, J. Neurochem. 4:487 (1973).
35. E. L. Engel, J. G. Wood, and F. I. Byrd, J. Neurobiol. 10:429 (1979).
36. L. N. Irwin and C. C. Irwin, Dev. Neurosci. 2:129 (1979).
37. P. Panzetta, H. J. Maccioni, and R. Caputto, J. Neurochem. 35:100 (1980).
38. H. Rösner, J. Neurochem. 24:815 (1975).
39. L. N. Irwin, H. Chen, and R. A. Barraco, Dev. Biol. 49:29 (1976).
40. H. Rösner, Roux's Arch. Dev. Biol. 188:205 (1980).

41. G. D. Hunter and A. Dunn, presented at Meeting of Intl. Sco. Neurochem., Vancouver, B. C., 15 July 1983.
42. K. Suzuki, J. Neurochem. 12:969 (1965).
43. M. T. Vanier, M. Holm, R. Ohman, and L. Svennerholm, J. Neurochem. 18:581 (1971).
44. C. Alling and I. Karlsson, J. Neurochem. 21:1051 (1973).
45. A. Merat and J. W. T. Dickerson, J. Neurochem. 20:873 (1973).
46. K. Suzuki, J. F. Poduslo, and S. E. Poduslo, Biochim. Biophys. Acta 152:576 (1968).
47. R. K. Yu and K. Iqbal, J. Neurochem. 32:293 (1979).
48. H. K. M. Yusuf and J. W. T. Dickerson, Biochem. J. 174:655 (1978).
49. K. H. Chou, C. E. Nolan, and F. B. Jungalwala, J. Neurochem. 39:1547 (1982).
50. M. Holm, J.-E. Mansson, M.-T. Vanier, and L. Svennerholm, Biochim. Biophys. Acta 280:356 (1972).
51. G. Tettamanti, in: "Chemistry and Brain Development," R. Paoletti and A. N. Davison, eds., Plenum, New York (1971).
52. E. Yavin and Z. Yavin, Dev. Neurosci. 2:25 (1979).
53. L. N. Irwin, D. B. Michael, and C. C. Irwin, J. Neurochem. 34:1527 (1980).
54. R. Hilbig, H. Rösner, G. Merz, K. Segler-Stahl, and H. Rahmann, Roux's Arch. Dev. Biol. 191:281 (1982).
55. M. J. Katz, R. J. Lasek, and I. R. Kaiserman-Abramof, Proc. Natl. Acad. Sci. USA 78:397 (1981).

Receptors and Function

GANGLIOSIDE RECEPTORS: A BRIEF OVERVIEW AND INTRODUCTORY REMARKS

Sen-itiroh Hakomori

Fred Hutchinson Cancer Research Center and
The University of Washington
1124 Columbia Street
Seattle, Washington 98104

Gangliosides are typical amphipathic molecules organized in plasma membranes with their hydrophobic ceramide moiety inserted into the lipid bilayer and their hydrophilic, ionogenic group faced towards the outer environment of cells.[1] This model of ganglioside organization in membranes, as well as the occurrence of a "specific" interaction of gangliosides with various bioactive factors, favors the idea that gangliosides are acting as receptors for toxins, hormones, and other bioactive factors. The idea has been endorsed by the specific interaction of cholera toxin subunit with GM1 and by the ability of GM1 to inactivate the toxin's biological activities and to furnish cells the toxin-dependent adenylate cyclase response when GM1 is exogenously added to the insusceptible cells. Bioactive factors which show interaction with various kinds of glycolipids are listed in Table 1, and general implications of such interaction involving ganglioside receptors have been reviewed repeatedly;[2-5] however, if one is allowed to include any protein, regardless of the absence of clear biologic activity, the list of proteins that interact with gangliosides would be much longer. The fact that gangliosides interact with bioactive factors may justify an interpretation of receptor activity in some cases; nevertheless, the interaction *per se* does not necessarily lead us to assume that gangliosides are the receptors in general. The major conflict in accepting the idea of ganglioside receptors can be summarized in four items as described below:

1) Gangliosides interact with a large variety of proteins including serum albumin,[6] amphipathic membrane proteins,[7] etc.; therefore, interactions such as those listed in Table 1 may

Table 1. Interaction of Glycolipids With Biofactors.

	Biofactor	Major interactant claimed	Ref.
1.	Bacterial toxins		
	cholera toxin	GM1	8,9
	tetanus toxin	GT1b,GQ1b	10-12
	Botulinus toxin	GT1b	13,14
	Staphylococcus toxin	*SPG	15
2.	Glycoprotein hormones		
	thyrotropin	GD1b	16
	chorionic gonadotropin	GT1	17
	luteinising hormone	GT1	18
3.	Sendai virus	GT1a,GQ1b	19,20
4.	Interferon (type 1 only)	GM2,GT1	21-23
5.	Fibronectin	GT1b	24
6.	Lymphokines		
	macrophage migration inhibitor	Fucosyl GgOse$_4$Cer	25-27
7.	Opiate and morphine	sulfatide	28,29
8.	Serotonin	GD3	30

* sialosylparagloboside

merely represent a few of many widely occurring ionic interactions whose biological significance is yet undetermined. The interaction of GM1 with cholera toxin is, of course, highly specific.

2) The common gangliosides (particularly GT1 and GD1b) inhibit the binding of various glycoprotein hormones (thyrotropin, chorionic gonadotropin and luteinizing hormone) to target tissue membranes.[16-18] Various gangliosides (GM2 and GT) inhibit both mouse and human interferon activity.[21-23] Similarly, the same gangliosides interact with different bacterial toxins (see Table 1). These non-specific interactions of gangliosides with hormones, interferon, and toxins conflict with the fact that toxic or hormonal activities of these factors are highly specific to target cells. Furthermore, the interferon activity is highly species-specific, human interferon acting on human cells but not on mouse cells and <u>vice versa</u>.

3) Some natural target cells for a specified bioactive factor are lacking in the expected ganglioside which showed the strongest interaction with the specified bioactive factor. Human intestinal epithelia, which is the primary target of cholera toxin, had an extremely low content of GM1, while the intestinal epithelia of some other animals which are not susceptible to cholera infection had a large quantity of GM1. Normal rat thyroid cell lines with high-affinity functional receptors for thyrotropin had none of the higher gangliosides postulated as the thyrotropin receptor.[31] Some fibroblasts, such as BHK and NIL, which have a high content of fibronectin at the cell surface had no GT or GD1b ganglioside, while other fibroblasts, such as 3T3 cells, which have a low quantity of fibronectin have a high quantity of higher gangliosides.

4) In several cases high affinity receptors for those biological factors have been isolated from target cells and characterized as protein, and the binding to gangliosides is generally of much lower specificity and affinity (with the exception of GM1 for cholera toxin.

What would be the most relevant explanation for the ganglioside interaction with various bioactive factors? A few possibilities are described briefly as follows (see also Table 2). (i)

Table 2. Perspectives For a Few Possibilities For Ganglioside Interaction With Various Proteins.

---

1. Gangliosides acting as receptors

2. Gangliosides as secondary auxiliary receptors; primary receptors could be "ganglioproteins" or other high affinity proteins

3. Gangliosides acting as co-factors of protein receptors (in analogy with phosphatidylcholine for Rh antigen and GM3 for Paul-Bunnell antigen)

4. Gangliosides as modulators of protein receptors (on the assumption that receptors are allosteric and have gangliophilic sites)

5. Ganglioside interactions with bioactive factors are non-specific, incidental phenomena

---

Some of these interactions may indeed represent receptor function and supporting data have been reported (e.g., cholera toxin and some other bacterial toxins), although conflicts of the type described above need to be answered. (ii) The same carbohydrate sequence and structure present in glycoprotein ("ganglioprotein") could be the primary receptor. Gangliosides may function as a secondary receptor. Glycoproteins with high affinity binding to Sendai virus have been isolated and characterized from cells (e.g., bovine erythrocytes) susceptible to Sendai virus infection,[33] although Sendai virus itself interacts clearly with some gangliosides, particularly those having the NeuAcα2-8NeuAcα-2-3Gal terminal structure.[19,20] Many tumor cells which are highly susceptible to Sendai virus infection do not usually contain polysialogangliosides; therefore, glycoprotein must be the major receptor in such cells. In fact, high affinity glycoproteins have been detected in HeLa cells[32] as well as bovine erythrocytes (see above). Thyrotropin interacts with both glycoprotein and gangliosides. A two component model for the thyrotropin receptor unit has been proposed.[34] The unit consists of both gangliosides and glycoprotein which are assumed to be acting in concert. Protein was believed to provide the high affinity recognition site and gangliosides the low affinity component which induces the necessary conformation in the ligand for its entry into the lipid bilayer. (iii) Gangliosides and glycolipids could be non-specific cofactors of some protein receptors analogous to phosphatidylcholine acting as the cofactor of Rh polypeptide antigen. Consistent with this idea is the recent success in demonstrating an opiate receptor (μ receptor) activity reconstituted from acidic lipids and an inactive protein (fraction A) eluted from an affinity column (succinylmorphine-Sepharose).[35] The acidic lipid was once claimed as sulfatide (see Table 1). (iv) Gangliosides may simply modulate receptor function specifically or non-specifically. For this possibility, an assumption that receptors are essentially allosteric proteins having a site to interact with gangliosides directly or indirectly would be appropriate. Thus, ligand receptor interaction and possibly a subsequent internalization could be regulated by gangliosides.

The mechanism of cellular response to each bioactive factor is significantly different for each ligand and for each type of cell. The idea for ganglioside receptor is reasonably well tested for cholera toxin and for a few other bacterial toxins although some important questions still remain to be answered. For other types of receptors, much debate seems to be warranted. Generalization is always misleading at this premature stage. In the subsequent presentation, each speaker will provide some evidence to support one of the views and the possibilities as discussed above.

REFERENCES

1. This organization is based on surface labeling data with inside-out vesicles as compared to normal erythrocytes. No extensive evidence for various types of plasma membranes has been provided. Alternative arrangement is also possible [cf. S. Hakomori, Glycosphingolipids in cellular interaction, differentiation and oncogenesis, in: "Sphingolipid Biochemistry," J. N. Kanfer and S. Hakomori, eds., Plenum Press, New York (1983).
2. W. E. van Heyningen, Gangliosides as membrane receptors for tetanus toxin, cholera toxin and serotonin, Nature 249:415 (1974).
3. P. H. Fishman and R. O. Brady, Biosynthesis and function of gangliosides, Science 194:906 (1976).
4. R. W. Ledeen, Gangliosides in: "Handbook of Neurochemistry," Vol. 3, A. Lajtha, ed., Plenum Press, New York (1983).
5. J. N. Kanfer, Glycosphingolipids as receptors, in: "Sphingolipid Biochemistry," J. N. Kanfer and S. Hakomori, eds., Plenum Press, New York (1983).
6. B. Venerando, S. Roberti, S. Sonnino, A. Fiorilla, and G. Tettamanti, Interactions of ganglioside GM1 with human and fetal calf sera. Formation of ganglioside-serum albumin complexes, Biochim. Biophys. Acta 692:18 (1982).
7. K. Watanabe, S. Hakomori, M. E. Powell, and M. Yokota, The amphipathic membrane proteins associated with gangliosides: The Paul-Bunnell antigen is one of the gangliophilic proteins, Biochem. Biophys. Res. Commun. 92:638 (1980).
8. W. E. van Heyningen, C. C. L. Carpenter, N. F. Pierce, and W. B. Greenough III, Deactivation of cholera toxin by ganglioside, J. Infect. Disease 124:415 (1971).
9. J. Holmgren, I. Lonnroth, and L. Svennerholm, Tissue receptor for cholera exotoxin: Postulated structure from studies with GM1 ganglioside and related glycolipids, Infect. Immun. 8:208 (1973).
10. W. E. van Heyningen, The fixation of tetanus toxin, strychnine, serotonin and other substances by ganglioside, J. Gen. Microbiol. 31:375 (1963).
11. F. D. Ledley, G. Lee, L. D. Kohn, W. H. Habig, and M. C. Hardegree, Tetanus toxin interaction with thyroid plasma membranes, J. Biol. Chem. 252:4049 (1977).
12. J. Holmgren, H. Elwing, P. Fredman, Ö. Strannegard, and L. Svennerholm, Gangliosides as receptors for bacterial toxins and Sendai virus, Adv. Exp. Med. 125:453: (1980).
13. L. L. Simpson, and M. M. Rapport, The binding of botulinum toxin to membrane lipids: sphingolipids, steriods and fatty acids, J. Neurochem. 18:1757 (1971).
14. M. Kitamura, M. Iwamori, and Y. Nagai, Interaction between clostridium botulinum neurotoxin and gangliosides,

Biochim. Biophys. Acta 628:328 (1980).
15. I. Kato and M. Naiki, Ganglioside and rabbit erythrocyte membrane receptor for staphylococcal alpha-toxin, Infect. Immunol. 13:289 (1976).
16. B. R. Mullin, P. H. Fishman, G. Lee, S. M. Aloj, F. D. Ledley, R. J. Winand, L. D. Kohn, and R. O. Brady, Thyrotropin-ganglioside interactions and their relationship to the structure and function of thyrotropin receptors, Proc. Natl. Acad. Sci. (USA) 73:842 (1976).
17. G. Lee, S. M. Aloj, R. O. Brady, and L. D. Kohn, The structure and function of glycoprotein hormone receptors: Ganglioside interaction with human chorionic gonadotropin, Biochem. Biophys. Res. Commun. 73:370 (1976).
18. G. Lee, S. M. Aloj, and L. D. Kohn, The structure and function of glycoprotein hormone receptors: Ganglioside interaction with luteinizing hormone, Biochem. Biophys. Res. Commun. 77:434 (1977).
19. A. M. Haywood, Characteristics of Sendai virus receptors in a model membrane, J. Mol. Biol. 83:427 (1974).
20. J. Holmgren, L. Svennerholm, H. Elwing, P. Fredman, and Ö. Strannegard, Sendai virus receptor: Proposed recognition structures based on binding to plastic-absorbed gangliosides, Proc. Natl. Acad. Sci. (USA) 77:1947 (1980).
21. F. Besancon and H. Ankel, Binding of interferon to gangliosides, Nature 252:478 (1974).
22. V. E. Vengris, F. H. Reynolds, Jr., M. D. Hollenberg, and P. M. Pitha, Interferon action: Role of membrane gangliosides, Virology 72:486 (1976).
23. H. Ankel, C. Krishnamurti, F. Besancon, S. Stenfanos, E. Falcoff, Mouse fibroblast (type I) and immune (type II) interferons: Pronounced differences in affinity gangliosides and in anti-viral and anti-growth factors on mouse leukemia L-1210R cells, Proc. Nat. Acad. Sci. (USA) 77:2528 (1980).
24. H. K. Kleinman, G. R. Martin, P. H. Fishman, Ganglioside inhibition of fibronectin-mediated cell adhesion to collagen, Proc. Natl. Acad. Sci. (USA) 76:3367 (1979).
25. G. Poste, R. Kirsh, and I. J. Fidler, Cell surface receptors for lymphokines I. The possible role of glycolipids as receptors for macrophage migration inhibiting factor (MIF) and macrophage activation factor (MAF), Cell Immunol. 44:71 (1979).
26. G. Poste, H. Allen, and K. L. Matta, Cell surface receptors for lymphokines II. Studies on the carbohydrate composition of the MIF receptor macrophage using synthetic saccharides and plant lectins, Cell Immunol. 44:89 (1979).
27. T. Miura, S. Handa, and T. Yamakawa, Specific inhibition of macrophage migration inhibition factor by fucosylated glycolipid, J. Biochem. (Tokyo) 86:773 (1979).
28. H. H. Loh, T. M. Cho, Y. C. Wu, R. A. Harris, and E. L. Way,

Opiate binding to cerebroside sulfate: A model system for opiate-receptor interaction, Life Science 16:1811 (1975).
29. F. B. Craves, B. Zalc, L. Leybin, N. Baumann, and H. H. Loh, Antibodies to cerebroside sulfate inhibit the effects of morphine and β-endorphin, Science 207:75 (1980).
30. D. W. Wooley and B. W. Gommi, Serotonin receptors VII. Activities of various pure gangliosides as the receptors, Proc. Natl. Acad. Sci. (USA) 53:959 (1965).
31. S. Beckner, R. O. Brady, and P. H. Fishman, Reevaluation of the role of gangliosides in the binding and action of thyrotropin, Proc. Natl. Acad. Sci. (USA) 78:4848 (1981).
32. P.-S. Wu, R. W. Ledeen, S. Udem, and Y. A. Isaacson, Nature of the Sendai virus receptor: Glycoprotein versus ganglioside, J. Virol. 33:304 (1980).
33. Y. Suzuki, S. Takahashi, and M. Matsumoto, Isolation and characterization of receptor sialoglycoproteins of human hemagglutinating virus of Japan (Sendai virus) from bovine erythrocyte membranes, J. Biochem. 93:1621 (1983).
34. L. D. Kohn, E. Consiflio, M. J. S. DeWolf, E. F. Grollman, F. D. Ledley, G. Lee, and N. P. Morris, Thyrotropin receptors and gangliosides, in: "Structure and Function of Gangliosides," L. Svennerholm, P. Mandel, H. Dreyfus, and P.-F. Urban, eds., Plenum Press, New York (1980).
35. T. M. Cho, B. L. Ge, C. Yamato, A. P. Smith, and H. H. Loh, Isolation of opiate binding components by affinity chromatography and reconstitution of binding activities, Proc. Natl. Acad. Sci. (USA) 80:5176 (1983).

GANGLIOSIDES AS MODULATORS OF THE COUPLING OF NEUROTRANSMITTERS TO ADENYLATE CYCLASE

Glyn Dawson and Elizabeth Berry-Kravis

Depts. Biochemistry and Pediatrics, Joseph P. Kennedy Jr. Mental Retardation Research Center, Univ. of Chicago, Chicago, Ill

INTRODUCTION

Polysialogangliosides are characteristic components of nervous tissue, enriched in neurons, but little is known of their biological role in the CNS. Since the oligosaccharide moiety in G_{M1} ganglioside has been shown to confer binding specificity towards cholera toxin,[1] this has become the model for gangliosides to act as receptors for a wide variety of biologically active compounds including hormones and neurotransmitters. However, in most of these cases the receptor has been subsequently characterized as a protein and the binding to ganglioside has turned out to be of rather low affinity and specificity. Examples of this are the thyrotropin[2] and opiate receptors.[3] Thus the precise function of gangliosides in the CNS remains problematical but of intense interest.

In membrane structural terms, the most important carbohydrate components of gangliosides are the negatively-charged N-acetylneuraminic acid residues, whose effect is to increase both the molecular area and fluidity of the membrane as a consequence of electrostatic repulsion.[4] Because of this, the rigid, hydrophobic, ceramide moiety has considerable freedom of movement and, for example, may play a key role in stabilizing the pentameric G_{M1} cholera toxin receptor. Gangliosides could also play a similar role in regulating the surface stereochemistry of protein receptors, and this possibility has been the subject of recent studies in this laboratory.

Of all the neurotransmitter receptors, the serotonin receptor has been the one most implicated as a ganglioside receptor. However, despite claims of specific 5HT binding to disialohematoside (G_{D3})[5,6] the low binding affinity for 5HT has precluded gangliosides from being seriously considered as candidates for the serotonin receptor. Recently we have characterized $5HT_1$ receptors in a neuroblastoma x Chinese hamster 18 day embryonic brain explant hybrid clonal cell line NCB-20.[7] The order of drug potency for inhibition of high affinity [^{3}H]5HT binding was essentially the same as that observed in CNS tissue (5,6-dihydroxytryptamine = 5HT = methysergide = 5-methoxytryptamine > cyproheptadine = clozapine = mianserin > spiperone > dopamine > ketanserin) and a 30 min exposure of cells to 10 μM 5HT gave a two- to five-fold stimulation of cyclic AMP synthesis. At first, we were puzzled by the fact that the affinity of the receptor for 5HT was only a fiftieth that of the CNS receptor. However we report here that increasing the ganglioside content of these cells, either by direct preincubation or by prolonged exposure to enkephalin, increases the affinity of the 5HT receptors almost to the level observed in CNS tissues. These observations suggest a novel role for gangliosides in neuronal function.

RESULTS

Confluent cultures of mouse neuroblastoma cells typically express the simpler gangliosides G_{M3}, G_{M2}, G_{M1} and G_{D1a} with only trace quantities of the more complex tri- and tetra-sialogangliosides associated with central nervous system (Fig. 1B, 1C) and, by inference, with neurons. A hybrid cell line, NCB-20, the clonal product of fusion between mouse neuroblastoma N18TG2 and an 18 day embryonic Chinese hamster brain neuronal explant culture,[8] grown in monolayer culture,[7] expresses a complex ganglioside pattern (Fig. 1B, 1C) which has been associated with the binding of tetanus toxin.[9,10] The TLC system used was a modification of Hunter et al,[11] using Whatman HP-KF high performance silica gel TLC plates (10 x 10 cm plates with a 200 μ thick coating), and the solvent system chloroform - methanol - 0.2% $CaCl_2$ (55-45-10), dry, rotate 90° followed by two developments in the same dimension in chloroform - methanol - 2.5M ammonia in 0.2% KCl (50-40-10) (Figs. 1B and 1C). It can be seen that the pattern in NCB-20 cells is simpler than that in human brain (Fig. 1A), that fatty acid heterogeneity (each ganglioside resolves into a long chain fatty acid (C_{22}, C_{24}) spot and a short chain (C_{16}, C_{18}) fatty acid spot on TLC) make interpretation of NCB-20 maps more complex, and that chronic exposure to $_D$Ala^2$_D$Leu^5 enkephalin (DADLE) increases the content of gangliosides, including polysialogangliosides.

Our previous studies on neurotumor cell lines N4TG1 and NG108-15 have indicated that certain ganglioside glycosyltransferases, most notably the UDP GalNAc:G_{M3}-N-acetylgalactosaminyltransferase, can be stimulated by cyclic AMP and inhibited by lowering cellular levels of cyclic AMP with the help of neuromodulators such as DADLE.[3,12] The initial effect of DADLE is to inhibit cyclic AMP synthesis and block both the stimulation of cyclic AMP synthesis by ligands such as PGE_1 and serotonin and the incorporation of [^{3}H] GlcN into gangliosides such as G_{M2} and G_{M1}[14] (Fig. 2). However, after approximately 5h of exposure of NCB-20 cells to DADLE, 5HT-stimulated cyclic AMP levels have returned to normal and after 30h,

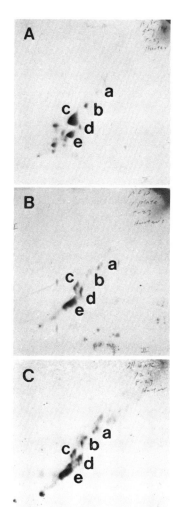

Fig. 1. 2-Dimensional TLC of gangliosides isolated from human brain (A), NCB-20 cells (B) and NCB-20 cells following chronic (72h) exposure to DADLE, (C) separated as described in the text. Gangliosides are identified as G_{M3} (a), G_{M2} (b), G_{M1} (c), G_{D3} (d), G_{D1a} (e) and polysialogangliosides (f).

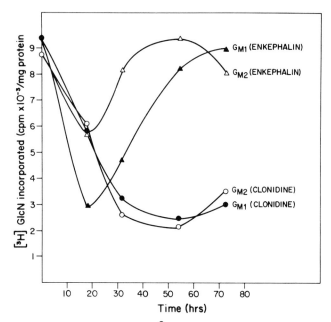

Fig. 2 Initial inhibition of [^{3}H]GlcN incorporation into G_{M1} and G_{M2} by both DADLE and clonidine is reversed by prolonged exposure (50h or more) to DADLE but not clonidine.

PGE_1-stimulated cAMP levels, as well as ganglioside synthesis have returned to normal levels (Fig. 2). During the next 10h the cells appear to undergo a further change wherein the sensitivity of cAMP synthesis to serotonin but not PGE_1 is substantially enhanced. When cells were exposed similarly to clonidine, the inhibition of both cyclic AMP synthesis (via the α_2-adrenergic receptor) and ganglioside synthesis persisted (Fig. 2). Cells exposed chronically (48h) to clonidine became tolerant but not supersensitive to 5HT.

Scatchard analysis of [^{3}H]5HT binding to membranes prepared from NCB-20 cells continuously exposed to DADLE for up to 60h revealed a dramatic (7-fold) increase in binding affinity by 40h which corresponded to the onset of cyclase supersensitivity to 5HT (Fig. 3A). The correlation of this finding with gangliosides was emphasized by the fact that a similar increase in binding affinity could be achieved by simply pre-incubating untreated NCB-20 cells with human brain ganglioside G_{M1} (Fig. 3B). Exposure of cells to clonidine for 48h produced no change in 5HT receptor affinity (Fig. 3A).

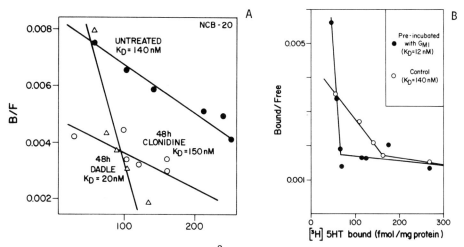

Fig. 3 Scatchard analysis of [^{3}H]5HT binding to NCB-20 membranes following chronic (48h) exposure to DADLE (△) or CLONIDINE (O) (A) or pre-incubation (24h) with ganglioside G_{M1} (B).

Specificity of ganglioside amplification of 5HT-stimulated adenylate cyclase

Most sialoglycosphingolipids tested were able to enhance the serotonin-mediated stimulation of cyclic AMP synthesis but the greatest potency was seen with two polysialoganglioside fractions - one predominantly G_{Q1b} and the other a mixed G_Q fraction. Other lipids had very little effect (Table 1). Of the non-sialylated lipids, only sphingomyelin gave an enhancement of 5HT-stimulated cAMP synthesis, but even this was well below the level seen with even simple gangliosides such as G_{M2}. Of the gangliosides tested, $G_{Q1b} > G_{D1a} > G_{M1} > G_{M2} > G_{T1b}$ was the order of potency, suggesting, perhaps, the importance of negatively charged residues in the stimulation phenomenon. The order of potency was approximately the same in both the [^{3}H]5HT membrane binding assay and the cyclic AMP stimulation studies employing whole cells. For maximum effect on adenylate cyclase, the potency order was $G_{Q1b} > G_{M1} > G_{D1a} = G_{T1b} = G_{M2}$. In these studies, the presence or absence of added gangliosides in the culture media had no effect on the stimulation, suggesting that the gangliosides did not bind <u>directly</u> to 5HT. Further, cells previously made supersensitive to 5HT by prolonged (48h) exposure to DADLE did not show any response to added gangliosides such as G_{M1} (Fig. 4).

The stimulation was typically saturated at around 0.5-1 µM, which is below the critical micelle concentration for all the gangliosides tested (Fig. 5).

Table 1 Effect of Lipid Pre-Incubation on Basal and 5HT-Stimulated Levels of Cyclic AMP in NCB-20 Cells. Basal Activity Was Measured in the Presence of IBMX (10^{-4}M), Added for 30 min Prior to Harvesting; Cells Were Stimulated with 10^{-7}M Serotonin (5HT).

Lipid added	Basal cAMP	Net 5HT Stimulation of cAMP synthesis
Cholesterol	110	130
Stearic acid	125	45
Phosphatidylcholine	90	105
Phosphatidylethanolamine	120	60
Phosphatidylinositol	115	120
Sphingomyelin	100	190
Sulfo Gal Cer	125	110
LacCer	110	90
GbOse$_3$Cer	90	95
GbOse$_4$Cer	105	95
II^{3}NeuAcGgOse$_3$Cer (G_{M2})	95	250
II^{3}NeuAcGgOse$_4$Cer (G_{M1})	90	300
II^{3}IV^{3}NeuAcGgOse$_4$Cer (G_{D1a})	100	250
G_{T1}	100	250
G_{Q1b}	80	400

The most potent ganglioside effects altered the sensitivity of the cells to 5HT by approximately one order of magnitude (Fig. 6) (maximum stimulation being to the same level but occurring at submicromolar 5HT concentrations rather than the 10 μM in untreated cells). This is in general agreement with the binding studies (Fig. 3B).

Evidence for actual uptake and incorporation of gangliosides into cell membranes was obtained by preparing [^{3}H]G_{M1} of high specific activity (20 x 10^6 cpm/ μmole) with the D-galactose oxidase/NaB[^{3}H]$_4$ procedure and performing binding assays (Fig. 7). It can be seen that saturation occurred after approximately 2h at 37° and that the concentration curve shows evidence of the biphasic phenomenon described by Toffano et al.[15]

<u>Specificity of the effect for the serotonin - GTP binding protein (Ns) - adenylate cyclase complex</u>

We have carried out preliminary studies on the effects of

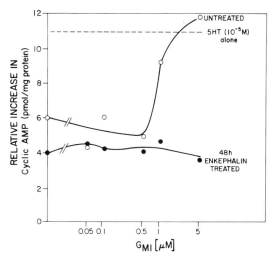

Fig. 4 Effect of ganglioside G_{M1} pretreatment on the ability of 100 nM 5HT to stimulate cyclic AMP synthesis in cells previously exposed to DADLE for 48h.

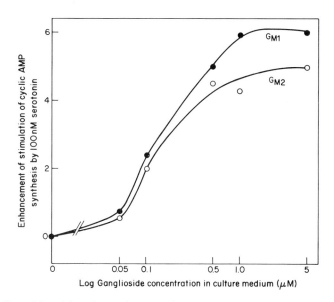

Fig. 5 Ganglioside-dependent enhancement of 100nM 5HT-mediated stimulation of cyclic AMP synthesis in NCB-20 cells.

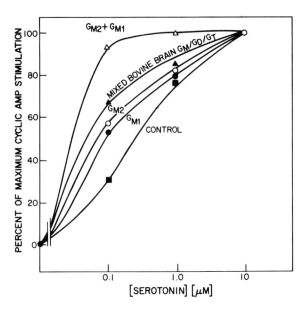

Fig. 6 Gangliosides decrease the EC_{50} for 5HT stimulation of cyclic AMP synthesis in NCB-20 cells.

exogenously added gangliosides on a number of other adenylate cyclase-linked systems in NCB-20 cells and thus far have found no other stimulatory effects. NCB-20 cells respond to added prostaglandins such as PGE_1 with a 300-fold stimulation of cyclic AMP synthesis over a 30 min period and pre-incubation with G_{M1} diminished this stimulation by up to 25% in some experiments (data not shown). In addition, pre-incubation with some gangliosides consistently diminished basal cyclic AMP levels by 10-20%. Since it is not known which exogenous ligands in calf serum determine the "basal stimulation" (but PGE_1 is a variable component of fetal or newborn (Bobby) calf serum) of cyclic AMP synthesis, the specificity of this effect is unknown. since we have shown that NCB-20 cells express enkephalin (delta opiate), dopamine (D-2) and clonidine (α_2-adrenergic) receptors which are coupled in an inhibitory manner to adenylate cyclase, we tested gangliosides in these three systems and found no major effect on the serotonin ($5HT_1$) receptor and that non-specific binding to biogenic amines is not a significant factor.

DISCUSSION

Wooley and Gommi[16] made the initial suggestion that gangliosides could be part of the gut serotonin (5HT) receptor following their observations that neuraminidase treatment abolished the 5HT

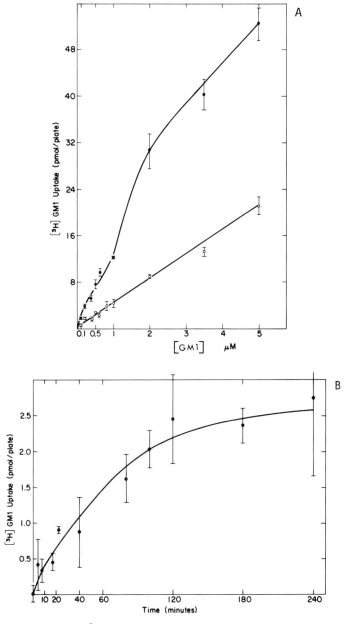

Fig. 7 Uptake of [^{3}H]G_{M1} by NCB-20 cells occurs in a biphasic manner, with the low concentration process showing saturation at 0.5-1 μM [^{3}H]G_{M1} added and the high concentration process at 10 μM (upper). Time course studies (lower) showed saturation of uptake by 2-4h.

response of rat stomach strips and brain ganglioside mixtures restored it. Subsequently, studies from several laboratories suggested that brain or adrenal medulla gangliosides bound 5HT with some degree of specificity, but in most cases the low affinity of the binding suggested a storage rather than a receptor function.[5,6] Tamir and associates[17] purified a specific serotonin binding protein (SBP) from both serotonergic neurons and non-neuronal tissue and showed that several gangliosides, most notably G_{D3} ganglioside, enhanced serotonin binding by increasing the number of binding sites rather than the affinity of 5HT binding.

Serotonin appeared to have a much higher affinity (nmolar) for SBP compared to gangliosides (negligible at 200 nM 5HT) and the dependence of 5HT binding on the presence of SBP, Fe^{2+} and egg phosphatidylcholine was interpreted as involving a conformational change in SBP to reveal new binding sites.[17] It was further suggested that the interaction of gangliosides such as G_{D3} with SBP regulated the concentration of 5HT in the synapse[17] rather than being connected with the function of either the $5HT_1$ or $5HT_2$ receptors.

The data presented here strongly support the concept that gangliosides form an intrinsic part of the high affinity neuronal surface $5HT_1$ receptor and facilitate its coupling via a GTP binding protein (N_s) to adenylate cyclase. Some ganglioside specificity was observed in that a tetrasialoganglioside fraction (G_Q) was more potent than either disialoganglioside G_{D1a} or monosialo-gangliosides (G_{M1} or G_{M2}). It was also clear from our studies that the addition of ganglioside to NCB-20 cells increased the binding affinity rather than increasing the number of binding sites. Gangliosides had no effect on the binding by supersensitive (chronically exposed to DADLE) cells presumably because the appropriate gangliosides had already been synthesized de novo. Further, at no time have we obtained any evidence for high affinity binding of 5HT to gangliosides alone - in agreement with Tamir et al.[17]

At present, we know very little of the biochemistry of the brain $5HT_1$ receptor. Our own studies on NCB-20 cells indicate that it is sensitive to trypsin and Pronase and moderately sensitive to dithiothreitol, but its resistance to tunicamycin sets it apart from the typical glycoprotein receptors for acetylcholine, enkephalin, glucagon and insulin. Electron inactivation studies in this laboratory, which showed the target size of the enkephalin receptor to be around 200,000 daltons,[18] have thus far provided evidence for a similarly large protein size for the $5HT_1$ receptor. Work is currently in progress to see if gangliosides are involved in modulating the action of other receptors which are positively coupled to adenylate cyclase (adenosine, secretion, β-adrenergic etc.). Thus far, we have seen no effect on receptors which are coupled negatively to adenylate cyclase (enkephalin, dopamine

(D-2), α_2-adrenergic), which suggests some form of specificity in ganglioside action - perhaps for the unique 45K subunit which couples to the common 35K subunit of N_s and activates adenylate cyclase.[19] Since we have previously shown that serum lipoproteins contain a complex array of glycosphingolipids and that certain cells can acquire these glycosphingolipids,[20] modulation of receptor function by sialoglycosphingolipids could have far-reaching physiological consequences.

SUMMARY

Cultured NCB-20 mouse neuroblastoma x Chinese hamster brain clonal hybrid cells express an adenylate cyclase-coupled receptor for serotonin (5HT) which corresponds pharmacologically to the $5HT_1$ receptor in whole brain, except for its much lower affinity for serotonin. Studies showed that the affinity of the NCB-20 receptor could be increased to near that of the whole brain receptor and the potency of 5HT in elevating cyclic AMP levels increased by pre-incubating NCB-20 cells for at least 3 hours with submicromolar concentrations of brain gangliosides. Tetrasialoganglioside (G_{Q1b}) was found to be the most potent ganglioside tested, producing a ten-fold increase in affinity. However, the actual 5HT binding site is a protein and we have obtained no evidence that serotonin binds directly to gangliosides at the concentrations at which it labels the receptor. The receptor-mediated inhibition of adenylate cyclase by biogenic amines such as dopamine and clonidine through dopamine (D_2) and α-adrenoreceptors was unaffected by pre-incubation of the NCB-20 cells with gangliosides.

Enkephalin was also found to acutely supress both the ability of 5HT to stimulate adenylate cyclase activity and the synthesis of polysialogangliosides in NCB-20 cells. After 6 hours of exposure, the cells became tolerant to enkephalin and after 36 hours the cells became supersensitive to 5HT in terms of adenylate cyclase activation and 5HT binding. The affinity of the receptor for 5HT increased the same 10-fold magnitude as achieved by G_{Q1b} pre-incubation in comparison with untreated cells. This increase in receptor affinity appeared to coincide chronologically with the increase in ganglioside synthesis observed in enkephalin tolerant cells, further suggesting an important role of polysialogangliosides in the function of the serotonin ($5HT_1$) receptor.

ACKNOWLEDGEMENTS

We would like to thank Mr. J. Y. Oh for help in purifing the gangliosides, Debbie Graves for cell culture and many colleagues, especially Dr. P. Hoffman for the cAMP antibody. Supported by USPHS Grants HD-04583, HD-06426, HD-09402 and GM 07281.

REFERENCES

1. P. Cuatrecasas, Interaction of vibrio cholerae enterotoxin with cell membranes Biochemistry 12:3547 (1973).
2. S. K. Becker, R. O. Brady, and P. H. Fishman, Reevaluation of the role of gangliosides in the binding and action of the role of gangliosides in the binding and action of thyrotropin, Proc. Natl. Acad. Sci. USA 78:4848 (1981).
3. R. W. McLawhon, G. S. Schoon, and G. Dawson, Glycolipids and opiate action, Europ. J. Cell Biol. 25:353 (1981).
4. B. Maggio, F. A. Cumar, and R. Caputto, Molecular behavior of glycosphingolipids in interfaces, Biochim. Biophys. Acta 650:69 (1981).
5. W. Gielen, Uber die Funktion von Gangliosiden. Ein serotonin und Ca^{++} receptor, Z. Naturforsch. (B) 23b:117 (1981).
6. H. C. Price, C. Byard, W. Sims, and R. Wilson, Gangliosides and other lipid micelles. A study of amine binding by dialysis/fluorescence method, Neurochem. Res. 4:63 (1979).
7. E. Berry-Kravis and G. Dawson, Characterization of an adenylate cyclase-linked serotonin ($5HT_1$) receptor in a neuroblastoma x brain explant hybrid cell line (NCB-20), J. Neurochem, 40:977 (1983).
8. J. McDermott, H. Higashida, S. Wilson, H. Matsuzawa, J. Minna, and M. Nirenberg, Adenylate cyclase and acetylcholine release regulated by separate serotonin receptors of somatic cell hybrids, Proc. Natl. Acad. Sci. USA 76:1135 (1979).
9. T. B. Rogers and S. H. Snyder, High affinity binding of tetanus toxin to mammalian brain membranes, J. Biol. Chem. 256: 2402 (1981).
10. Z. Yavin, E. Yavin, and L. Kohn, Sequestration of tetanus toxin in developing neuronal cell cultures, J. Neurosci Res. 7:267 (1982).
11. G. D. Hunter, V. M. Wiegant, and A. J. Dunn, Interspecies comparison of brain ganglioside patterns studied by two-dimensional thin-layer chromatography, J. Neurochem. 37:1025 (1981).
12. R. W. McLawhon, G. S. Schoon, and G. Dawson, Possible role of cyclic AMP in the receptor-mediated regulation of glycosyltransferase activities in neurotumor cell lines, J. Neurochem. 37:132 (1981).
13. S. K. Sharma, W. A. Klee, and M. Nirenberg, Dual regulation of adenylate cyclase accounts for narcotic dependence and tolerance, Proc. Natl. Acad. Sci. USA 72:3092 (1975).
14. G. Dawson, R. W. McLawhon, and R. J. Miller, Inhibition of sialoglycosphingolipid (ganglioside) biosynthesis in mouse clonal lines N4TG1 and NG108-15 by -endorphin, enkephalins and opiates, J. Biol. Chem. 255:129 (1980).
15. G. Toffano, D. Benvegnu, A. C. Bonetti, L. Facci, A. Leon, P. Orlando, R. Ghidoni, and G. Tettamanti, Interactions of G_{M1}

ganglioside with crude rat brain neuronal membranes, J. Neurochem. 35:861 (1980).
16. D. W.. Woolley and B. W. Gommi, Serotonin receptors-VII. Activities of various pure gangliosides as the receptors, Proc. Natl. Acad. Sci. USA 53:959 (1965).
17. H. Tamir, W. Brunner, D. Casper, and M. M. Rapport, Enhancement by gangliosides of the binding of serotonin to serotonin binding protein, J. Neurochem. 34:1719 (1980).
18. R. W. McLawhon, J. C. Ellory, and G. Dawson, Molecular size of opiate (enkephalin) receptors in neuroblastoma - glioma hybrid cells as determined by radiation inactivation analysis, J. Biol. Chem. 258:2102 (1983).
19. E. Hanski, P. C. Sternweis, J. K. Worthup, A. W. Dromerick, and A. G. Gilman, The regulatory component of adenylate cyclase, J. Biol. Chem. 256:12911 (1981).
20. G. Dawson, B. C. P. Kwok, M. Nishigaki, and B. W. Shen, Role of serum lipoproteins in the pathogenesis of Gaucher Disease, in: "Gaucher Disease," R. J. Desnick, J. Grabowski and S. Gatt, eds., Alan R. Liss, Inc., N.Y. pp. 253-265 (1982).

GANGLIOSIDES, THE THYROTROPIN RECEPTOR, AND AUTOIMMUNE THYROID DISEASE

Paolo Lacetti, Donatella Tombaccini, Salvatore Aloj, Evelyn F. Grollman and Leonard D. Kohn

Section on Biochemistry of Cell Regulation, Laboratory of Biochemical Pharmacology, National Institute of Arthritis, Diabetes, and Digestive and Kidney Diseases National Institutes of Health, Building 4, Room B1-31 Bethesda, Maryland 20205, U.S.A.

INTRODUCTION

The interaction of thyrotropin (TSH)* with a specific receptor on the thyroid cell surface induces changes in adenylate cyclase activity which result in enhanced iodide uptake, thyroglobulin biosynthesis, iodination of thyroglobulin, degradation of iodinated thyroglobulin to form thyroid hormone, and the release of thyroid hormone (T_3 and T_4) into the blood stream. Although changes in all of the above activities have been and can be used to define receptor function, definition of the structure of the TSH receptor on a molecular level, has required the identification of specific membrane components using binding studies and ^{125}I-labeled hormone.

The sum of the binding studies identified a membrane ganglioside, as well as a membrane glycoprotein as two potentially important components in the cell surface recognition event.[1-5] TSH binding to the glycoprotein component of the membrane was proposed to be the initial high affinity recognition event on the cell

---

*Abbreviations: TSH, thyrotropin; FRTL-5 cells, functioning rat thyroid cells derived from normal Fisher rat thyroid glands; T_3, triiodothyronine; T_4 thyroxine; CBA, cytochemical bioassay; TSAb, thyroid-stimulating antibody.

surface, i.e., the necessary first step in receptor recognition; however, a full functional response was postulated to require the ganglioside.[1-5] The ganglioside was suggested to contribute to the following receptor functions.[1-5] It completed specificity by distinguishing among glycoprotein hormones and related ligands such as tetanus toxin, cholera toxin, and interferon. It modulated the apparent affinity and capacity of the glycoprotein receptor component and induced a conformational change in the hormone believed necessary for subsequent message transmission. It allowed the ligand to perturb the phospholipid bilayer through alterations in lipid order and contributed to the ability of TSH to alter the ion flux across the membrane. Finally, the model proposed that after the glycoprotein receptor component trapped the TSH, the ganglioside acted as an emulsifying agent to allow the hormone to interact with other membrane components within the hydrophobic environment of the lipid bilayer and thereby initiate the signal processes. The physical basis of the emulsification process was the formation of a hydrogen-bonded, anhydrous complex between the hormone and the oligosaccharide moiety of the ganglioside. By excluding water from the interface of the ligand-receptor complex, membrane penetration was facilitated, and interactions involving membrane components of the adenylate cyclase ensued.

The present report summarizes the results of two different approaches which validate the model and define the ganglioside role in particular. The primary discussion will center on the use of monoclonal antibodies to identify receptor components and their link to message signals. The second, a reconstitution approach, will complement and amplify these conclusions. The data will also provide major insight into the antigenic determinants of antibodies which are believed etiologic in the pathologic expression of autoimmune thyroid diseases.[6] In particular, it will be shown that thyroid-stimulating antibodies (TSAbs) in Graves' patients are directed against gangliosides and are the autoimmune equivalents of cholera toxin.

RESULTS AND DISCUSSION

Monoclonal antibodies to the TSH receptor were made by injecting crude solubilized thyroid membrane preparations into mice followed by fusion, the identification of hybridomas secreting antibodies related to TSH receptor function and, finally, characterization of the antigenic determinants of the antibodies with particular respect to the already identified TSH binding components.[4,5] An alternative approach which directly related TSH receptor structure to the autoantibodies in Graves' sera, involved the formation of heterohybridomas between lymphocytes from patients with active Graves' disease and a non-IgG secreting mouse myeloma cell line, followed by the same identification procedure above.[4,5] A two-stage screening procedure with thyroid membranes was utilized.[4,5] Hybridoma antibodies reactive with thyroid membranes in the first screening assay, but blocked by 1 x 10^{-6} M unlabeled TSH in the second, were chosen as potential TSH receptor antibodies. As a first approximation the following criteria were used to further identify a monoclonal antibody to the TSH receptor. (i) The antibody had to inhibit TSH binding to thyroid membranes or, conversely, be itself prevented from binding to thyroid membranes by TSH. (ii) Inhibition had to be <u>specific</u> and had to be <u>competitive</u> as opposed to noncompetitive or uncompetitive. (iii) The antibody had to competitively inhibit TSH-stimulated functions or, conversely, had to mimic TSH activity and exhibit competitive agonism.

Most of the monoclonal antibodies to the TSH receptor thus far identified, can be broadly grouped as "inhibitors", "stimulators", or "mixed antibodies" based on these criteria.[4,5] An "inhibitor" has the following characteristics: competitive inhibition of ^{125}I-TSH binding; competitive inhibition of TSH-stimulated adenylate cyclase activity; inhibition of TSH-stimulable thyroidal iodine release or uptake; and no direct stimulation of thyroid adenylate cyclase activity, T_3/T_4 release, or iodide uptake. A "stimulator" (TSAb) in general has the following characteristics: significantly weaker inhibition of ^{125}I-TSH binding than the first group but equally potent competitive inhibition of TSH-stimulable adenylate cyclase activity; direct TSH-like, stimulatory action with respect to adenylate cyclase activity, as well as both iodide uptake and T_3/T_4 release by the thyroid; and competitive agonism when included with low concentrations of TSH in measuring adenylate cyclase activity. An antibody with "mixed" activity is a stimulatory antibody (TSAb) which also has the ability to inhibit ^{125}I-TSH binding to a significant degree. The significance of the existence of antibodies with "mixed" properties will be discussed with respect to the organization of receptor components and their active site determinants.

In essence, all inhibitor antibodies have been found to be directed at the glycoprotein receptor component, stimulators at the ganglioside, and mixed at both components.[4,5]

All of the monoclonal antibodies selected on the basis of the specific ability of TSH to inhibit their binding to membranes were able to inhibit ^{125}I-TSH binding to thyroid membranes; however, significant differences existed in their relative potencies. The antibodies in the most potent "inhibitory" group were able to not only significantly block ^{125}I-TSH binding to human, bovine, or rat thyroid membranes, but also to liposomes embedded with the high affinity glycoprotein component of the thyroid membranes (Table 1, "Inhibitors"). In contrast, antibodies in the low potency group (Table 1, "Stimulators") using membranes had no significant activity

TABLE 1. Ability of antibodies to prevent ^{125}I-TSH binding to liposomes containing the glycoprotein component of TSH receptor [a] or to react with various ganglioside preparations [a]

Antibody	^{125}I-TSH bound (cpm)			Ganglioside reactivity measured as ^{125}I-Protein A or ^{125}I-anti-human F(ab)$_2$ binding to antibody ganglioside complexes in a solid phase assay (cpm)				
	Bovine Glycoprotein Component	Human	Rat	No added lipid	Bovine thyroid gang.	Human thyroid gang.	Rat thyroid gang.	Mixed bovine brain gang.
Controls								
Monoclonal control	18,200	14,800	29,200	150	160	160	155	155
Nl mouse IgG	17,940	14,700	29,500	128	185	185	125	135
Nl human IgG	18,400	14,800	31,200	149	144	144	160	145
Inhibitors								
13D11	4,100	3,800	5,400	141	235	235	175	215
11E8	2,100	400	2,050	156	206	206	210	204
59C9	6,200	1,200	5,900	121	195	195	195	198
60F5	4,100	1,050	4,200	110	210	210	210	200
129H8	1,400	980	2,800	101	285	285	256	170
122G3	6,400	4,200	9,200	146	268	268	235	155
Stimulators								
22A6	14,800	13,600	26,800	137	870	870	982	410
206H3	15,500	12,200	24,500	156	2,889	2,889	1,210	128
307H6	16,200	12,800	26,400	129	4,210	4,210	2,450	240
Mixed								
52A8	9,800	5,400	15,300	135	1,480	1,480	1,210	205
208F7	12,400	6,100	14,400	136	920	920	1,140	180

[a] Assays were performed as detailed in References (4, 5).

in the liposome assays. Inhibition was evident under many of the in vitro experimental conditions historically used to measure ^{125}I-TSH binding. Inhibition was competitive with respect to TSH whether measured using ^{125}I-TSH and unlabeled antibody or unlabeled TSH and unlabeled antibody, binding of antibody being measured with ^{125}I-labeled protein A. No antibody had anti-TSH activity.

The antibodies from the inhibitory group were reactive with intact functioning rat thyroid cells, as well as thyroid membranes.[4,5] Their interaction with the functioning thyroid cells was specifically inhibited by TSH but not hCG (human chorionic gonadotropin). Using these cells, the antibodies noted as inhibitors were in no case able to mimic TSH as direct stimulators of adenylate cyclase activity.[4,5] They were, however, able to inhibit TSH-stimulated adenylate cyclase activity,[4,5] inhibition in each case was competitive. The antibodies also inhibited TSH-stimulated radioiodine uptake in thyroid cells[4,5] and TSH-stimulated T_3/T_4 release in a mouse bioassay used to measure in vivo TSH activity.[4,5]

In short, the monoclonal antibodies classified as inhibitors in these studies, i.e., those acting as good competitive inhibitors of TSH binding and TSH-stimulated function, were antibodies directed at the glycoprotein receptor component of the TSH receptor. Yet, despite the importance of the glycoprotein receptor component as a portion of the physiologic TSH receptor on the surface of the thyroid cell, it was not the primary antigen against which auto-immune stimulating antibodies (TSAbs) are directed, since in no case were antibodies to the glycoprotein receptor component stimulating.

The several antibodies (22A6, 206H3, and 307H6) which were not significant inhibitors of TSH binding by comparison to the inhibitory antibodies (Table 1), were, however, equipotent inhibitors of TSH-stimulated adenylate cyclase activity.[4,5] Resolution of this contradiction evolved with the observation that each of these antibodies was a potent direct stimulator of human and rat thyroid cell adenylate cyclase activity[4,5] and that their inhibitory action with respect to TSH was an example of pseudo-inhibition resultant from their lower direct stimulatory activity. Thus, at low TSH levels, 22A6, 206H3, 307H6, 52A8, and 208F7 were more than additive competitive agonists, and at high TSH concentrations they were competitive antagonists.[4,5] The antibodies were active in the mouse bioassay measuring thyroid hormone release[4,5] and were able to enhance iodide uptake by thyroid cells in a manner identical to TSH,[4,5] i.e., under conditions where cAMP-dependent iodide uptake was the measured response.

The stimulating antibodies were all able to interact with ganglioside preparations of thyroid membranes (Table 1), whereas, with two exceptions, most were poorly reactive with the glycoprotein component (Table 1). The significance of the two exceptions, the

208F7 and 52A8 antibodies designated as mixed (Table 1) and which exhibit reactivity in both assays, will be discussed below. The reactivity with the ganglioside preparations was reasonably thyroid-specific when using total ganglioside extracts (Table 1); the stimulating antibody reactivity with ganglioside preparations was lost if the glycolipid preparations were pretreated with neuraminidase (Table 2); and reactivity was highest in disialoganglioside fractions obtained by column chromatographic techniques (Table 2). The highest reactivity is, however, evident with a single minor component of the disialoganglioside preparation (see below and references 4 and 5).

In sum, then, the monoclonal antibodies establish that a ganglioside is a component of the TSH receptor; that it is a minor component of the total ganglioside pool; and that it appears to be vital in linking the recognition process to cAMP signal generation.

The activities of the antibodies and TSH were compared when they were mixed together using a cytochemical bioassay (CBA). In the CBA, 22A6, 307H6, and 208F7 are stimulators, whereas 11E8 is inactive as a stimulator over a wide dose range. 11E8 does, however, inhibit TSH stimulation in the CBA at a concentration where 22A6 is a stimulator. In contrast to its effect on TSH, 11E8 shows relatively low potency (> 10,000-fold lower) when inhibiting stimulation of the thyroid-stimulating antibodies 22A6 and 307H6. Antibody 208F7 is inhibited at a 10,000-fold dilution of 11E8. Differences in the effect of 11E8 on TSH as opposed to 22A6, 307H6, and 208F7 are not the result of a 10,000-fold difference in binding constants. Since separate experiments indicate that the 11E8 mono-

Table 2. Ability of Monoclonal Antibodies to React With Modified or Partially Purified Thyroid Ganglioside Preparations.[a]

Ganglioside Preparation	Ganglioside Reactivity (cpm)	
	22A6	307H6
Human thyroid gangliosides	930	3,955
+ neuraminidase	210	410
Rat thyroid gangliosides	884	2,290
+ neuraminidase	195	290
Human neutral glycolipid	900	800
Monosialogangliosides	1,100	2,000
Disialogangliosides	8,050	10,700
Rat FRTL-5 disialogangliosides	11,400	22,600

[a]Prepared and assayed as detailed in references 4 and 5.

clonal antibody interacts predominantly with a membrane glycoprotein, whereas the 22A6 and 307H6 monoclonal antibodies interact with a ganglioside (see above), the simplest way of reconciling the above observations is to apply the two component receptor model suggested in receptor binding studies (Fig. 1). Thus, TSH can be envisaged to interact first with the glycoprotein component of the TSH receptor, which exhibits high affinity binding properties. Its biological action, however, required an additional or subsequent interaction with a ganglioside. In contract, TSAbs represented by 307H6 and 22A6 can be envisaged to bypass the glycoprotein receptor component and interact with the ganglioside to initiate the hormone-like signal. Since 11E8 is an antibody binding to the glycoprotein receptor component, the two TSAbs, 307H6, and 22A6 are minimally affected by 11E8 since they are directed at the next sequential step, the ganglioside, which is vital in ligand message transmission. Antibody 208F7 raises another issue in this respect. It can interact with both ganglioside and glycoprotein receptor components. It is a good inhibitor of TSH binding and a potent stimulator of adenylate cyclase activity. It is also sensitive to 11E8 inhibition, albeit at very high concentrations by comparison to TSH. These findings suggest that there is an "overlap" or "interaction" between the two receptor components and that 208F7 interacts at this common site (Fig. 1).

This model in Fig. 1 has been supported by reconstitution studies. Membrane preparations from the 1-8 rat thyroid tumor with a TSH receptor defect are coincidentally devoid of higher order gangliosides.[1-5] The higher order ganglioside defect in the 1-8 tumor line has in turn been associated with a defect in the synthesis

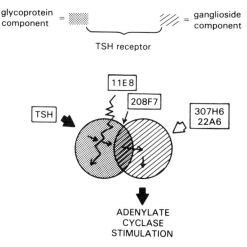

Fig. 1. Hypothetical model of TSH receptor as suggested by studies with monoclonal antibodies to the TSH receptor.

of higher order gangliosides (i.e., only G_{M3} could be detected in its membrane fraction by thin layer chromatographic analysis. The TSH receptor defect is expressed as low TSH binding and no TSH-stimulated adenylate cyclase activity despite normal thyroid functional responses to prostaglandins or dibutyryl cyclic AMP.[1-5] The original 1-8 tumor also had no cholera toxin stimulated adenylate cyclase response despite the ability of forskolin to activate adenylate cyclase activity.[1-5]

Incubation of a total ganglioside extract from the FRTL-5 thyroid cells or the disialoganglioside fraction from the FRTL-5 thyroid cells (Fig. 2) with primary cultures of the original 1-8 tumor resulted in return of TSH, TSAb, and cholera toxin stimulated adenylate cyclase activity (Table 3). No reconstitution of TSH receptor function occurred in control reconstitution incubations containing a ganglioside extract from the original 1-8 tumor (G_{M3}); mixed brain gangliosides; or the FRTL-5 ganglioside extracts treated with a mixture of neuraminidases capable of converting 87% to 94% of the sialic acid residues from a lipid bound to free form.

Incorporation of the gangliosides from the FRTL-5 thyroid cell preparations into the 1-8 original tumor cells was monitored by reactivity with the 22A6 or 307H6 stimulating monoclonal anti-receptor antibodies; these antibodies have been shown above to react with thyroid ganglioside preparations (see above). Thus, for example, as measured by ^{125}I-protein A, 22A6 binding to the no addition cells, to cells incubated with G_{M3}, or to cells incubated with neuraminidase treated FRTL-5 gangliosides was 150 ± 60 cpm membrane protein. The 22A6 binding to cells incubated with FRTL-5 ganglio-

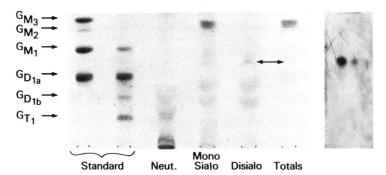

Fig. 2. (Left) Thin layer chromatogram of lipid extracts (Table 2) from human thyroid evaluated by resorcinol staining. The minor disialoganglioside (arrow) is identified by autoradiography (right) after sequential treatment of plates with 307H6 or 208F7 and with ^{125}I-labeled anti-human IgG.

Table 3. Reconstitution of TSH receptor expression in 1-8 rat thyroid tumors.

Thyroid Membrane or Cell Prep.	Adenylate cyclase activity[a] p moles cAMP/µg DNA			
	Basal	Cholera Toxin 1×10^{-9} M	TSH 1×10^{-9} M	307H6 20 µg/ml
FRTL-5 thyroid cell	0.5	6.4	18	5.4
1-8 tumor (original)	0.8	0.4	0.8	0.7
+ FRTL-5 thyroid cell gangliosides	0.7	3.8	5.2	2.6
+ FRTL-5 thyroid cell disialogangliosides	0.6	1.1	7.8	5.4
+ NDase treated FRTL-5 gangliosides	0.7	1.2	0.9	0.8
+ NDase treated FRTL-5 disialogangliosides	0.7	0.5	1.9	1.2
+ mixed brain gangliosides	0.8	3.1	1.1	0.8
+ 1-8 tumor gangliosides	0.5	0.7	0.5	--
+ G_{M3}	0.8	0.7	0.7	0.8
+ FRTL-5 thyroid cell gangliosides followed by trypsin treatment[b]	0.5	--	0.9	1.9

[a] Assays were performed in triplicate; results are the average of at least three separate experiments. In no case did the standard deviation of any value exceed ± 10%.

[b] Trypsin treatment of cell preparations used the procedures referred to in Reference 1.

sides or disialogangliosides was 520 ± 40 and 830 ± 50 cpm, respectively (p values < 0.01 compared to the controls above).

Again, the reconstitution data should in no way be construed to negate the importance of the glycoprotein component of the TSH receptor as noted from the trypsin sensitivity data in Table 3. Thus 1-8 cells, reconstituted with FRTL-5 gangliosides, and then exposed to trypsin, lost their ability to respond to TSH. It is, however, notable that the TSAb and cholera toxin stimulation activity were significantly less altered by the trypsin treatment (Table 3), i.e., TSAb behaves as an autoimmune equivalent of cholera toxin.

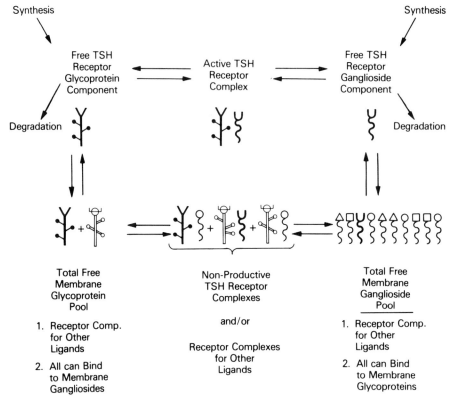

Fig. 3. Hypothetical model of active TSH receptor complex in equilibrium with its free component parts which are in turn in equilibrium with other glycoproteins and gangliosides of its membrane.

If a model of a receptor complex of both components is postulated (Fig. 1), it is an intuitive extrapolation that the complex is in equilibrium with free components (Fig. 3). Further, the gangliosides can be viewed as a group or series of similar structure with greater or lesser affinities for a particular ligand but each with the potential for interacting with a membrane glycoprotein, of which the TSH receptor glycoprotein is only one. The implications of such a model are far-reaching.

The TSH receptor has been related to receptor structures for cholera toxin, tetanus toxin, and interferon.[1-5] Thus, each of these ligands has been associated with a receptor structure, involving gangliosides, albeit structures with different ganglioside specificities. All four, TSH, cholera, tetanus and interferon can influence the binding and bioactivity of the other.[1] For example, in the absence of exogenous NAD cholera toxin can first enhance TSH binding and function, then inhibit, as its concentration is increased. That these phenomena involve interactions with components of receptor structure is illustrated by the fact that the cholera toxin interaction results in enhanced surface exposure of higher order gangliosides in the G_{M1} to G_{D1b} area, where the ganglioside with the highest affinity to TSH migrates.[1] A possible simple explanation for these phenomena resides in the equilibrium model of Fig. 3. Thus, cholera toxin at low concentrations (Fig. 3) interacts with G_{M1} and favors the interaction of the more relevant TSH receptor ganglioside with the TSH receptor glycoprotein, since the competing free ganglioside pool is smaller and its affinity is best. At high concentrations, cholera toxin would also interact with the TSH receptor glycoprotein[1] and decrease its availability for forming the active TSH receptor complex. This model would predict that cholera toxin does indeed interact with the glycoprotein receptor component and that the G_{M1} is merely the kinetically detected step because of the unusual affinity and multivalent binding state of the β-subunit.

Using the theory of antiidiotype and anti-antiiodiotype antibodies to receptors, one can presume that the monoclonal TSAb which stimulates adenylate cyclase activity via the ganglioside component of the receptor is not only anti-receptor, but equivalent to hormone. Under this circumstance, determinants involved in the antireceptor "active" site should be identical to those of TSH and the antiidiotype species of anti-TSH should interact with this and inhibit TSAb activety. This is true, i.e., a human anti-TSH preparation has been shown to inhibit 307H6 and a Graves' serum TSAb.[4,5] It also inhibited cholera toxin suggesting the model in Fig. 3 is valid, since anti-TSH has no G_{M1} interactivity.[5]

SUMMARY

The thyrotropin (TSH) receptor has been proposed to be composed of a membrane glycoprotein and a membrane ganglioside, the former important in high affinity recognition, the latter vital for message coupling to the adenylate cyclase system. The present study used two approaches, formation of antireceptor monoclonal antibodies and reconstitution, to validate the model and further examine the role of the ganglioside. Three kinds of monoclonal antireceptor antibodies are defined. One group which inhibits TSH binding and TSH functions, i.e., TSH-stimulated adenylate cyclase activity, iodide uptake, and thyroid hormone release, is shown to be directed against the glycoprotein component of the receptor. The second group includes antibodies which mimic TSH in all stimulatory actions, are competitive agonists of TSH, are equivalent to thyroid stimulating antibodies in the sera of patients with Graves' disease, and are directed against the ganglioside component of the receptor. These stimulating monoclonal antibodies are directed against a minor ganglioside membrane component which fractionates as a disialoganglioside. When this ganglioside is incorporated into 1-8 thyroid cells which have a correlated ganglioside deficiency and TSH receptor defect, reconstitution of TSH stimulated adenylate cyclase activity occurs. Whereas the first group of antibodies inhibits TSH-stimulated function, they do not inhibit the stimulatory antibodies which mimic TSH, an observation consistent with the 2 component hypothesis of the receptor model. The third group of antibodies have a mix of properties from the first two groups and suggests that the TSH receptor in situ is an actual complex of the two components or that there are common carbohydrate determinants in the functional sites of each receptor component. Implications of a TSH receptor structure in which its ganglioside and glycoprotein components are in equilibrium with pools of free components and, in turn, components important for cholera toxin, tetanus toxin and interferon receptors are discussed. In regard to the pathogenesis of Graves' disease, the data indicate that thyroid stimulating autoantibodies are autoimmune equivalents of cholera toxin with respect to the importance of ganglioside function. Since antiidiotype studies of antibodies against TSH confirm a structural relationship between receptors for thyrotropin, cholera toxin, and thyroid stimulating autoantibodies, the data establish an unequivocal role for the ganglioside in TSH receptor structure which facilitates interpretation of in vitro experiments aimed at understanding the mechanism of ganglioside-ligand interactions.

REFERENCES

1. L. D. Kohn, Relationships in the structure and function of receptors for glycoprotein hormones, bacterial toxins, and interferon, in: "Receptors and Recognition", P. Cuatrecasas and M. F. Greaves, eds., Chapman and Hall, London, Vol. 5, Series A (1978).
2. L. D. Kohn, S. M. Aloj, F. Beguinot, P. Vitti, E. Yavin, Z. Yavin, P. Laccetti, E. F. Grollman, and W. A. Valente, Molecular interactions at the cell surface: role of glycoconjugates and membrane lipids in receptor recognition processes, in: "Membranes and Genetic Diseases", J. Shepard, ed., Alan R. Liss, New York, Vol. 97 (1982).
3. L. D. Kohn and S. Shifrin, Receptor structure and function: an exploratory approach using the thyrotropin receptor as vehicle, in: "Horizons in Biochemistry and Biophysics, Hormone Receptors", J. Wiley and Sons, New York, Vol. 6, (1982).
4. L. D. Kohn, W. A. Valente, P. Laccetti, C. Marcocci, M. DeLuca, P. A. Ealey, N. J. Marshall, and E. F. Grollman, Monoclonal antibodies as probes of thyrotropin receptor

structure, in: "Receptor Biochemistry and Methodology - Monoclonal and Antiidiotypic Antibodies: Probes For Receptor Structure And Function", J. C. Venter, C. M. Fraser and J. M. Lindstrom, eds., Alan R. Liss, New York, Vol 4 (in press).

5. L. D. Kohn, E. Yavin, Z. Yavin, P. Laccetti, P. Vitti, E. F. Grollman, and W. A. Valente, Autoimmune thyroid disease studied with monoclonal antibodies to the thyrotropin receptor, in: "Monoclonal Antibodies: Probes For Study of Autoimmunity And Immunodeficiency", G. Eisenbarth and R. Haynes, eds., Academic Press, New York (in press).

6. A. Pinchera, G. F. Fenzi, E. Macchia, P. Vitti, F. Monzani, and L. D. Kohn, Immunoglobulines thyreostimulantes antigenes correspondants, Annales d'Endocrinologie (Paris), 43:520 (1982).

SPECIFIC GANGLIOSIDES ARE RECEPTORS FOR SENDAI VIRUS

Proteins in Lipid Samples Can Mask Positive Biological Effects

Mary Ann K. Markwell, Pam Fredman* and Lars Svennerholm*

Molecular Biology Institute, Department of Microbiology
University of California, Los Angeles, CA 90024, USA
*Department of Neurochemistry, Psychiatric Research
Centre, University of Goteborg, S-422 03 Hisings Backa
3, Sweeden

RECEPTOR GANGLIOSIDES

Sendai virus (also known as the Hemagglutinating Virus of Japan, HVJ) is a member of the family Paramyxoviridae. This virus along with select other members of the same family such as measles and Newcastle disease viruses have the remarkable ability to induce membrane fusion in the neutral to slightly alkaline pH range normally found at the cell surface. This ability is used by Sendai virus to initiate infection. Infectious viral particles (virions) recognize specific receptors on the host cell surface and, through interaction with these receptors, enable the viral membrane to fuse with the surface membrane of the host, thereby releasing the viral nucleocapsid directly into the cytoplasm to continue the infectious process.[1] Noninfectious viral particles also bind and enter the cell, but do so by adsorptive endocytosis which results in their subsequent degradation.[2]

The use of membrane fusion as an infectious mode of entering the cell is, of course, limited to those viruses which can cause membrane fusion in the neutral to slightly alkaline pH range encountered at the cell surface. Other enveloped viruses, such as influenza A of the family Myxoviridae which are lacking this ability, rely on adsorptive endocytosis for infectious entry into the cell.[3,4] A functional host cell receptor for any virus must, therefore, not only present specific sites on the cell surface to which the virus can bind, but also must allow entry of the virion in a manner which leads to the subsequent steps of infection.

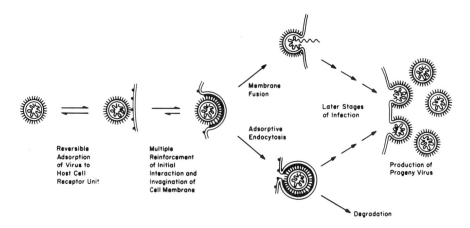

Fig. 1. The mode of infectious entry for an enveloped virus, membrane fusion or adsorptive endocytosis, depends upon its ability to cause fusion in the pH range encountered at the cell surface.

Because both noninfectious and infectious viral particles can bind to and enter the cell, and because the former constitute the majority of particles in a typical mammalian virus preparation, it is essential that the binding of virus to putative receptors be shown to be relevant to the infectious process.

Our receptor assay was specifically designed to rely on the ultimate indicator of infection, the production of progeny virus. By defining conditions to limit the initial period of contact between inoculum and host cell, and to limit the replication of the virus to a single cycle, the number of progeny virus produced was found to be directly proportional to the number in the inoculum. Thus the natural amplification of viral particles which occurs during infection was used to quantitate successful virion-cell interactions at the initial phase of infection. Virus production was measured by hemagglutination titer (HAU/ml) of the medium from infected cells 48 hours after inoculation.

By using this functional receptor assay, the nature of the endogenous host cell receptors for Sendai virus was investigated. Native HeLa cells produced an HAU/ml of 256 within 48 hr. after being infected (Table 1). Treatment of the cell surface with proteases such as trypsin or pronase consistently enhanced the susceptibility of these cells to infection. In contrast, treatment of the cell surface with Vibrio cholerae sialidase destroyed the endogenous receptors for Sendai virus as evidenced by the resistance the cells acquired to infection.

Table 1. The Nature of Sendai Virus Receptors*

Cell State	Addition	Virus Production
Native	---	256 HAU/ml
Trypsin-treated	---	512
Pronase-treated	---	512
Sialidase-treated	---	<2
Sialidase-treated	GM1 (1 nmol)	<2
Sialidase-treated	GD1a (1 nmol)	256
Sialidase-treated	GQ1b (0.1 nmol)	512

*HeLa cells growing in monolayer culture in 35-mm dishes were treated with 0.5% trypsin, 0.05% pronase, or 50 mU of Virbio cholerae sialidase before inoculation with Sendai virus, Z strain. The conditions used were the maximum treatment with protease during which >95% of the cells remain viable. The virus was protected from the action of protease adsorbed to the cell by adding the inoculum in 10% fetal calf serum.

These data indicated that the receptors were sialoglycoconjugates. Previously we had demonstrated that resialylation of sialidase-treated (receptor-deficient) cells with a sialytransferase preparation that elaborated NeuAcα2,3Galβ1,3GalNAc as the terminal sequence on glycoconjugates fully restored the susceptibility of host cells to infection, but use of a second sialyltransferase preparation which elaborated NeuAcα2,6Galβ1,3GalNAc as a terminal sequence had no effect.[5] Concurrently, Holmgren et al. (1980) had independently demonstrated the ability of Sendai virus to bind to specific gangliosides when homogeneous preparations of these were immobilized on plastic.[6] All of the gangliosides recognized by the virus contained NeuAcα2,3Galβ1,3GalNAc or NeuAcα2,8NeuAcα2,3Galβ1,3GalNAc as their terminal sequence. The actual functioning of these putative receptors was first demonstrated by their addition to receptor-deficient MDBK cells.[7]

In Table 1 the results of adding GM1, GD1a, and GQ1b to receptor-deficient HeLa cells are shown. Gangliosides of the gangliotetraose series containing a terminal sialic acid functioned as receptors; those lacking the terminal sialic acid had no effect. The same results were obtained for three different strains of Sendai virus (Z, Enders, and ESW$_5$) all of which are highly fusogenic.

In order to finally designate specific gangliosides as receptors it remained to be established that the gangliosides endogenously present in host cells function as receptors. The ganglioside content of three different types of host cells--MDBK, HeLa, and MDCK cells--were analyzed (Table 2). It was found that the ganglioside content of each cell line varied depending on the passage number and the surface on which the cells were grown. MDCK cells grown in glass roller bottles were 16-fold less susceptible to infection than cells of the same passage grown in plastic roller bottles. This difference in susceptibility directly correlated to a lower content of oligosialogangliosides.

In all three cell lines (MDBK, HeLa, and MDCK) GM3 was the major ganglioside. The presence of GM1 and the more complex homologs of the gangliotetraose series was also established. For each of the three lines examined, receptor gangliosides GD1a, GT1b, and GQ1b were present in more than sufficient quantities to fully account for the susceptibility of these cells to infection on the day they were harvested (Table 2).

Sendai virus receptors are macromolecules naturally occurring on the host cell surface which specifically bind the virion and

Table 2. Composition of Membrane Components of MDBK, HeLa, and MDCK Cells*

CELL TYPE:	MDBK	HeLa	MDCK
CONTENTS:	(nmol/mg protein)		
Phospholipids	353	402	340
Protein-bound sialic acid	6.8	12.1	10.4
Gangliosides	7.7	2.5	9.4
Individual Gangliosides			
GM3	4.0	1.6	6.7
LM1	1.5	0.2	0.1
GM1	1.3	0.2	0.5
GD3	0.2	0.1	0.4
GD1a	0.3	0.2	0.9
GD2	0.1	0.1	0.2
GD1b	0.1	0.1	0.3
GT1b	0.4	0.1	0.4
GQ1b	trace	trace	trace

*Cells were grown in monolayer in plastic dishes or roller bottles. The identity of each line was established by karyotyping. The equivalent of 100-200 10-cm dishes was used to analyze the ganglioside content of each cell line.

Table 3. The Functioning of Endogenous MDBK, HeLa, and MDCK Cell Gangliosides as Receptors for Sendai Virus*

Addition to Sialidase-Treated Cells	Virus Production
None	<2 HAU/ml
GM1	<2
GD1a	256
MDBK cell oligosialogangliosides	256-512
HeLa cell oligosialogangliosides	256-512
MDCK cell oligosialogangliosides	256

*Oligosialoganglioside fractions were prepared from MDBK, HeLa, and MDCK cells as described previously.[8] An aliquot of the oligosialoganglioside fraction from each cell line containing 0.3 nmoles of GD1a plus GT1b or standards containing 0.3 nmoles of GM1 or GD1a were added to sialidase-treated HeLa cells before inoculation with Sendai virus.

through this binding facilitate the subsequent events of infection. The oligosialoganglioside fractions from all three host lines examined were shown to function as receptors when added to receptor-deficient cells (Table 3). The demonstration that the gangliosides endogenously present in host cells function as receptors completes the first identification of receptors for a mammalian virus.

Our data thus far emphasize the importance of gangliosides in virus adsorption. In addition gangliosides may also have a role in membrane fusion. In envisioning the initial contact between the virus and its ganglioside receptors on the cell surface it would be logical to assume that the binding of a highly multivalent ligand such as Sendai virus which has hundreds of binding spikes on its surface would temporarily stabilize a clustering of its receptors at the region of virus-cell contact. It is at this point where membrane fusion must occur if the virion is to continue its infectious process. Oligosialogangliosides such as GD1a and GT have been shown to act as natural membrane fusogens, causing extensive cell-cell fusion when their concentration is increased by exogenous addition to cells.[9] Such a concentration increase would be achieved naturally in a localized area on the cell surface by the binding of Sendai virus. It is therefore tempting to speculate that viruses such as Sendai virus which rely on membrane fusion for their entry into cells have evolved to recognize oligosialogangliosides to allow not only their adsorption but to facilitate the fusion event.

Because of the dependence of Sendai virus on membrane fusion as the infectious mode of entering the host, it will be of interest to determine if sialoglycoproteins are also recognized by the virus as additional receptors. A number of studies including the early ones of Burnet[10] and the more recent ones of Haywood[11] and Ledeen[12] suggest that Sendai virus may have some affinity for sialoglycoproteins, but give no indication if it is in the range that would make them biologically significant. For some ligands such as cholera toxin the difference in affinities between glycolipids and glycoproteins is so great that the ganglioside GM1 appears to be its only natural receptor (see reference 13 for review). A second consideration is the distance of the oligosaccharide from the lipid bilayer. For membrane fusion to occur it is generally estimated that the viral membrane and host cell membranes must approach within 10-15A of each other.[14] It has already been demonstrated that Sendai virus can fuse with liposomes containing receptor gangliosides but no host proteins.[15] It remains to be established that the binding of Sendai virus to any sialoglycoprotein on the cell surface can produce an approach proximal enough to result in membrane fusion in the absence of receptor gangliosides.

PROTEINS IN LIPID SAMPLES CAN MASK POSITIVE BIOLOGICAL EFFECTS

It has been shown that some membrane proteins can be extracted with the organic solvent mixtures typically used for ganglioside isolation[12] and that particular amphipathic proteins can co-purify with gangliosides.[16] Although the ganglioside preparations used in our studies were highly purified (<0.02% of the dry weight was protein[7]), we were initially concerned that the very minor amount of protein remaining could cause false positives. Therefore the specificity of interaction of Sendai virus was confirmed by using gangliosides from a second, independent source. GD1a and GQ1b were prepared and generously supplied by Robert W. Ledeen.

A second approach to this problem was to purposely use gangliosides which were known to contain substantial amounts of protein but lacked receptor gangliosides to determine if these would produce positive results in our receptor assay system. Commercial preparations of GM1, GD1a, and GT1b (Fig. 2) were found to contain as much as 1-5% of their dry weight as protein as determined by radioiodination.[7] This ultrasensitive technique produced not only a quantitative estimate of the amount of protein contamination when compared to BSA standards, but also provided a means of analyzing the species present when the radiolabeled samples were electrophoresed on SDS polyacrylamide gels. In general the commercially prepared GM1, GD1a, and GT1b samples contained the same or similar species of proteins except for a band migrating at 40K (indicated by an arrow) which was more prominent in GM1. Only the samples of GD1a and GT1b (Lanes B and C) produced positives in our receptor

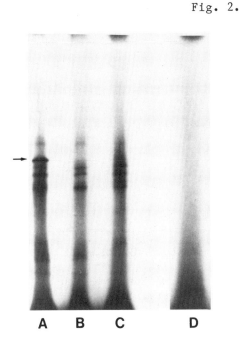

Fig. 2. Autoradiographic analysis of proteins in ganglioside samples. Purified individual gangliosides were prepared by Supelco from a Folch extract of bovine brain by successive purification on three silicic acid columns (Iatrobeads). Aliquots (50-100 μg) of GM1 (Lane A), GD1a (B), and GT1b (C) were radioiodinated[7] and then applied with nonradioactive molecular weight standards to a 5-12.5% polyacrylamide gel. One half of the iodinated GT1b sample was incubated with a 1% trypsin solution (porcine pancreas, Sigma) for 60 min at room temperature before being applied to the gel (Lane D). The gels were electrophoresed, fixed, stained, and dried as described.[17] The autoradiographs shown are the results of a 2-day exposure.

assay system, and these results were not altered by protease treatment even though the latter extensively hydrolysed the contaminating proteins (Lane D).

In addition to testing those proteins present in commercial preparations we also investigated sialoglycoproteins known to have sialyloligosaccharide structures similar to but not identical with those found on receptor gangliosides (Fig. 3). Fetuin, a known hemagglutination inhibitor for Sendai virus,[11] was incubated with Sendai virus at 10 mg/ml to test its ability to act as an inhibitor of viral infection of native cells (Table 4). Glycophorin A, an amphipathic protein which is thought to be the main component on human erythrocytes responsible for the binding of the virus during hemagglutination,[18] was incubated with receptor-deficient cells to test its ability to function as a receptor.

Neither sialoglycoprotein had any effect. The fetuin preparation which did inhibit hemagglutination did not inhibit viral infection. The second result, that glycophorin A showed no evidence of functioning as a receptor in our receptor assay system, is not surprising because our assay depends upon the spontaneous insertion of the putative receptor into a preformed membrane. These data do not rule out the possibility that sialoglycoproteins could be receptors in addition to gangliosides for Sendai virus infection or inhibitors of the infective process by virtue of their being receptor

Table 4. Preliminary Examination of Sialoglycoproteins as Inhibitors or Receptors for Sendai Virus Infection*

Cell State	Addition	Virus Production
Native	---	256 HAU/ml
Native	Fetuin (60 nmol)	256
Sialidase-treated	---	<2
Sialidase-treated	GD1a (1 nmol)	128-256
Sialidase-treated	GM1 (1 nmol)	<2
Sialidase-treated	glycophorin A (2 nmol)	<2
Sialidase-treated	glycophorin A (2 nmol) plus GM1 (20 nmol)	<2

*Fetuin (10 mg/ml, 32 hemagglutination inhibition units/ml) was preincubated with the virus inoculum for 30 min at 4°C to examine its ability to act as an inhibitor of virus infection in native cells. Glycophorin A and gangliosides were preincubated with receptor-deficient MDBK cells for 20 min at 37°C before addition of virus to examine their ability to function as receptors.

analogs. They do, however, indicate that protein contaminants in ganglioside samples are highly unlikely to be the cause of false positives in our receptor assay system.

However, a new problem presented by protein contaminants in ganglioside samples became apparent when the total ganglioside fractions isolated from MDBK, HeLa, and MDCK cells were reinserted in receptor-deficient cells to test for receptor function of endogenous gangliosides. Samples derived from MDBK and MDCK cells produced the anticipated results. But with samples derived from HeLa cells, although analysis of the gangliosides had demonstrated that more than sufficient amounts of GD1a and GT1b were present in the samples, only partial restoration of susceptibility occurred. Upon further purification of the gangliosides to produce an oligosialoganglioside fraction, full receptor capability was manifested. The removal of monosialogangliosides seemed an unlikely explanation of the observed phenomenon because the addition of purified GM3 to the oligosialoganglioside fraction had no effect on the ability of the receptor gangliosides in that fraction to biologically function. Examination of the radioiodinated gels did indicate the presence of protein in the samples (0.1 to 1%). Gangliosides are known to bind to proteins such as bovine serum albumin.[19] Therefore purified GD1a was added to a 1% solution of BSA in Tris-saline and incubated with receptor-deficient cells. The inclusion of BSA completely interfered with the biological functioning of GD1a as a

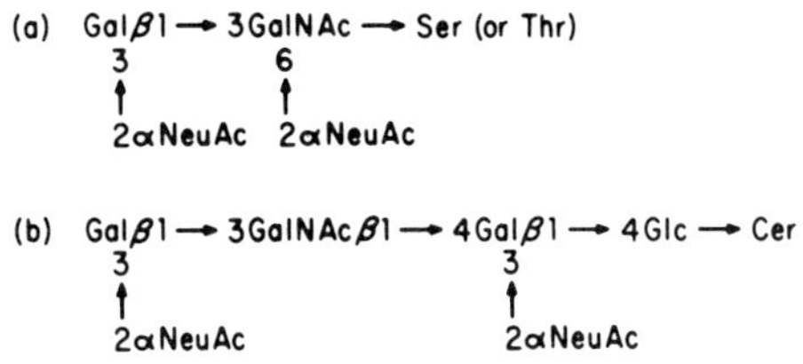

(a) Galβ1 → 3GalNAc → Ser (or Thr)
    3$$6
    ↑$$↑
    2αNeuAc 2αNeuAc

(b) Galβ1 → 3GalNAcβ1 → 4Galβ1 → 4Glc → Cer
    3$$3
    ↑$$↑
    2αNeuAc2αNeuAc

Fig. 3. Sialyloligosaccharide structures found on proteins such as fetuin and glycophorin A (a) are similar to but not identical with those found on receptor gangliosides such as GD1a (b).

receptor. The substitution of 10% fetal calf serum for the BSA gave identical results when added simulataneously with the GD1a (Table 5). Neither BSA nor fetal calf serum had any effect if added after GD1a had inserted into the membrane. It thus appears that the presence of proteins can interfere with the demonstration of the biological functioning of gangliosides by possibly preventing their proper insertion into the receptor-deficient membrane.

In conclusion we would like to emphasize that (1) specific gangliosides are natural receptors for Sendai virus, (2) that these

Table 5. Masking of Positive Biological Effects of Gangliosides By Contaminants in the Sample*

Cell State	Addition	Virus Production
Native	---	256 HAU/ml
Sialidase-treated	---	<2
Sialidase-treated	Total ganglioside extract	64
Sialidase-treated	Oligosialoganglioside fraction	256-512
Sialidase-treated	GM3 (3 nmol) plus oligosialo-ganglioside fraction	256
Sialidase-treated	GD1a (0.3 nmol)	256
Sialidase-treated	GD1a plus 1% BSA	<2
Sialidase-treated	GD1a plus 10% fetal calf serum	<2

*Samples containing 0.3 nmoles of GD1a standard, or of GD1a plus GT1b in the total ganglioside extract or oligosialoganglioside fraction from HeLa cells were incubated for 20 min at 37°C with receptor-deficient cells.

gangliosides could have an additional role in membrane fusion, (3) that the question of whether sialoglycoproteins can act either as receptors or inhibitors of Sendai virus infection needs further investigation, and (4) that protein contaminants in lipid samples can mask biological effects of gangliosides.

ACKNOWLEDGMENTS

The authors would like to thank Drs. Debi P. Nayak and Arnold J. Berk for initial cell stocks, and Drs. Anne M. Haywood and Allen Portner for initial virus stocks. In addition we would like to acknowledge Dr. Kathryn E. Kronquist for her help in karyotyping the cells, Mrs. Birgitta Dellheden for lipid determinations and Dr. Alfred T. H. Burness for preparing the glycophorin A. This research was supported by research grants from the National Institutes of Health, the Kroc Foundation, and the Swedish Medical Research Council.

REFERENCES

1. D. P. Fan and B. M. Sefton, The entry into host cells of Sindbis virus, Vesicular Stomatitis virus and Sendai virus, Cell 15: 985 (1978).
2. Y. Yasuda, Y. Hosaka, Y. Fukami, and K. Fukai, Immunoelectron microscopy study on interactions of noninfectious Sendai virus and murine cells, J. Virol. 39:273 (1981).
3. K. S. Matlin, H. Reggio. A. Helenius, and K. Simons, Infectious entry pathway of influenza virus in a canine kidney cell line, J. Cell Biol. 91:601 (1981).
4. A. Yoshimura, K. Kuroda, K. Kawasaki, S. Yamashina, T. Maeda, and S. -I. Ohnishi, Infectious cell entry mechanism of influenza virus, J. Virol. 43:284 (1982).
5. M. A. K. Markwell and J. C. Paulson, Sendai virus utilizes specific sialyloligosaccharides as host cell receptor determinants, Proc. Natl. Acad. Sci. USA 77: 5693 (1980).
6. J. Holmgren, L. Svennerholm, H. Elwing, P. Fredman, and O. Strannegard, Sendai virus receptor: Proposed recognition structure based on binding to plastic-adsorbed gangliosides, Proc. Natl. Acad. Sci. USA 77:1947 (1980).
7. M. A. K. Markwell, L. Svennerholm, and J. C. Paulson, Specific gangliosides function as host cell receptors for Sendai virus, Proc. Natl. Acad. Sci. USA 78:5406 (1981).
8. P. Fredman, O. Nilsson, J. -L. Tayot, and L. Svennerholm, Separation of gangliosides on a new type of anion-exchange resin, Biochim. Biophys. Acta 618:42 (1980).
9. B. Maggio, F. A. Cumar, and R. Caputto, Induction of membrane fusion by polysialogangliosides, FEBS Lett. 90:149 (1978).

10. C. H. Andrews, F. B. Bang, and F. M. Burnet, A short description of the myxovirus group (influenza and related viruses), Virology 1:176 (1955).
11. A. M. Haywood, Characteristics of Sendai virus receptors in a model membrane, J. Mol. Biol. 83:427 (1974).
12. P. -S. Wu, R. W. Ledeen, S. Udem, and Y. A. Isaacson, Nature of the Sendai virus receptor: Glycoprotein versus ganglioside, J. Virol. 33:304 (1980).
13. P. H. Fishman, Role of membrane gangliosides in the binding and action of bacterial toxins, J. Mem. Biol. 69:85 (1982).
14. D. Gingell and L. Ginsberg, Problems in the physical interpretation of membrane interaction and fusion, in: "Membrane Fusion", G. Poste and G. L. Nicolson, eds., North-Holland Publishing Company, New York, p. 791 (1978).
15. A. M. Haywood and B. P. Boyer, Sendai virus membrane fusion: Time course and effect of temperature, pH, calcium, and receptor concentration, Biochemistry 24:6041 (1982).
16. K. Watanabe, S. Hakomori, M. E. Powell, and M. Yakota, The amphipathic membrane proteins associated with gangliosides: The Paul-Bunnell antigen is one of the gangliophilic proteins, Biochem. Biophys. Res. Commun. 92:638 (1980).
17. M. A. K. Markwell and C. F. Fox, Surface-specific iodination of membrane proteins of viruses and eucaryotic cells using 1,3,4,6-tetrachloro-3 ,6 -diphenylglycoluril, Biochemistry 17:4807 (1978).
18. C. Howe, J. E. Coward, and T. W. Fenger, Viral invasion: Morphological, biochemical, and biophysical aspects, in: "Virus-Host Interactions", H. Frankel-Conrat and R. R. Wagner, eds., Plenum Press, New York, p. 1 (1980).
19. G. Tettamanti, A. Preti, B. Cestaro, M. Masserini, S. Sonnino, and R. Ghidoni, Gangliosides and associated enzymes at the nerve-ending membranes, in: "Cell Surface Glycolipids", C. C. Sweeley, ed., American Chemical Society, Washington, D. C., p. 321 (1980).

GANGLIOSIDES AS RECEPTOR MODULATORS

Eric G. Bremer and Sen-itiroh Hakomori

Fred Hutchinson Cancer Research Center and
The University of Washington
1124 Columbia Street
Seattle, Washington 98104

INTRODUCTION

A number of studies have postulated cell surface gangliosides to be receptors for various bioactive factors such as glycoprotein hormones,[1-3] interferon,[4-6] and bacterial toxins,[7-9] based on the interaction of gangliosides with these bioactive factors or ganglioside modification of the cellular effect of these factors. These observations, however, do not necessarily prove or support the idea that gangliosides function as receptors for these factors. In fact, in most of these cases, the receptor has been subsequently characterized as a protein, and the binding to gangliosides is generally of much lower affinity and specificity. With the exception of cholera toxin binding to the oligosaccharide moiety of GM1,[10] few examples of ganglioside receptors have been demonstrated. Our recent studies on the effect of gangliosides on cell growth have suggested that gangliosides may alter the binding of polypeptide growth factors to their receptors in an indirect way.

The possibility that gangliosides may be regulators of cell growth was suggested by changes in glycosphingolipid metabolism closely related to density-dependent growth inhibition of cultured cells (for review, see 11). Addition of exogenous glycosphingolipids has been one approach utilized to test the possibility that glycolipids influence the control of cell proliferation.[11] Hamster NIL cells cultured in media containing $GbOse_4Cer$ showed increased cell adhesiveness, and NIL cells transformed by polyoma virus reverted to normal NIL cell morphology. The prereplicative period of these cells also lengthened two-fold with added glycolipid.[12]

Addition of ganglioside to a human neural cell line resulted in a 40% lengthening of the G1 phase.[13] Similarly, the addition of several gangliosides to culture media reduced both the growth rate and saturation density of SV40 transformed and untransformed 3T3 cells.[14] With the availability of purified growth factors, gangliosides, and serum-free culture conditions, we have examined the phenomenon of ganglioside-induced growth inhibition in greater detail. We have examined the effect of exogenously-added gangliosides on cells in serum-free medium, on the response of the cells to individual polypeptide growth factors, and on the cellular binding of growth factors.

RESULTS

Cell growth inhibition by gangliosides

Swiss mouse 3T3 cells (Swiss 3T3) or baby hamster kidney cells (BHK) were cultured as previously described[15,16] in serum-free hormone-supplemented medium consisting of a nutrient medium plus insulin, transferrin, hydrocortisone, linoleic acid, bovine serum albumin, and fibronectin. In addition to this medium, 3T3 cells required platelet-derived growth factor (PDGF) and epidermal growth factor (EGF), and BHK cells required fibroblast growth factor (FGF). Addition of 50 nmoles per ml of GM3 inhibited cell growth of both 3T3 (Fig. 1) and BHK cells,[20] while 50 nmoles per ml GM1 strongly inhibited only 3T3 cells (Fig. 1). NeuAc-nLcOse$_4$Cer, which has the same terminal disaccharide as GM3, did not inhibit cell growth at the concentrations used (data not shown).

Growth inhibition by exogenous gangliosides was further examined by measurement of the mitogenicity of individual growth factors. The mitogenicity of PDGF, EGF, and FGF was determined for cells cultured in the presence or absence of exogenous ganglioside by [^{3}H]-thymidine incorporation.[17,18] The results of the [^{3}H]-thymidine incorporation assay are shown in Fig. 2. Exogenously-added GM1 and GM3 both inhibited PDGF-stimulated thymidine incorporation. GM1, which was the most potent inhibitor of 3T3 cell division (Fig. 1), was also the strongest inhibitor of PDGF-stimulated thymidine incorporation (Fig. 2a). As with its effect on cell growth, NeuAc-nLcOse$_4$Cer had no effect on PDGF-stimulated thymidine incorporation (data not shown). These data suggest that GM1 and GM3 may inhibit cell growth through inhibition of PDGF-stimulated DNA synthesis. When EGF-stimulated thymidine incorporation was examined, GM3 was the strongest inhibitor (Fig. 2b), while GM1 was only marginally inhibitory and NeuAc-nLcOse$_4$Cer had no effect. Specificity is implied, since GM3 is the stronger inhibitor of EGF mitogenicity and GM1 is the stronger inhibitor of PDGF mitogenicity. FGF-stimulated thymidine incorporation did not appear to be significantly altered by any of the gangliosides tested.

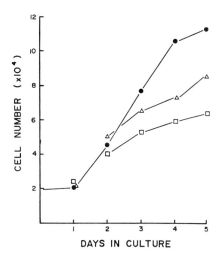

Fig. 1. Cell growth inhibition by exogenously added gangliosides. Swiss 3T3 cells were seeded into 24-well plates at a concentration of 2×10^4 cells/well, and grown in serum-free DME/F12 medium containing no gangliosides, ●; 50 nmoles per ml GM1, □; or 50 nmoles per ml GM3, △. The medium was changed on the second and fourth days. Each day cells were harvested with trypsin and counted with a Model Z_{B1} Coulter counter. Each data point represents the average of six determinations.

Cellular binding of growth factors

Since exogenously-added ganglioside appeared to inhibit the mitogenicity of PDGF and EGF, the binding of these growth factors to their receptors was also examined. Binding assays were performed on cells that had been cultured in the presence of GM3, GM1, NeuAc-nLcOse$_4$Cer, or no ganglioside. Fig. 3a shows the concentration-dependence of [^{125}I]-PDGF binding for 3T3 cells cultured with GM1 or with no ganglioside added. The number of binding sites is the same for both the GM1-fed and control cells (no ganglioside added); however, the concentration-dependence of PDGF binding by GM1 differs from control. Scatchard[19] analysis of the binding data confirms this observation (Table 1), showing an apparent K_d of 12 pmolar for control 3T3 and an apparent K_d of 7 pmolar for GM1-fed cells with no significant difference in the number of PDGF receptors per cell (6.1×10^4 and 5.8×10^4, respectively). Exogenous addition of GM3 to cell cultures also decreased the apparent K_d of [^{125}I]-PDGF binding, in a similar manner to GM1. NeuAc-nLcOse$_4$Cer, on the other hand, had no effect on the binding of [^{125}I]-PDGF. The effect of exogenous gangliosides on [^{125}I]-PDGF binding therefore correlates with their effect on PDGF-stimulated cell division: GM1 and GM3 inhibited 3T3 cell growth, while NeuAc-nLcOse$_4$Cer had no effect (Figs. 1 and 2).

The increased binding affinity for PDGF by GM1- or GM3-fed cells appears to be reversible. Cells were grown in medium to which 50 nmoles per ml GM1 were added for 4 days and then the medium was replaced with fresh medium to which no ganglioside was added. The cells were allowed to recover for an additional 3 days in normal medium and then the [^{125}I]-PDGF binding was determined as described above. In this case, the binding affinity and receptor number were both similar to those of the control cultures (Table 1).

The influence of exogenously-added gangliosides on EGF binding was also examined. Fig. 3b shows the 4°C binding of EGF to 3T3 cells. GM1 and GM3 did not appear to influence the equilibrium binding of EGF. Since 3T3 cells respond to FGF ([^{3}H]-thymidine incorporation assay, Fig. 2c), we also examined FGF binding to Swiss 3T3 cells. FGF binding was carried out at 4°C as previously described by Bremer and Hakomori.[20] Fig. 3c shows the [^{125}I]-FGF binding to 3T3 cells cultured in the presence or absence of gangliosides. GM3-fed cultures bound twice as much FGF as did cultures with no added ganglioside. GM1- and NeuAc-nLcOse$_4$Cer-fed cultures were indistinguishable from the control cultures.

Interaction between gangliosides and growth factors

PDGF and FGF have both been shown to be very cationic peptides with isoelectric points of 9-10.[17,26] Gangliosides that contain sialic acid, on the other hand, are anionic. This suggests a possible direct ionic interaction between the cationic growth factors and the added gangliosides. The lack of any influence by gangliosides on EGF binding is consistent with this

---

Fig. 2. (Next Page). Growth factor stimulated [^{3}H]-thymidine incorporation. Swiss 3T3 cells were cultured in DME/F12 plus 5% FCS for 3 days with no ganglioside, ● ; 50 nmoles per ml GM1,□ ; or 50 nmoles per ml GM3,△ . This medium was then replaced with DME/F12 plus 1% human plasma derived serum, but with ganglioside present or absent as listed above for 48 hours. Cultures were incubated for 18 hours with increasing concentrations of PDGF, A ; EGF, B ; or FGF, C . The incorporation of [^{3}H]-thymidine into trichloroacetic acid-insoluble material was measured as previously described.[18] The results are plotted (A and B) as fold stimulation above a no growth factor added control and are the average of four determinations.

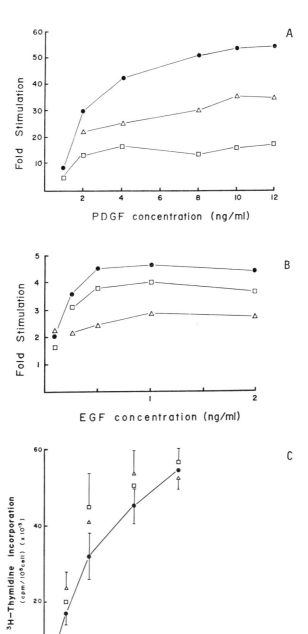

Table 1. Scatchard Analysis of 4°C [^{125}I]-PDGF Binding

	Apparent K_d (pmolar)	Number of receptors per cell
No ganglioside added	12.0	6.1×10^4
GM1	7.0	5.8×10^4
GM3	8.0	6.0×10^4
NeuAc-nLcOse$_4$Cer	12.0	6.3×10^4
72 hours after GM1 removal	11.0	5.4×10^4

interpretation. However, since NeuAc-nLcOse$_4$Cer has no effect on PDGF or FGF binding and GM1 has no effect on FGF binding, electrostatic binding alone seems unable to account for our results.

The possibility of direct interaction between the growth factors and gangliosides was further investigated by competition between cells and ganglioside liposomes for [^{125}I]-PDGF and [^{125}I]-FGF binding. Single bilayer ganglioside liposomes were prepared by injecting an ethanol solution containing 25 nmoles per ml lecithin, 50 nmoles per ml cholesterol, and ganglioside into phosphate-saline.[21] Liposomes containing increasing amounts of ganglioside (G1 or GM3) were mixed with growth factors and then added to cells. As indicated in Table 2, there does not appear to be any competition between ganglioside liposomes and cells for PDGF or FGF[20] binding. These competition data and the

Fig. 3. (Next Page). Binding of [^{125}I]-growth factors to Swiss 3T3 cells. Cells were grown in the presence or absence of 50 nmoles per ml ganglioside as described for Fig. 2. Binding was determined by incubation with 1 ml binding medium (DME/F12 plus 1mg/ml BSA) containing increasing amounts of [^{125}I]-PDGF, A ; [^{125}I]-EGF, B ; or [^{125}I]-FGF, C . Cultures were incubated for four hours at 4°C with gentle agitation before measurement of cell-bound [^{125}I] as previously described.[18,20] Non-specific binding was determined by addition of cold competitor and has been subtracted. Cell numbers were determined by Coulter Counter. ●, no ganglioside added; □, 50 nmoles per ml GM1 added; Δ, 50 nmoles per ml GM3 added.

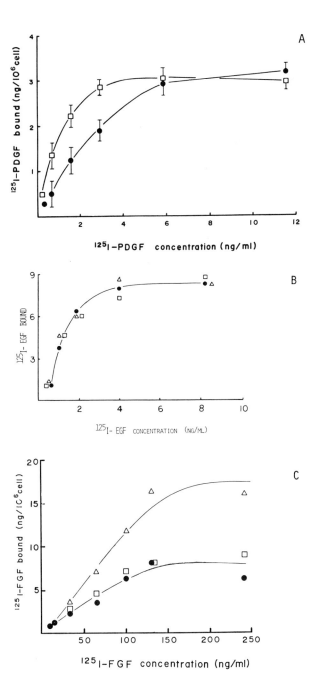

observation that not all gangliosides affect $[^{125}I]$-PDGF or $[^{125}I]$-FGF binding make direct interaction between the gangliosides and growth factors seem even less likely.

Incorporation of exogenously-added gangliosides into 3T3 cells

The differential effects on growth factor mitogenicity and binding could be explained by differences in the cellular incorporation of exogenous gangliosides. Considerable evidence suggests that at least some exogenous glycolipids become oriented with the ceramide moiety in the lipid bilayer.[22-24] We examined the incorporation of exogenous gangliosides into 3T3 cells with tritiated gangliosides labeled in the ceramide moiety by catalytic hydrogenation.[25] $[^{3}H]$-GM3, $[^{3}H]$-GM1, and $[^{3}H]$-NeuAc-nLcOse$_4$Cer were added to 3T3 cell cultures and the incorporation of radioactivity was measured after trypsinization of the cells. All of these gangliosides were incorporated into the cell to about the same extent (Table 3). The trypsin-sensitive, or supernatant, counts and the trypsin-insensitive, or cell pellet, counts were also compared (Table 3). Of the cell-associated material, only 10% was trypsin-insensitive after 6 hours, but 60-70% of the cell-associated radioactivity was trypsin-insensitive after 48 hours. Since ganglioside inserted into the cell bilayer should not be trypsin-sensitive, these data suggest that it takes up to 48 hours for the majority of the cell-associated

Table 2. Competition for $[^{125}I]$-PDGF Binding by Ganglioside Liposomes

Ganglioside Added to Liposome*	μg Ganglioside Added to Liposomes	ng PDGF/10⁶ cell**		
		No Ganglioside Added	GM1-Fed Cells	GM3-Fed Cells
None		2.3	3.4	3.0
GM1	14	2.1	3.2	3.0
	70	2.2	3.2	2.9
GM3	10	2.2	3.4	3.2
	70	2.0	3.3	3.1

* Single bilayer liposomes were prepared according to Batzri and Korn[21].
**Determined from the binding of 3 ng/ml $[^{125}I]$-PDGF.

exogenously-added ganglioside to become inserted into the cell membrane rather than adsorbed to the cell surface. These incorporation data may also explain the lag time before growth inhibitory effects can be measured (Fig. 1).

DISCUSSION

Our results indicate that growth inhibition by exogenously-added gangliosides may reflect ganglioside-induced changes in the way cells respond to growth factors. PDGF is considered to be the major mitogen for 3T3 cells.[18] After addition of GM1, PDGF-stimulated thymidine incorporation is reduced almost five-fold (Fig. 2). GM3 appears to decrease the mitogenic responses to both PDGF and EGF. FGF-stimulated thymidine incorporation was unaffected by addition of ganglioside. These data argue for selectivity in the effects of added gangliosides, since they differentially affect these growth factor receptor systems.

We are currently examining possible mechanisms for ganglioside-induced growth inhibition by studying the binding kinetics of growth factors to the cells. Gangliosides do not act as receptors for growth factors, since PDGF and EGF receptors have been shown to be proteins with molecular weights of 180K and 170K, respectively.[27,28] Furthermore, competition experiments revealed little or no interaction between gangliosides and iodinated growth factors (Table 2). Binding data, however, suggest that gangliosides can influence growth factor binding. PDGF

Table 3. Incorporation of Exogenously-Added Gangliosides into Swiss 3T3 Cells

Time	Cell-associated Radioactivity*			
	Total CPM/10^5 Cell		Trypsin Insensitive (%)	
	GM3	GM1	GM3	GM1
6 h	1656	1921	7	12
24 h	1894	1950	30	47
48 h	3262	2888	62	70

*[^{3}H]-Labeled gangliosides were added to the medium at 45 nmoles/ml (1500 cpm/nmole).

bound to its receptor more efficiently on the GM1- and GM3-treated cells. GM3 alone increases accessible FGF binding sites on both 3T3 (Fig. 3c) and BHK cells.[20] This suggests that gangliosides may interact with specific membrane proteins which may directly or indirectly affect the function of that protein. The data for PDGF and FGF support this interpretation, since there is an alteration of binding, and the gangliosides do not directly compete for PDGF or FGF binding.

Glycolipid interaction with membrane proteins has also been suggested by electron spin resonance spectra of spin-labeled glycolipids[29] and by cross-linking studies.[30] Studies using monoclonal antibodies to the thyrotropin receptor have suggested

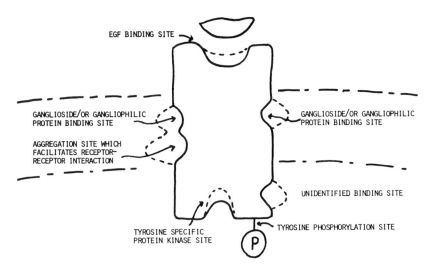

Fig. 4. "Allosteric Regulator" model of the EGF receptor as adapted from Schlessinger et al.[38] The primary adaptation has been to include ganglioside or gangliophilic protein binding sites. The binding of EGF to the receptor may affect (i) tyrosine-specific protein kinase which is independent from cyclic AMP-dependent kinase, and (ii) receptor-receptor aggregation which facilitates clustering and internalization of receptor. In addition, it is proposed that the receptor should have binding sites to gangliosides and gangliophilic proteins. EGF binding to the receptor could be allosterically regulated through binding of gangliosides or gangliophilic protein to the receptor.

that gangliosides may be involved in receptor function.[31] One antibody that can stimulate thyrotropin function is very active against a preparation of human thyroid gangliosides. This same antibody does not compete with [^{125}I]-thyrotropin binding. A second antibody directed against the thyrotropin receptor appears to bind a glycoprotein and competes for [^{125}I]-thyrotropin binding. These data further support the notion that gangliosides may affect receptor function in a non-competitive manner.

The binding properties of individual growth factor receptors have been shown to be affected by other receptor systems. Phorbol esters,[32] phospholypase C,[33] dexamethasone,[34] and vasopressin[35] have all been shown to alter the K_d of EGF binding to its receptor. The binding of PDGF has also been shown to down-regulate the EGF receptor at 37°C[36] and alter its affinity at 4°C.[37] These data suggest that receptor activity is modulated by other membrane components. Recently, Schlessinger has proposed an "allosteric regulator" model for the EGF receptor.[38] This model proposes several binding sites on the receptor protein in addition to the EGF binding site (Fig. 4). The chemical level and the organization of gangliosides may affect the allosteric configuration of the receptor. It is possible that one of the additional sites may bind gangliosides directly or through an intermediate gangliophilic protein. Alternatively, exogenous gangliosides may mask another interaction site required for microclustering or interaction with other receptors.

The differences observed in the [^{125}I]-growth factor binding kinetics, however, do not explain how gangliosides can decrease the mitogenic response, but rather they point where the explanation may be found. Tyrosine autophosphorylation of the receptor after growth factor binding has been suggested to be an important step initiating the mitogenic response after binding of PDGF[39] or EGF.[27] If gangliosides alter this process, it could imply that the gangliosides more or less directly affect the receptor. There are, however, other non-receptor, membrane-associated events which gangliosides may inhibit. An amiloride-sensitive Na^+/H^+ exchange system has been reported to be activated by growth factor binding and would be an important event in the mitogenic response of cells.[40] It is possible that the gangliosides may inhibit this type of process instead of directly affecting the growth factor receptor. These possibilities are currently under investigation.

ACKNOWLEDGEMENTS

This investigation was supported by a research grant from the National Institutes of Health, CA20026, and E.G.B. is the recipient of an Otsuka Research Foundation fellowship.

REFERENCES

1. B. R. Mullin, P. H. Fishman, G. Lee, Sm. M. Aloj, F. D. Ledley, R. J.Winand, L. D. Kohn, and R. O. Brady, Thyrotropin-ganglioside interactions and their relationship to the structure and function of thyrotropin receptors. Proc. Natl. Acad. Sci. USA 73:842 (1976).
2. G. Lee, S. M. Aloj, R. O. Brady, and L. D. Kohn, The structure and function of glycoprotein hormone receptors: Ganglioside interactions with human chorionic gonadotropin, Biochem. Biophys. Res. Commun. 73:370 (1976).
3. G. Lee. S. M. Aloj, and L. D. Kohn, The structure and function of glycoprotein hormone receptors: Ganglioside interactions with luteinizing hormone, Biochem. Biophys. Res. Commun. 77:434 (1977).
4. F. Besancon, and H. Ankel, Binding of interferon to gangliosides, Nature (Lond.) 252:478 (1974).
5. V. E. Vengris, R. H. Reynolds, M. D. Hollenberg, and P. M. Pitha, Interferon action: Role of membrane gangliosides, Virology 72:486 (1976).
6. H. Ankel, C. Krishnamaurti, F. Besancon, S. Stefanos, and E. Falcoff, Mouse fibroblast (type I) and immune (type II) interferons: Pronounced differences in affinity for gangliosides and in antiviral and antigrowth effects on mouse leukemia L-1210R cells, Proc. Natl. Acad. Sci. USA 77:2528 (1974).
7. A. M. Haywood, Characteristics of Sendai virus receptors in a model membrane, J. Mol. Biol. 83:427 (1974).
8. J. Holmgren, L. Svennerholm, H. Elwing, P. Fredman, and O. Strannegard, Sendai virus receptor: Proposed recognition structure based on binding to plastic-adsorbed gangliosides, Proc. Natl. Acad. Sci. USA 77:1947 (1980).
9. F. D. Ledley, G. Lee, L. D. Kohn, W. H. Habig, and M. C. Hardegree, Tetanus toxin interactions with thyroid plasma membranes, J. Biol. Chem. 252:4049 (1977).
10. P. Cuatrecasas, Interaction of Vibrio cholerae enterotoxin with cell membranes, Biochemistry 12:3547 (1973).
11. S. Hakomori, Glycosphingolipids in cellular interaction, differentiation, and oncogenesis, Annu. Rev. Biochem. 50:733 (1981).
12. R. A. Laine, and S. Hakomori, Incorporation of exogenous glycosphingolipids in plasma membranes of cultured hamster cells and concurrent change of growth behavior, Biochem. Biophys. Res. Commun. 54:1039 (1973).
13. I. Icarel-Liepkalns, V. A. Liepkalns, A. J. Yates, and R. E. Stephens, Cell cycle phases of a novel human neural cell line and the effect of exogenous gangliosides, Biochem. Biophys. Res. Commun. 105:225 (1982).
14. T. W. Keenan, E. Scmid, W. W. Franke, and H. Wiegandt, Exogenous glycosphingolipids suppress growth rate of

transformed and untransformed 3T3 cells, Exp. Cell Res. 92:259 (1975).
15. D. Barnes, and G. Sato, Methods for growth of cultured cells in serum-free medium, Anal. Biochem. 102:255 (1980).
16. T. Macaig, B. Kelley, J. Cerundolo, S. Ilsley, P. R. Kelley, J. Gaudreau, and R. Forand, Hormonal requirements of baby hamster kidney cells in culture, Cell Biol. Int. Reports 4:43 (1980).
17. E. Raines, and R. Ross, Platelet-derived growth factor. I. High yield purification and evidence for multiple forms, J. Biol. Chem. 257:5154 (1982).
18. D. F. Bowen-Pope, and R. Ross, Platelet-derived growth factor. II. Specific binding to cultured cells, J. Biol. Chem. 257:5161 (1982).
19. G. Scatchard, The attraction of proteins for small molecules and ions, Ann. N. Y. Acad. Sci. 51:660 (1949).
20. E. G. Bremer, and S. Hakomori, GM3 ganglioside induces haster fibroblast growth inhibition in chemically-defined medium: Gangliosides may regulate growth factor receptor function, Biochem. Biophys. Res. Commun. 106:711 (1982).
21. S. Batzri, and E. D. Korn, Single bilayer liposomes prepared without sonication, Biochim. Biophys. Acta 298:1015 (1973).
22. S. Kanda, K. Inone, S. Nojima, H. Utsumi, and H. Wiegandt, Incorporation of spin-labeled ganglioside analogues into cell and liposomal membranes, J. Biochem. 91:1707 (1982).
23. P. H. Fishman, J. Moss, and M. Vaughan, Uptake and metabolism of gangliosides in transformed mouse fibroblasts: Relationship of ganglioside structure to choleragen response, J. Biol. Chem. 251:4490 (1976).
24. D. M. Marcus, and L. Cass, Glycosphingolipids with Lewis blood group activity: Uptake by human erythrocytes, Science 164:553 (1969).
25. G. Schwarzmann, A simple and novel method for tritium labeling of gangliosides and other sphingolipids, Biochim. Biophys. Acta 529:106 (1978).
26. D. Gospodorowicz, Purification of a fibroblast growth factor from bovine pituitary, J. Biol. Chem. 250:2515 (1975).
27. S. Cohn, G. Carpenter, and L. King, Jr., Epidermal growth factor-receptor-protein kinase interactions. Co-purification of receptor and epidermal growth factor-enhanced phosphorylation activity, J. Biol. Chem. 255:4834 (1979).
28. K. Glenn, D. F. Bowen-Pope, and R. Ross, Platelet-derived growth factor. III. Identification of platelet-derived growth factor receptor by affinity labeling, J. Biol. Chem. 257:5172 (1982).
29. F. J. Sharom, and C. W. M. Grant, A model for ganglioside behavior in cell membranes, Biochim. Biophys. Acta 507:280 (1978).

30. C. A. Longwood, S. Hakomori, and T. H. Ji, A glycolipid and its associated proteins: Evidence by crosslinking of human erythrocyte surface components, FEBS Lett. 112:265 (1980).
31. W. A. Valente, P. Vitti, Z. Yavin, E. Yavin, E. F. Grollman, R. S. Toccafondi, and L. D. Kohn, Monoclonal antibodies to the thyrotropin receptor: Stimulating and blocking antibodies derived from the lymphocytes of patients with Graves disease, Proc. Natl. Acad. Sci. USA 79:6680 (1982).
32. L. S. Lee, and I. B. Weinstein, Tumor-promoting phorbol esters inhibit the binding of epidermal growth factor to cellular receptors, Science 202:313 (1978).
33. M. Shoyab, and G. T. Todaro, Perturbation of membrane phospholipids alters the interaction between epidermal growth factor and its membrane receptors, Arch. Biochem. Biophys. 296:222 (1981).
34. J. B. Baker, G. S. Barsh, D. H. Carney, and D. D. Cunningham, Dexamethasone modulates binding and action of epidermal growth factor in serum-free cell culture, Proc. Natl. Acad. Sci. USA 75:1882 (1978).
35. E. Rozengurt, K. D. Brown, and P. Petticum, Vasopressin inhibition of epidermal growth factor binding to cultured mouse cells, J. Biol. Chem. 256:716 (1981).
36. C. Heldin, A. Wasteson, and B. Westemark, Interaction of platelet-derived growth factor with its fibroblast receptor: Demonstration of ligand degradation and receptor modulation, J. Biol. Chem. 257:4216 (1982).
37. D. F. Bowen-Pope, P. E. DiCorleto, and R. Ross, Interactions between the receptors for platelet-derived growth factors and epidermal growth factor, J. Cell Biol. 96:679 (1983).
38. J. Schlessinger, A. B. Schreiber, A. Levi, I. Lax, T. Libermann, and Y. Yarden, Regulation of cell proliferation by epidermal growth factor, Critical Reviews in Biochemistry 14:93 (1983).
39. J. Nishimura, J. S. Huang, and T. F. Duel, Platelet-derived growth factor stimulates tyrosine-specific protein kinase activity in Swiss mouse 3T3 cell membranes, Proc. Natl. Acad. Sci. USA 79:4303 (1982).
40. J. Pouyssegur, J. C. Chambard, A. Franchi, S. Paris, and E. Van Obberghen-Schilling, Growth factor activation of an amiloride-sensitive Na^+/H^+ exchange system in quiescent fibroblasts: Coupling to ribosomal protein S6 phosphorylation, Proc. Natl. Acad. Sci. USA 79:3935 (1982).

BRAIN GANGLIOSIDES AND THERMAL ADAPTATION IN VERTEBRATES

Hinrich Rahmann, Reinhard Hilbig, Wolfgang Probst and Martin Muhleisen

Institute of Zoology, University of Stuttgart-Hohenheim
D-7000 Stuttgart 70 (Hohenheim), West Germany

INTRODUCTION

The ability of the vertebrates to adapt to fluctuations in their environment is mainly based upon adaptive changes within the CNS, in which the synapses have been shown to be the most sensitive structures. Therefore it is assumed that changes in the physicochemical properties of the synaptic membrane probably are responsible for maintaining adequate transmission and possibly related kinds of adaptive neuronal processes.

There is no doubt that almost every stage of neuronal activity, especially that of synaptic transmission, can be modified by changes in extracellular calcium. With regard to this phenomenon it must be considered that it is not only the transmitter release itself which might be responsible for synaptic transmission, but that it is of great importance to take into account the modulation of this process. This should be the focus of future study in order to understand the adaptive mechanisms of the transmission of information.

With regard to modulatory properties of synaptic components we have proposed the hypothesis of a functional implication of gangliosides in synaptic transmission and adaptive neuronal processes, such as thermal adaptation.[1-3]

In order to obtain further evidence for our hypothesis we have adopted several experimental approaches concerning changes in brain gangliosides of vertebrates which have developed during phylogeny different strategies for survival in the cold. In addition we have begun in vitro experiments measuring the influence of different temperatures on the interaction of Ca^{2+} ions with monolayers prepared from single ganglioside fractions or ganglioside mixtures derived from the brain of warm or cold-adapted animals. Our goal was to elucidate a possible causal mechanism for the modulatory function of these glycosphingolipids in the process of neuronal transmission.

MATERIALS AND METHODS

Gangliosides were extracted from whole brain or single brain regions of about 80 vertebrate species belonging to 7 different classes (lampreys, cartilaginous fish, bony fish, amphibians, reptiles, bird and mammals). Care was taken to collect all animals from their normal habitats. Quantitative estimations of ganglioside-bound neuraminic (sialic) acid (NeuAc) were carried out according to Svennerholm and Fredman.[4] Individual ganglioside fractions were separated by HP-TLC[5], visualized with resorcinol reagent and quantified by spectrophotometric scanning at 580 nm. From identified peak areas the percent distribution and concentration of single ganglioside fractions were calculated. For comparative calculations the different ganglioside fractions were arbitrarily arranged into three groups: —either according to their content of NeuAc-residues (3 NeuAc, and respectively more or less than 3), which roughly corresponds with the polarity of gangliosides, —or according to one of the three possible pathways of biosynthesis.[6] The ganglioside nomenclature was that of Svennerholm.[7]

Furthermore, the influence of different temperatures ($11°$, $20°$ and $37°C$) on physico-chemical properties (surface pressure, area/molecule) of single ganglioside fractions (GD1a and GD1b) and of brain ganglioside mixtures derived from summer or winter acclimatized dsungarian hamsters (Phodopus sungorus) was measured by means of monolayers in a teflon trough with triethanolamine-HCl buffer (pH 7.4) as subphase.

RESULTS

1. Neuronal Gangliosides and Thermal Adaptation Phenomena in Vertebrates

The comparative investigation of about 80 different vertebrate species reveals an extraordinary variety in concentration and composition of brain gangliosides between cold-blooded or ectothermic vertebrates (all classes of fish, amphibians, reptiles) and warm-blooded or homeothermic birds and mammals. The total amount of ganglioside-bound sialic acid is obviously correlated with the

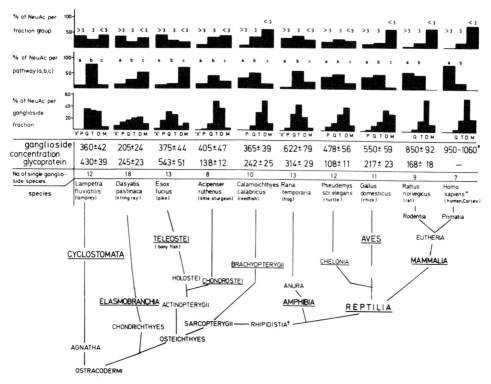

Fig. 1. Brain gangliosides in vertebrates including the phylogenetical dendrogram of representative species. Relative proportions of different gangliosides containing 3, more than or less than 3 NeuAc-residues, and their affiliation to one of the three pathways of biosynthesis, and to increasing numbers of NeuAc-residues (from mono-(M) to penta-(P) and higher sialylated fractions ('X')) is given. Number of single ganglioside fractions and concentration of ganglioside- or glycoprotein-bound NeuAc (μg/g fresh wt.) is indicated.

level of nervous organization: about 110 to 800 μg per g fresh wt. in lower vertebrates, and about 500 to 1000 μg/g in higher evolved vertebrates. When the ratio of neuronal gangliosides to sialoglycoproteins is compared across phyletic lines, it becomes evident that a drastic decrease in the glycoproteins from about 700-800 μg/g in teleost fish to less that 200 μg/g in mammals took place. This phenomenon of the changing ratio between these two sialoglycomacro-molecules has not been explained up to now.

For demonstration of anagenetic changes, representatives were selected from all classes of vertebrates (Fig. 1) and their ganglioside compositions were compared to their approximate position in the evolutionary stambrush. Besides the differences in the concentration, there is a large diversity in the molecular character of the vertebrate brain gangliosides, especially with regard to the number of sialic acid residues, i.e. polarity. The ganglioside pattern is composed of a large number of individual fractions (up to 17 in a ray) in all classes of fish, amphibians and reptiles. The CNS of adult warm-blooded vertebrates, on the other hand, generally revealed ganglioside patterns showing only small numbers of relatively less polar fractions (only 9 in mammals). It can be concluded that, during phylogeny, not only the concentration of gangliosides increased, and the number of single ganglioside fractions decreased, but that changes in the preponderance of one or two of the three possible pathways of ganglioside biosynthesis occurred as well (mainly "b"-pathway in lampreys, "b" + "c" in cartilaginous fish and amphibiams, "c" in bony fish, "b" + "a" in reptiles and birds and "a" in mammals).

When arranging the brain gangliosides of all vertebrates investigated up to now according to their polarity (3 NeuAc, more or

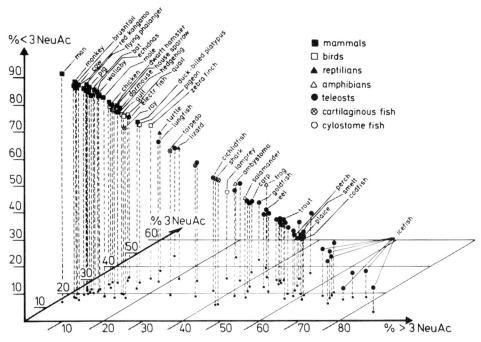

Fig. 2. Three dimensional arrangement of brain gangliosides from 81 vertebrate species in relation to their polarity (3 NeuAc-residues, more or less than 3). For reasons of clarity only some representatives have been assigned.

less than 3 NeuAc-residues; Fig. 2) distinct correlations between the ganglioside composition and the state of thermal adaptation became evident. The general tendency is that "the lower the environmental (- body) temperature, the higher is the polarity of brain gangliosides. Those species being adapted to warm habitats (tropic fish) or having developed thermoregulation ability with the phenomenon of homeothermy (adult birds and mammals) have relatively less polar brain gangliosides. Those vertebrates living under cold temperature conditions have gangliosides which are the most polar (Fig. 2).

This general result has been proved experimentally (Fig. 3) for different vertebrate species naturally adapted to habitats with extreme temperature conditions (for instance cold stenothermic icefish from the poles, gurnard codfish, rainbow trout, warm-stenothermic angle fish, or eurythermic ectotherms as for instance carp, newt and turtle, which can tolerate larger thermal fluctuations in their environment). Especially in hibernating mammals (fat dormouse, Glis glis) we succeeded in showing structure-specific changes in the brain ganglioside pattern depending on their state of seasonal adaptation. During hibernation a poly-sialylation of gangliosides, expecially in the pons, medulla and spinal cord occurred; in normothermic active dormice a less polar fraction accumulated (Fig. 4). This fraction, according to two-dimensional TLC, was shown to be O-Ac-GT1b, a ganglioside which is absent in hibernating dormice.[8]

Quite similar results were obtained when comparing the developmental profile of brain gangliosides in birds and mammals during early ontogeny in their heterothermic phase. Parallel to the ontogenetic development of homeothermy the pattern of brain gangliosides changes, thus indicating distinct correlation between the acquisition of the ability to thermoregulate and the degree of polarity of neuronal membranes. These results could be followed in two developmentally different types of birds[9]: the nidifugous type, characterized by a functional brain nearly reaching adult level at birth (e.g. chicken, quail), and the nidiculous type, showing at birth a brain with reduced functions (zebrafinch, pigeon, house sparrow).

2. Ganglioside-Calcium-Interactions

Since gangliosides possess the ability to complex strongly and specifically with Ca^{2+}, such complexes are assumed to reflect an efficient mechanism for modulating the sensitivity of the membrane-mediated processes of transmission and by this of functional synaptic plasticity.[3] In previous studies we investigated the influence of temperature changes on Ca^{2+}-ganglioside interactions by means of ion-selective electrodes[10] or by equilibrium dialysis using $^{45}Ca^{2+}$.[11] In continuation of these experiments we have now

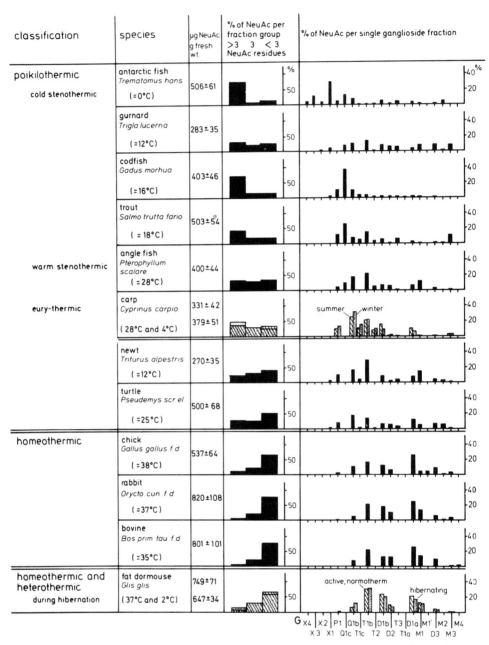

Fig. 3. Relative composition of major polarity groups of brain gangliosides (cf. Fig. 1 and 2) and single fractions of several vertebrates according to their thermal classification and adaptation properties to changes in the environmental temperature.

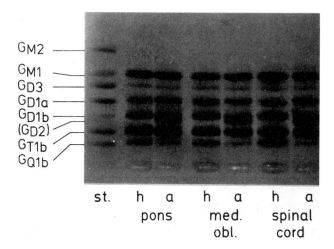

Fig. 4. Brain ganglioside pattern of 3 different brain structures of hibernating (h) and normothermic, active (a) fat dormouse (Glis glis).

analyzed variations of surface pressure-area-isotherms caused by different temperatures[12] and/or by addition of Ca^{2+} to the subphase of monolayers. These were prepared from single ganglioside fractions (e.g. GD1a, GD1b) or ganglioside mixtures derived from the brains of summer and winter adapted dsungarian hamsters which show significant differences in their composition and hence polarity (%-composition of main ganglioside fractions at 22° versus -4°C: GM1:16/13%; GD1a:11/9%; GD1b:17/19%; O-AcGT1b:14/0%; GT1b:27/30%; GQ1b:7/20%; Fig. 5).

The changes of the isotherms were most distinct for GD1a as compared with GD1b, and in case of the mixtures, for the less polar gangliosides from warm-adapted hamsters as compared with those from cold-adapted counterparts. (Lecithin and phosphatidyl-serine tested in the same system did not show any significnat variations.) In other words the data revealed a condensing effect of Ca^{2+} at low temperatures but not at warm temperatures where the area per molecule is partly increased in the presence of Ca^{2+}. Increased Ca^{2+}-binding in the cold might be responsible for this effect.

These results can be taken as evidence that synaptic membranes from cold-adapted animals with a high portion of polar gangliosides in the cold may have similar Ca^{2+}-binding abilities and hence viscosities as those from warm-adapted animals with a higher amount of less polar gangliosides.

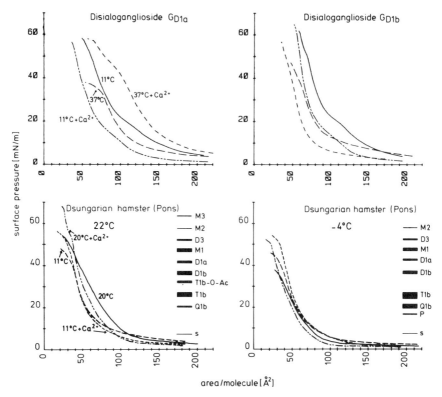

Fig. 5. Surface pressure/area – isotherms of monolayers from disialogangliosides (GD1a and GD1b) and from ganglioside mixtures pons-region of summer and winter adapted dsungarian hamsters (Phodopus sungorus) at two different temperatures each with or without Ca^{2+} (10^{-5}M) in the subphase.

DISCUSSION

The great variability in concentration and composition of gangliosides among vertebrates is traceable to different thermal adaptations: ectothermic lower vertebrates are phylogenetically adapted to distinct habitats or to seasonal fluctuations whereas higher evolved birds and mammals maintain high and constant endogenous body temperatures (with the exception of their neonatal phase of heterothermy or hibernation). The differences in gangliosides in relation to the respective thermal state demonstrate that the molecular mechanism of thermal adaptation in the CNS of all vertebrates is based upon a similar principle, which is that changes in the polarity of neuronal gangliosides reflect alterations in the environmental temperature. This principle obviously has had great evolutionary advantage insofar as it probably enabled the synaptic transmission process to proceed effectively even under changed temperature conditions. The latter is supported by ultrastructural changes of synapses following thermal adaptation.[13]

The in vitro experiments with different gangliosides demonstrate the pronounced adaptive abilities of these membraneous lipids to complex with Ca^{2+}-ions, and the high thermosensitivity of those complexes. We therefore assume, that under natural in vivo conditions the neuronal gangliosides obviously act as modulatory substances at the very local zone of the synaptic membrane by changing their polarity (sialylation and de-sialylation or N- or O-acetylation) in connection with the extent of Ca^{2+}-binding. By these mechanisms an undisturbed transmission process may be guaranteed. Since the physico-chemical results complement one another with the evolutionary and eco-physiological data, gangliosides may provide an excellent tool in our hands for obtaining in the not too distant future a satisfactory understanding of the functional events during chemical synaptic transmission of information.

SUMMARY

1. Comparative studies on brain gangliosides of about 80 species belonging to all classes of vertebrates reveal: a: distinct increases in concentration with phylogenetical progress of nervous organization, b) decreases in number of single fractions, c) changes in the polarity (degree of sialylation, N- or O-acetylation of sialic acids), d) alterations in the preponderance of one of the three possible pathways of biosynthesis.

2. In addition to these phylogenetical trends, clear correlations between the brain ganglioside composition and the state of thermal adaptation were shown: "The lower the environmental (- body) temperature, the higher the polarity of brain gangliosides". This principle was proved for ectotherms being adapted to habitats with extreme temperatures, during seasonal acclimatization and for homeotherms during early neonatal heterothermic development or during hibernation.

3. Surface pressure-area isotherms of monolayers from single ganglioside fractions (GDla, GDlb) or differently composed ganglioside mixtures from brains of warm- or cold-adapted hamsters as physico-chemical parameters show significant differences in their variability concerning temperature and/or Ca^{2+}-influences.

4. The results are taken as evidence that variations in the composition of synaptic-bound gangliosides may induce alterations in physico-chemical properties of the neuronal membrane, thus modulating synaptic transmission during temperature adaptation.

REFERENCES

1. H. Rahmann, H. Rosner, and H. Breer, A functional model of sialoglyco-macromolecules in synaptic transmission and memory formation, J. Theor. Biol. 57:231 (1976).
2. H. Rahmann, W. Probst, and M. Muhleisen, Gangliosides and synaptic transmission (Review), Japan. J. Exp. Med. 52:275 (1982).
3. H. Rahmann, Functional implication of gangliosides in synaptic transmission, Neurochem. Intern., 5:539 (review) (1983).
4. L. Svennerholm and P. Fredman, A procedure for the quantitative isolation of brain gangliosides, Biochim. Biophys. Acta 617:97 (1980).
5. H. Rosner, A new thin-layer chromatigraphie approach for separation of multisialogangliosides, Analyt. Biochem. 109:437 (1980).
6. R. K. Yu and S. Ando, Structures of some new complex gangliosides of fish brain, Adv. Exp. Med. Biol. 125:33 (1980).
7. L. Svennerholm, Ganglioside designation, Adv. Exp. Med. Biol. 125:11 (1980).
8. H. Rahmann, M. Muhleisen, R. Hilbig, S. Sonnino, and G. Tettamanti, in preparation.
9. H. Rahmann, U. Seybold, and V. Seybold, Comparative developmental profiles of brain gangliosides in fishes and birds, in: "Ganglioside structure, function and biomedical potential". Intern. Soc. Neurochem. Satellite Meeting. Vancouver (abstract) (1983).
10. W. Probst and H. Rahmann, Influence of temperature changes on the ability of gangliosides to complex with Ca^{2+}, J. Therm. Biol. 5:243 (1980).
11. M. Muhleisen, W. Probst, K. Hayashi, and H. Rahmann, Calcium binding to liposomes composed of negatively charged lipid moieties, Japan Exp. Med. 53:103 (1983).
12. W. Probst, D. Mobius, and H. Rahmann, Surface behavior of gangliosides at different temperatures, Hoppe-Seyler's Z. Physiol. Chem. 363:1313 (1983).
13. W. Probst, A. Choms, and H. Rahmann, Ultrastructural differences of synapses in the optic tectum of teleosts following seasonal acclimatization, J. Therm. Biol. 8:442 (1983).

Gangliosides and Neurological Diseases

GANGLIOSIDES AND DISEASE: A REVIEW

Kunihiko Suzuki

The Saul R. Korey Department of Neurology, Department
of Neuroscience, and the R. F. Kennedy Center for
Research in Mental Retardation and Human Development
Albert Einstein College of Medicine, Bronx, N.Y. 10461

INTRODUCTION

From the beginning, research on gangliosides has been closely associated with studies of diseases. Although there had been a few articles which may well have described the compounds which we now know as gangliosides,[1,2] discovery of gangliosides is justifiably credited to Klenk, who first discovered them in 1935 in the brain of patients with Niemann-Pick disease[3] and then in a larger quantity in the brain of Tay-Sachs disease patients.[4] Tay-Sachs disease became the first and then the only genetic disorder of abnormal ganglioside metabolism. During the next three decades, fundamental studies of chemistry and metabolism of gangliosides and studies of genetic diseases of ganglioside metabolism advanced hand in hand to the present state of enormous complexity. The synergistic interactions between basic and clinical sciences have proved to be highly beneficial to both sides. We now know several major groups of genetic gangliosidosis of known enzymatic causes and several more of yet to be characterized genetic variants.

Despite the progress in the chemistry and enzymology of gangliosides, their functional roles remained largely unknown until the 1970's, a few earlier pioneering studies notwithstanding.[5-7] This area of research has been gathering momentum in recent years. With this came the realization that some pathological conditions might be related to biological functions of gangliosides and that the course of some other disorders might be influenced by administration of gangliosides as a therapeutic agent. As in any rapidly moving field, controversies and uncertainties abound. One can confidently predict, however, that it will not be too long before a clearer picture of their biological functions and of their potential as therapeutic agents would emerge. One of the important purposes of this symposium was to accelerate this process. This brief review is intended to provide the general background for this border-area of the ganglioside-disease relationship.

The ganglioside-disease relationship can be divided into several distinct categories; (a) disorders that are caused by primary abnormality in ganglioside metabolism, (b) diseases in which ganglioside abnormalities occur non-specifically and secondarily to the disease process, (c) conditions in which ganglioside might be a potential pathogenetic factor, and (d) diseases for which ganglioside could serve as a therapeutic agent.

PRIMARY GENETIC ABNORMALITIES OF GANGLIOSIDE METABOLISM

The classical Tay-Sachs disease was the only known genetic gangliosidosis until 20 years ago. The situation has changed dramatically since with advances in basic knowledge and technologies in isolation, chemical structures, metabolism, enzymology, and physiological functions of gangliosides. Table 1 lists the major categories of well-characterized genetic disorders caused by primary abnormalities in ganglioside metabolism.

Catabolic enzyme defect

Three major categories of genetic diseases are known, caused by defective enzymes that normally catabolize gangliosides: GM2-gangliosidosis due to defective β-N-acetylhexosaminidases, GM1-gangliosidosis due to abnormal β-galactosidase, and mucolipidosis IV due to ganglioside sialidase deficiency. These diseases belong to the inborn lysosomal diseases of Hers[8] in that they are caused by genetic mutations that result in defective activities of lysosomal acid hydrolases.

The classical Tay-Sachs disease, the prototype of all ganglioside storage diseases, is caused by almost total lack of β-N-acetylhexosaminidase A activity.[9,10] The hexosaminidase B activity is either normal or, more commonly, higher than normal. The A isozyme appears to be exclusively responsible for degradation of GM2-ganglioside in vivo. This results in a several-fold increase in total brain gangliosides in affected patients. Since GM2-ganglioside constitutes 80-90% of total ganglioside in patients' brains and since it is a very minor component in normal brain, the increase of GM2-ganglioside in a patient's brain can be 100 times normal. When the hexosaminidase deficiency is partial rather than complete, the juvenile form of GM2-gangliosidosis results with a slower clinical course and milder pathological and compositional abnormalities.[11,12] The enzymatic deficiency in the classical Tay-Sachs disease and in juvenile GM2-

Table 1. Genetic Gangliosidosis in Man.

1. Catabolic Enzyme Defect

    a. β-N-Acetylhexosaminidase defect (GM2-gangliosidosis)

        i) β-N-Acetylhexosaminidase A defect

            Tay-Sachs disease (complete defect)
            Juvenile GM2-gangliosidosis (partial defect)
            Defect toward natural substrate (GM2)
            Miscellaneous variants

        ii) β-N-Acetylhexosaminidase A & B defect (Sandhoff)

    b. β-Galactosidase defect (GM1-gangliosidosis)

    c. Ganglioside sialidase defect (mucolipidosis IV)

2. Activator Defect

    a. Hexosaminidase-GM2 activator defect (GM2-gangliosidosis AB variant)

3. Anabolic Enzyme Defect?

---

gangliosidosis can be detected with chromogenic or fluorogenic artificial substrates. Within the past few years, a new form of hexosaminidase A mutation has been discovered.[13-16] In this disease, the mutation appears to have occurred in such a way that hexosaminidase A has lost its catalytic activity toward GM2-ganglioside but not toward the commonly used artificial substrates, such as p-nitrophenyl or 4-methylumbelliferyl substrates. The defective activity however can be detected with p-nitrophenyl β-N-acetylglucosamine-6-sulfate as substrate. Since hexosaminidase activities are normal when tested with the conventional artificial substrates, this disease was originally classified as the AB variant of GM2-gangliosidosis. In view of the nature of the mutation, it should now be classified as a hexosaminidase A defect.

As the technology of molecular genetics advances, more and more "variants" of hexosaminidase abnormalities are being discovered. Some of them manifest themselves as phenotypically atypical forms.[17-19] Some others may only be distinguished from the more common forms by the unique molecular abnormalities.[20] Atypical phenotypes can also result from genetic compounds (double heterozygotes).

The lysosomal β-N-acetylhexosaminidase A is a hetero-dimer consisting of α and β chains. Hexosaminidase B on the other hand is a homo-dimer of β+β. The above hexosaminidase A deficiencies occur when the mutation is on the α chain. When the mutation is on the β chain, both hexosaminidase A and B are affected, as expected. When a mutation inactivates both hexosaminidase A and B, it causes Sandhoff disease (O variant).[10] Due to the in vivo substrate specificities of the A and B isozymes, degradation of two related glycosphingolipids with terminal β-N-acetylgalactosamine, globoside and asialo-GM2-ganglioside (GA2), is blocked, in addition to the defective GM2-ganglioside degradation.

In contrast to GM2-gangliosidosis, GM1-gangliosidosis is caused by genetic abnormalities of one of the two lysosomal β-galactosidases, GM1-ganglioside β-galactosidase.[21] As the result, hydrolysis of the terminal galactose of GM1-ganglioside is blocked. This β-galactosidase has a broad substrate specificity and also hydrolyzes β-galactose moieties from oligosaccharide chains of glycoproteins. Thus, fragments of partially hydrolyzed glycoproteins, oligosaccharides, and keratan sulfate-like high-molecular-weight materials also accumulate in the tissues of patients.[22,23] The compositional abnormalities in the brain are analogous to those in GM2-gangliosidosis. The total brain ganglioside can be several times normal and GM1-ganglioside constitutes 80-90% of total ganglioside.

Bach et al.[24] suggested that mucolipidosis type IV was caused by deficiency of a specific sialidase which hydrolyzed N-acetylneuraminic acid from polysialogangliosides and GM3-ganglioside. At least two other laboratories have recently confirmed the finding.[25,26] The sialidase involved in this disease is clearly different from the one which hydrolyzes sialic acid from glycoproteins and oligosaccharides, since the genetic deficiency of the latter results in an entirely different disease, sialidosis. Although the biochemical data appear reliable, this reviewer has not been completely convinced by the deficient sialidase activity as the primary cause of the disease, mainly because accumulation of gangliosides in patients' tissues seems to be much milder than one would expect for a genetic condition in which the ubiquitous ganglioside, GM3, cannot be degraded. Furthermore, if polysialogangliosides cannot be degraded in the brain, one might expect exceedingly severe neurological manifestations, but patients with mucolipidosis IV are generally less severely affected neurologically than, for example, Tay-Sachs disease patients. Whether this can be explained by the relatively high residual activity of ganglioside sialidase reported in patients' tissues remains to be clarified.

Activator defect

A patient with clinical, pathological and compositional features essentially indistinguishable from those of classical Tay-Sachs disease but with apparently normal hexosaminidase A and B activities was first recognized in 1969.[10] Hexosaminidase A in such a patient was found to be completely normal not only toward the artificial substrates but also toward the natural substrates.[27] This specific form of GM2-gangliosidosis is known as "AB variant", a name that indicates normal hexosaminidases. The genetic cause of this disease was later identified as lack of a specific natural activator protein which is required for normal in vivo catabolism of GM2-ganglioside by hexosaminidase A.[28] Although this disease appears identical clinically and pathologically with the classical Tay-Sachs disease, except for its pan-ethnic distribution, it must be recognized that conceptually it belongs to an entirely different disease category. GM2-gangliosidosis AB variant is not an inborn lysosomal disease as defined by Hers. The conventional chromogenic or fluorogenic substrates will not distinguish this disease from the hexosaminidase A mutation which abolishes catalytic activity toward the natural substrate only (see above). Use of p-nitrophenyl β-N-acetylglucosamine-6-sulfate should prove useful for differentiation of the two diseases.[15,16]

Anabolic defect?

Several years ago, a patient was described who had a rapidly progressive course with macroglossia and generalized seizures leading to death.[29] The white matter showed severe spongy changes. Analytically, the brain and liver showed marked decreases of higher gangliosides beyond GM3. GM3-ganglioside was the main component in these organs. Consistent with this compositional abnormality, UDP-N-acetylgalactosamine:GM3 N-acetylgalactosaminyltransferase activity was reduced.[30] It was thus considered that this patient represented an entirely new category of genetic disorders of ganglioside metabolism, caused by an anabolic defect. As of this writing, however, existence of this disease must be viewed with caution, because the biochemical findings in the first case were not replicated in the second patient in the same family, who showed clinical and pathological pictures identical with the first patient.[31] Confirmation with additional cases would be required to establish this disease entity.

Animal models

Genetic disorders of ganglioside metabolism are also known in several species of mammals. For the most part, they are equivalent to human diseases in genetics, clinical and pathological features, compositional abnormalities, and enzymatic defects.[32,33] GM1-gangliosidosis is known for cat, dog and cattle. They are all characterized by genetic defect in acid β-galactosidase, as in the human disease. On the other hand, GM2-gangliosidosis is known in cat, dog and swine. The feline disease is enzymatically similar to Sandhoff disease in humans in that both hexosaminidase A and B are deficient. Due to the different pattern of hexosaminidase isozymes in the dog, the nature of the enzymatic defect has not been clarified, except that the total activity of hexosaminidase does not appear deficient.

Total hexosaminidase activity was reported to be approximately 30% of normal in the porcine GM2-gangliosidosis.

These "authentic animal models" provide extremely useful tools for research in genetic gangliosidoses. The human diseases are rare, and the nature of experiments permissible with human patients is severely restricted for ethical reasons. The animal models can overcome many of the inherent restrictions in working with human patients. It is anticipated that these animal models will be used extensively for various approaches for treatment of these genetic diseases, including those that utilize the recombinant DNA technology.

NONSPECIFIC ABNORMALITY

In many neurological disorders, compositional abnormalities of brain gangliosides occur not due to primary abnormality in ganglioside metabolism but nonspecifically and secondarily to the pathological alterations.

The most common and perhaps most nonspecific is relative increases in monosialogangliosides, GM1, GM2, and GM3. In the normal brain, GM1-ganglioside is the major monosialoganglioside, constituting 10-20% of total ganglioside sialic acid, but GM2- and GM3-gangliosides are present in very minor amounts. In diseases of diverse etiology and pathology, such as Niemann-Pick disease,[34] metachromatic leukodystrophy,[35] mucopolysaccharidosis,[35] Batten disease,[36] and Lafora disease,[37] the monosialogangliosides are proportionately increased, particularly the normally minor GM2- and GM3-gangliosides. This abnormality is so often seen in so many different pathological conditions that it is difficult to ascribe to it any particular histopathology. While the significance of this abnormality is not clear, its nonspecific nature must be recognized. It cannot be considered as evidence for abnormality in ganglioside metabolism unless supported by additional data. There are case reports in the literature in which the authors appear to have fallen into this error.

Another nonspecific abnormality was first reported in 1966 in the white matter of a patient who died of subacute sclerosing panencephalitis.[38] Similar abnormality was later discribed in white matter of adrenoleukodystrophy patients.[39,40] The abnormality involves conspicuous increases in normally very minor disialogangliosides, GD2 and GD3. In many other pathological conditions, relatively minor increases of these disialogangliosides have been observed. Some authors consider that this abnormality is a chemical counterpart of reactive astrocytic gliosis.

Secondary abnormality of brain ganglioside also occurs as a result of histopathological deletion of certain tissue components, which have unusual ganglioside compositions. A dramatic example is found in the plaque areas of the white matter of patients with multiple sclerosis.[41] Human myelin has a unique ganglioside composition with GM1- and GM4-gangliosides being the predominant species. Due to the total loss of myelin, the areas of plaques are almost devoid of GM4-ganglioside, and GM1-ganglioside is greatly reduced.

GANGLIOSIDE AS PATHOGENETIC AGENT

We have so far examined disease conditions in which gangliosides themselves are abnormal in one way or another. In an entirely different type of ganglioside-disease relationship, gangliosides can be the causative agent either directly or indirectly.

At least two groups of investigators reported occurrence of an autoimmune disease resembling experimental allergic encephalomyelitis (EAE) in animals immunized with gangliosides. Nagai and coworkers[42] described such a disease ("ganglioside syndrome") in rabbits immunized with either a total brain ganglioside mixture, GD1a- or GM1-ganglioside. The animals developed EAE-like signs, including weight loss, muscular weakness, and hindleg paralysis. The pathological lesions were more severe in the peripheral nervous system than in the central nervous system. The guinea pig was also susceptible to this experimental disease. Similarly, Sela and colleagues reported an EAE-like syndrome in rabbits immunized with gangliosides[45] (also see the chapter by Sela et al. in this volume). The clinical and pathological findings of these autoimmune conditions are somewhat unexpected in the sense that, despite the predominant localization of gangliosides in neuronal elements, pathological lesions are primarily in the white matter and peripheral nerves. From the localization of the antigen alone, one would expect more severe cortical involvement. Accesibility of the antibody to the tissue antigens, or presence of a very minor ganglioside in the mixture, which is primarily localized in the white matter, could explain this apparent discrepancy between the antigen localization and the distribution of pathological lesions.

In another entirely different approach, Rapport, Karpiak and coworkers described effects of anti-ganglioside serum on brain function. They found that direct application of anti-ganglioside antiserum to the rat brain surface induced recurrent epileptiform discharges. The effect was abolished by absorption of the antiserum with GM1-ganglioside.[44] They also showed inhibition of learned avoidance response by anti-GM1-ganglioside.[45]

GANGLIOSIDE AS THERAPEUTIC AGENT

For a long time, the physiological function of ganglioside remained a matter of speculation. In recent years, however, gangliosides have been implicated in an increasing number of cellular and organ functions. The space limitation does not permit this review to go into details here but many chapters in this volume address various aspects of the physiological function of gangliosides. To list just a few, gangliosides have been reported to be the receptor for serotonin, tetanus toxin, cholera toxin, botulinum toxin, thyroid stimulating hormone, leutenizing hormone, interferon, and Sendai virus. They have been reported to promote cellular differentiation, to release dopamine from synaptosomes, to accelerate axonal transport, to activate Na^+,K^+-ATPase, to stimulate lymphocytes, and to interact with fibronectin. Among the implicated functions is the ability of gangliosides to promote axon sprouting, nerve regeneration and neuronal cell regeneration.[46-49] These observations led to therapeutic trials with ganglioside first in experimental conditions, such as induced or genetic diabetes mellitus in rats, with encouraging results.[50] These developments then prompted clinical trials of ganglioside with human patients with disorders, such as diabetic neuropathy, amyotrophic lateral sclerosis, and others. The results reported in the literature so far have not been consistent, with some very encouraging observations and others with negative results. It is this reviewer's impression that this particular field is too young at this time for a cohesive review and critical assessment of the results. Some series involve too small a number of subjects with inadequate controls. Some of the positive results are merely anecdotal. What is needed at this time is a few well-controlled studies with a large enough number of patients that would allow rigorous statistical evaluation of the findings. One of the important purposes of this symposium was to have an opportunity to assess objectively the current status of ganglioside as a therapeutic agent. Several active groups are contributing their results in the following pages for the reader's scrutiny. One hopes that, from these well-designed trials, rigorous and objective conclusions would be forthcoming, whatever they may turn out to be. I believe it is simply premature to draw any conclusion, positive or negative, concerning efficacy of ganglioside for treatment of any human neurological disorders.

REFERENCES

1. K. Landsteiner and P. A. Levene, On the heterogenetic haptene, Proc. Soc. Exp. Biol. Med. 23:343 (1925).
2. E. Walz, Ueber das Vorkommen von Kerasin in normaler Rindermilz, Z. Physiol. Chem. 166:210 (1927).
3. E. Klenk, Ueber die Natur der Phophatide und anderer Lipoide im Gehirn und Leber, Z. Physiol. Chem. 235:24 (1935).

4. E. Klenk, Beiträge zur Chemie der Lipidosen. 3. Niemann-Picksche Krankheit und amaurotische Idiotie, Z. Physiol. Chem. 262: 128 (1939).
5. W. E. van Heyningen, Tentative identification of the tetanus toxin receptor in nervous tissue, J. Gen. Microbiol. 20:310 (1959).
6. W. E. van Heyningen and P. A. Miller, The fixation of tetanus toxin by ganglioside, J. Gen. Microbiol. 24:107 (1961).
7. D. W. Woolley and B. W. Gommi, Serotonin receptors, VII. Activities of various pure gangliosides as the receptors, Proc. Natl. Acad. Sci. USA 53:959 (1965).
8. H. G. Hers, Inborn lysosomal disease, Gastroenterology 48:625 (1966).
9. S. Okada and J. S. O'Brien, Tay-Sachs disease: Generalized absence of a beta-D-N-acetylhexosaminidase component, Science 165:698 (1969).
10. K. Sandhoff, Variation of β-N-acetylhexosaminidase pattern in Tay-Sachs disease, FEBS Lett. 4:351 (1969).
11. K. Suzuki, K. Suzuki, I. Rapin, Y. Suzuki, and N. Ishii, Juvenile GM2-gangliosidosis. Clinical variant of Tay-Sachs disease or a new disease, Neurology 20:190 (1970).
12. Y. Suzuki and K. Suzuki, Partial deficiency of hexosaminidase component A in juvenile GM2-gangliosidosis, Neurology 20:848 (1970).
13. S.-C. Li, Y. Hirabayashi, and Y.-T. Li, A new variant of type-AB GM2-gangliosidosis, Biochem. Biophys. Res. Commun. 101:479 (1981).
14. Y. Hirabayashi, Y.-T. Li, and S.-C. Li, The protein activator specific for the enzymatic hydrolysis of GM2-ganglioside in normal human brain and brains of three types of GM2-gangliosidosis, J. Neurochem. 40:168 (1983).
15. H.-J. Kytzia, U. Hinrichs, I. Maire, K. Suzuki, and K. Sandhoff, Variant of GM2-gangliosidosis with hexosaminidase A having a severely changes substrate specificity, EMBO J. 2:1201 (1983).
16. Y.-T. Li, Y. Hirabayashi, and S.-C. Li, Differentiation of two variants of type-AB GM2-gangliosidosis using chromogenic substrates, Am. J. Human Genet. 35:520 (1983).
17. I. Rapin, K. Suzuki, K. Suzuki, and M. P. Valsamis, Adult (chronic) GM2-gangliosidosis - Atypical spinocerebellar degeneration in a Jewish sibship, Arch. Neurol. 33:120 (1976).
18. W. G. Johnson and A. M. Chutorian, Inheritance of the enzyme defect in a new hexosaminidase deficiency disease, Ann. Neurol. 4:399 (1978).
19. W. G. Johnson, C. S. Cohen, A. F. Miranda, S. P. Waren, and A. M. Chutorian, α-Locus hexosaminidase genetic compound with juvenile gangliosidosis phenotype: Clinical, genetic and biochemical studies, Am. J. Human Genet. 32:508 (1980).

20. R. L. Proia and E. F. Neufeld, Synthesis of β-hexosaminidase in cell-free translation and in intact fibroblasts: An insoluble precursor chain in a rare form of Tay-Sachs disease, Proc. Natl. Acad. Sci. USA 79:6360 (1982).
21. S. Okada and J. S. O'Brien, Generalized gangliosidosis: beta-galactosidase deficiency, Science 160:1002 (1968).
22. K. Suzuki, Cerebral GM1-gangliosidosis: chemical pathology of visceral organs, Science 159:1471 (1968).
23. L. S. Wolfe and N. M. K. Ng Ying Kin, Storage and excretion of oligosaccharides and glycopeptides in the gangliosidosis, in: "Current Trends in Sphingolipidoses and Allied Disorders", B. W. Volk and L. Schneck, eds., p. 15, Plenum Press, New York (1976).
24. G. Bach, M. Zeigler, T. Schaap, and G. Kohn, Mucolipidosis type IV: Ganglioside sialidase deficiency, Biochem. Biophys. Res. Commun. 90:1341 (1979).
25. Y. Ben-Yoseph, T. Momoi, L. C. Hahn, and H. L. Nadler, Catalytically defective ganglioside neuraminidase in mucolipidosis IV, Clin. Genet. 21:374 (1982).
26. L. Caimi, G. Tettamanti, B. Berra, F. O. Sale, C. Borrone, R. Gatti, P. Durand, and J. J. Martin, Mucolipidosis IV, A sialidosis due to ganglioside sialidase deficiency, J. Inherit. Metab. Dis. 5:218 (1982).
27. E. Conzelmann, K. Sandhoff, H. Nehrkorn, B. Geiger, and R. Arnon, Purification, biochemical and immunological characterization of hexosaminidase A from variant AB of infantile GM2-gangliosidosis, Europ. J. Biochem. 84:27 (1978).
28. E. Conzelmann and K. Sandhoff, AB variant of infantile GM2-gangliosidosis: Deficiency of a factor necessary for stimulation of hexosaminidase A-catalyzed degradation of ganglioside GM2 and glycolipid GA2, Proc. Natl. Acad. Sci. USA 75:3979 (1978).
29. J. Tanaka, J. H. Garcia, S. R. Max, J. E. Viloria, Y. Kamijo, N. K. McLaren, M. Cornblath, and R. O. Brady, Cerebral sponginess and GM3-gangliosidosis: Ultrastructure and probable pathogenesis, J. Neuropath. Exp. Neurol. 34:249 (1975).
30. S. R. Max, N. K. McLaren, R. O. Brady, R. M. Bradley, M. B. Rennels, J. Tanaka, J. H. Garcia, and M. Cornblath, GM3 (hematoside) sphingolipodystrophy, N. Engl. J. Med. 291:929 (1974).
31. R. O. Brady, Inherited metabolic diseases and pathogenesis of mental retardation, Ann. Biol. Clin. 36:113 (1978).
32. H. J. Baker, J. A. Mole, J. R. Lindsey, and R. M. Creel, Animal models of human ganglioside storage diseases, Fed. Proc. 35: 1193 (1976).
33. K. Suzuki, "Authentic animal models" for biochemical studies of human genetic diseases, in: Proc. 4th Int. Symp. Developmental Disabilities, Y. Suzuki, ed., University of Tokyo Press, Tokyo, in press.

34. S. Kamoshita, A. M. Aron, K. Suzuki, and K. Suzuki, Infantile Niemann-Pick disease: A chemical study with isolation and characterization of membranous cytoplasmic bodies and myelin, Am. J. Dis. Child. 117:379 (1969).
35. K. Suzuki, Ganglioside patterns of normal and pathological brains, in: "Inborn Disorders of Sphingolipid Metabolism", S. M. Aronson and B. W. Volk, eds., p. 215, Pergamon Press, Oxford, (1966).
36. P. E. Duffy, M. Kornfeld, and K. Suzuki, Neurovisceral storage disease with curvilinear bodies, J. Neuropath. Exp. Neurol. 27:351 (1968).
37. R. Janeway, J. R. Ravens, L. A. Pearch, L. Odor, and K. Suzuki, Progressive myoclonus epilepsy with Lafora inclusion bodies. I. Clinical, genetic, histopathologic and biochemical aspects, Arch. Neurol. 16:565 (1967).
38. W. T. Norton, S. E. Poduslo, and K. Suzuki, Subacute sclerosing leukoencephalitis II. Chemical studies including abnormal myelin and an abnormal ganglioside pattern, J. Neuropath. Exp. Neurol. 25:5826 (1966).
39. Y. Suzuki, S. H. Tucker, L. B. Rorke, and K. Suzuki, Ultrastructural and biochemical studies of Schilder's disease II. Biochemistry, J. Neuropath. Exp. Neurol. 29:405 (1970).
40. M. Igarashi, D. Belchis, and K. Suzuki, Brain gangliosides in adrenoleukodystrophy, J. Neurochem. 27:327 (1976).
41. R. K. Yu, R. W. Ledeen, and L. Eng, Ganglioside abnormalities in multiple sclerosis, J. Neurochem. 23:169 (1974).
42. Y. Nagai, T. Momoi, M. Saito, E. Mitsuzawa, and S. Ohtani, Ganglioside syndrome, a new autoimmune neurologic disorder, experimentally induced with brain gangliosides, Neurosci. Lett. 2:107 (1976).
43. G. Konat, H. Offner, V. Lev-Ram, O. Cohen, M. Schwartz, I. R. Cohen, and B. Sela, Abnormalities in brain myelin of rabbits with experimental autoimmune multiple sclerosis-like disease induced by immunization to gangliosides, Acta Neurol. Scand. 66:568 (1982).
44. S. E. Karpiak, L. Graf, and M. M. Rapport, Antiserum to brain gangliosides produces recurrent epileptiform activity, Science 194:735 (1976).
45. S. E. Karpiak, L. Graf, and M. M. Rapport, Antibodies to GM1 ganglioside inhibit a learned avoidance response, Brain Res. 151:637 (1978).
46. A. Gorio, G. Carmignoto, L. Facci, and M. Finesso, Motor nerve sprouting induced by ganglioside treatment. Possible implications for gangliosides on neuronal growth, Brain Res. 197: 236 (1980).
47. M. M. Rapport and A. Gorio, (eds.), Gangliosides in Neurological and Neuromuscular Function, Development, and Repair, Raven Press, New York (1981).
48. J. R. Sparrow and B. Grafstein, Sciatic nerve regeneration in ganglioside-treated rats, Exp. Neurol. 77:230 (1982).

49. A. Gorio, P. Marini, and R. Zanoni, Muscle reinnervation - III. Motoneuron sprouting capacity, enhancement by exogenous gangliosides, Neuroscience 8:417 (1983).
50. F. Norido, R. Canella, and A. Gorio, Ganglioside treatment of neuropathy in diabetic mice, Muscle and Nerve 5:107 (1982).

CEREBRAL AND VISCERAL ORGAN GANGLIOSIDES AND RELATED GLYCOLIPIDS IN

GM1-GANGLIOSIDOSIS TYPE 1, TYPE 2 AND CHRONIC TYPE

Tamotsu Taketomi, Atsushi Hara, and Tooru Kasama

Department of Biochemistry, Institute of Adaptation
Medicine, Shinshu University School of Medicine
Matsumoto 390, Japan

INTRODUCTION

GM1-gangliosidosis is an autosomal recessive hereditary disease caused by deficiency of lysosomal acid β-galactosidase. Recent studies[1,2] have indicated that most of GM1-gangliosidoses can be clinically classified into three types: infantile Type 1, juvenile Type 2 and adult Chronic Type. Also, a genetic and biochemical study[3] has suggested that the phenotypic variation found in GM1-gangliosidosis results from different alleic mutations affecting the GM1 β-galactosidase locus, and that different combinations of these mutations account for the clinical heterogeneity of this disease. However, it seems that further biochemical investigation still remains to be carried out on storage substances in cerebral and visceral organs of patients with GM1-gangliosidosis. We have recently studied gangliosides and related glycolipids in brain, liver, spleen and kidney of Type 1, Type 2 and Chronic Type patients. These experimental results which showed chemically characteristic properties in each Type will be presented and discussed here.

MATERIALS AND METHODS

Case reports

A female patient (Type 1) had severe mental retardation, decerebrate rigidity and hepatomegaly, and died at 4 years and 2 months of age because of bronchopneumonia which repeated frequently since 1 year and 5 months of age. Her parents were cousins. A male patient (Type 2) had also severe mental retardation and decerebrate rigidity, but slight hepatomegaly later. He died of bronchopneumonia at 5 years and 5 months of age. These patients showed acid β-galactosidase deficiency in the leukocytes. A male patient with dystonia musculorum deformans could not be diagnosed as GM1-gangliosidosis before his death at 54 years and 1 months of age. After his death, it was found that his three brothers with the same clinical symptom showed the deficiency of acid β-galactosidase in the leukocytes. Thus, he was finally diagnosed as GM1-gangliosidosis Chronic Type together with histopathological and biochemical results.

Pathological examination

Almost no myelination and atrophic cortex were found in the brain (780 g, weight) of the Type 1 patient. Ballooning of nerve cells and numerous MCB were observed. Vacuolization of liver parenchymal cells, renal glomerular epithelia and vascular endothelia were also found. The ultrastructure of fibrillar materials accumulating in Kupffer cells was markedly different from MCB in neurons. The Type 2 patient brain (750 g, weight) showed enlargement of ventricle and atrophy in brain stem and cerebellum. Ballooning of nerve cells and numerous MCB were found together with dysmyelination and marked gliosis in the grey matter. Swollen Kupffer cells were found, but no marked change in visceral organs. The Chronic Type patient had an edematous and atrophic brain (1190 g, weight). Ballooning of nerve cells was found mainly in basal ganglia including caudate nucleus and putamen. Shrinkage was also found in the third layer of the cortex. No accumulation of abnormal materials was observed in atrophic visceral organs.

Experimental procedures

Fresh brain tissue of Type 1 and Type 2 patients and various regions dissected from fresh brain tissue of Chronic Type patient together with fresh liver, spleen and kidney tissues were available for the present investigation. Lipids were extracted from the different tissues with C-M (2:1) and then with a mixture of C-M (1:2) containing 5% water according to Suzuki.[4] Cholesterol, cholesterol ester and free fatty acids were separated by TLC and analyzed by GLC. Various phospholipids were separated by two-dimensional TLC with the first developing solvent of C-M-2.5M Ammonia (60:35:8) and the second solvent of C-M-Acetone-Acetic acid-W (75:15:30:15:7.5); the phosphorus content for each spot was determined by the method of Bartlett.[5] For glycolipid analysis, the total lipid fraction was subjected to mild alkaline hydrolysis to eliminate glycerophospholipids. The saponified fraction was dialyzed against distilled water and then applied to a column of DEAE-Sephadex A-25 (acetate form) according to Ledeen et al.[6] for separation of neutral and acidic glycolipid fractions. Following acetylation,

the neutral glycolipid fraction was separated from sphingomyelin by
a column chromatography and then deacetylated with ammonia in dry
methanol. The neutral glycolipids thus obtained were subjected to
preparative TLC and developed with C-M-W (65:25: 4). The acidic
glycolipids were separated into sulfatide and gangliosides on a
column of Silica gel 40. The ganglioside fraction was separated
into individual gangliosides by preparative TLC with C-M-0.25% KCl
(60:35:8). The total lipids of basal ganglia of Chronic Type were
directly applied to preparative TLC to separate individual gangliosides by developing with C-M-0.25% KCl (60:35:8) for 2.5 hr at room
temperature. Fatty acid and sugar components of glycolipids were
analyzed by GLC rountinely as already reported by us.[7] Chemical
structure of glycolipids was determined by the methylation procedure using dimsyl potassium according to Phillips and Fraser.[8] The
partially methylated hexitol or hexosaminitol acetates were
analyzed by GLC with FQ OV-101 capillary column (o 0.2mm x 25m),
programmed from 150° to 220° at 2° per min. Spots of simple lipids,
phospholipids and glycolipids on TLC were located by spraying brom-
othymol blue, anthrone reagent, resorcinol reagent and cupric
phosphoric acid or with iodine vapor. α-L-fucosidase of Charonia
lampas, β-galactosidase and β-N-acetylhexosaminidase of Jack bean
(Seikagaku Kogyo Co. Ltd., Tokyo) were used for sequential degradation of unknown fucolipids. Immunological activity of asialo GM1
(GA1) was tested by Ouchterlony procedure using rabbit anti-GA1
antiserum and the distribution of GA1 in cerebral and visceral
organs was checked by indirect immunofluorescence technique.

RESULTS

Lipids in brain tissues

A characteristic TLC pattern of the total lipids extracted
from the brain tissue of Type 2 patient is shown in Fig. 1. A
similar pattern was also found in the Type 1 patient. As shown in
Fig. 2, marked increase in GM1 was detected by TLC even in the
Chronic Type patient, as already reported by Kobayashi and Suzuki.[2]
Analytical results of the total lipids including simple lipids,
phospholipids, galactolipids and gangliosides and related neutral
glycolipids are summarized in Table 1. It was noted that the Type
1 brain showed almost no myelination in comparison to the Type 2,
because it contained very small amounts of myelin lipids including
ethanolamine-plasmalogen, cerebroside and sulfatide. The Type 1
brain contained free fatty acids in addition to cholesterol as
simple lipids, whereas the Type 2 brain contained cholesterol ester
together with cholesterol. Then, as shown in Table 1 and Table 2,
it was recognized that large amounts of gangliosides, particularly
GM1 and GA1, were accumulated in the brain tissues of Type 1 and
Type 2 and that other neutral glycolipids including glucosylceram-

ide, lactosylceramide, globotriaosylceramide (CTH) and globoside increased relatively in comparison to myelin galactolipids. N-acetylglucosamine-containing sialosylparagloboside (LM1) and paragloboside were present in the Type 1 brain, but not clearly in the Type 2. The chemical structure of accumulated GM1 and GA1 was confirmed by the methylation procedure, that is, GLC analysis of partially methylated alditol acetates of sugars (Fig. 3). The Chronic Type brain showed marked increase in GM1, GA1 and lactosylceramide in basal ganglia including caudate nucleus and putamen, but slight increase in the grey matter (Fig. 2 and Table 2).

The findings that ethanolamine-plasmalogen values were 2 - 3 times higher than those for PE for both specimens of the Chronic Type seemed to be unusual in the brain, but the reason is unfortunately unknown. Of course, it is found that the ethanolamine-plasmalogen values are usually more than the amount of PE in the myelin or other tissues like skeletal muscle.

Immunological activity of GA1 was tested by the Ouchterlony procedure using rabbit anti-GA1 antiserum and GA1 was detected in ballooned neurons by indirect immunofluorescence technique. Fatty acid compositions of gangliosides are similar in all Types of GM1-gangliosidosis, and the fatty acid compositions of other glycolipids in Type 1 and Type 2 are similar to each other but fairly different from those in Chronic Type.

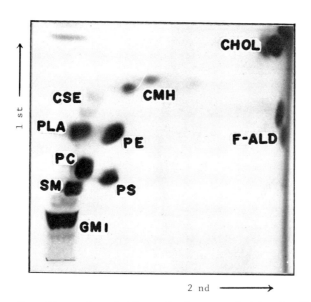

Fig. 1. Two-dimensional TLC of the total lipids (Type 2).

Table 1. Lipid Analysis of Brain Tissues.

mol/g wet weight	Type 1	Type 2	Chronic Type*	**
Free Fatty Acid	2.5			
Cholesterol	6.8	29.6	44.8	39.3
Cholesterolester		17.5		
Phospholipids				
PC	11.7	13.7	16.0	20.2
PS	0.2	3.3	7.2	6.7
PE	5.8	5.6	4.9	5.9
Plasmalogen	1.2	4.9	12.9	12.3
Sphingomyelin	2.8	5.2	7.9	8.4
Cerebroside	0.1	4.7	18.8	16.8
Sulfatide	trace	2.1	5.2	5.7
Total Gangliosides	4.0	6.2	1.8	2.5

*Caudate nucleus,   **Putamen

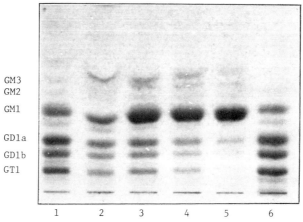

Fig. 2. TLC of gangliosides (Chronic Type). 1, Grey matter; 2, White matter; 3, Caudate nucleus; 4, Putamen; 5, Type 2; 6, Normal adult.

Table 2. Ganglioside and Related Glycolipid Contents in Brain Tissues.

nmol/g wet weight	Type 1	Type 2	Chronic Type	
			*	**
GM3	90	440	265	366
GM2	120	250	132	129
GM1	3430	4870	1049	1494
GD3	140		48	56
GD1a	100	390	107	153
GD1b	70	170	99	147
GT1	60	80	51	79
Glucosylceramide	310	2140	17	48
Lactosylceramide	370	1780	450	402
CTH	150	1100[a]	24	20
Globoside	220[b]		38	18
GA1	1670	4270	312	268

*Caudate nucleus, **Putamen,
[a]Contained globoside, [b]Contained paragloboside

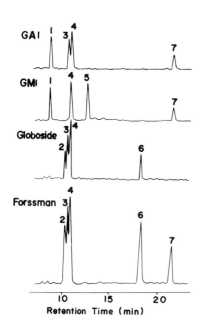

Fig. 3. GLC of partially methylated hexitol or hexaminitol acetate.
1, 2,3,4,6-Gal;
2, 2,4,6-Gal;
3, 2,3,6-Gal;
4, 2,3,6-Glc;
5, 2,6-Gal;
6, 3,4,6-GalNAc;
7, 4,6-GalNAc.

Glycolipids in visceral organs

Liver, spleen and kidney of Type 1 patient were subjected to analysis of neutral and acidic glycolipids in comparison to Type 2. TLC patterns of neutral glycolipids are shown in Fig. 4. These analytical results are summarized in Table 3. It was of interest that the visceral organs of the Type 1 patient contained fairly large amounts of unusual glycolipids, for example, galactosylceramide, sulfatide, digalactosylceramide, paragloboside and fucolipids in the liver, digalactosylceramide in the kidney and paragloboside in the spleen. However, these visceral organs did not show any increase in gangliosides except the normal presence of GM3, LM1, GM1 and GD3. The Type 2 patient showed almost normal presence of neutral glycolipids and gangliosides in the liver and spleen as compared to normal subjects.

We paid attention to two kinds of unknown fucolipids in the liver of Type 1. They were isolated and purified as shown in Fig. 5. The sugar components of fucolipid I and fucolipid II consisted of glucose:galactose:N-acetylglucosamine:fucose (1:2:1:1) and (1:3:2:2), respectively. The enzymatic sequential degradation of

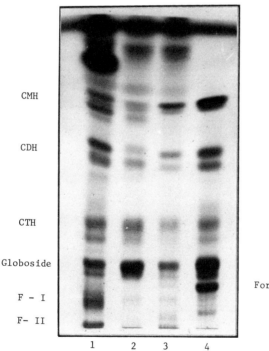

Fig. 4. TLC of neutral glycolipids (Type 1). 1, Liver; 2, Kidney; 3, Spleen; 4, Authentic samples.

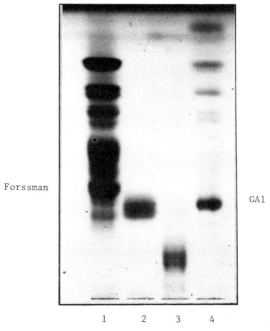

Fig. 5. TLC of purified fucolipids (Type 1). 1, Authentic samples; 2, Fucolipid I; 3, Fucolipid II; 4, Authentic sample.

Table 3. Glycolipid Contents in Visceral Organs.

nmol/g wet weight	Liver Type 1	Liver Type 2	Spleen Type 1	Kidney Type 1
Galactosylceramide	37.7			49.7
Glucosylceramide	31.6	51.8	24.0	75.5
Sulfatide	68.9			48.7
Lactosylceramide	71.5	57.7	17.3	36.8
Digalactosylceramide	17.1			55.4
CTH	54.0	27.8	72.9	120.8
Globoside	36.2	22.7	98.1	450.7
Paragloboside	15.1		62.0	
Fucolipid I	105.0		trace	trace
Fucolipid II	16.9			
Hematoside	61.5	210.2	144.3	37.3
LM1	6.8		9.3	4.7
GM1	9.8		13.0	3.5
GD3	12.7	15.5	13.0	10.5

fucolipid I with various combinations of α-L-fucosidase, β-galactosidase and β-N-acetylhexosaminidase showed that the chemical structure of oligosaccharide moiety of it may be identical with Lacto-N-fucopentaose II or Lacto-N-fucopentaose III. The methylation study for the confirmation of the structure is in progress. The fucolipid II may be proposed to be Lacto-N-difucooctaose, but its chemical structure still remains to be solved.

DISCUSSION

It was reported by some workers[10,11] that the visceral histiocytosis in generalized GM1-gangliosidosis appeared to be chiefly due to the storage of mucopolysaccharides rather than of gangliosides and that galactose, N-acetylglucosamine and mannose containing glycoproteins accumulated in the liver and spleen. It has been found recently by Yamashita et al.[12] that different excretion patterns of urinary oligosaccharides released from glycoproteins by an endo-β-N-acetylglucosaminidase are obtained by Type 1 and Type 2 patients. It has also been suggested by Farrell and MacMartin[13] that the multiplicity of GM1-gangliosidosis phenotypes can be explained by different mutations of the structural gene of an acidic β-galactosidase responsible for the catabolism of sugar chains of both glycolipids and glycoproteins. As already mentioned above, the Type 1, Type 2 and Chronic Type are similar in the accumulation of GM1 and GA1 in the nervous system, but the sites of accumulation are localized in the basal ganglia, particularly in the Chronic Type. Also, it is noted that the Type 1 visceral organs contain fairly large amounts of unusual glycolipids rather than gangliosides and that they are contrasted to those of the Type 2. The abnormal increase in some of these glycolipids may be due to the reduced activity of β-galactosidase, but the increase of other glycolipids may be due to a secondary effect of exoglycosidase inhibition caused by accumulation of glycoproteins which are histopathologically observed, but not clarified biochemically in the present paper. However, as indicated in GM1-gangliosidosis and Krabbe's disease,[14] the deficiency of β-galactosidase brings up some difficult problems. These include substrate specificity, activating agents such as activator proteins and detergents, inhibitors and localization in the nervous system, visceral organs and reticuloendothelial system. These problems still remain to be elucidated with the further investigation of the enzymatic variation and genetic mutation.

SUMMARY

Biochemical analyses of the samples of GM1-gangliosidosis Type 1, Type 2, and Chronic type at autopsy showed that GM1 and asialo-GM1 are markedly increased in the whole cerebral tissues of patients with Type 1 and Type 2, but mainly in the basal ganglia including caudate nucleus and putamen in the Chronic Type as already reported by Kobayashi and Suzuki in 1981. On the other hand, the finding that the Type 1 visceral organs contained unusual glycolipids rather than gangliosides was contrasted to those of Type 2 which seemed less abnormal. The unusual glycolipids included particularly fucolipid I and II consisting of glucose:galactose:N-acetylglucosamine:fucose (1:2:1:1) and (1:3:2:2), respectively. The chemical structure of oligosaccharide moiety of fucolipid I may be identical with Lacto-N-fucopentaose II or Lacto-N-fucopentaose III. The fucolipid II may be proposed to be Lacto-N-difucooctanose.

REFERENCES

1. J. S. O'Brien, Molecular genetics of GM1 -galactosidase, Clin. Genet. 8:303 (1975).
2. T. Kobayashi and K. Suzuki, Chronic GM1 gangliosidosis presenting as dystonia: II Biochemical Studies, Ann. Neurol. 9:476 (1981).
3. D. F. Farrell and U. Ochs, GM1 gangliosidosis: Phenotypic variation in a single family, Ann. Neurol. 9:225 (1981).
4. K. Suzuki, The pattern of mammalian brain gangliosides II. Evaluation of the extraction procedure, post-mortem changes and the effect of formalin preservation, J. Neurochem. 12:629 (1965).
5. G. R. Bartlett, Phosphorus assay in column chromatography, J. Biol. Chem. 234:466 (1959).
6. R. W. Ledeen, R. K. Yu, and L. F. Eng, Gangliosides of human myelin: Sialosylgalactosylceramide (G7) as a major component, J. Neurochem. 21:829 (1973).
7. T. Taketomi and N. Kawamura, Chemical and immunological characterization of caprine erythrocyte glycolipids in comparison with porcine erythrocyte glycolipids. J. Biochem. 72:799 (1972).
8. L. R. Phillips and B. A. Frasser, Methylation of carbohydrates with dimsyl potassium in dimethyl sulfoxide, Carbohydrate Res. 90:149 (1981).
9. K. Stellner, H. Saito, and S. Hakomori, Determination of amino-sugar linkages in glycolipids by methylation, Arch. Biochem. Biophys. 155:464 (1973).
10. K. Suzuki, Y. Suzuki, and S. Kamoshita, Chemical pathology of GM1-gangliosidosis (generalized gangliosidosis), J. Neuropath. Exp. Neurol. 28:25 (1969).
11. L. S. Wolfe, R. G. Senior, and N. M. K. Ng Ying Kim, The structure of oligosaccharides accumulating in the liver of GM1-gangliosidosis Type 1, J. Biol. Chem. 249:1828 (1974).
12. K. Yamashita, T. Ohkura, S. Okada, H. Yabuuchi, and A. Kobata, Urinary oligosaccharides of GM1-gangliosidosis, J. Biol. Chem. 256:4789 (1981).

13. D. F. Farrell and M. P. MacMartin, GM1 gangliosidosis: Enzymatic variation in a single family, Ann. Neurol. 9:232 (1981).
14. K. Suzuki and Y. Suzuki, Globoid cell leucodystrophy (Krabbe's Disease): Deficiency of galactocerebroside β-galactosidase, Proc. Natl. Acad. Sci. USA 66:302 (1970).

CANINE GM_2-GANGLIOSIDOSIS: CHEMICAL AND ENZYMATIC FEATURES

Yoshikatsu Eto*, Lucila Autilio-Gambetti** and John T. McGrath***

*Department of Pediatrics
The Tokyo Jikei University School of Medicine
Tokyo, Japan
**Institute of Pathology
Case Western Reserve University
Cleveland, Ohio
***Veterinary School
University of Pennsylvania
Philadelphia, USA

INTRODUCTION

GM2-gangliosidosis in animals has been observed in cat, dog and swine.[1-4] In 1968, McGrath et al.[4] found a new strain of GM2-gangliosidosis in male German short hair pointer dog which had shown signs of progressive neural dysfunction. Onset of the disease was 10-11 months. Clinical signs initiated by a deterioration of his previously keen response to a whistle, included in sequence: incoordination and postual difficulties, progressive impairment of hearing and vision, psychic changes simulating idiocy, terminal convulsions and prostration. Histological observations showed typical ballooning of the cell bodies of many neurons throughout the nervous system. The intracytoplasmic materials was Sudan black B and PAS positive. Electron microscopic examination revealed typical MCB as found in human Tay-Sachs disease.[4] Clinical and pathological features in dog GM2-gangliosidosis appear to be similar to human Tay-Sachs disease. Brief biochemical features of canine GM2-gangliosidosis were previously reported by Autilio-Gambetti.[4] The present report concerns more detail on chemical and enzymatic analysis in tissues of canine GM2-gangliosidosis.

MATERIALS AND METHODS

Canine GM2-gangliosidosis tissues frozen under $-20^\circ C$ were obtained from Autilio-Gambetti. The pathological and brief biochemical studies of these tissues have been described elsewhere. Lipid extraction was essentially carried out by the method of Folch et al.[5] Ganglioside fraction was obtained by partition as described by Suzuki.[6] Ganglioside sialic acid was estimated by the method of Svennerholm.[7] Quantitative analysis of gangliosides was carried out by densitometric analysis as described by Ando et al.[8] The compositions of glycosphingolipids were analyzed by thin-layer chromatography and subsequent densitometric analysis. Thin-layer chromatography with silica gel G plates (Merck, HPTLC) for gangliosides was performed with the solvent system chloroform-methanol-water containing 0.2% $CaCl_2$ (50:45:10, v/v/v) or chloroform-methanol-2.5N ammonia (50:45:10, v/v/v) and visualized by resorcinol spraying reagent. Thin-layer chromatography for neutral glycosphingolipids was performed with the solvent system chloroform-methanol-water (65:25:4, v/v/v) and visualized by spraying 50% sulphuric acid. Enzyme digestions of gangliosides with neuraminidase and beta-hexosaminidase were essentially carried out according to Li et al.[9] DEAE-Sepharose column chromatography of gangliosides was carried out as described by Momoi et al.[10] FD-MS or SI-MS for identification of gangliosides was carried out by Hitachi M-8022 type. The enzyme assays of 4MU beta-glactosidase and N-acetyl-beta-hexosaminidase were performed as described previously.[11] Isoelectric focusing polyacrylamide gel of beta-hexosaminidase was performed by the method of Wringley.[12] DEAE-Sepharose column chromatography of N-acetyl-beta-hexosaminidase was carried out as described by Robinson et al.[13]

RESULTS

Table 1 shows ganglioside sialic acid content in brain and liver of canine GM2-gangliosidosis. Affected dog brain contained a 3-10 fold increase of ganglioside sialic acid and liver a 3-5 increase over control.

Table 2 shows the quantitative analysis of gangliosides in brains of canine GM2-gangliosidosis. Affected dog brain contained about 40-54% GM2-ganglioside of total gangliosides, whereas in control this ganglioside consisted of 1-5% of total gangliosides. Furthermore, GD1a-galNAc consisted of 4-6% of total gangliosides in affected dog brain, while in control this ganglioside was practically absent. Other polysialogangliosides were greatly reduced in affected dog.

Table 1. Total Ganglioside Sialic Acid Content in Brain and Liver From Canine GM2-Gangliosidosis.

		Brain	Liver
Control	No 1	700	128
	2	778	141
	3	658	60
	4	1000	-
	5	796	-
	6	536	-
	7	686	-
Affected	No 1	6587	276
	2	4131	350
	3	2724	218

Values are expressed as µg per g wet tissue.

Further characterization of gangliosides in dog brain was carried out by DEAE-Sepharose column chromatography using ammonium acetate gradient, according to the method of Momoi et al.[10] Fig. 1 shows the thin-layer chromatogram of gangliosides isolated from control and affected dog brain after DEAE-Sepharose column chromatography. As shown in Fig. 1, the exclusive monosialoganglioside was GM2, with a significant increase of GD1a-galNAc ganglioside. Mild acid hydrolysis of GM2-ganglioside produced CMH, CDH and asialo GM2, which indicates that the major accumulated compound is GM2. Sialidase treatment produced asialo GM2 and also beta-hexosaminidase treatment with Jack bean produced GM3 ganglioside. The molar ratio of glucose, galactose and galactosamine by GLC was 1.0, 1.02 and 0.91, respectively. These analytical data suggested that the accumulated compound was GM2 ganglioside. Further confirmation of the structure of canine GM2-ganglioside purified by DEAE-Sepharose and Iatrobead column chromatography was carried out by FD-MS (type 8022). The characteristic signals of GM2-ganglioside were 750 m/e, 912 m/e, 1115 m/e which are coincident to glucosylceramide, lactosylceramide, and asialo GM2 fragments, respectively. This MS pattern was completely identical to that of human Tay-Sachs GM2-ganglioside (Fig. 2).

Fig. 3 shows the thin-layer chromatogram of neutral glycolipids isolated from canine GM2-gangliosidosis brain. As shown, ceramide dihexoside and asialo GM2-ganglioside were accumulated

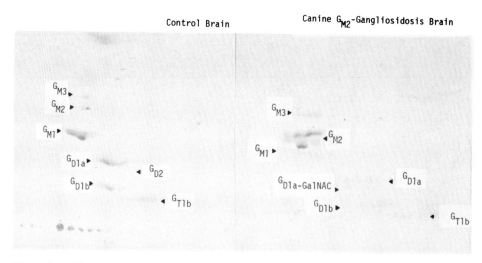

Fig. 1. Thin-layer chromatogram of gangliosides isolated from control and canine GM2-gangliosides after DEAE-Sepharose column chromatography. The plate was developed with chloroform-methanol-water containing 0.2% $CaCl_2$ (50:45:10, v/v/v) and spots were stained with resorcinol agent.

Table 2. Ganglioside Compositions in Canine GM2-gangliosidosis Brain.

	Control								Affected dog		
	1	2	3	4	5	6	7	8	1	2	3
GM3	4.4	2.7	1.7	2.6	2.0	3.8	1.8	2.8	9.3	7.9	6.4
GM2	5.4	0.3	0.3	1.8	1.3	1.7	2.5	0.4	54.6	49.6	40.0
GM1	18.2	18.3	21.1	21.0	25.5	15.8	20.5	17.9	7.1	9.6	10.1
GD3	tr	tr	tr	tr	tr	5.5	5.6	1.8	tr	tr	4.6
GD1a	22.3	26.1	24.4	27.3	28.2	26.1	12.1	16.3	7.3	8.0	9.1
GD1a-galNAc	−	−	−	−	−	−	−	−	3.9	5.5	5.7
GD1b	18.4	16.6	22.1	16.4	16.0	20.6	22.1	19.3	6.7	9.2	10.2
GT1b	20.4	17.1	28.7	20.9	22.3	22.9	17.2	20.2	9.5	6.9	9.6
GQ	1.4	1.9	1.4	1.6	4.9	5.0	5.1	1.5	1.6	3.3	3.3

No. 1: Control human brain grey matter. Values expressed as percentage of total NANA.

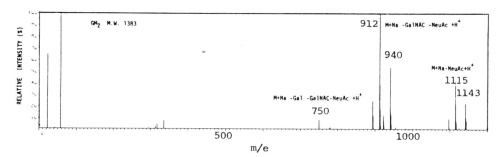

Fig. 2. FD-MS of GM2 ganglioside

in affected dog brain, but their amounts were much less as compared with those of human GM2- gangliosidosis. Furthermore, the levels of cerebroside and sulfatide in affected dog brain were well preserved, compared to those of human Tay-Sachs brain in which these glycolipids were greatly diminished. Fig. 4 shows a two-dimentional thin-layer chromatogram of neutral glycolipids from affected dog brain. There was some accumulation of ceramide lactoside and asialo GM2 which confirmed previous Fig. 3.

Next, we studied enzymatic features in canine GM2-gangliosidosis. Fig. 5 shows isoelectric focusing acrylamide gel electrophoresis of canine kidney hexosaminidase compared with that of human beta-hexosaminidase, using the technique of Wringley. Human hexosaminidase was separated into A and B fractions and their isoelectric points were 5.2 and 7.2, respectively. However, there were no A and B fractions in canine kidney beta-hexosaminidase. The isoelectric point of canine kidney beta-hexosaminidase was 6.0. Each sub-component was separated, but there was no apparent difference for each component between control and affected dog tissues.

Further characterization of beta-hexosaminidase in canine tissues was carried out by DEAE-Sepharose column chromatography.

Fig. 6 shows DEAE-Sepharose column chromatography of brain, liver and kidney beta-hexosaminidase obtained from control and affected dogs. On DEAE-Sepharose column chromatography dog hexosaminidases were separated into absorbed and non-absorbed fractions. There was no apparent difference between control and affected dog liver. However, in affected dog kidney the brain A-like fractions which absorbed in DEAE-Sepharose column chromatography were significantly decreased as compared with those of control tissues.

Table 3 shows the total activity of beta-hexosaminidase and also quantitative data of absorbed and non-absorbed fractions on DEAE-Sepharose column chromatography which are tentatively

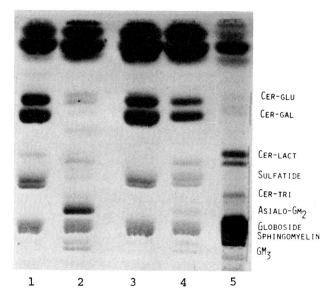

Fig. 3. Thin-layer chromatogram of neutral glycolipids in canine GM2-gangliosidosis brain. The plate was developed with chloroform-methanol-water (65:25:4, v/v/v). Lane 1: normal human white matter; lane 2: human Tay-Sachs disease brain; lane 3: control dog brain; lane 4: dog GM2-gangliosidosis; lane 5: erythrocyte glycolipids.

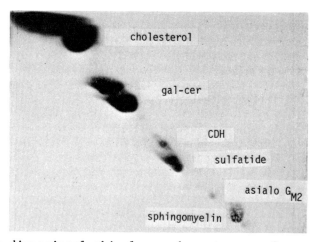

Fig. 4. Two dimensional thin-layer chromatogram of neutral glycolipids in dog GM2-gangliosidosis brain. The plate was developed first with chloroform-methanol-water (65:25:4, v/v/v) and second with chloroform-methanol-2.5N-ammonia (60:40:8, v/v/v).

Table 3. Lysosomal Enzyme Activities in Canine GM2-Gangliosidosis.

Tissues	4MU-beta-glu-cosaminadase	% A	4MU-beta-galac-tosaminidase	% A
Control				
brain	789	44.7	185	49.6
liver	3414	40.4	1051	41.0
kidney	3174	43.0	907	34.5
spleen	1294	44.6	382	44.6
Affected				
brain	2195	25.9	770	17.1
liver	2427	44.6	682	34.6
kidney	1994	15.7	579	7.6
spleen	1516	37.0	382	24.5

Activities are expressed as nmoles/mg protein/hour.

assigned as A and B, in tissues of control and canine GM2-gangliosidosis. Total hexosaminidase activity in affected dog tissues was not decreased. In affected dog brain and kidney, the A-like fraction seems to be decreased to about 40-50% of controls, but this finding was not observed in affected dog liver.

DISCUSSION

Analytical features in canine GM2-gangliosidosis are interesting, since the accumulation of GM2-ganglioside was not exclusive as compared with that of human Tay-Sachs disease in which more than 90% of total ganglioside in brain was GM2. The pattern of gangliosides in canine GM2-gangliosidosis was similar to that of human juvenile GM2-gangliosidosis. The structure of GM2-ganglioside accumulated in canine GM2-gangliosidosis brain was confirmed by chemical and enzymatic analyses, and also by FD-MS pattern. Furthermore, the accumulation of GD1a-galNAc in canine GM2-gangliosidosis brain was also observed, as shown in human Tay-Sachs disease. Neutral glycolipids in canine GM2-gangliosidosis brain showed increased ceramide lactoside and asialo GM2, but no increase of globoside was found. The degree of accumulation of ceramide lactoside and asialo GM2 was much less than those of human Tay-Sachs disease. These analytical data indicated that canine GM2-gangliosidosis was equivalent to human juvenile GM2-gangliosidosis.

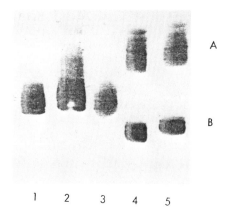

Fig. 5. Isoelectric focusing polyacrylamide gel electrophoresis of N-acetyl-beta-hexosaminidase obtained from canine GM2-gangliosidosis kidney. Lane 1: control dog kidney; lanes 2,3: Tay-Sachs dog kidney; lanes 4,5: human kidney. Ampholyte used: 3.0-10.0.

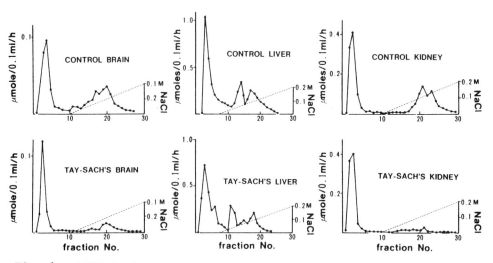

Fig. 6. DEAE-Sepharose column chromatography of beta-hexosaminidase.

Enzymatic features of canine GM2-gangliosidosis have unique characteristics, since total beta-hexosaminidase activity in affected dog tissues was not apparently reduced and also beta-hexosaminidase A or B like fractions were not separated in isoelectric focusing or cellulose acetate electrophoresis. However, the use of DEAE-Sepharose column chromatography permitted separation of beta-hexosaminidase A and B like fractions. Hexosaminidase A like fractions in canine brain, kidney and spleen were deficient, but not in liver. In cat GM2-gangliosidosis,[14] beta-hexosaminidase was separated into A and B fractions on cellulose acetate electrophoresis and total beta-hexosaminidase activity in affected cat tissues was reduced to less than 10% of control. In heterozygotes of cat GM2-gangliosidosis, hexosaminidase activity was intermediate. Therefore, cat GM2-gangliosidosis is considered to be analogous to human Sandhoff disease.[14]

In conclusion, this animal model of canine GM2-gangliosidosis is probably caused by either partial deficiency of beta-hexosaminidase A or activiator deficiency against GM2-ganglioside beta-hexosaminidase. Further characterization of beta-hexosaminidase in canine tissues is in progress.

SUMMARY

The chemical and enzymatic features in tissues of GM2-gangliosidosis are characterized by the analysis of glycolipids and FD-MS, and also by enzymatic analysis with DEAE-Sepharose column chromatography. The results suggest that canine GM2-gangliosidosis is equivalent to human juvenile GM2-gangliosidosis.

REFERENCES

1. J. S. O'Brien, S. Okada, M. W. Ho, D. L. Fillerup, M. L. Veath, and K. Adams, Ganglioside storage disease, Fed. Proc. 30:956 (1971).
2. H. J. Baker, J. Mole, J. R. Lindsey, and R. M. Creel, Animal models of human ganglioside storage diseases, Fed. Proc. 35:1193 (1976).
3. H. Bernheimer, H. and E. Karbe, Morphologische und neurochemische Untersuchungen von 2 Formen der amourotischen Idiotie des Hundes: Nachweis einer GM2-gangliosidose Acta Neuropathol. 16:243 (1970).
4. J. H. McGrath, A. M. Kelly, L. Autilio-Gambetti, and S. A. Steinberg, Nervous system: Storage diseases in dogs and cats, in: "International Encyclopedia of Psychiatry, Psychology, Psychoanalysis and Neurology," Vol. 8, B. B.

Wolman, ed., Human Sciences Press, New York (1977).
5. J. Folch, M. G. Lees, and G. H. Sloan Stanley, A simple method for the isolation and purification of total lipids from animal tissues, J. Biol. Chem. 226:497 (1957).
6. K. Suzuki, The pattern of mammalian brain gangliosides. III. Regional and developmental differences, J. Neurochem. 11:169 (1965).
7. L. Svennerholm, Quantitative estimation of sialic acids. II. A colorimetric resorcinol-hydrochloric acid method, Biochim. Biophys. Acta 24:604 (1957).
8. S. Ando, N.-C. Chang, and R. K. Yu, High performance thin-layer chromatography and densitometric determination of brain ganglioside composition of several species, Anal. Biochem. 89:437 (1978).
9. Y.-T. Li, M. J. King, and S.-C. Li, Enzymic degradation of gangliosides, in: "Structure and Function of Gangliosides," Vol. 125, p. 93, Plenum Press, New York (1980).
10. T. Momoi, S. Ando, and Y. Nagai, High resolution preparative column chromatographic system for gangliosides using DEAE-Sephadex and a new porous silica, Iatrobeads, Biochim. Biophys. Acta 441:488 (1976).
11. Y. Z. Frohwein and S. Gatt, Isolation of N-acetyl-beta-hexosaminidase from calf brain, Biochemistry 6:2775 (1967).
12. C. W. Wringley, Analytical fractionation of plant and animal proteins by gel electrofocusing, J. Chromat. 36:362 (1968).
13. D. Robinson, R. G. Price, and N. B. Dance, Separation and properties of beta-glucosidase, beta-glucuronidase and N-acetyl-beta-glucosaminidase from rat kidney, Biochem. J. 102:525 (1967).
14. L. C. Cork, J. F. Munnell, M. D. Lorenz, J. V. Murphy, H. J. Baker, and M. C. Rattazzi, GM2-ganglioside lyosomal storage disease in cats with beta-hexosaminidase deficiency, Science 196:1014 (1977).

IMMUNOLOGICAL EXPRESSION OF GANGLIOSIDES IN MULTIPLE SCLEROSIS AND
IN A DEMYELINATING MODEL DISEASE IN RABBITS

Ben-Ami Sela,[a] Halina Offner,[b] Gregory Konat,[b]
Varda Lev-Ram,[c] Oded Cohen[c] and Irun R. Cohen[c]

Departments of Biophysics[a] and Cell Biology[c], The
Weizmann Institute of Science, Rehovot, Israel and The
Neurochemical Institute[b], Copenhagen, Denmark

INTRODUCTION

Multiple sclerosis (MS) is a neurological disease of an unknown aetiology characterized by the destruction of myelin in the central nervous system (CNS). One possible mechanism implicated in the process of demyelination and plaque formation is an autoimmune response to a nervous system antigen(s), and the search continues to identify myelin components which might provoke such postulated autoimmune reactions. T-cell-dependent cellular immunity has been suggested as playing a role in the pathogenesis of MS,[1,2] as shown in numerous reports on the specific sensitization of peripheral blood lymphocytes from MS patients to myelin basic protein. Similarly, a humoral (B cell) response against myelin glycolipid components, such as galactocerebroside[3] and galactosyl diglyceride,[4] were also investigated. Studies on the possible involvement of myelin gangliosides as potential target antigens in the pathological autoimmune process in MS are presently summarized. These studies were initiated following reports on the emergence of anti-ganglioside antibodies in sera of MS patients[5,6] and by the consideration that gangliosides are exclusively located on the external (intraperiod line) apposition of myelin, and are thus rendered accessible for an autoimmune attack.

Two independent approaches were undertaken. (i) Since active E-rosetting had been proposed as a sensitive marker of T-cell-mediated immunocompetence,[7] highly purified brain gangliosides were tested for their potential triggering of peripheral blood lymphocytes from MS patients to actively rosette sheep red blood cells (SRBC). The use of this assay indicated that poly-sialylated gangliosides are potent and specific stimulators of active E-rosetting MS lymphocytes to a degree not encountered hitherto with other tested myelin antigens. (ii) It was reported by Nagai and his associates[7] that rabbits immunized with gangliosides developed lesions of demyelination in their peripheral nervous system. This pathological phenomenon was referred to as ganglioside syndrome which showed characteristics of experimental allergic neuritis. It was of interest to study systematically whether pathological features involving CNS myelin could be induced in rabbits under varying methods of ganglioside immunization. Data are summarized suggesting that some of the immunized rabbits develop a chronic, partially remitting disease, with clinical and pathological features reminiscent of demyelinating diseases of the CNS.

MATERIALS AND METHODS

Gangliosides

The total ganglioside fraction was extracted from either bovine or human brain. The latter was from a 44-year-old female, afflicted with MS for 30 years, who died of bronchopneumonia. Extraction was performed with chloroform/methanol, followed by alkaline treatment and phase partition. Further purification was achieved by chromatography on DEAE-Sephadex, DEAE-Sepharose and activated silicic acid (Unisil) columns,[8] which yielded a fraction that contained 28 percent sialic acid. Individual gangliosides were further fractionated on a porous silica gel column (Iatrobeads, GRS-8060) as described.[9] These individual gangliosides were found by amino acid analysis to contain less than 0.05% protein contamination, and their purity was assessed on high performance thin layer chromatography plates developed in chlofoform/methanol/ 0.2% $CaCl_2$ (50:40:10).

Myelin

The isolation of myelin from forebrains of rabbits by the flotation method on sucrose gradients was as previously reported.[10] The fraction of the light and heavy myelin species on discontinuous gradients, and the characterization of lipid and protein components of myelin were established recently.[11,12]

Active E-rosette test (AER)

Venous blood was defibrinated with glass beads, and leukocytes were isolated on Ficoll-Hypaque gradients. The mononuclear cell layer was washed with medium 199 or Hanks' balanced salt solution and resuspended at 5×10^6 cells/ml in RPMI 1640 without serum. SRBC in Alsever's solution were washed in medium 199 and resuspended at a concentration of 1%. Ganglioside fractions were dissolved in saline, and incubated with the lymphocytes at concentrations of 0.05 to 5 picogram/0.5 ml, for periods ranging from 30 min to 2 hrs. After the repeated wash of the lymphocytes, an equal volume of SRBC was added (0.25 ml), the cells were then mixed and centrifuged at 200g for 3 min, followed by incubation for 5 min at 4°C, under constant rotation. After additional centrifugation, the cells were counted directly. Each sample was run in triplicate, and at least 200 cells were counted per determination. Lymphocytes binding three or more SRBC were considered as positive. To overcome the technical problem of standardized resuspension of lymphocytes and SRBC in a relatively short period of incubation, the rosetting assay was performed with a special rotating apparatus,[13] which allowed optimal swinging and rotatory motions of the cell-tube holder and afforded consistent rosetting determinations.

Lymphocyte donors

(a) Forty six MS patients meeting the criteria of McAlpine et al.[14] for definite MS. The median age was 42 years, and none of the patients received immunosuppressive therapy during the course of this study. (b) Eight patients with unilateral optic neuritis, with a median age of 36. (c) Forty two patients with a median age of 46 years with other neurological diseases (OND), including polyneuropathy, Guillain-Barre syndrome, brain tumors (astrocytoma and meningioma), spinal tumor, cerebrovascular accidents, disc prolapse, epilepsy, Parkinson's disease, dementia, acute transverse myelitis, acute disseminated encephalitis, and plexitis. (d) Thirty two healthy blood donors with a median age of 39 years.

Immunization of rabbits

Outbred rabbits of both sexes, weighing 2.5 to 3.0 kg, were obtained from the breeding colony of the Weizmann Institute Animal Center. The composition of the administered mixture of ganglioside-protein carrier-adjuvant was apparently crucial, yet another important factor was the dose of the inoculated gangliosides. Preliminary immunization of rabbits with ganglioside-methylated bovine serum albumin (mBSA) (Sigma, USA) at a ratio of 1:1, emulsified with complete or incomplete Freund's adjuvant, yielded very few cases of neurological disorders in the tested rabbits. We have found empirically that an effective immunizing mixture was achieved with the ganglioside-mBSA complex emulsified in incomplete adjuvant into which 5 mg/ml of Mycobacterium tuberculosis H37Ra powder (Difco, USA) was ground and dissolved thoroughly. Each rabbit was inoculated with 5 mg of mixed brain gangliosides conjugated non-covalently to an equal weight of mBSA (Sigma, USA) emulsified in the adjuvant in a total volume of 2 ml. Inoculation was divided into three sites in the scapular on the back, while control rabbits received 5 mg of mBSA in adjuvant without gangliosides. The rabbits were observed twice weekly for periods up to several months, animals on the verge of death were sacrificed and autopsied. Histological sections of the brain, spinal cord and sciatic nerve were stained with hematoxylin and eosin or with luxol fast blue. The rabbits were bled before immunization and at weekly intervals thereafter. Using solid phase radioimmunoassay, the sera were studied for antibodies to brain gangliosides. Cerebrospinal fluid was obtained from the cisterna magna of control and test rabbits without trauma or bleeding and was centrifuged at 900g at 4°C for 20 min and stored at -20°C. Cerebrospinal fluid was studied for immunoglobulin content by SDS polyacrylamide gel electrophoresis, and sera were applied to glass plates containing 1% noble agar overlaid to a thickness of 1 mm. Electrophoresis was carried out in 0.05 M barbital buffer pH 8.6, after which gels were fixed, dried and stained with 1% Coomassie blue.

RESULTS AND DISCUSSION

Active E-rosette test (AER)

The in-vitro active rosetting was suggested to correlate with in-vivo delayed hypersensitivity,[15] and numerous attempts were made to demonstrate sensitivity to myelin basic protein in MS and other syndromes of the CNS. Such studies have indicated that lymphocytes primed by basic protein in-vivo were not restricted to MS.[16,17] On the other hand, sensitivity to gangliosides was demonstrated as a more specific property of MS lymphocytes.[19]

The priming of T lymphocytes is expressed as the ratio of the percentage of a patient's lymphocytes rosetting in the presence of gangliosides (Gang.-AER) to the percentage rosetting in their absence (AER). Cellular sensitivity to the presented antigen was implicated when an increase of 15% or more was measured in the ratio of Gang.-AER/AER. The results of AER assays with peripheral MS lymphocytes presented with mono-, di-, tri- and tetra-sialo gangliosides are shown in Table 1. A clear pattern is indicated in the table whereby an increased active rosetting by MS lymphocytes is proportional to the number of sialic acid residues of the tested glycosphingolipids. The average percentage of positive cases of the two tested ganglioside concentrations is 7% with GM1, 10% with GD3, 30% with GD1a, 52% with GD1b, 55% with GT1b, and 94% with GQ1b. It should be noted that increasing the concentration of the tested ganglioside by 10-100 fold did not affect significantly the indicated rosetting percentages with the various gangliosides. The lowest dose of the tetra-sialo-ganglioside that was effective in the AER corresponds to an extremely low concentration of 10^{-13}M, whereas other relevant lipids (such as ceramide, sulfatide, or lactosyl ceramide) did not show any AER inducing effect on the tested lymphocytes.[19]

In control experiments none of the lymphocytes collected from 32 healthy blood donors gave any significant reaction. Of greater significance is the finding that of over 40 patients with other neurological diseases (OND) only 12% of the samples were stimulated by gangliosides to a ratio greater than 1.15. The lack of reaction to gangliosides with healthy and OND lymphocytes was very recently confirmed by Ilyas and Davison,[20] who have also shown that in comparison to gangliosides, myelin basic protein is a weak stimulator of the AER. The latter study has also indicated that cellular hypersensitivity to myelin basic protein is not specific to MS but is also found in patients with leucoencephalopathy, and in most of the patients with cerebrovascular accidents. This supports the assumption that in non-related diseases with brain damage, some of

Table 1. In Vitro Effect of Fractionated MS Brain Gangliosides on Active T-Cell Rosetting of SRBC by Peripheral Lymphocytes From MS Patients.

MS Brain Ganglioside Added (pg)		Mean Antigen Response Ratio[a]	Number of Individuals Tested	Positive Cases[b]	
				Number	%
GM1	0.5	0.90 ± 0.04	36	4	11
	0.05	1.02 ± 0.06	36	1	3
GM3	0.5	1.05 ± 0.07	28	4	14
	0.05	0.95 ± 0.05	28	2	7
GD1a	0.5	1.07 ± 0.06	28	10	35
	0.05	1.05 ± 0.07	28	7	25
GD1b	0.5	1.13 ± 0.08	24	11	49
	0.05	1.09 ± 0.06	24	13	54
GT1b	0.5	1.10 ± 0.06	42	20	47
	0.05	1.14 ± 0.07	42	26	62
GQ1b	0.5	1.30 ± 0.09	42	39	92
	0.05	1.22 ± 0.08	42	40	95

[a]The results represent the mean antigen response ratios (Gang.-Aer/AER ± S.D.)

[b]Positive cases are those with a ratio > 1.15. All individuals tested were clinically stable with varying degrees of disability afflicted with MS for at least 5 years, with a median age of 41 years (range 26-50).

the abundant myelin basic protein might be released or exposed to sensitize circulating lymphocytes.

One exception to the rather generalized non-reaction of OND lymphocytes to gangliosides was found while examining cells from patients with unilateral optic neuritis (ON). This syndrome could be considered as an early manifestation of MS in a significant number of cases, as the frequency with which these patients later develop MS varies between 13 and 85%.[21,22] The interval between the first bout of ON and the first AER test of ON lymphocytes varied between 2-4 weeks. Five out of 8 tested ON patients responded to the tetrasialoganglioside GQ1b similarly to the MS patients 2 weeks to 1 month after the appearance of the first symptoms, whereas the lymphocytes from 3 ON patients remained negative for a period of 1 year. The relatively high response to gangliosides on AER test of 5 out of 8 patients is compatible with the rather high proportion of ON cases that develop to MS. Most intriguing is the observation that the reaction to GQ1b in early optic neuritis was apparent several weeks before the response to myelin basic protein in such rosetting assay became positive.

Noteworthy in the recent report from Davison's group is the observation that sensitivity of lymphocytes to gangliosides was found only in MS patients in acute relapse or with progressive disease, but not in patients during remission. This finding, that ought to be substantiated with a large group of patients, suggests that autoimmunity to gangliosides is expressed during the period of active disease, and decays at the stable phase of remission. A longitudinal survey is underway now to test the suggested correlation of lymphocyte sensitivity to gangliosides and the clinical stage of the disease.

Considerable current attention is focused on the possible combination of lipid haptens and non-lipid myelin antigens in enhancing demyelination in autoimmune disease.[3] At this point we cannot assess whether the observed enhanced lymphocyte response to multisialo gangliosides in MS is initiated by a sudden rise in these molecules in the CNS, nor can we relate it to any postulated autoimmune reaction leading to the pathological lesions in myelin. It should be emphasized that the active rosetting test is only one measure of in vivo lymphocyte priming by a given antigen, and its mechanism is still not resolved. It is assumed that after contact with the specific antigen, lymphokine-like factors are released from the antigen specific T-cells that increase the affinity of other T lymphocytes for SRBC.[23] The potentiation by gangliosides of T cells to active rosetting could indeed be blocked by cyclosporin A.[24] Current studies in our laboratory are aimed at the application of other assays to test the significance of the in vivo priming of MS lymphocytes by multisialo gangliosides, such as measuring the appearance of T killer cells in peripheral blood that

Chronic, demyelinating experimental disease of the CNS induced in the rabbit immunized with gangliosides

Numerous attempts have been reported aimed at the induction of a demyelinating disease that could be a reliable experimental model of multiple sclerosis. Various components of myelin served for immunizing with galactocerebroside[25] or gangliosides,[7] or as in experimental allergic encephalitis with myelin basic protein as immunogen. As the classical form of experimental allergic encephalitis was an acute, monophasic short disease which did not resemble the chronic nature of MS, a chronic, relapsing variant of experimental allergic encephalitis was developed by injecting rabbits with the total homogenate of rabbit spinal cord.[26] Many factors are likely to dictate the type or form of neurologic disease induced such as the type of brain antigen employed, its dose, the route of administration and most importantly the nature of the adjuvant and the ratio of antigen/adjuvant.[27]

Rabbits of a mixed breed were inoculated with a mixture of purified bovine brain gangliosides and mBSA emulsified in Mycobacterium tuberculosis - fortified adjuvant. Up to 70 percent of the rabbits developed clinical signs of neurological dysfunction starting not earlier than 3 weeks from injection. Delayed onset of up to 40 days was found in some rabbits, while none of control rabbits injected with mBSA + adjuvant showed signs of illness. The clinical symptoms consisted of progressive weakness of the trunk and limbs, paralysis of one or more limbs, incontinence and occasional righting-reflex, ataxia, tremors and even rotatory nystagmus was

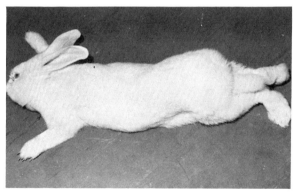

Fig. 1. Rabbit immunized with gangliosides developing neurologic lesions in the central nervous system. Note rigid paralysis of fore and hind limbs.

diagnosed in a small fraction of the sick rabbits. The mild and transient symptoms of weakness were observed in over 50% of the tested rabbits whereas the more severe and consistent signs were found in 8 to 16% of the sick animals, and 7 to 14% of the rabbits manifested behavioral abnormalities such as extreme aggression or catatonic-like state. The constellation of symptoms varied in each particular rabbit but the clinical course in the sick rabbits was marked by remission and exacerbations of variable duration. In some of the more severely paralyzed rabbits that were supported through the acute phase of the disease, a partial recovery occurred to a state of continuous weakness which developed in some cases to hindleg paralysis or frank paralysis (Fig. 1).

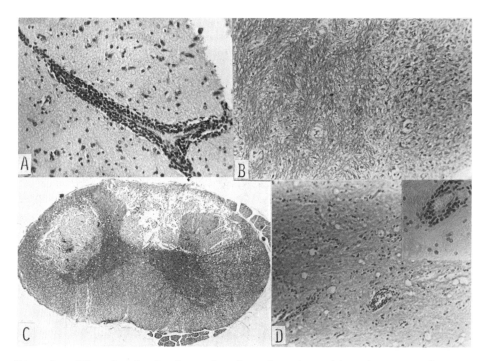

Fig. 2. Histologic lesions in chronic, demyelinating neurologic disease of rabbits immunized with gangliosides
(A) Perivascular infiltration of mononuclear cells in the white matter. Hematoxylin and eosin x 200. (B) Plaque of demyelination in optic nerve. Right part of section shows lack of myelin and increased cellularity. Luxol fast blue x 300. (C) Lumbar section stained for myelin showing advanced demyelinating lesion in the right lateral and posterior columns x 75. (D) Perivascular cuffing in the area of a plaque. Hematoxylin and eosin x 250.

Histological examination of the CNS revealed a variety of pathological lesions consisting of plaques of demyelination in the white matter of brains or in optic nerve with mononuclear cell infiltration around blood vessels in the plaque area or beneath the meninges and ventricular lining cells (Fig. 2A, B, D). Massive demyelination was also apparent in the spinal cord (Fig. 2C), yet sciatic nerve showed no lesions. The exclusive histological damage to the CNS, and lack of lesions in sciatic nerve is distinguishable from the histopathological findings in the ganglioside syndrome characterized by damage to the peripheral nervous system.[7] Such massive difference in the appearance of demyelination induced in rabbits by gangliosides in two independent systems, underscores the importance of the composition of the complex ganglioside-auxiliary substances and the scheme of immunization. It is also assumed that the different genetic background of the rabbits in these two studies is relevant to their different susceptibilities to the immunogen and the type of symptoms induced. The exclusive damage to CNS in the immunized rabbits in our system was also confirmed by electromyographic studies of selected ill rabbits that have shown normal conduction in peripheral nerves and normal transmission across the neuromuscular junction. No apparent block of the acetylcholine receptor (typical of myasthenia gravis) was detected in sick rabbits, as the Tensilon test performed on these rabbits was negative.

Cerebrospinal fluid obtained from test and control rabbits by cisternal puncture was subjected to electrophoresis on acrylamide SDS gels, followed by incubation with either ^{125}I-wheat-germ agglutinin (Fig. 3A) or with goat anti-rabbit immunoglobulins. Fluid of ill rabbits gave distinct staining of an immunoglobulin band (glycoslyated heavy chain of IgG) that was not found in cerebrospinal fluid of healthy rabbits, whether or not they had been immunized against gangliosides. Of 9 rabbits developing severe or mild clinical signs, 8 showed the new immunoglobulin band in serum 60 to 90 days after immunization. Electrophoresis on agarose according to Link,[28] gave with serum from sick rabbits a pattern typical of oligoclonal banding, whereas control serum gave much weaker staining in that region (Fig. 3B). The increase in oligoclonal banding in serum of MS patients has been reported recently.[26]

Another consistent trait that is characteristic of ganglioside-immunized rabbits that developed paralysis and other neurological symptoms was found while examining the distribution of myelin components derived from forebrain myelin at the time of sacrifice.[11] Myelin from sick and control rabbits was fractionated on sucrose gradients into light and heavy species, with buoyant densities smaller or larger than 0.625 M, respectively. Myelin from sick rabbits had 31% less "light myelin" and 39% more "heavy myelin" compared to control myelin (Fig. 4). This abnormal pattern is similar to that observed with myelin isolated from MS brains

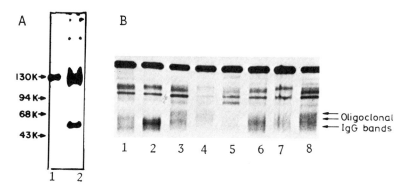

Fig. 3. Increased immunoglobulin level in cerebrospinal fluid and serum of rabbits showing chronic, demyelinating, neurologic disease. (A) Autoradiogram of acrylamide-SDS gel showing ^{125}I-wheat germ agglutinin binding to glycosylated proteins in cerebrospinal fluid from control rabbit (1) and paralyzed rabbit (2). 50K band was shown by goat anti-rabbit IgG to be the glycosylated heavy chain of immunoglobulin G. (B) Agar gel electrophoresis showing increased oligoclonal IgG in the serum of rabbits immunized with gangliosides. Lanes 1,2,3,6,7,8 refer to sick rabbits; 4 and 5 are control rabbits that were not sensitized with gangliosides. Electrophoresis was basically according to Link and Muller.[28]

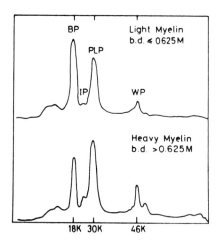

Fig. 4. Densitometric profiles of proteins in myelin fractions isolated from rabbit forebrain. Upper and lower scans represent light myelin (LM) and heavy myelin (HM) respectively. BP, basic protein; IP, intermediate protein; PLP, proteolipid protein; WP, Wolfgram protein.

due to the relative decrease in the 18K basic protein in the latter myelin.[10]

In conclusion, the symptoms induced by the immunization of rabbits with gangliosides under specific conditions exhibited features of chronic, relapsing neurological disease. It is assumed that premeditated immunization of rabbits against gangliosides led to an immune response which could be responsible for initiating a primary lesion in myelin. Future studies will determine the relevance of autoimmunity to gangliosides in the pathogenesis of MS and the mechanisms underlying the development of demyelination and other CNS lesions in ganglioside-immunized rabbits.

SUMMARY

Accumulating evidence suggests that the process of demyelination in MS might involve an autoimmune response to one or more myelin components. A combination of myelin basic protein and myelin haptens was considered as possibly enhancing a cellular or humoral autoimmune reaction in MS. In line with this notion we have used an in-vitro active E-rosette assay that correlates with in-vivo delayed hypersensitivity to demonstrate specific immunologic sensitivity of lymphocytes from MS patients to polysialogangliosides. A recent report that only lymphocytes from patients in relapse, but not in remission, are primed by gangliosides, underscores the relevance of the antigenic expression of gangliosides during the active pathological phase of the disease. The antigenic capacity of gangliosides to induce upon immunization a neurological disorder featured by demyelination in the CNS was demonstrated in rabbits. This and previous reports on the induction of peripheral demyelination in rabbits immunized with gangliosides will be further analyzed to gain insight on the possible role of these myelin lipid components as targets for an autoimmune mechanism in MS.

ACKNOWLEDGMENT

These studies were supported by the Israel Multiple Sclerosis Society.

REFERENCES

1. W. Sheremata, J. B. R. Cosgrove, and E. H. Eylar, Cellular hypersensitivity to basic myelin (A_1) protein and clinical multiple sclerosis, N. Eng. J. Med. 291:14 (1974).
2. S. C. Knight, Cellular immunity in multiple sclerosis, Brit. Med. Bull. 33:45 (1977).

3. C. S. Raine and U. Traugott, The pathogenesis and therapy of multiple sclerosis is based upon the requirement of a combination of myelin antigens for autoimmune demyelination, J. Neuroimmunol. 2:83 (1982).
4. H. E. Hilde and M. E. Parks, Serological reactions against glycolipid-sensitized liposomes in multiple sclerosis, Nature 264:785 (1976).
5. R. Arnon, E. Crisp, R. Kelley, G. W. Ellison, L. W. Myers, and W. W. Tourtellotte, Anti-ganglioside antibodies in multiple sclerosis, J. Neurol. Sci. 46:179 (1980).
6. B. R. Mullin, A. J. Montanaro, J. D. Reid, and R. N. Nishimura, Interaction of multiple sclerosis serum with liposomes containing ganglioside GM1, Ann. Neurol. 7:587 (1980).
7. Y. Nagai, T. Momoi, M. Saito, F. Mitsuzawa, and S. Ohtano, Ganglioside syndrome, a new autoimmune neurologic disorder, experimentally induced with brain gangliosides, Neurosci. Lett. 2:107 (1976).
8. R. W. Ledeen and R. K. Yu, Gangliosides: structure, isolation and analysis, in: "Methods in Enzymology", Vol 83, V. Ginsburg, ed., Academic Press, Inc., New York and London (1982).
9. T. Momoi, S. Ando, and Y. Nagai, High resolution preparative column chromatography system for gangliosides using DEAE-Sephadex and a new porous silica, Iatrobeads, Biochim. Biophys. Acta 441:488 (1976).
10. G. Konat and H. Offner, Density distribution of myelin fragments isolated from control and multiple sclerosis brains, Neurochem. Int. 4:241 (1982).
11. G. Konat, H. Offner, V. Lev-Ram, O. Cohen, M. Schwartz, I. R. Cohen, and B. Sela, Abnormalities in brain myelin of rabbits with experimental autoimmune multiple sclerosis-like disease induced by immunization to gangliosides, Acta Neurol. Scand. 66:568 (1982).
12. G. Konat and J. Clausen, The effect of long term administration of triethyllead on the developing rat brain, Envir. Physiol. Biochem. 4:236 (1974).
13. H. Offner, T. Fog, S. C. Rastogi, G. Konat, and J. Clausen, The enhancing effect of multiple sclerosis brain homogenates on the active E-rosette forming lymphocytes in neurological disorders, Acta Neurol. Scand. 59:49 (1979).
14. O. McAlpine, C. E. Lumsden, and E. D. Acheson, Multiple sclerosis - A reappraisal, Livingstone, London, pp. 142-148 (1965).
15. P. J. Felsburg and R. Edelman, The active E-rosette test: a sensitive in vitro correlate for human delayed type hypersensitivity, J. Immunol. 118:62 (1977).
16. U. Trougott, L. C. Scheinberg, and C. S. Raine, Multiple sclerosis circulating antigen-reactive lymphocytes, Ann. Neurol. 6:425 (1979).
17. H. Offner, G. Konat, N. E. Raun, and J. Clausen, E-rosette-forming lymphocytes in multiple sclerosis patients - Basic

protein stimulation of rosette-forming cells, Acta Neurol. Scand. 57:380 (1978).
18. H. Offner and G. Konat, Stimulation of active E-rosette forming lymphocytes from multiple sclerosis patients by gangliosides and cerbrosides, J. Neurol. Sci. 46:101 (1980).
19. H. Offner, G. Konat, and B. Sela, Multi-sialo brain gangliosides are powerful stimulators of active E-rosetting lymphocytes from multiple sclerosis patients, J. Neurol. Sci. 52:279 (1981).
20. A. A. Ilyas and A. N. Davison, Cellular hypersensitivity to gangliosides and myelin basic protein in multiple sclerosis, J. Neurol. Sci. 59:85 (1983).
21. G. D. Perkin, Optic neuritis and multiple sclerosis. An immunological comparison, in: "Clinical Neuroimmunology", A. S. Rose, ed., Blackwell, Oxford (1979).
22. A. S. Rose, G. W. Ellison, L. W. Myers, and W. W. Tourtellotte, Criteria for the clinical diagnosis of multiple sclerosis, Neurology 26:20 (1976).
23. A. I. Agabata and C. H. Kirkpatrick, Release of E-rosette augmenting factor (E-RAF) after stimulation of human leucocytes with mitogens or antigens, J. Immunol. 122:1080 (1979).
24. A. N. Davison and A. A. Ilyas, Cyclosporin A inhibits ganglioside-stimulated lymphocyte rosette formation in multiple sclerosis, Int. Arch. Allergy Appl. Immun. 69:393 (1982).
25. T. Saida, K. Saida, S. H. Dorfman, D. H. Silberberg, A. J. Summer, M. C. Manning, R. P. Lisak, and M. J. Brown, Experimental allergic neuritis induced by sensitization with galactocerebroside, Science 204:1103 (1979).
26. C. C. Whitacre, D. H. Wattson, P. Y. Paterson, R. P. Roos, D. J. Paterson, and E. G. W. Arnason, Cerebrospinal fluid and serum oligoclonal IgG bands in rabbits with experimental allergic encephalomyelitis, Neurochem. Res. 6:87 (1981).
27. F. D. Lublin, Delayed relapsing experimental allergic encephalomyelitis in mice: Role of adjuvants and pertussis vaccine, J. Neurol. Sci. 57:105 (1982).
28. H. Link and R. Muller, Immunoglubulins in multiple sclerosis and infections of the nervous system, Arch. Neurol. 25:326 (1971).

ANTIBODIES TO GLYCOSPHINGOLIPIDS IN PATIENTS WITH MULTIPLE

SCLEROSIS AND SLE

T. Endo, D. D. Scott, S. S. Stewart, S. K. Kundu and
D. M. Marcus

Departments of Medicine, Microbiology and Immunology, and
Neurology, Baylor College of Medicine, Houston, TX 77030

INTRODUCTION

Antibodies against gangliosides and neutral glycosphingolipids (GSLs) have been reported previously in the sera of patients with multiple sclerosis (MS)[1-5] and systemic lupus erythematosus (SLE).[6] The specificity of these antibodies has not been examined in detail, and in studies in which a liposome lysis assay was used the lytic factor in serum was not clearly identified as an immunoglobulin. The long term objectives of this study are to characterize the specificity of anti-GSL antibodies, to determine their incidence in immunological and non-immunological diseases that affect the nervous system, and to examine the temporal fluctuation of antibody levels in relation to disease activity. In this report we present our initial findings on the incidence and specificity of anti-GSL antibodies.

METHODS

The liposomes used in our assay contain the following proportions of lipids: sphingomyelin-cholesterol-dicetyl phosphate-GSL (1 µmol - 1 µmol - 0.1 µmol - 20 µg) and 4-methylumbelliferyl phosphate was incorporated as a marker. In order to obtain incorporation of a large amount of marker it is important to heat the lipids at 80° for 1 minute during formation of the liposomes. The total reaction mixture of 1 ml contained 10 µl of liposomes, 15 µl of human serum, 30 µl of guinea pig serum, 20 mM Tris buffer, pH 8.0, 150 mM NaCl, 0.15 mM Ca^{++}, and 0.5 mM Mg^{++}. The assay was performed for 1 hour at room temperature.

RESULTS

A summary of the data is presented in Table 1; release of more than 10% of the marker was considered to represent a positive test. No significant lysis occurred if the complement was inactivated by heating or if the GSL was omitted from the liposome. To demonstrate that the lytic factor was an immunoglobulin, aliquots of six positive sera were passed over an immunoadsorbent column that contained affinity-purified goat antibodies to human immunoglobulins, or over a control column that contained normal goat IgG. The lytic activity was completely removed by the anti-human immunoglobulin column and was unaffected by the control column (data not shown). The sera of approximately 75% of the MS patients and 56% of the SLE patients contained antibodies to one or more GSLs (Table 1), as did most of the patients who had extensive head trauma or large cerebrovascular accidents. Antibodies against G_{M1} and asialo G_{M1} were detected most frequently in all categories of patients, and the lytic activity of these antibodies is presented in Figures 1 and 2.

Data on the frequency with which antibodies to G_{M1} and asialo G_{M1} occurred in the same sera or separately are presented in Table 2. Among patients who had antibodies to either GSL, antibodies to asialo G_{M1} only were found in 2/35 MS patients and 10/22 SLE patients, a significant difference (p = 0.002). Conversely, antibodies to G_{M1} only occurred in 9/35 MS patients and in only 1/22 SLE patients (p = 0.002).

Fine specificity of antibodies

We noted previously[7,8] that rabbit IgG antibodies against asialo G_{M1} were highly specific, whereas IgG anti-G_{M1} antibodies crossreacted extensively with asialo G_{M1}, G_{D1b}, and to a lesser extent with other gangliosides. To further define the specificity of patients' sera that contained antibodies against both GSLs, inhibition studies were performed by preincubating sera with liposomes that contained one of these glycolipids without any trapped marker, and then adding target liposomes.

Table 1. Incidence of Antibodies to Glycosphingolipids

Disease	# Samples	# Pos	G_{M1}	AsG_{M1}	G_{M4}	Gal-Cer	G_{D3}	G_{D1a}
MS	47	35	30	23	3	2	6	2
SLE	46	26	13	21	3	1	9	0
Trauma	18	14	9	8	1	0	4	2
Stroke	6	6	4	6	2	1	1	0

(Antigen in liposome)

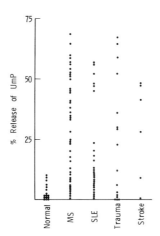

Fig. 1. Release of marker from liposomes that contain G_{M1} by sera from patients with MS, SLE, cranial trauma or cerebrovascular accidents.

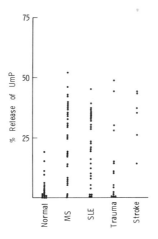

Fig. 2. Release of marker from liposomes that contain asialo G_{M1} by sera from patients with MS, SLE, cranial trauma or cerebrovascular accidents.

Table 2. Frequency of Antibodies to G_{M1} and Asialo G_{M1}

Disease	# Samples	# Pos	Both Pos	G_{M1} only	AsG_{M1} only
MS	47	35	24	9	2
SLE	46	22	11	1	10
Trauma	18	11	4	5	2
Stroke	6	6	3	1	2

As seen in Fig. 3B, asialo G_{M1} completely inhibited the antibodies that reacted with G_{M1} in two sera and substantially inhibited these antibodies in two other sera, but G_{M1} did not inhibit the anti-asialo G_{M1} in any of the four sera (Fig. 4A). These data indicate that sera which react with both GSLs contain a mixture of specific antibodies directed against asialo G_{M1} and antibodies that crossreact with both GSLs.

DISCUSSION

The incidence of antibodies to G_{M1} in patients with MS in this study is considerably higher than in previous reports;[4,5] antibodies to asialo G_{M1} were not examined in these investigations. Our data also differ from previous publications in other aspects. In contrast to Mullin et al.[4] we did not observe significant release of trapped marker from the liposomes in the absence of complement, nor did we find an extensive overlap between normal and MS sera. We observed more extensive release of marker than Arnon et al.,[5] most of whose positive sera released less than 20%. We feel certain technical aspects of liposome preparation and composition are crucial for obtaining stable liposomes that contain a large quantity of trapped marker. Liposomes that contain sphingomyelin exhibit less spontaneous release than those made with lecithin, and in forming the liposomes it is necessary to heat them above the transition temperature of the lipids in order to trap a large amount of marker.

Our studies of anti-asialo G_{M1} antibodies also differ from a previous report[6] that they were restricted to a subset of patients with SLE. Hirano et al. used a liposome agglutination technique that is much less sensitive than the liposome lysis assay that we used, but the apparent discrepancy between our reports cannot be attributed solely to this technical difference. We found comparable titers of antibodies to asialo G_{M1} and G_{M1}, but no anti-G_{M1} antibodies were noted by Hirano et al. We did note, however, that the occurrence of antibodies to asialo G_{M1} only was significantly more frequent in patients with SLE than with MS.

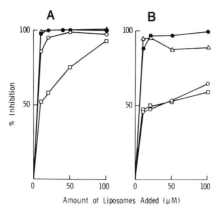

Fig. 3. Inhibition of the release of marker from liposomes that contain G_{M1} by sera from 4 different patients by (A) liposomes containing G_{M1} and (B) liposomes containing asialo G_{M1}.

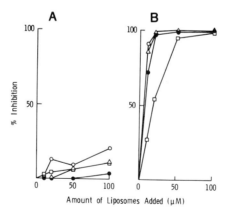

Fig. 4. Inhibition of the release of marker from liposomes that contain asialo G_{M1} by sera from 4 different patients by (A) liposomes containing G_{M1} and (B) liposomes containing asialo G_{M1}.

There are some interesting similarities between these data and previous studies of rabbit antisera. Rabbits immunized with mixed brain gangliosides responded primarily to G_{M1} and asialo G_{M1},[9] which were the commonest specificities detected in our human subjects. Both human and rabbit antibodies to asialo G_{M1} are highly specific, whereas anti-G_{M1} crossreacted extensively with asialo G_{M1}. G_{M1} is not the most abundant brain ganglioside, and asialo G_{M1} is present only in trace amounts in brain ganglioside preparations. The more abundant compounds, G_{D1a}, G_{D1b}, and G_{T1b} may be hydrolyzed by neuraminidases in inflamed tissues, and/or polysialo compounds may be less immunogenic than monosialo compounds or neutral GSLs.

Several factors might contribute to the appearance of these antibodies in our study population. The frequency of antibodies in the sera of patients who have experienced traumatic or ischemic injury to the CNS suggests that the antibodies can arise in response to inflammation and damage of neural tissues. Patients with MS or SLE do not generally have as extensive or severe damage to their nervous system as the trauma and stroke patients. A more appropriate control group would be patients with similar patchy lesions. A second possibility is that these antibodies are part of a polyclonal hypergammaglobulinemia, but this is not a common occurrence in MS, in contrast to SLE. Finally, the antibodies may represent an autoimmune response that may be responsible for some of the neuropathology associated with these diseases. At this time we are unable to evaluate the relative contributions of these factors in eliciting anti-GSL antibodies, nor can we determine whether these antibodies play a role in initiating or perpetuating the neural lesions of patients with MS or SLE. The local production of antibodies within the CNS would be more effective in producing tissue damage. We have had the opportunity of examining only three samples of cerebrospinal fluid from MS patients whose sera contained antibodies, and only one of the spinal fluid samples was positive. We hope to obtain more samples of spinal fluid for analysis, and we are performing longitudinal studies of patients to determine if the antibody titers fluctuate in relation to disease activity.

REFERENCES

1. M. Yokoyama, E. G. Trams, and R. O. Brady, Sphingolipid antibodies in sera of animals and patients with central nervous system lesions, Proc. Soc. Exp. Biol. Med. 111:350 (1962).
2. H. E. Hirsch and M. E. Parks, Serological reactions against glycolipid-sensitised liposomes in multiple sclerosis, Nature 264:785 (1976).
3. B. Ryberg, Multiple specificities of antibrain antibodies in multiple sclerosis and chronic myelopathy, J. Neurol. Sci. 38:357 (1978).
4. B. R. Mullin, A. J. Montanaro, J. D. Reid, and R. N. Nishimura,

Interaction of multiple sclerosis serum with liposomes containing ganglioside G_{M1}, Ann. Neurol. 7:587 (1979).
5. R. Arnon, E. Crisp, R. Kelley, G. W. Ellison, L. W. Myers, and W. W. Tourtellotte, Anti-ganglioside antibodies in multiple sclerosis, J. Neurol. Sci. 46:179 (1980).
6. T. Hirano, H. Hasimoto, Y. Shiokawa, M. Iwamori, Y. Nagai, M. Kasai, Y. Ochiai, and K. Okumura, Anti-glycolipid autoantibody detected in the sera from systemic lupus erythematosus patients, J. Clin. Invest. 66:1437 (1980).
7. M. Naiki, D. M. Marcus, and R. Ledeen, Properties of antisera to ganglioside G_{M1} and asialo G_{M1}, J. Immunol. 113:84 (1974).
8. S. K. Kundu, D. M. Marcus, and R. W. Veh, Preparation and properties of antibodies to G_{D3} and G_{M1} gangliosides, J. Neurochem. 34:184 (1980).
9. M. M. Rapport, L. Graf, Y. L. Huang, W. Brunner, and R. K. Yu, Antibodies to total brain gangliosides: Titer and specificity of antisera, in: "Structure and Function of Gangliosides", L. Svennerholm, P. Mandel, H. Dreyfus, and P. F. Urban, eds., Plenum Press, New York (1980).

Neuritogenesis and Regeneration

NEURITOGENESIS AND REGENERATION IN THE NERVOUS SYSTEM: AN OVERVIEW OF THE PROBLEM AND ON THE PROMOTING ACTION OF GANGLIOSIDES

Alfredo Gorio, Damir Janigro and Renzo Zanoni

Fidia Research Laboratories
Dept. of Cytopharmacology
Via Ponte della Fabbrica 3/A
35031 Abano Terme, Italy

INTRODUCTION

Promoting restoration of functions after injury to the nervous system is one of the major goals in neuroscience and in the whole field of medical research. Such an achievement is, however, quite difficult unless the basic mechanisms underlying neuronal plasticity, neuritogenesis, sprouting and regrowth of injured axons are better understood. Very few molecules have been shown experimentally to be active in promoting such processes; however careful examinations made at a later time showed that several results were misinterpretated due to the poor basic knowledge of the experimental model employed.[1] A good example of the difficulties which can be encountered in this field is the Nerve Growth Factor (NGF) story. As early as 30 years ago Rita Levi-Montalcini and co-workers showed that mouse sarcoma tumors, transplanted into the chorioallantoic membrane of chick embryos, induced a massive invasion of blood vessels and viscera by sympathetic nerve fibers. Such an interesting finding was further investigated and the active factor (NGF) of the mouse sarcoma purified.[2] When large quantities of NGF became available an increasing number of laboratories began to investigate its action and the pioneering work of Rita Levi-Montalcini was broadened. It is known that NGF is a single protein which is capable of inducing neurite outgrowth and regeneration in several types of neurones in vitro and in vivo.[2] Despite years of effort and the large number of scientists involved, the mode of action of NGF remains an open question. NGF stands as the first molecule isolated from animal tissue capable of promoting neurite growth and as an example of how difficult the challenge is.

SPECIFICATION OF NEURONAL MORPHOLOGY

Understanding of the role played by extrinsic factors in neurite formation or regeneration is difficult to pursue in vivo. In vitro experiments, on the other hand, offer the advantage of direct visualization of neurite outgrowth and a control of the bathing medium composition. Using neurons in culture, Silvio Varon has shown the existence of at least two types of endogenous substances active on neurons. One family supposedly released from the territory of innervation may contain substances named Neurotrophic Factors (NTF), which may play a role during development, growth and maintenance, while the other one contains neurite promoting factors (NPF) and may be related to neurite outgrowth and binding to extracellular surfaces.[3,4] The likely presence of several kinds of external factors, playing a regulatory role for various neurons, raises questions about the respective relevance of external and internal determinants of neuronal morphology and plasticity. It is well known that growth cones and neurites distinguish among different substrata and will reproducibly move toward one and avoid another.[5,6] This substratum specificity is clearly demonstrated by regenerating motor axons during muscle end-plate reinnervation. Sciatic nerve axons of the rat cease regrowing and differentiate into nerve terminals capable of releasing transmitter as soon as they make contact with the denervated EDL muscle end-plates. However, a few microns away the same axons keep sprouting and growing undisturbed.[7,8] The information for triggering such recognition processes seems to be specifically part of the end-plate basal lamina.[9] Therefore it is reasonable to conclude that mechanisms of external guidance[5] together with specific growth factors[2,10,11] are capable of regulating neuronal morphology.

However, it is also known that neurons contain information that specifies their unique and detailed morphology. Such hypotheses are difficult to establish in vivo, since the possibility remains that in spite of an accurate destruction of the environment, relevant bits may remain and influence regeneration. Again tissue culture has offered great assistance in solving such problems. For instance, sister neuroblastoma cells in culture are exact mirror images or identical twins of another one. If neuroblastoma microtubules are destroyed with Nocadozole there is a disassembly of the cell processes followed, after drug removal, by reorganization of a cell morphologically identical to the previous one.[12,13] Probably the exact reformation of cell processes is not conditioned by external guidance material deposited previously by the cell, since neuroblastoma cells move or roll away from the original position during drug incubation. Both these observations show that neurons retain specific information about their morphology even when their structure has been suppressed. Such a memory must not reside in the cytoskeleton geometry, in fact neuroblastoma cells express the capacity of recapitulating the previous shape when the expression of such information is suppressed.[12,13]

SPROUTING AND REGENERATION OF NEURONS

Indeed the cellular mechanisms of nervous system regeneration are dictated by neuronal structure and shape. A transected axon must elongate to reconnect with the peripheral territory again, and then, by sprouting, form several new processes for reinnervating various target neurons. In such a way the divergence of neuronal information is restored. Axonal elongation and sprouting are the two basic cellular mechanisms utilized by neurons for recovering a lesioned neuronal network. Central and peripheral neurons may respond to axotomy by chromatolysis, increased RNA and protein synthesis, changes in various enzyme activities, loss of synapses from soma and dendrites.[1] However the major changes occurring in the cell body are related to synthesis of proteins specific for regeneration such as growth associated proteins[14,15] and a change in the rate of tubulin transport.[16,17] In fact, it has been shown that the rate of axonal elongation is similar to the rate of axonal transport of a fraction, slow component B (SCb), that during axonal regeneration also contains tubulin, which usually travels slower.[16,17] We have obtained results which indirectly confirm this concept. We found that decreased slow transport corresponds to a decreased rate of axonal regeneration (see Gorio et al. this volume). These data indicate that the mechanisms of neurite reformation and extension are regulated at the gene level. However axonal sprouting, which occurs at the site of injury and into the target, is regulated by external influences such as factors released from the denervated territory.[1] Motoneurones have a great capacity for sprouting and are capable of expanding their peripheral field of innervation up to 4-5 fold.[18] This property is not affected by a decreased slow transport (see Gorio et al. this volume).

In conclusion, there is sufficient evidence that neuronal regeneration is dependent on both intrinsic properties of neurons and extrinsic signals from the denervated territory. These well demonstrated capacities for regeneration expressed by motoneurons are not clear in CNS where regeneration may not occur, and if it does occur, it is not clear if recovery of anatomical structure means restoration of functions. However, in spite of these difficulties it is quite clear that CNS neurons of adult mammals can sprout and even regrow over long distances, making connections with appropriate targets.[19] In addition, partial denervation of the septum induces collateral sprouting from the surviving monoaminergic axons and reinnervation of the vacated synaptic sites.[20,21] Unlike the peripheral nervous system, one of the major problems in CNS neuronal regeneration after trauma is caused by the glia scar formed after mechanical lesion. This inhibits neurite outgrowth. Another problem is that the primary effect of the lesion, i.e. Wallerian degeneration of the distal stump after axotomy, is followed by a slower retrograde degeneration which can induce neuronal cell death. Therefore, experiments oriented to evaluate the effects of molecules promoting regeneration in CNS should, at first, assess precisely the specificity and the extent of the lesions.

PROMOTING REGENERATION

By definition, molecules promoting regeneration of the nervous system should either increase sprouting or elongation rate of axons. On the other hand, equally important for CNS neurons is the protective action against retrograde degeneration. In such a case a prevented cell death may result in enhanced regeneration. At the same time that Rita Levi-Montalcini was describing NGF promoting activity, Hoffmann described molecules and methods capable of affecting motor neuron sprouting and regeneration. For instance Pyronin G, given to rats in drinking water, enhanced muscle reinnervation after partial denervation.[22] However recent results indicate that Pyronin G accelerates Wallerian degeneration, thereby preparing a proper pathway for sprouts.[23] Hoffman also described the action of "Neurocletin," a mixture of unsaturated lipids extracted from nerves and CNS white matter capable of stimulating sprouting in intact and partially denervated muscles.[24] This work was popular then but there was no follow up, and no further information on Neurocletin is available. Perhaps it might be part of the "products of nerve degeneration" which are suspected to promote sprouting and several postsynaptic changes after denervation.[1] Also electrical stimulation has been claimed to be active in sprouting,[22] but the effect, if present, may be different according to which parts of the nervous system are under examination (further information in ref. 1). The conditioning lesion paradigm is a method utilized by some neurosurgeons to improve peripheral regeneration. It was found that a preliminary lesion enhances elongation and sprouting of regenerating peripheral nerves if the second lesion is made 2 weeks later.[25]

The above examples are the best known attempts to improve regeneration of the nervous system; they are both quite old and have not fulfilled completely the hopes raised. The most recent development in this field is the discovery that exogeneously applied gangliosides can improve regeneration of the nervous system. The contents of this book are witness to the growing interest in this approach. Gangliosides were discovered some 50 years ago and chemically characterized some 20 years ago; their presence in the nervous system has been correlated with biological functions such as neuronal development and maturation only in recent years.[26,27] However, similar to the NGF problem, the most important and earlier experiments on gangliosides and regeneration were performed in vivo. Ceccarelli and co-workers[28] showed that if a cat nictitating membrane was denervated by cutting either pre- or post-ganglionically the sympathetic nerve, the reinnervation process was faster and more complete in the animals treated daily with 50 mg/Kg of gangliosides.

We decided to investigate such an effect in order to examine the cellular action of gangliosides. The rat extensor digitorum longus muscle was denervated by crushing the sciatic nerve and the reinnervation monitored electrophysiologically and morphologically.[7,8] The treatment with gangliosides strongly enhanced the dynamics of sprouting and polyinnervation without alterating synaptic repression[29] or maturation.[30] In particular it was also shown that the treatment did not alter the elongation rate of regenerating axons. These results yielded clear evidence of an interaction between gangliosides and specific sprouting factors. In fact, their action was solely exerted during the reinnervation phase, and when synaptic repression began their action became negligable. The lack of effect upon the innervation of the non-denervated contralateral muscle shows that these molecules were active specifically during the dynamics of nerve regeneration.[31] Partial denervation of rat soleus muscle confirmed the capacity of gangliosides to enhance sprouting. Following resection, L_5 mixed-nerve motoneurones are able to expand the motor-unit size up to a maximum of 4.5 fold. Daily treatment with 5 mg/Kg of gangliosides raised the maximum expansion to 6.3 fold.[18] Gangliosides can enhance by some 50% the sprouting capacity of motoneurones.

The in vivo experiments strongly suggested that gangliosides require the presence of specific growth factors for stimulating sprouting. Confirmation of this evidence was achieved by using pheochromocytoma PC12 cells, which stop reduplication and form neurites after addition of NGF to the culture medium. The results indicated that treatment with either ganglioside mixture or purified GM_1 and GT_{1b} enhanced neurite outgrowth and regeneration; no effect was observed if the bathing medium lacked NGF. On the other hand, it was also found that the addition of gangliosides did not alter NGF binding and incorporation.[32] Antibody to GM_1 added to the culture medium decreased PC12 neurite outgrowth and regeneration, again without alteration of NGF binding (unpublished observation, G. Ferrari, M. Fabris and A. Gorio). The capacity of both gangliosides and their antibodies to influence neuritogenesis and regeneration of neurites shows that the neuronal cell membrane plays an important role in determining neuronal morphology, independent of the action of growth factor. In particular, it is clear from both in vitro and in vivo experiments that exogenous gangliosides are capable of increasing the responsiveness of neurones to specific factors without selectivity for the type of neurones.

Dr. Oderfeld-Nowak and co-workers[33] showed that gangliosides also stimulated sprouting in the central nervous system. Lesions of the septum caused a decrease in cholinergic enzymes in the hippocampus, followed by a slow recovery which was accelerated by treatment with gangliosides. In this book there are two more articles showing that GM_1 enhances recovery in the striatum, after hemitransection of the nigro-striatal pathway, and in the hippocampus after an entorhinal cortex lesion. The latter report is very important since it shows that the enhanced recovery is probably due to protection against neuronal damage as well as sprouting. We could confirm this conclusion in a more precise manner.[34] Rats were injected subcutaneously at birth with 5,7 HT, which depletes serotonin in the CNS. At first we observed a very sharp drop which occured a few hours after toxin injection (primary degeneration) followed by a slower degeneration which proceeded for several days (secondary degeneration). Animals treated for the first 4 days of life with GM_1 showed the same kind of primary degeneration while the secondary was fully prevented. This effect was, then, followed by a faster regeneration of such pathways and reinnervation of the frontal and occipital cortex, so that 4 weeks after birth rats treated with GM_1 showed full recovery of serotonin levels in the frontal cortex while untreated ones were still at about the 20% level.[34]

CONCLUSION

The processes of neuritogenesis, sprouting and regeneration are regulated at different levels in the nervous system. Several components of neuronal regeneration are regulated at the soma level (memory, elongation, neuritogenesis), others at the peripheral level (sprouting and survival), although the two systems may be interrelated. Exogenous gangliosides are capable of interacting with both regulatory processes of regeneration by enhancing sprouting, regeneration and neuronal survival to injury.

REFERENCES

1. A. Gorio, Sprouting and regeneration of peripheral nerve, in: "Advances in Cellular Neurobiology", J. Zagoren, and S. Fedoroff, eds., Academic Press, New York, in press (1984).
2. R. Levi-Montalcini, The nerve growth factor-target cells in-

teraction: A model system for the study of directed axonal growth and regeneration, in: "Nervous System Regeneration", B. Haber, J. R. Perez-Polo, G. A. Hashim, and A. M. Giuffrida Stella, eds., Birth defects: Original articles series (1983).
3. S. Varon and R. Adler, Nerve growth factors and control of nerve growth, Curr. Topics Develop. Biol. 16:207 (1980).
4. S. Varon and R. Adler, Trophic and specifying factors directed to neuronal cells, Adv. Cell Neurobiol. 2:115 (1981).
5. P. Weiss, Cell contact, Int. Rev. Cytol. 7:391 (1958).
6. S. B. Carter, Haptotaxis and the mechanism of cell motility, Nature 213:256 (1967).
7. G. Carmignoto, M. Finesso, R. Siliprandi and A. Gorio, Muscle reinnervation I. Restoration of transmitter release mechanisms, Neuroscience 8:393 (1983).
8. A. Gorio, G. Carmignoto, M. Finesso, P. Polato and M. G. Nunzi, Muscle Reinnervation II. Sprouting, synapse formation and repression, Neuroscience 8:403 (1983).
9. J. R. Sanes, L. M. Marshall and U. J. McMahan, Reinnervation of muscle fiber basal lamina after removal of myofibers. Differentiation of regenerating axons at original synaptic sites, J. Cell Biol. 78:176 (1978).
10. R. B. Campenot, Local control of neurite development by nerve growth factor, Proc. Natl. Acad. Sci. (USA) 74:4516 (1977).
11. P. C. Letorneau, Chemotactic response of nerve fiber elongation to nerve growth factor, Develop. Biol. 66:183 (1978).
12. F. Solomon, Detailed neurite morphologies of sister neuroblastoma cells are related, Cell 16:165 (1979).
13. F. Solomon, Neuroblastoma cells recapitulate their original neurite morphologies after reversible microtubule depolymerization, Cell 21:333 (1980).
14. J. H. Pate Skene and M. Willard, Changes in axonally transported proteins during axon regeneration in toad retinal ganglion cell, J. Cell Biol. 89:86 (1981).
15. J. H. Pate Skene and M. Willard, Characteristics of growth-associated polypeptides in regenerating toad retinal ganglion cell axons, J. Neuroscience 1:419 (1981).
16. R. J. Lasek and P. N. Hoffman, The neuronal cytoskeleton, axonal transport and axonal growth, in: "Cell motility. Book C. Microtubules and Related Proteins", R. Goldman, T. Pollard, and J. Rosenbaum, eds., Cold Spring Harbor Laboratory, New York (1976).
17. R. J. Lasek, I. G. McQuarrie and J. R. Wujek, The Central Nervous System Regeneration: Neuron and Environment, in: "Post Traumatic Peripheral Nerve Regeneration: Experimental Basis and Clinical Implications", A. Gorio, H. Millesi, and S. Mingrino, eds., Raven Press, New York (1981).

18. A. Gorio, P. Marini and R. Zanoni, Muscle Reinnervation III. Motoneuron sprouting capacity, enhancement by exogenous gangliosides, Neuroscience 8:417 (1983).
19. A. Bjorklund and U. Stenevi, Regeneration of monoaminergic and cholinergic neurons in the mammalian central nervous system, Physiol. Rev. 59:62 (1979).
20. G. Raisman, Neuronal Plasticity in the septal nuclei of the adult rat, Brain Res. 14:25 (1969).
21. G. Raisman and P. M. Field, A quantitative investigation of the development of collateral reinnervation after partial deafferentiation of the septal nuclei, Brain Res. 50:241 (1973).
22. H. Hoffman, Acceleration and retardation of the process of axon-sprouting in partially denervated muscle, Aust. J. Exp. Biol. Med. Sci. 30:541 (1952).
23. R. J. Keynes, The effects of pyronin on sprouting and regeneration of mouse motor nerves, Brain Res. 253:13 (1982).
24. H. Hoffman and P. H. Springell, An attempt at the chemical identification of "neurocletin" (the substance evoking axon sprouts), Aust. J. Exp. Biol. Med. Sci. 29:417 (1951).
25. A. Gorio, H. Millesi and S. Mingrino, "Posttraumatic Peripheral Nerve Regeneration: Experimental Basis and Clinical Implications", Raven Press, New York (1981).
26. R. W. Ledeen, Ganglioside structure and distribution: are they localized at the nerve ending?, J. Supramol. Struct. 8:1 (1978).
27. M. M. Rapport and A. Gorio, "Gangliosides in Neurological and Neuromuscular Function, Development and Repair", Raven Press, New York (1981).
28. B. Ceccarelli, F. Aporti and M. Finesso, Effects of brain gangliosides on functional recovery in experimental regeneration and reinnervation, in: "Advance in Experimental Medicine and Biology", G. Porcellati, B. Ceccarelli, and G. Tettamanti, eds., Plenum Press, New York (1976).
29. A. Gorio, G. Carmignoto, L. Facci and M. Finesso, Motor nerve sprouting induced by ganglioside treatment. Possible implications for gangliosides on neuronal growth, Brain Res. 197:236 (1980).
30. A. Gorio, G. Carmignoto, M. Finesso, A. Leon, P. Marini, L. Tredese and R. Zanoni, Electrophysiological correlates of the re-innervation of rat neuromuscular junction: implications on the role of membrane components such as gangliosides in the motor nerve sprouting, in: "Cholinergic Mechanisms", G. Pepeu, and H. Ladinsky, eds., Plenum Publishing Corp., New York/London (1981).
31. A. Gorio, G. Carmignoto and G. Ferrari, Axon sprouting stimulated by gangliosides: A new model for elongation and sprouting, in: "Gangliosides in Neurological and Neuromuscular Function, Development and Repair", M. M.

Rapport, and A. Gorio, eds., Raven Press, New York (1981).
32. G. Ferrari, M. Fabris and A. Gorio, Gangliosides enhance neurite outgrowth in PC12 cells, Develop. Brain Res. 8:215 (1983).
33. W. Wojcik, J. Ulas and B. Oderfeld-Nowak, The stimulating effect of ganglioside injections on the recovery of choline acetyltransferase and acetylcholinesterase activities in the hippocampus of the rat after septal lesions, Neuroscience 7:495 (1982).
34. G. Jonsson, H. Kojima and A. Gorio, GM_1 ganglioside has a counteracting effect on neurotoxin induced alteration of the postnatal development of central serotonin (5-HT) neurons, Abstract ENA Meeting, Hamburg (1983).

EFFECTS OF GANGLIOSIDES ON THE FUNCTIONAL RECOVERY OF DAMAGED BRAIN

G. Toffano, G. Savoini, C. Aldinio, G. Valenti,
R. Dal Toso, A. Leon, L. Calza*, I. Zini*, L. F.
Agnati*, and K. Fuxe**

Department of Biochemistry, Fidia Research Laboratories, 35031 Abano Terme, Italy; *Institute of Human Physiology, University of Modena, 41100 Modena, Italy; **Department of Histology, Karolinska Institutet, 104 01 Stockholm, Sweden

INTRODUCTION

Treatment of brain and spinal cord injuries is usually undertaken with the tacit assumption that anatomical repair is not possible. This view implies serious limitations for the comprehension of phenomena such as neuronal plasticity and functional recovery of mature CNS. Currently however neuronal circuitries are envisaged as highly adaptable structures, intrinsically capable of remodeling and establishing new functional connections not only during development but also in the mature brain in response to perturbations such as lesions.[1,2] The failure of regeneration in the mature central nervous tissues may then rather be due to the incapability of growing axon sprouts to elongate and find a pathway to the denervated target.[3]

Transplantation experiments in mammals seem to be a promising strategy for understanding conditions that influence regeneration in the CNS. Lesioned neurons of the brain can grow into transplanted grafts from the peripheral nervous system, while embryonic transplanted central neurons receive and project fibers in adult deafferented central regions with restoration of the physiological functions. In this context an attempt has been made to correct the dopamine deficiency in parkinsonian patients by transplanting their own adrenal tissue into the caudate nucleus.[4]

An alternative approach is to modify, by means of

pharmacological agents, the response of central neurons to injury. The mechanisms involved in neuronal plasticity may represent an ideal target for a pharmacological treatment aiming to restore function of both central and peripheral nervous system after injury, and for defining the etiopathology of several neurological diseases.[5] Reactive phenomena such as axon sprouting and synaptogenesis may represent a major factor in the restoration of function in damaged brain.[6] To pursue this goal, fundamental prerequisites are: (i) knowledge of the molecular mechanisms underlying neuronal plasticity, (ii) availability of drugs modifying the molecular process, (iii) parallelism between molecular effects and physiological functions and finally, (iv) correspondence between physiological and therapeutic effects in vivo.

FACTORS AFFECTING NEURONAL REGENERATION

Studies with neuronal tissue culture techniques have indicated that neuronal ontogeny is highly dependent on the occurrence of an orderly sequence of extrinsic neuronotrophic, neurite-promoting, and guidance signals.[7] These signals may either derive from the humoral environment or be associated with the surrounding target cell surfaces or extracellular matrices, and apparently operate within a fixed period of time with remarkable precision. Similarly, CNS regeneration depends on the availability of suitable environmental factors.[3] Factors with trophic effect on the neuronal cells have been extracted from adult brain at the side of the lesion (see ref. 6 and 8). Depressive factors and/or processes, such as scar formation, autoimmune response, inflammatory reaction, lack of blood supply and others, may presumably inactivate or counteract the action of these neuronotrophic factors. It is conceivable that the functional responsiveness of neurons to neuronotrophic factors ultimately depends on their intrinsic capabilities of regulating signal recognition and transduction. Hence the plasma membranes and their constituents ought to play a crucial role in regulating neuritogenesis and presumably also in maintaining correct intercellular relationships.

NEURONAL DEVELOPMENT, REGENERATION, AND CELL DIFFERENTIATION: A ROLE FOR GANGLIOSIDES

The functional role of intrinsic neuronal membrane components in the regulation of cell surface responses may be relevant for the processes regulating surface-related cellular activities of both developing and mature neurons. Among them a particular role has been attributed to gangliosides, components particularly abundant in the CNS.[9] Because of their physical-chemical characteristics and asymmetrical distribution in the outer leaflet of the membrane bilayer they have been assumed to play a role in a variety of cell surface events including cell-cell recognition[10] and transduction of biological signals.[11]

Gangliosides stimulate the morphological and biochemical differentiation of clonal and primary cells in culture.[8,12-18] Particularly in neuroblastoma cells, gangliosides enhance neurite formation, increase cAMP content and decrease ^{3}H-thymidine incorporation, phenomena which reflect an elongation of G_1 phase of the cell cycle.[12] Binding studies with ^{3}H-GM$_1$ monosialoganglioside labelled on the terminal galactose[19] indicate that exogenous gangliosides can be stably inserted with defined kinetics into the lipid phase of cell membranes.[20] The insertion is accompanied by an enhancement of the adenylate-cyclase and (Na^+, K^+) ATPase activities, the structural and functional nature of which is still unclear.[21,22] Although the cause-effect relationship between modification of cell surface ganglioside content and neurite growth is still obscure, a participation of endogenous gangliosides is indicated by the observation that GM$_1$ expression parallels neurite outgrowth in the cerebellum during development,[23] and that ganglioside complexity and content change according to the neuronal developmental stages. Conversely administration of anti-ganglioside antibodies to developing animals produces behavioral dysfunction with morphological abnormalities in dendritic arborization[28] and inhibits neurite outgrowth.[27]

Finally an enhancement of axonal sprouting during peripheral reinnervation[24,25] and an accelerated recovery of choline acetyltransferase and acetylcholinesterase activities in rat hippocampus after medioventral septal lesion[26] have been reported.

GM$_1$ ganglioside facilitates the differentiation of both clonal and primary cells in culture, while antibodies to GM$_1$ inhibit the neurite outgrowth from regenerating retinal explants, from dorsal root ganglia as induced by NGF, and from embryonic dissociated mesencephalic cells.[13,18,20,27] These observations together with those done on neuron storage disorders[29] and on development[23] suggest an important role of GM$_1$ in neuronal differentiation and have been the starting point for experiments directed to ascertain a possible effect of GM$_1$ in the recovery of brain after injuries.

GM$_1$ MONOSIALOGANGLIOSIDE STIMULATES THE FUNCTIONAL RECOVERY OF DOPAMINERGIC NEURONS IN ADULT NERVOUS SYSTEM

We have recently demonstrated that repeated intraperitoneal administrations of GM$_1$ produces a significant recovery of those biochemical, immunohistochemical and behavioral paramaters which characterize the dopaminergic nigro-striatal pathways in unilaterally hemitransected rats.[6,30,31]

In the lesioned side of control animals, striatal tyrosine hydroxylase (TH) activity, striatal TH-positive nerve terminals (stained with specific fluorescent antibodies against TH) and HVA content decrease after surgery. In GM_1-treated rats a significant increase of TH activity, the number of TH-positive nerve terminals and of HVA content occur in the lesioned striatum as compared with the saline-treated group. GM_1 treatment stimulates the collateral sprouting of dopaminergic axons, and the new sprouts functionally impinge on the denervated target cells, as indicated by the decreased degeneration of striatal dopaminergic binding site-containing cells and by the decreased sensitivity to apomorphine.[30]

The effects elicited by GM_1 in the lesioned striatum are paralleled by a significant increase of TH activity and TH immunoreactivity in the cell bodies and dendrites in the substantia nigra ipsilateral to the lesion. Not only the entity of TH-immunoreactivity, but also the survival capabilities of substantia nigra neurons, as demonstrated by Nissl staining,[31] have been affected. An example of the GM_1 effect is reported in Fig. 1 (L. Agnati, L. Calza, and G. Toffano, unpublished results). These findings suggest that GM_1 favors the recovery of dopaminergic synaptic function in the striatum and increases the survival of lesioned dopaminergic neurons in hemitransected rats.

The mechanism by which GM_1 favors the collateral sprouting of dopaminergic axons in the striatum and maintains the number of dopaminergic cell bodies in the substantia nigra after axotomy is still under investigation. Collateral sprouting of mesolimbic and nigro-striatal dopaminergic neurons has been observed after axonal electrolytic injury.[32,33] Spontaneous sprouting of some nerve fibers can be elicited by trophic factors unmasked by lesion and specific trophic dopaminergic factors are present in striatal membranes.[34] A possible correlation between trophic factors and gangliosides has been suggested.[8,13,27,35] With regard to the mechanism underlying GM_1-induced survival of TH-positive cell bodies in the substantia nigra, we have observed that in the control group, there exists a reduction of TH-positive dendrites in the substantia nigra of the lesioned side as compared with that of the unlesioned side. GM_1 treatment restores the density of TH-positive dendrites and stimulates their elongation into substantia nigra pars reticulata.[31] Moreover GM_1 stimulates the formation of collateral fibers at the site of the lesion. After hemitransection the dopaminergic neurons react by elongating the TH-positive collaterals going back to the substantia nigra pars reticulata (Fig. 2a). Such a reaction is markedly enhanced by GM_1 treatment (Fig. 2b).

Thus it is possible that dopaminergic cells through dendrites and collaterals maintain their trophic interaction with adjacent cells, this allowing them to survive. The effect could be achieved through potentiation of the action of neuronotrophic factors released

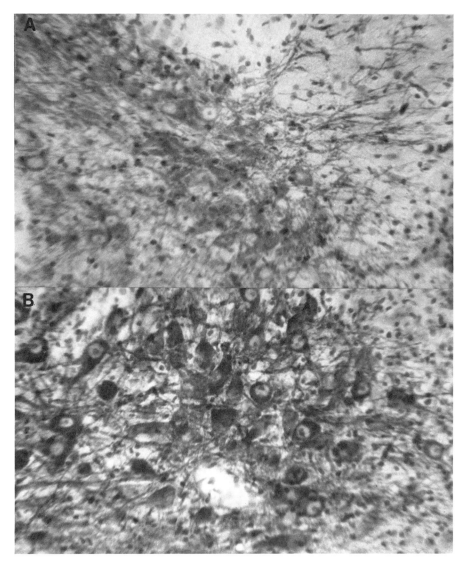

Fig. 1. PAP (peroxidase anti-peroxidase) immunocytochemical staining of TH-positive cell bodies in the substantia nigra (A 2180, Konig & Klippel Atlas). The immunocytochemical procedure and Nissl staining have been performed as already described.[31] The upper part (A) is an example of TH-positive cell bodies in the substantia nigra of the lesioned side in the control group, while the lower part (B) refers to the same anatomical region of the GM_1-treated group.

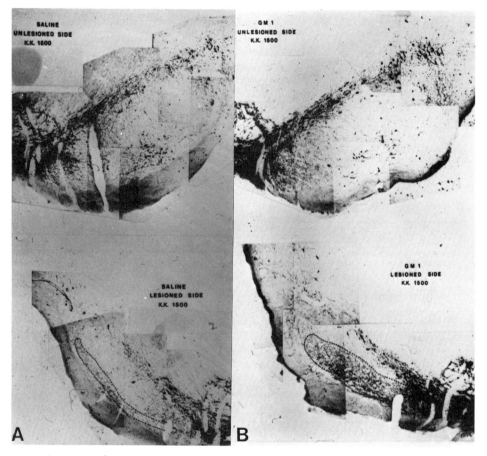

Fig. 2. PAP (peroxidase anti-peroxidase) immunocytochemical staining of TH-positive collaterals in a transversal section (A 1500, Konig & Klippel Atlas) of (A) a saline-treated rat and (B) a 10 mg/kg GM_1-treated rats 21 days after hemitransection. (L. Calza, L. Agnati, K. Fuxe, G. Toffano, and L. Giardino, submitted for publication).

at the level of the lesion. Support to this interpretation is given by the fact that the effect of GM_1 on the survival is maximal close to the lesion (Fig. 3), probably because of the higher concentration of neuronotrophic factors in that area. Moreover when the dopaminergic nigro-striatal system is destroyed by 6OH-DA, GM_1 becomes inactive. It is possible that the 6OH-DA oxidation products, by binding covalently to neuronal elements, may cause denaturation of neuronotrophic factors.[6]

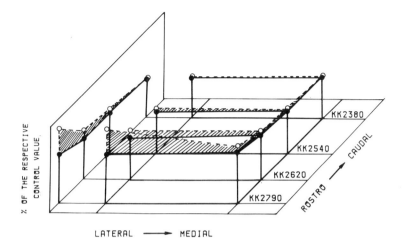

Fig. 3. Evidence for the existence of a "gradient" in GM1 effect (optical density). Computerized immunofluorescence analysis of TH-positive cell bodies in the substantia nigra ipsilateral to hemitransection of a 10 mg/kg GM_1-treated rat. The analysis was done on 4 consecutive rostro-caudal or 3 lateral-medial sections as indicated in the scheme. (L Agnati, L. Calza, K. Fuxe, F. Benfenati, and G. Toffano, submitted for publication).

EFFECT OF GM_1 GANGLIOSIDE ON THE MATURATION OF EMBRYONIC DOPAMINERGIC MESENCEPHALIC CELLS IN CULTURE

In order to obtain information useful for the understanding of the effect of GM_1 in vivo on neuronal regeneration, the GM_1 effect on the maturation of embryonic dopaminergic mesencephalic neurons cultured with or without striatal target cells was studied. This model is suitable for studying neuronal regeneration since cultured dissociated cells share common features with in vivo lesioned neurons.[36]

Rostral mesencephalic tegmentum and corpus striatum from brains of 13 and 15 day old mouse embryos respectively were dissected, mechanically dissociated and plated in serum free medium (Dal Toso et al., manuscript in preparation). In these conditions 90 to 95% of counted cells (Fig. 4) were positively stained with the anti-neurofilament monoclonal antibody and identified as neurons, while only 3 to 5% of cultured cells were reactive to GFAP antiserum indicating a limited presence of astroglial cells. Morphologically, the addition of GM_1 facilitates neurite outgrowth (referred to overall neuronal population present in the culture) (Table 1). GM_1 addition to mesencephalic cultures induces an increase in mean neurite length per cell, whereas it doubles mean neurite number and length per cell with respect to non-treated mesencephalic striatal co-cultures.

Biochemical evaluation of GM_1 effects on maturation of dopaminergic neurons indicated that GM_1 increases ^{3}H-DA uptake in mesencephalic cells, cultured in the absence of striatal cells (100% increase with 10^{-7}M GM_1). Since striatal cells have been suggested to possess factors which affect the maturation of mesencephalic dopaminergic neurons we investigated whether GM_1 was able to potentiate the effect of the target. Fig. 5 shows that in control samples the addition of striatal cells to a fixed number of plated mesencephalic cells induces a significant increase of ^{3}H-DA uptake. The effect is dependent on the number of striatal cells present in the culture. This effect is additive to that produced by GM_1 directly on mesencephalic cells.

DISCUSSION

Repeated administration of GM_1 monosialoganglioside stimulates the functional recovery of the dopaminergic nigro-striatal system as indicated by a number of biochemical, immunohistochemical and behavioral parameters. GM_1 favors the collateral sprouting of

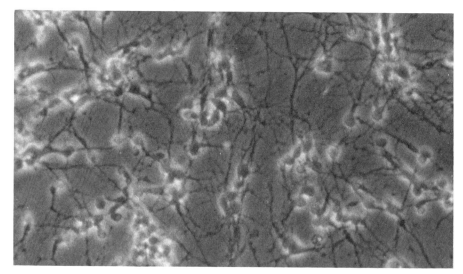

Fig. 4. Mesencephalic-striatal dissociated cells after 4 days in serum free medium culture. In this condition 90 to 95% of the cultured cells were identified as neurons on the basis of the reaction with the anti-neurofilament monoclonal antibody.

Table 1. Effect of GM_1 on Mesencephalic-Striatal Cultures: Morphometric Evaluation

Type of cell	Addition	Neurites per cell	Neurite length per cell (μM)	Cell Body circumference
Mesencephalic Cells (10^6)	Control	0.751 ± 0.167	5.81 ± 2.61	7.23 ± 1.96
	GM_1 7.5 x 10^{-8}M	0.724 ± 0.153	7.29 ± 3.24	8.53 ± 2.03
Mesencephalic cells (10^6) + Striatal cells (10^6)	Control	0.769 ± 0.182	7.28 ± 1.41	8.60 ± 2.41
	GM_1 7.5 x 10^{-8}M	1.575 ± 0.425*	13.05 ± 3.27*	9.38 ± 3.2

* $p < 0.01$ by Student's t test

Morphological analysis was done after 4 days in culture on at least 500 cells with a Leitz ASM Digitizer interfaced with a HP-9845 computer. Total number of neurites, neurite length and cell body circumference were referred to the total number of cells counted in the same photogram. Mesencephalic and striatal dissociated cells were obtained from 13 and 15 day old mouse embryos respectively and plated in a serum free medium (R. Dal Toso, A. Leon, and G. Toffano, submitted for publication).

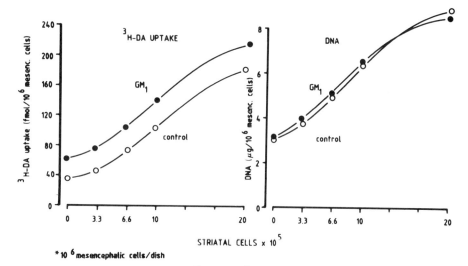

Fig. 5. Effect of GM_1 ($10^{-7}M$) on ^{3}H-DA uptake and DNA content in dissociated mesencephalic cells as a function of the number of striatal cells. 10^6 mesencephalic cells/dish were plated with or without different amounts of striatal cells. Biochemical analysis was done after 4 days in culture.

dopaminergic axons in the striatum and maintains the dopaminergic cell bodies in the substantia nigra after hemitransection. The effect of GM_1 does not occur when dopaminergic neurons are destroyed by injecting 6OH-DA directly into the substantia nigra and it is not specific for a single neuronal system. In fact GM_1 enhances also the recovery of cholinergic parameters in the hippocampus after septal lesion.[37] The mechanism by which GM_1 exerts the above effect is still a matter of investigation. The mechanical lesion may unmask trophic factors which elicit nerve sprouting[38] of intact surviving dopaminergic axons, stimulate the pruning process at the level of the lesion, as indicated by the formation of TH-positive collaterals, and the formation of longer TH-positive dendrites in the substantia nigra. GM_1 apparently interferes with these processes and favors the formation of new functional circuitries. Support to an involvement of neuronotrophic factors in the action of GM_1 is provided by a number of observations in vitro[8,13,27,35] and by the failure of GM_1 to produce sprouting and cell maintenance after lesioning the dopaminergic nigro-striatal neurons with 6OH-DA. The chemical neurotoxin is taken up and accumulates in catecholaminergic neurons. The oxidation products bind covalently to neuronal elements leading to protein denaturation, presumably also of trophic factors, and degeneration.

Finally, the effect of GM_1 was studied in vitro on the maturation of dissociated embryonic dopaminergic neurons in culture. This model is suitable for studying neuronal regeneration since cultured dissociated cells have common features with the in vivo lesioned neurons. In this system GM_1 enhances the maturation of dopaminergic cells by acting directly on mesencephalic cells and the effect is additive to, but not potentiated by, the presence of striatal target cells.

In conclusion the present results support the working hypothesis that it is possible to stimulate pharmacologically those compensatory mechanisms which seem to be involved in the restoration of the brain function after lesion. GM_1 may be envisaged not only as a pharmacological agent active on these processes but also as a useful tool for understanding the mechanism which underlies neuronal plasticity and leads to central regeneration.

SUMMARY

The effect of GM_1 ganglioside on the recovery of dopaminergic nigro-striatal neurons was studied in rats after unilateral hemitransection. GM_1 treatment favoured the collateral sprouting of dopaminergic axons in the striatum as indicated by the induced increase of tyrosine hydroxylase (TH) activity and immunofluorescence. Concomitantly GM_1 partially prevented the decrease of TH activity caused by the hemitransection in the substantia nigra ipsilateral to the lesion. A significant increase of TH immunoreactivity was also detected in the substantia nigra: GM_1 prevented the disappearance of TH-positive cell bodies and increased the formation of TH-positive collaterals and dendrites with respect to the saline treatment. The addition of GM_1 to embryonic dissociated mesencephalic cell cultures stimulates the expression of dopaminergic characteristics as suggested by the increase of ^{3}H-DA uptake.

REFERENCES

1. A. Björklund and U. Stenevi, Regeneration of monoaminergic and cholinergic neurons in the mammalian central nervous system, Physiol. Rev. 59:62 (1979).
2. C. W. Cotman, M. Nieto-Sampedro, and E. W. Harris, Synapse replacement in the nervous system of adult vertebrates, Physiol. Rev. 61:684 (1981).
3. M. Benfey and A. J. Aguayo, Extensive elongation of axons from rat brain into peripheral nerve grafts, Nature 296:150 (1982).
4. G. Kolata, Grafts correct brain damage, Science 217:342 (1982).
5. S. H. Appel, A unifying hypothesis for the cause of amyotrophic lateral sclerosis parkinsonism and Alzheimer disease,

Ann. Neurol. 10:499 (1981).
6. G. Toffano, G. Savoini, C. Aldinio, G. Valenti, F. Moroni, G. Lombardi, L. F. Agnati, and K. Fuxe, A role for exogenous gangliosides in the functional recovery of adult lesioned nervous system, in press.
7. S. Varon and R. Adler, Trophic and specifying factors directed to neuronal cells, in: "Advances in Cellular Biology", Vol. 2, Academic Press, New York (1981).
8. G. Toffano, D. Benvegnu, R. Dal Toso, L. Facci, and A. Leon, Neuronal development and regeneration: a role for gangliosides, in: "Neural Membranes", G. Y. Sun, N. Bazan, J.-Y. Wu, G. Porcellati, and A. Y. Sun, eds., The Humana Press, Clifton (1983).
9. R. W. Ledeen, Ganglioside structure and distribution: are they localized at the nerve ending?, J. Supramol. Struct. 8:1 (1978).
10. T. Yamakawa and Y. Nagai, Glycolipids at the cell surface and their biological functions, Trends Biochem. Sci. 3:128 (1978).
11. G. Toffano, G. E. Savoini, F. Moroni, M. G. Lombardi, L. Calza, and L. F. Agnati, GM_1 ganglioside stimulates the regeneration of dopaminergic neuron in the central nervous system, Brain Res. 261:128 (1983).
12. A. Leon, L. Facci, D. Benvegnu, and G. Toffano, Morphological and biochemical effects of gangliosides in neuroblastoma cells, Develop. Neurosci. 5:108 (1982).
13. G. Ferrari, M. Fabris, and A. Gorio, Gangliosides enhance neurite outgrowth in PC12 cells, Develop. Brain Res. (in press.
14. F. J. Roisen, H. Bartfeld, R. Magelle, and G. Yorke, Ganglioside stimulation of axonal sprouting in vitro, Science 214: 577 (1981).
15. U. Mengs, H. L. U. Tullner, R. Goldschmidt, and F. K. Pieran, Influence of gangliosides on neurite sprouting and arborization in vitro, Int. J. Tiss. Reac. 4:277 (1982).
16. K. Obata, M. Oide, and S. Handa, Effects of glycolipids on in vitro development of neuromuscular juction, Nature 266:369 (1977).
17. J. I. Morgan and W. Seifert, Growth factors and gangliosides: a possible new perspective in neuronal growth control, J. Supramol. Struct. 10:111 (1979).
18. M. Schwartz and N. Spirman, Sprouting from chicken embryo dorsal root ganglia induced by nerve growth factor is specifically inhibited by affinity purified antiganglioside antibodies, Proc. Natl. Acad. Sci. USA 79:6080 (1982).
19. G. Toffano, D. Benvegnu, A. C. Bonetti, L. Facci, A. Leon, P. Orlando, R. Ghidoni, and A. C. Tettamanti, Interactions of GM_1 ganglioside with crude rat brain neuronal membranes, J. Neurochem. 35:861 (1980).
20. L. Facci, A. Leon, G. Toffano, S. Sonnino, R. Ghidoni, and G.

Tettamanti, Promotion of neuritogenesis in mouse neuroblastoma cells by exogenous gangliosides. Relationship between the effect and the cell association of ganglioside GM_1, J. Neurochem. in press.
21. A. Leon, L. Facci, G. Toffano, S. Sonnino, and G. Tettamanti, Activation of (Na^+,K^+) ATPase by nanomolar concentrations of GM_1 ganglioside, J. Neurochem. 37:350 (1981).
22. C. R. Partington and J. W. Daly, Effect of gangliosides on adenylate cyclase activity in rat cerebral cortical membranes, Mol. Pharmacol. 15:484 (1979).
23. M. Willinger and M. Schachner, GM_1 ganglioside as a marker for neuronal differentiation in mouse cerebellum, Develop. Biol. 74:101 (1980).
24. B. Ceccarelli, F. Aporti, and M. Finesso, Effects of brain gangliosides on functional recovery in experimental regeneration and reinnervation, Adv. Exp. Med. Biol. 71:275 (1976).
25. A. Gorio, G. Carmignoto, L. Facci, and M. Finesso, Motor nerve sprouting induced by ganglioside treatment. Possible implication for gangliosides neuronal growth, Brain Res. 197:236 (1980).
26. M. Wojcik, J. Ulas, and B. Oderfeld-Nowak, The stimulating effect of ganglioside injection on the recovery of choline acetyltransferase and acetylcholinesterase activities in the hippocampus of the rat after septal lesions, Neurosci. 7:495 (1982).
27. N. Spirman, B. A. Sela, and M. Schwartz, Antiganglioside antibodies inhibit neuritic outgrowth from regenerating goldfish retinal explants, J. Neurochem. 39:874 (1982).
28. S. Karpiak, Y. L. Huang, and M. M. Rapport, Immunological model of epilepsy, J. Neurochem. 3:15 (1982).
29. D. P. Purpura, G. D. Pappas, and H. J. Baker, Meganeurites and aberrant processes of neurones on GM_1-gangliosidosis. A Golgi study, Brain Res. 145:13 (1977).
30. G. Savoini, K. Fuxe, L. F. Agnati. L. Calza, F. Moroni, M. G. Lombardi, M. Goldstein, and G. Toffano, Effect of GM_1 ganglioside on the recovery of dopaminergic nigro-striatal neurons after lesion, Study Group on Neuroplasticity and Repair in Central Nervous System, Geneva (1982).
31. K. Fuxe, L. F. Agnati, L. Calza, I. Zini, F. Benfenati, G. Toffano, M. Goldstein, and C. Kohler, Chemical morphometry as tool to study gangliosides as growth regulation factor in central catecholamine neurons. Study Group on Neuroplasticity and Repair in Central Nervous System, Geneva (1983).
32. G. M. Gilad and D. J. Reis, Collateral sprouting in cerebral mesolimbic dopamine neurons: biochemical and immunocytochemical evidence of changes in the activity and distribution of tyrosine hydroxylase in terminal fields and in cell bodies, Brain Res. 160:17 (1979).

33. D. J. Reis, G. Gilad, V. M. Pickel, and T. H. Joh, Reversible changes in the activities and amounts of tyrosine hydroxylase in dopamine neurons of the substantia nigra in response to axonal injury as studied by immunochemical and immunocytochemical methods, Brain Res. 144:325 (1978).
34. A. Prochiantz, M. C. Daguet, A. Herbert, and J. Glowinski, Specific stimulation of in vitro maturation of mesencephalic dopaminergic neurones by striatal membranes, Nature 293:570 (1981).
35. W. Seifert, Gangliosides in nerve cell cultures, in: "Gangliosides in neurological and neuromuscular function, development and repair," M. M. Rapport and A. Gorio, eds., Raven Press, New York (1980).
36. B. Berger, U. Di Porzio, M. C. Daguet, M. Gay, A. Virginy, J. Glowinski, and A. Pochiantz, Long term development of mesencephalic dopaminergic neurons of mouse embryos in dissociated primary cultures: morphological and histochemical characteristics, Neuroscience 7:193 (1982).
37. B. Oderfeld-Nowak, J. Ulas, M. Jezierska, M. Skup, M. Wojcik, and K. Domanska, Role of GM_1 ganglioside in repair processes after hippocampal deafferentation in rats. Study Group on Neuronal Plasticity and Repair in Central Nervous System, Geneva (1982).
38. A. Björklund and U. Stenevi, In vivo evidence for a hippocampal adrenergic neuronotrophic factor specifically released on septal deafferentation, Brain Res. 229:403 (1981).

EXOGENOUS GANGLIOSIDES ENHANCE RECOVERY FROM CNS INJURY

Stephen E. Karpiak

Division of Neuroscience, New York State Psychiatric
Institute and the Dept. of Psychiatry, College of
Physicians & Surgeons, Columbia Univ., N.Y., N.Y.

INTRODUCTION

Since gangliosides are found in high concentration in the CNS[1,2] and are topographically localized to the outer surface of neuronal membranes[3,4], they have been implicated as receptor molecules. Some hypotheses suggest that gangliosides participate in the regulation of neurogenesis,[5] synaptogenesis,[6] regeneration[7] as well as cell-cell interaction.[8] Reports that antibody to ganglioside can inhibit neurite outgrowth in vitro[9,10] and can interfere with synaptogenesis and neurogenesis in vivo[11] indirectly support these hypotheses.

Other evidence which suggests that gangliosides play a critical role in neuronal development and regeneration, is based upon recent studies which report that exogenous ganglioside administration can accelerate neurite outgrowth in vitro[12,13] and enhance peripheral nerve regeneration after injury.[14,15] Although the mechanism underlying these effects is unknown, it has been reported that incubation[16,17] of rat brain neuronal membranes with [^{3}H] GM1 ganglioside results in plasma membrane incorporation. These "inserted" gangliosides are termed "functional" since cells incubated with GM1 ganglioside show increased choleragen binding,[18,19] which is constant for 4 days, indicating no increase in metabolism or degradation of the incorporated ganglioside. NMR studies have also confirmed this "insertion" phenomenon.[20]

The following study was undertaken to determine if exogenous ganglioside administration could affect CNS processes, i.e., CNS regeneration, following injury. To begin to examine this question, a well-studied model was chosen[21-23] where it had been shown in rat that a unilateral entorhinal cortical (EC) lesion results in rapid denervation of the outer molecular layer of the hippocampal dentate gyrus.[24] This denervated dendritic region is reinnervated (collateral sprouting) by afferents from the contralateral EC, cholinergic septal afferents, commissural and associational afferents.[25-27] After this unilateral EC lesion, rats trained on an alternation task show a loss of the behavior, followed by a gradual recovery. The rate of functional recovery correlates with the rate of reinnervation of the previously denervated dentate gyrus. Lesioning the contralateral EC will again cause a loss of the behavior, but no functional recovery occurs. It is concluded that the recovery of function depends on reinnervating fibers that originate from the contralateral EC.[28]

The following experiment was designed to assess the effect of exogenous ganglioside treatment on functional recovery (learned alternation behavior) following a unilateral EC lesion.

METHODS

Animals

Male Sprague-Dawley rats were used (180-200 gr).

Behavioral Testing

Both the testing apparatus and procedures for the learned alternation task are modifications of those reported in the literature.[29] A t-maze painted flat black, with 10 cm wide alleys, a main arm of 45 cm, side arms of 25 cm, with walls 20 cm in height was used. Food wells were located in the end of each arm. After 72 hrs of food deprivation (rats were maintained at 80-90% of original body weight) daily training began on the alternation behavior. On days when the rats were being initially food-deprived they were given a 10 minute habituation period in the maze.

Training and subsequent testing consisted of placing the rat in the base of the main arm of the maze, and then immediately releasing the animal. On the first trial, in order to receive a food reward, the rat was required to move into the left arm of the maze where two 37 mg food pellets were located. After eating the food pellets, or if the animal turned into the right (incorrect) arm of the maze, it was then picked-up by the experimenter and held for ten secs (inter-trial interval) and returned to the base of the main arm and released. On the second trial the food pellets were located in the right maze arm, on the third in the left. On each day this training continued until the rat made 20 correct choices. Scores were based upon the number of incorrect responses made on each day. In addition, runway latencies (secs to traverse the main arm) were monitored to assess motor activity.

Rats received a unilateral entorhinal cortical lesion in the afternoon on the 4th day of training on the alternation behavior. The animals had been previously tested in the A.M. on that day.

Entorhinal Cortical Lesions

Rats were anesthetized using Ketamine (100 mg/kg) supplemented with xylazine. For these studies, aspiration lesions were used,[30] although similar data have been obtained using electrolytic lesions. After all testing, rat brains were removed and examined for placement of the lesion. Data were eliminated from the study when the extent of the entorhinal lesion was smaller than 2 cubic mm or larger than 3 cubic mm, and when any part of the hippocampus or brain stem was involved.

Ganglioside Injections

In the first set of experiments rats received daily i.m. injections (50 mg/kg) of total bovine brain ganglioside (average composition: GD1b 16%; GT1 19%; GM1 21%; GD1a 39%). Subsequent experiments examined the effectiveness of varying doses of total ganglioside (5, 25, 50, 75 mg/kg) and of the individual ganglioside species GM1 (10 mg/kg). All injections were begun on the day before lesioning and continued until the last day of behavioral testing. Control rats received saline injections.

RESULTS

The data clearly demonstrate that there is a marked behavioral deficit after the unilateral EC lesion (saline group) followed by a gradual recovery of the behavior and return to baseline 8 days later (Fig. 1). As compared to controls, rats treated with total ganglioside or GM1 ganglioside showed the following (Fig. 1): 1) a reduction ($p<0.01$) in the extent of behavioral deficit seen 24 hrs after the lesion; 2) consistently fewer errors on each subsequent day of testing; 3) faster return to baseline performance; 4) an improved final net performance that was almost error-free (ANOVA: days x treatment $p<0.01$). Rats treated with GM1 ganglioside returned to baseline performance (days 7-8) earlier than rats injected with total ganglioside (days 8-9). No statistical difference between rats injected with 50 mg/kg of total ganglioside or rats injected with 10 mg/kg of GM1 was seen.

In addition it was found that the mortality rate after the entorhinal lesion was reduced by almost 50% ($p<0.01$) when rats were pretreated with total ganglioside or GM1 ganglioside (Fig. 2). Runway latencies were identical for all tested groups.

In assessing the dose response of total ganglioside administration, no difference was found between doses of 50 mg/kg, or 75 mg/kg. There was a slight ($p<0.05$) decrease in the effectiveness of a 25 mg/kg dose as compared to 50 mg/kg, although rats which received the 25 mg/kg dose still showed a facilitated functional recovery when compared to controls ($p<0.01$). Rats which received 5 mg/kg were not different from controls.

DISCUSSION

Our experiments show that exogenous ganglioside treatment facilitates functional recovery of alternation behavior after an entorhinal cortical lesion. This facilitation is characterized by a reduction in the extent of behavioral deficit seen one day after damage to the CNS and subsequent improved recovery from the lesion. Though the actual rates of recovery for both ganglioside treated animals and controls would seem to be identical, the final performance of the ganglioside-treated animals was better that that of controls. A dose response was seen with total ganglioside, where 75 mg/kg and 50 mg/kg were identical in effectiveness, and 25 mg/kg was less effective, but still significantly facilitated the functional recovery as compared to controls. A 5 mg/kg dose was not effective.

Equally as facilitatory as total ganglioside injections, if not slightly more so, was GM1 ganglioside at 10 mg/kg. In fact this dose is equivalent to the amounts of GM1 being injected when total ganglioside was injected. It will be of interest to

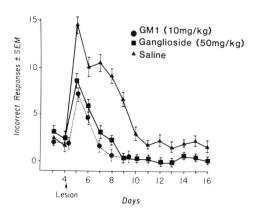

Fig. 1. Functional recovery on alternation behavior seen after a unilateral entorhinal cortical lesion. Daily injections (i.m.) of either total brain ganglioside (N=41) or GM1 ganglioside (N=18) significantly reduced 1) the behavioral deficit seen 24 hrs after lesioning; 2) the rate of recovery and 3) the final performance level.

determine if other major ganglioside species are also effective. Gangliosides are a well-characterized class of glycosphingolipids, whose individual molecular variation makes it suitable to study the differential effectiveness of different but related molecules. Such differences in effectiveness may result from structural differences.

Runway latencies for all groups were identical on each day of testing. Therefore the facilitatory effects of the gangliosides cannot easily be attributed to enhanced motor ability.

The reduced mortality rate as well as the behavioral deficit seen 24 hrs after the entorhinal cortical lesion suggests that pretreatment with gangliosides provides some form of "protection". The action of gangliosides on cell membranes may in some way "stabilize" these membranes (e.g. through alterations in membrane fluidity) in such a way as to prevent the death of cells at the perimeter of the lesion.

The mechanisms by which gangliosides exert their effects in this system is open to speculation. Based upon reports that exogenously administered gangliosides can spontaneously insert into membranes[18-20], two hypotheses seem reasonable at this time. One hypothesis is[14,31] that gangliosides incorporated into the membranes of regenerating neurons (neurites) may act as receptors for "growth factors". A correlation exists between the levels of certain gangliosides and various stages of neuronal maturation in some studies.[32] Tissue culture studies provide some support for such hypotheses since neurite outgrowths of regenerating retinal or dorsal root ganglia are inhibited by antibody to GM1.[9,10] Similarly, we have shown (in vivo),[11] that antibodies to GM1 ganglioside administered neonatally caused subtle behavioral dysfunctions in the adult associated with CNS loss of gangliosides as well as morphological abnormalities in dendritic arborization and spine morphology. It was hypothesized that the antibody to GM1 interfered with synapse formation which led to impaired dendrogenesis.

A second hypothesis may be that exogenous gangliosides are acting quite differently as compared to endogenous molecules. "Incorporated gangliosides may induce changes in membrane permeability. Gangliosides can cluster and cause permeability changes in membranes, and at the site of clustering, induce membrane modifications (i.e. changes in fluidity)[14,33] which make that membrane locus a site suitable for forming a new membrane branch. In the extreme example (i.e. an endogenous increase in ganglioside concentration) Purpura theorizes[34] that the increase in gangliosides as seen in the gangliosidoses appears to be associated with an array of morphological alterations which include meganeurite formation, secondary neurite outgrowth as well as aberrant dendrite differentiation. These processes are analogous to those seen in vitro

Fig. 2: Mortality rate seen 24 hrs after a unilateral entorhinal lesion was significantly reduced when rats were pretreated either with total brain ganglioside or GM1 ganglioside.

when exogenous gangliosides are administered.

The facilitatory effects which the gangliosides are exerting can be hypothesized to be due to enhanced collateral sprouting into the denervated dendritic region of the granular cells in the dentate gyrus. The enhanced functional recovery would seem to support this hypothesis. Preliminary data[35,36] show that with ganglioside treatment there are increased levels of acetylcholinesterase activity in the dentate, indicating enhanced cholinergic sprouting. However, morphological evidence will be necessary to confirm the hypothesis that gangliosides enhance neuronal sprouting in the dentate after the entorhinal lesion.

SUMMARY

Reports indicate that exogenous gangliosides can accelerate neurite outgrowth in vitro and facilitate peripheral nerve regeneration in vivo. An experiment was designed to assess whether ganglioside administration alters functional recovery and neuronal regeneration after a CNS lesion. Rats trained on an alternation behavior and subjected to a unilateral entorhinal cortical lesion were given daily (i.m.) injections of either total brain ganglioside or GM1 ganglioside.

Results show that ganglioside administration reduces the extent of behavioral deficit caused by the lesions and facilitated the course of functional recovery. It is hypothesized that gangliosides are enhancing hippocampal sprouting which occurs subsequent to the entorhinal lesion.

REFERENCES

1. J. A. Skrivanek, R. W. Ledeen, R. U. Margolis, and R. K. Margolis, Gangliosides associated with microsomal sub-

fractions of brain: Comparison with synaptic plasma membranes, J. Neurobiol. 13:95 (1982).
2. P. Fishman and R. Brady, Biosynthesis and function of gangliosides, Science 194:904 (1976).
3. S. P. Mahadik, B. Hungund, and M. M. Rapport, Topographic studies of glycoproteins of intact synaptosomes from rat brain cortex, Biochimica & Biophys. Acta 515:240 (1978).
4. B. Hungund and S. P. Mahadik, Topographic studies of gangliosides of intact synaptosomes from rat brain cortex, Neurochem. Res. 6:183 (1981).
5. M. Willinger and M. Schachner, GM1 ganglioside as marker for neuronal differentiation in mouse cerebellum, Dev. Biol. 74:101 (1980).
6. K. Obata, M. Oide, and S. Handa, Effects of glycolipids on in vitro development of neuromuscular junction, Nature 266:369 (1977).
7. A. Gorio, P. Marini, and R. Zanoni, Muscle reinnervation - III. Motoneuron sprouting capacity, enhancement by exogenous gangliosides, Neurosci. 3:417 (1983).
8. T. Yamakawa and Y. Nagai, Glycolipids at the cell surface and their biological function, TIBS 3:128 (1979).
9. M. Schwartz and N. Spirman, Sprouting from chicken embryo dorsal root ganglia induced by nerve growth factor is specifically inhibited by affinity-purified antiganglioside antibodies, Proc. Natl. Acad. Sci. USA 79:6080 (1982).
10. N. Spirman, B. Sela, and M. Schwartz, Antiganglioside antibodies inhibit neuritic outgrowth from regenerating goldfish retinal explants, J. Neurochem. 39:874 (1982).
11. E. Kasarskis, S. Karpiak, M.M. Rapport, R. Yu, N. Bass, Abnormal maturation of cerebral cortex and behavior in adult rats after neonatal administration of antibodies to GM1 ganglioside, Dev. Brain Res. 1:1 (1980).
12. F. J. Roisen, H. Bartfeld, R. Nagele, and G. Yorke, Ganglioside stimulation of axonal sprouting in vitro, Science 214:577 (1981).
13. A. Leon, L. Facci, D. Benvegnu, and G. Toffano, Morphological and biochemical effects of gangliosides in neuroblastoma cells, Dev. Neurosci. 5:108 (1982).
14. A. Gorio, P. Marini, and R. Zanoni, Muscle reinnervation - III. Motoneuron sprouting capaticy, enhancement by exogenous gangliosides, Neurosci. 3:417 (1983).
15. D. Kleinebeckel, Acceleration of muscle re-innervation in rats by ganglioside treatment: An electromyographic study, Eur. J. Pharm. 80:243 (1982).
16. G. Tettamanti, B. Veerando, S. Roberti, V. Chigorno, S. S. Sonnino, R. Ghidoni, P. Orlando, and P. Masari, The fate of exogenously administered brain gangliosides. In "Gangliosides in Neurological and Neuromuscular Function, Development and Repair", M. Rapport and A. Gorio, eds., Raven Press, New York, pp. 225-240 (1981).

17. G. Toffano, D. Benvegnu, A. Bonetti, L. Facci, A. Leon, P. Orlando, R. Ghidoni, and G. Tettamanti, Interaction of GM1 ganglioside with crude brain neuronal membranes, J. Neurochem. 35:861 (1980).
18. P. Fishman, J. Moss, and V. Manganiello, Synthesis and uptake of gangliosides by choleragen-responsive human fibroblasts, J. Biochem. 16:1871 (1977).
19. E. O'Keefe and P. Cuatrecasas, Persistence of exogenous, inserted ganglioside GM1 on the cell surface of cultured cells, Life Sci. 21:1649 (1977).
20. S. Kanda, K. Inoue, S. Nojima, H. Utsumi, and H. Wiegandt, Incorporation of spin-labeled ganglioside analogues into cell and liposomal membranes, J. Biochem. 91:1707 (1982).
21. O. Steward, C. Cotman, and G. Lynch, A quantitative autoradiographic and electrophysiological study of the reinnervation of the dentate gyrus by the contralateral entorhinal cortex following ipsilateral entorhinal lesions, Brain Res. 114:181 (1976).
22. O. Steward, Assessing the functional significance of lesion-induced neuronal plasticity, Int. Rev. Neurobiol. 23:197 (1982).
23. A. Caceres and O. Steward, Dendritic reorganization in the denervated dentate gyrus of the rat following entorhinal cortical lesions: a golgi and electron microscopic analysis, J. Comp. Neurol. 214:387 (1983).
24. J. Loesche and O. Steward, Behavioral correlates of denervation and reinnervation of the hippocampal formation of the rat: Recovery of alternation performance following unilateral entorhinal cortex lesions, Brain Res. 2:31 (1977).
25. O. Steward, Reinnervation of dentate gyrus by homologous afferents following entorhinal cortical lesions in adult rats, Science 194:426 (1976).
26. O. Steward and S. Vinsant, Collateral projections of cells in the surviving entorhinal area which reinnervate the dentate gyrus of the rat following unilateral entorhinal lesions, Brain Res. 149:216 (1978).
27. O. Steward and S. Scoville, Cells of origin of entorhinal cortical afferents to the hippocampus and fascia dentata of the rat, J. Comp. Neurol. 169:347 (1976).
28. O. Steward and S. Vinsant, Identification of the cells of origin of a central pathway which sprouts following lesions in mature rats, Brain Res. 147:223 (1978).
29. S. Karpiak, M. M. Rapport, and F. Bowen, Immunologically induced behavioral and electrophysiological changes in the rat, Neuropsychologia 12:313 (1973).
30. S. Karpiak, Ganglioside treatment improves recovery of altenation behavior after unilateral entorhinal cortex lesion, Exp. Neurol. 81:330 (1983).
31. E. Bremer and L. Hakomori, GM3 ganglioside induces hamster fibroblast growth inhibition in chemically-defined medium:

Ganglioside may regulate growth factor receptor function, Biochem. Bioph. Res. Comm. 106:711 (1982).
32. L. Irwin, D. Michael, and C. Irwin, Ganglioside patterns of fetal rat and mouse brain, J. of Neurochem. 34:1527 (1980).
33. F. Cumar, B. Maggio, and R. Capputo, Dopamine release from nerve endings induced by polysialogangliosides, Biochem. and Biophys. Res. Comm. 84:65 (1978).
34. D. Purpura and H. Baker, Neurite induction in mature cortical neurones in feline GM1-ganglioside storage disease, Nature 266:553 (1977).
35. S. Karpiak, F. Vilim, and S. Mahadik, GM1 ganglioside facilitates functional recovery after an entorhinal lesion; increase in AChE in dentate gyrus, Soc. Neurosci Abstr. 9:699 (1983).
36. B. Oderfeld-Nowak, M. Jezierska, J. Ulas, K. Mitros, and A. Wieraszko, Plastic responses of cholinergic parameters in the hippocampus induced by entorhinal cortex lesions are intensified by GM1 ganglioside treatment, Abstr from the Cell Biology of Neuronal Plasticity, Sardinia, Italy, p. 115 (1983).

Acknowledgement. Gangliosides used in these experiments were generously supplies by FIDIA Research Labs., Abano Terme, Italy.

GANGLIOSIDE INDUCED SURFACE ACTIVITY AND NEURITE FORMATION OF

NEURO-2A NEUROBLASTOMA CELLS

Fred J. Roisen, David A. Spero, Susan J. Held,
Glee Yorke and Harry Bartfeld

Department of Anatomy, Rutgers Medical School, UMDNJ
Piscataway, NJ 08854, and ALS Res. Center, St. Vincent's
Hospital and Medical Center, NY, NY 10011, U.S.A.

INTRODUCTION

Gangliosides are membrane-associated acidic glycolipids that have been implicated in the processes of differentiation,[1-4] growth,[5,6] and regeneration.[7-11] Their relative abundance in the nervous system[12] and the fact that both qualitative and quantitative changes occur in their composition during cephalogenesis and synaptogenesis suggests that they play a role in neuronal development. By virtue of their location within plasma membranes and contribution to the surface carbohydrate network, gangliosides may participate in a variety of surface-mediated regulatory activities including cell recognition,[13] contact inhibition,[14] and information transfer by acting as receptors for toxins[15] and hormones.[16]

The ganglioside GM1 has been shown to bind cholera toxin.[17] Taking advantage of this property, Fishman et al.[18] were able to show that when GM1-deficient fibroblasts, which are normally unresponsive to cholera toxin, are grown in medium supplemented with GM1, they are able to respond to the toxin, demonstrating that exogenously applied gangliosides are incorporated in a functional manner into cell membranes. This made it possible to test the effects of gangliosides on neuronal development. Obata et al.[7] have found that GM1 stimulates the formation of neuromuscular junctions in mixed cultures of skeletal muscle and spinal cord at low concentrations and has the opposite effect at high concentrations. We have shown previously that exogenously applied mixtures of bovine brain gangliosides (BBG) are incorporated into the plasma membranes of both murine Neuro-2a neuroblastoma and chick embryonic dorsal root ganglia (DRG) in vitro.[2] Media containing BBG have been reported to enhance neurite formation in primary neuronal cultures[19,20] and established neuronal lines.[2-4,21] We found that the enhanced neuritogenesis was accompanied by a corresponding two-fold increase in the ornithine decarboxylase activity in DRG and Neuro-2a cells 6h after BBG treatment.[19] ODC is the rate-limiting enzyme in the polyamine biosynthetic pathway and as such provides a convenient index of metabolic activity and growth. Furthermore, the dose-related increase in number and length of Neuro-2a cell processes produced by BBG-supplemented media was accompanied by prolongation of the G1 phase, an increase in intracellular cAMP levels and a decrease in thymidine incorporation into DNA.[3,4]

Our previous scanning electron microscopic (SEM) examination of Neuro-2a cells revealed relatively smooth surfaces on untreated cells in contrast to the heavy concentration of microvilli along the dorsal surfaces of BBG-treated cells. In the present study we have examined the morphological basis of BBG-mediated neuritogenesis. SEM demonstrated that BBG-treatment produces an initial burst of surface-related activity which results in the immediate (within 3 min) formation of numerous microvilli. Cytoskeletal disruptive agents (Colcemid and cytochalasin D) were employed to probe the role of microtubules and microfilaments in the BBG-enhanced neurite formation. SEM, whole-cell transmission electron microscopy (WCTEM) and thin-section TEM analysis of cultures treated simultaneously with gangliosides and these agents suggest that microtubules play a role in neurite elongation while microfilaments, the apparent dominant site of the ganglioside action, are responsible for spine formation and neurite branching in Neuro-2a cells.

MATERIALS AND METHODS

Cell Culture

Neuro-2a murine neuroblastoma (American Type Culture Collection - CCL 131) were grown in standard medium (SM) consisting of Minimum Essential Medium with Hanks' balanced salt solution (HBSS) (GIBCO, Grand Island, NY) supplemented with 10% heat-inactivated fetal calf serum (Irvine Scientific, Irvine, CA), 10 mg% gentamicin (Schering Corp., Kenilworth, NJ), 75 mg% additional $NaHCO_3$ and 0.1 mM nonessential amino acids at 35°C in a humidified atmosphere of 5% CO_2 and 95% air. Stock cultures were maintained routinely in Corning plastic flasks and passed via trypsin - EDTA (0.05% - 1 mM) every 5 days. For each study 2×10^5 cells were plated onto 35 mm dishes containing collagen-coated or poly DL-ornithine-coated coverslips. All substances tested were added to the culture medium at the indicated final concentrations: BBG at 250 μg/ml (a mixture containing 19% GM1, 44% GD1a, 16% GD1b, and 20% GT was generously supplied by Fidia Research Laboratories, Abano Terme, Italy); Colcemid at 0.5 µg/ml (GIBCO, Grand Island, NY); and cytochalasin D at 2 µg/ml (Sigma Chemical Co., St. Louis, MO).

For the study of early surface activity Neuro-2a cells were plated and allowed to attach to the collagen-coated coverslips 4h prior to addition of either fresh SM or SM supplemented with BBG and then fixed rapidly 0, 1, 3, 7, or 15 minutes after treatment. Coverslips representative of the various treatment periods were processed for SEM as described below, and screened at low magnification (800X) for comparable fields in a semi-random fashion. The degree of surface activity was evaluated semi-quantitatively on a scale of one to five at a magnification of 3500X: (1) relatively smooth surfaces, with occasional ridges; (2) few microvilli and numerous ridges; (3) moderate number of microvilli; (4) approximately 75% of surface covered with microvilli; and (5) perikaryon almost entirely covered with long microvilli. A minimum of 100 cells per treatment interval were evaluated from two separate experiments. Ultrastructural studies were carried out on cells maintained for 24-48h in SM alone or SM containing BBG. Cytoskeletal disruptive agents were added to SM in the presence or absence of BBG. A neuritogenic control was obtained by serum deprivation for 48h.

Electron Microscopy

Cells for TEM and SEM were fixed and processed as described previously.[2] For SEM the cells were critical-point-dried from liquid CO_2, coated with gold-palladium and examined with a Hitachi S-450 microscope. For WCTEM the cells were plated on Formvar-coated nickel grids previously treated with poly-DL-ornithine, grown for 24h, rinsed free of debris with HBSS and fixed for 1h at room temperature in 2.5% glutaraldehyde in 0.1 M cacodylate buffer (pH 7.3). They were rinsed in buffer, post fixed in 1% osmium tetroxide for 2 min at $4°C$ prior to ethanol dehydration, and critical-point-dried in CO_2. The cells were rotary-coated with carbon to enhance their stability and examined in a Philips 300 microscope at 80 or 100 kV.

RESULTS

Neuro-2a cells plated for 4h on collagen-coated coverslips in SM generally have round perikarya which lack distinctive surface modifications. Slender filopodia less than one cell diameter in length frequently extend from the perikarya and appear to anchor the cells to the underlying substratum (Fig. 1A). Treatment of these cells with BBG for three minutes resulted in a dramatic morphological change (Fig. 1B). The smooth somal membranes were transformed into surfaces that were heavily covered with microvilli. Several distinct surface modifications were observed: (1) ridge-like veils; (2) long microvilli; (3) small bulbous vesicles; and (4) occasional large

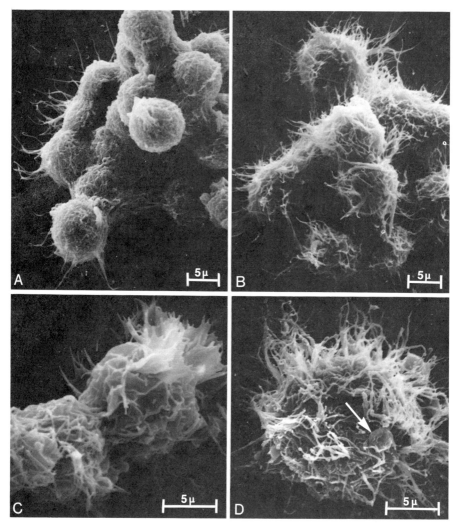

Fig. 1. Neuro-2a cells preplated for 4h in SM (A), then treated with SM and BBG for 3 min (B). BBG treatment induces rapid formation of microvilli on the relatively smooth untreated perikaryal surfaces. Several surface modifications were seen: (C) ridge-like veils; (D) long microvilli and blebs (arrow).

blebs (Fig. 1C & D). Semi-quantitative analysis revealed more than 50% of the untreated cells had surfaces totally devoid of microvilli and ridge-like veils (Fig. 2A & B). Within one minute after the addition of BBG, surface activity increased; only 25% of the cells remained inactive (Fig. 2C & D). After three minutes of ganglioside

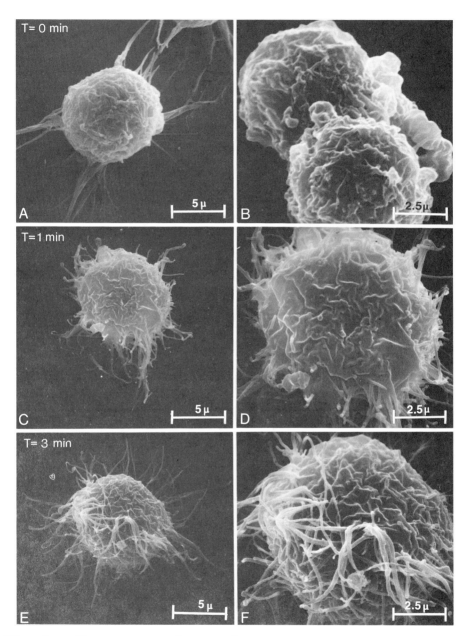

Fig. 2. The exposure of Neuro-2a cells to BBG initiates a rapid highly ordered sequence of surface-related activity resulting in microvillar formation.

exposure less than 2% of the cells appeared quiescent with surfaces lacking microvilli (Fig. 2E & F). Further exposure to BBG increased the number of microvilli per cell (Fig. 3). A small but constant number of cells (less than 2%) remained inactive throughout all the observation periods. A wide range of surface-related activity was observed at each time period (Fig. 3E). Maximum surface activity was observed after 7 minutes exposure to BBG, with most cells exhibiting microvilli; 50% of the cells were almost entirely covered with microvilli (Fig. 3A & B). By 15 minutes the number of microvilli per cell began to decrease although the length appeared equivalent to those found after 7 minutes exposure to BBG (Fig. 3C & D). The decrease was not complete since the dorsal surfaces of Neuro-2a cells were still covered with microvilli after 24h BBG exposure (Fig. 3F).

The cytoskeletal organization of Neuro-2a cells grown for 24h in SM or for 48h in serum-depleted medium consisted of a homogeneous lattice of fine cytoplasmic filaments (Fig. 4A). In contrast, BBG-treated cells displayed a more complex cytoskeleton consisting of numerous well-organized filament bundles, which appear to be distributed into the secondary and tertiary neuritic branches (Fig. 4B). These BBG-related differences were not a function of the degree of neuritogenesis, since control cells with extensive neuritic development (grown in serum-depleted SM) lacked a complex cytoskeleton. Thin sections through BBG-treated cells demonstrated that these bundles consisted of 50Å microfilaments (Fig. 4C). Low temperature pre-treatment prior to fixation produced no apparent changes in the cytoskeletal organization of BBG-treated Neuro-2a cells.

The effects of 24h exposure to SM supplemented with Colcemid alone or Colcemid and BBG were evaluated. Colcemid had no observable effect on the perikaryal surface but prevented filopodial formation (Fig. 5A). The simultaneous treatment with Colcemid and BBG resulted in cells with numerous spine-like projections but lacking filopodial extensions (Fig. 5B). Treatment with cytochalasin D alone for 24h produced cells without surface spines or processes (Fig. 5C). In contrast, the simultaneous exposure of cells to cytochalasin D and BBG for 24h resulted in cells devoid of spines, but exhibiting anomalous neuritic outgrowth consisting of many long, thin, unbranched neurites (Fig. 5D & E). These neurites lacked characteristic growth cones (Fig. 5F) and had a tendency to grow in circular fashion.

Fig. 3. A maximum number of microvilli per cell were present after 7 min BBG exposure (A & B); a decrease in number of microvilli was observed at 15 min (C & D). Frequency distribution histograms of the surface activity of 100 cells per interval was staged on a scale of 1-5 as indicated in the micrographs (E). Although the number of microvilli decreased after 7 min BBG exposure, many were present after 24h (F).

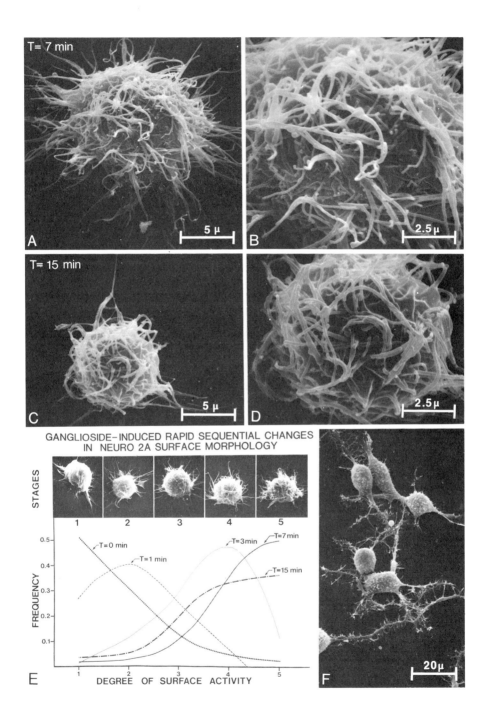

DISCUSSION

We have demonstrated that exogenously applied bovine brain gangliosides (BBG) increase the neuritogenic activity of Neuro-2a cells and initiate a rapidly onsetting sequence of changes in surface topography. These surface-related changes appear to be the most immediate responses to gangliosides yet detected and may reflect primary events in their uptake, lateral mobility, stimulation of neuritogenesis or numerous other functional factors.

Confluent synchronized cultures of Chinese hamster ovary cells have been reported to undergo sequential surface changes during the cell cycle.[22] The Neuro-2a cells used in our study were not synchronized by BBG-treatment and continued to divide, although at a reduced rate after extensive neuritogenesis. Thus the relatively

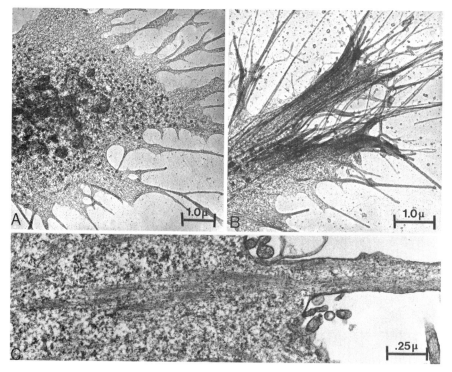

Fig. 4. WCTEM of growth cones: (A) 48h in serum-depleted medium; (B) 24h in BBG supplemented medium. Gangliosides induce the formation of bundles of filaments. (C) TEM of BBG-induced growth cone confirms the microfilamentous nature of the filament bundles.

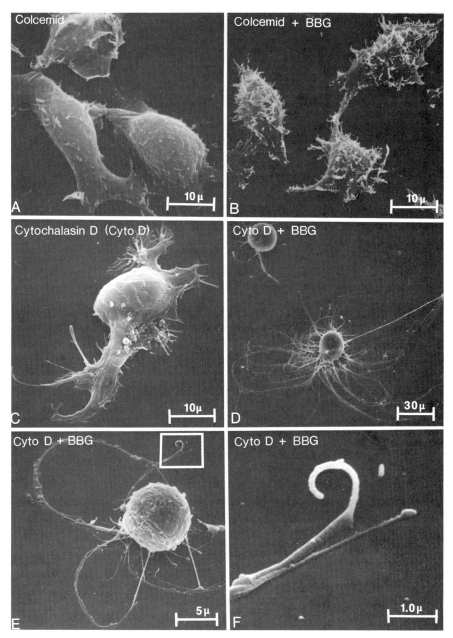

Fig. 5. Neuro-2a cells grown for 24h in SM supplemented with cytoskeletal disruptive agents and BBG as indicated: BBG (250 µg/ml); Colcemid (0.5 µg/ml) and cytochalasin D (2 µg/ml).

uniform responses produced by BBG-treatment are probably not related to changes in the cell cycle. Exposure of PC-12 pheochromocytoma cells to Nerve Growth Factor (NGF) induces an immediate and highly ordered burst of surface ruffling activity which precedes neuritogenesis by many hours.[23] It has been suggested that this series of sequential surface changes is a consequence of the binding and internalization of NGF. The addition of Epidermal Growth Factor to human glioma cells causes a rapid increase in membrane ruffling which is maximal within 5 minutes and then gradually declines but remains higher than that found in control cultures.[24] These studies raise the possibility that ganglioside-induced changes in surface morphology may reflect a general response of cells to growth factors. A further possibility is that the surface changes are more directly related to neurite formation. These changes may reflect an increased fluidity of the membrane either by ligand-induced binding,[25] or the incorporation of ganglioside into the membrane. The capacity of GM1 ganglioside to activate membrane-bound adenylate cyclase[26] and phosphodiesterase activity[27] may be an additional means of mediating these events. The activation of adenylate cyclase might be especially important in light of recent evidence demonstrating that increases in intracellular cAMP resulting in the release of cellular calcium can produce rapid and reversible changes in neurite orientation.[28] Therefore these early surface-associated changes may represent the initial steps in neuritogenesis.

The cytoskeletal organization of BBG-mediated neurite formation was shown via WCTEM to consist of a highly organized filamentous lattice which was absent in untreated controls and neurites formed in response to serum deprivation. The microfilamentous (50 to 70Å) nature of these filaments was revealed by their resistance to cold and by high resolution TEM. The role microtubules (MT) and microfilaments (MF) play in BBG-induced neuritogenesis was examined further with Colcemid[29] and cytochalasin D,[30] respectively. Simultaneous treatment with BBG and Colcemid produced Neuro-2a cells which lacked neurites (but had numerous spine-like projections) and demonstrated the dependence of BBG-induced neurites on MT. In contrast, treatment of these cells with BBG and cytochalasin D resulted in long, thin, circularly oriented, unbranched neurites without typical growth cones. Similar circular patterns of neuritic growth have been observed in cultures of cerebral cortex[31] and retinal explants[32] and have been suggested to reflect the helicity of microtubules and/or neurofilaments. Under these conditions the intrinsic properties of other cytoskeletal elements may emerge to produce the circular growth patterns.

SUMMARY

These studies demonstrate that while microtubules are essential for BBG-mediated neurite initiation and elongation, they are not involved in microfilament-dependent ganglioside-mediated surface activity. Microfilaments may be more directly altered by exogenous gangliosides than microtubules since they are the major structural elements of microvilli and are required for neurite branching. Our studies suggest that normal neuritogenesis requires a delicately balanced interaction between various cytoskeletal elements. Since there is a close relationship between membrane-associated lipid molecules and submembranous cytoskeletal elements,[33] the incorporation of gangliosides into membranes may alter this balance and result in neurite formation. The use of gangliosides to enhance neurite production provides a unique model for the study of nerve development. We have shown that bovine brain gangliosides stimulate an immediate sequence of surface-related changes as well as microtubule and microfilament dependent neurite formation in Neuro-2a cells. However, the precise molecular events by which gangliosides enhance neuritogenesis await further study.

REFERENCES

1. J. R. Moskol, D. A. Gardner, and S. Basu, Changes in glycolipid glycosyltransferase and glutamate decarboxylase and their relationship to differentiation in neuroblastoma cells, Biochem. Biophys. Res. Commun. 61:751 (1974).
2. F. J. Roisen, H. Bartfeld, and M. M. Rapport, Ganglioside mediation of in vitro neuronal maturation, in: "Gangliosides in Neurological and Neuromuscular Function, Development, and Repair", M. M. Rapport and A. Gorio, eds., Raven Press, New York (1981).
3. W. Dimpfel, W. Moller, and U. Mengs, Ganglioside-induced neurite formation in cultured neuroblastoma cells, Ibid.
4. A. Leon, L. Facci, D. Benvengnu, and G. Toffano, Morphological and biochemical effects of gangliosides in neuroblastoma cells, Dev. Neurosci. 5:108 (1982).
5. R. Langenback and S. Kennedy, Gangliosides and their cell density-dependent changes in control and chemically transformed C3H/10T1/2 cells, Exp. Cell Res. 112:361 (1978).
6. J. I. Morgan and W. Seifert, Growth factors and gangliosides: a possible new perspective in neuronal growth control, J. Supramol. Struct. 10:111 (1979).
7. K. Obata, M. Momoko, and S. Handa, Effects of glycolipids on in vitro development of neuromuscular junctions, Nature 266: 369 (1977).
8. M. R. Caccia, G. Meola, C. Cerri, L. Frattola, G. Scarlato, and F. Aporti, Treatment of denervated muscle by gangliosides, Muscle & Nerve 2:381 (1979).
9. A. Gorio, G. Carmignoto, L. Facci, and M. Finesso, Motor nerve sprouting induced by ganglioside treatment. Possible implications for gangliosides on neuronal growth, Brain Res. 197:236 (1980).

10. F. Norido, R. Canella, and A. Gorio, Ganglioside treatment of neuropathy in diabetic mice, Muscle & Nerve 5:107 (1982).
11. N. Spirman, B. A. Sela, and M. Schwartz, Antiganglioside antibodies inhibit neuritic outgrowth from regenerating goldfish retinal explants, J. Neurochem. 39:874 (1982).
12. R. W. Ledeen, Ganglioside structures and distribution: are they localized at the nerve ending?, J. Supramol. Struct. 8:1 (1978).
13. M. Willinger and M. Schachner, GM1 ganglioside as a marker for neuronal differentiation in mouse cerebellum, Dev. Biol. 74: 101 (1980).
14. D. R. Critchley and I. Macpherson, Cell density dependent glycolipids in NIL2 hamster cells, derived malignant and transformed cell lines, Biochim. Biophys. Acta 296:145 (1973).
15. E. Yavin, Z. Yavin, and L. D. Kohn, Temperature-mediated interaction of tetanus toxin with cerebral neuron cultures: characterization of a neuraminidase-insensitive toxin-receptor complex, J. Neurochem. 40:1212 (1983).
16. B. R. Mullin, P. H. Fishman, G. Lee, S. Aloj, F. D. Ledley, R. J. Winand, L. D. Korn, and R. O. Brady, Thyrotrophin-ganglioside interactions and their relationship to the structure and function of thyrotrophin receptors, Proc. Natl. Acad. Sci. USA 73:842 (1976).
17. P. Cuatrecasas, Gangliosides and membrane receptors for cholera toxin, Biochem. 12:3558 (1973).
18. P. H. Fishman, J. Moss, and V. C. Manganiello, Synthesis and uptake of gangliosides by choleragen-responsive human fibroblasts, Biochem. 16:1871 (1977).
19. F. J. Roisen, H. Bartfeld, R. Nagele, and G. Yorke, Ganglioside stimulation of axonal sprouting in vitro, Science 214:577 (1981).
20. J. J. Hauw, S. Fenelon, J. M. Boutry, Y. Nagai, and R. Escourolle, Effects of brain gangliosides on neurite growth in guinea pig spinal ganglia tissue cultures and on fibroblast cell cultures, in: "Gangliosides in Neurological and Neuromuscular Function, Development, and Repair", M. M. Rapport and A. Gorio, eds., Raven Press, New York (1981).
21. W. Seifert, Gangliosides in nerve cell cultures, Ibid.
22. K. Porter, D. Prescott, and J. Frye, Changes in surface morphology of Chinese Hamster ovary cells during cell cycle, J. Cell Biol. 57:815 (1973).
23. J. L. Connelly, L. A. Greene, R. R. Viscarello, and W. D. Riley, Rapid sequential changes in surface morphology of PC12 pheochromocytoma cells in response to nerve growth factor, J. Cell Biol. 82:820 (1979).
24. B. Westermark, A. Magnusson, and C.-H. Heldin, Effect of epidermal growth factor on membrane motility and cell locomotion in cultures of human clonal glioma cells, J. Neurosci. Res. 8: 491 (1983).

25. T. Revesz and M. Greaves, Ligand-induced redistribution of lymphocyte membrane ganglioside GM1, Nature 257:103 (1975).
26. J. W. Daly, The effect of gangliosides on the activity of adenylate cyclase and phosphodiesterase from rat cerebral cortex, in: "Gangliosides in Neurological and Neuromuscular Function, Development, and Repair", M. M. Rapport and A. Gorio, eds., Raven Press, New York (1981).
27. C. W. Davis and J. W. Daly, Activation of rat cerebral cortical 3',5'-cyclic nucleotide phosphodiesterase activity by gangliosides, Mol. Pharmacol. 17:206 (1980).
28. R. W. Gunderson and J. N. Barrett, Characterization of the turning response of dorsal root neurites toward nerve growth factor, J. Cell Biol. 87:546 (1980).
29. F. J. Roisen, W. G. Braden, and J. Friedman, Neurite development in vitro: III. The effect of several derivatives of cyclic AMP, colchicine and colcemid, Ann. N.Y. Acad. Sci. 253:545 (1975).
30. M. Schliwa, Action of cytochalasin D on cytoskeletal networks, J. Cell Biol. 92:79 (1982).
31. H. J. Romijn, M. T. Mud, P. S. Wolters, and M. S. Corner, Neurite formation in dissociated cerebral cortex in vitro: evidence for clockwise outgrowth and autotopic contacts, Brain Res. 192:575 (1980).
32. A. M. Heacock and B. W. Agranoff, Clockwise growth of neurites in retinal explants, Science 198:64 (1977).
33. G. L. Nicolson, Transmembrane control of the receptors of normal and tumor cells. I. Cytoplasmic influence over cell surface components, Biochim. Biophys. Acta 457:57 (1976).

ACKNOWLEDGMENTS

The authors thank Ms. Janet Baxter for her expert technical assistance and Ms. Jacqueline Salomon for typing the manuscript. This work was supported by NIH grants NS11299 and NS11605.

EFFECT OF EXOGENOUS GANGLIOSIDES ON THE MORPHOLOGY AND BIOCHEMISTRY OF CULTURED NEURONS

Henri Dreyfus, Blandine Ferret, Suzanne Harth, Alfredo Gorio*, Louis Freysz and Raphael Massarelli

Unite 44 INSERM and Centre de Neurochimie du CNRS, 5 rue Blaise Pascal
67084 Strasbourg Cedex, France
*Fidia Research Laboratories
Department of Cytopharmacology
Abano Terme, Italy

INTRODUCTION

The role of sialoglycoconjugates in various processes essential for the life and development of the neuron is becoming more frequently apparent. In particular, gangliosides have been involved in a large variety of phenomena ranging from cell to cell recognition and adhesion,[1,2] to differentiation[1,3] and from possible receptors for neurotransmitters and toxins[1,4] to modulators of the movements of solutes across the nerve membranes.[5]

The ensemble of these functions and roles sustain the hypothesis that gangliosides may act as "coding" molecules at the external site of the membrane for the sequence of events which ranges from the recognition of a target cell to the differentiation of the "recognizing" neuron. The presence of complex polysialogangliosides at the external side of the nerve cells is a further support to the hypothesis. A corollary to this is the finding that gangliosides play a major role in the regeneration of nerve fibers, stimulating in particular the "sprouting" of neuronal processes.[6]

A possible approach to the study of the role of gangliosides in nerve cells is the use of a suitable, simplified model of the nervous system represented by nerve cell cultures. Recent developments in the techniques involved in the growth of cells in vitro have led to the possibility of obtaining cultures of neurons grown and differentiated in the absence of supporting cells.[7]

The morphological (including synaptogenesis and related phenomena) and biochemical parameters of these neuronal cultures parallel those known for the neuronal elements of the nervous system in vivo.[8]

The addition of purified gangliosides (either as a mixture of different brain gangliosides or as individual components) to the growth medium of these cell cultures causes important changes in both neuronal morphology and biochemistry. The description of these modifications is the subject of the present report.

RESULTS

Morphology

The morphological aspects of neurons after the addition to the growth medium of a mixture of purified gangliosides (10^{-5} and 10^{-9}M composed of 19.8% GM1, 5.2% GD3, 39.6% GD1a, 14.6% GD1b, 17.6% GT1b, 3.2% GQ1b from Fidia Research Laboratories) is shown in Fig. 1.

Cells were treated with gangliosides at the 3rd day and at the 5th day of culture and the pictures were taken at the 7th day of culture. An increase in the number of cells, in the number of contacts and a "sprouting" effect were observed (and more objectively measured by a Leitz image analyzer). The results of the morphometric analysis showed a striking effect of gangliosides upon some cellular parameters (Table 1).
The number of cells per microscopic field was significantly increased both after 1 (results not shown) and 4 days of treatment while no difference was observed between the two concentrations utilized. The surface of the cell bodies did not change after 1 day of treatment and increased significantly after 4 days. It should be noted however that the control areas were significantly smaller at 7 days rather than in cells at 4 days. 10^{-9}M gangliosides had a more pronounced effect after 4 days of treatment than 10^{-5}M. The number of main processes did not significantly change after 1 day of treatment and increased at 4 days (more so with 10^{-5}M than with 10^{-9}M gangliosides). Again the latter effect was found because of a decrease in the number of main branches of the control cells. The number of secondary processes increased dramatically after 1 day of treatment (not shown) and showed no difference between 10^{-5}M and 10^{-9}M concentrations while a smaller increase was observed at 4 days of treatment with 10^{-5}M compared to the 10^{-9}M treated cells.

There was an increase in the length of the main processes with 10^{-9}M gangliosides after 4 days of treatment but no difference in the length of the secondary processes.

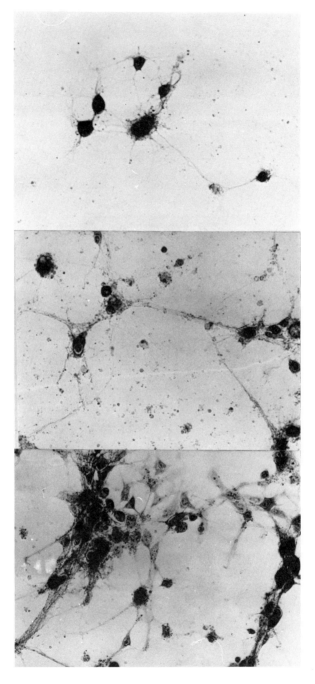

Fig. 1. Morphological aspects of 7-day-old neuronal cells examined with a phase contrast microscope. (a): control cells; (b,c): cells after 4 days treatment (addition at days 3 and 5 of culture) with 10^{-5}M and 10^{-9}M gangliosides, respectively.

Table 1. Morphometric Analysis of Chick Neurons.

	Control	10^{-5}M	10^{-9}M
		(4 days treatment)	
N (nph)	22.6 ± 6.05 (58)	29.79 ± 5.06** (52)	27.41 ± 7.26** (73)
S (μm^2) (n)	24.34 ± 9.73 (300)	31.48 ± 11.80** (300)	34.24 ± 12.81** (300)
Nmp (n)	2.53 ± 1.27 (72)	4.28 ± 1.16** (54)	3.43 ± 1.98** (60)
Nsp (n)	6.29 ± 2.80 (72)	9.76 ± 3.94** (54)	12.23 ± 4.19** (60)
λmp (μ) (n)	46.03 ± 21.86 (149)	40.35 ± 35.90* (232)	77.40 ± 70.35** (153)
λsp (μ) (n)	17.15 ± 8.6 (100)	22.23 ± 109.11* (224)	17.93 ± 10.47 (100)

N : number of cells per microscopic field (photo)
S : area of cell bodies
Nmp : number of main processes
Nsp : number of secondary processes
λmp : length of main processes
λsp : length of secondary processes
nph : number of photos analyses
n : number of determinations

**statistically(*not) significant (Student´s or Wilcoxon´s test performed at the Centre de Statistiques, Strasbourg by Dr. C. Nanopoulos).

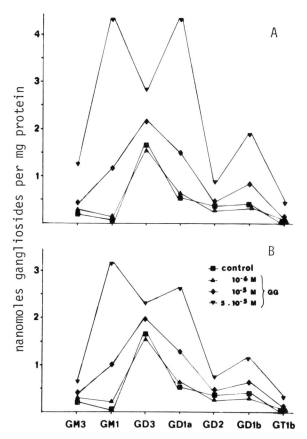

Fig. 2. Distribution of gangliosides of 7 day-old neurons after 4 days treatment (A) and 2 days treatment (B) by different concentrations of gangliosides. Cells were intensively washed with 9% NaCl and medium containing 10% fetal calf serum before ganglioside analysis.[3]

BIOCHEMISTRY

Ganglioside metabolism

The amount of individual gangliosides after incubation of the cells with a mixture of gangliosides changed as a function of the concentration of the exogenous gangliosides. Neurons were treated two days (from days 5 to 7 in culture, Fig. 2b) and 4 days (from day 3 to day 7, Fig. 2a), with 5×10^{-5}M, 10^{-5}M and 10^{-6}M gangliosides and the endogenous content of the individual species was measured on day 7. With 10^{-5}M the amount of GM1 increased 7-fold while the amount of GD1a and GD1b increased 2.5- and 1.5-fold respectively (Fig. 2). These values were doubled when the external concentration of added gangliosides was increased to 5×10^{-5} M (Fig. 2).

No difference was observed between 2 and 4 days of treatment. The effect of the addition of GM1, GD1a and GT1b on the synthesis of the endogenous gangliosides was studied by incubation of the neurons with N-acetyl-D-[U-^{14}C]mannosamine (N-Ac Mann) for 6 or 24 hours (Fig. 3). After 4 days of treatment with 10^{-5}M gangliosides less [^{14}C]-sialic acid was incorporated into gangliosides as could be judged from the decrease of the radioactive incorporation into the total gangliosidic fraction (Fig. 3). The decrease parallels the degree of sialylation of the exogenously added gangliosides and was already present after 6 hours of incubation with N-Ac-[^{14}C]Mann. The ratio of the radioactivity measured at 24 and 6 hours was similar after treatment with the various gangliosides indicating no additional effect of the degree of sialylation.

Influx of choline

An acid-soluble extract from 7 day-old-neuronal cultures grown in the presence of 10^{-5}M gangliosides for 4 days showed a linear incorporation of [^{14}C]choline for 80 min. An increase in

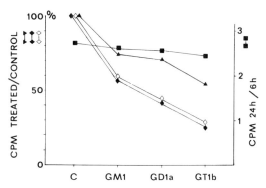

Fig. 3. Incorporation of N-acetyl-[U^{14}C]mannosamine in neuronal cells, control (c) and treated by different gangliosides (GM1, GD1a, GT1b). Gangliosides were added in the medium (◆◆, ◇◇ 10^{-5}M; ▲▲ 5×10^{-6}M) at days 3 and 5. Cells were intensively washed at day 6 and incubated at 37°C with N-Ac-[^{14}C]Mann (254 mCi/mmol, Amersham, 1 μCi/5 ml Sato Bottenstein medium/dish), during 6 h (◇◇) and 24 h (◆◆, ▲▲). Radioactivity linked to gangliosides was determined as described earlier.[5]

radioactive content was observed in treated cells after 30 min of incubation and a similar increase was observed in the radioactive content of the total lipid fraction. A more thorough analysis showed that such increase was present in the free choline, in the phosphorylcholine and acetylcholine compartments (results not shown).

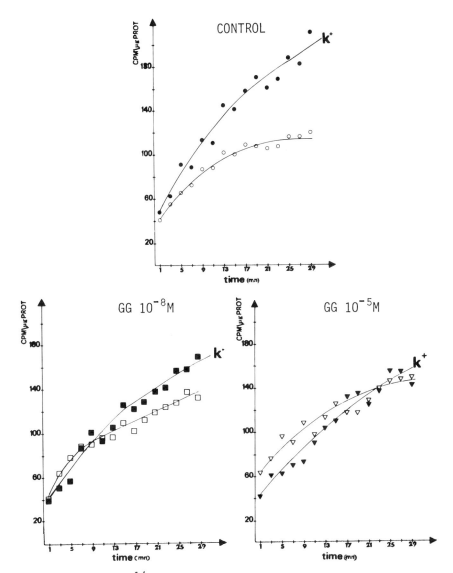

Fig. 4. Efflux of [^{14}C]choline. Cells were incubated in the presence (●,▼,■) or in the absence (○,▽,□) of 50 mM potassium. Incubations were performed as described earlier.[9]

Influx of dopamine

No apparent effect of 10^{-5}M gangliosides was observed on the influx of [^{3}H]dopamine at least from 3 to 30 min of incubation. A possible difference in the shape of the curve after ganglioside treatment, though not significant, might suggest the presence of an effect taking place before 3 min (results not shown). Such a possibility is presently under investigation.

Release of choline

The spontaneous release of choline from nerve cells can be stimulated by 50 mM K^+ (Fig. 4). After treatment with 10^{-5}M and 10^{-8}M gangliosides for 4 days the efflux of choline is not any longer stimulated by an excess of K^+ (Fig. 4).

Dopamine influx, choline influx and release, and separation of choline containing compounds were monitored as described elsewhere.[9]

Release of dopamine

The K^+ evoked release of dopamine observed in control cells, as well as the spontaneous release, was reduced after treatment of the cells with 10^{-5}M gangliosides (Fig. 5).

Release of GABA

Cells preincubated with [^{14}C]GABA released the neurotransmitter spontaneously and reached a plateau at around 9-10 min. Ganglioside treatment (10^{-5}M) increased this spontaneous release as well as the K^+ evoked release[10] while no effect was instead observed with lower concentration of gangliosides (10^{-8}M) (results not shown).

DISCUSSION

Effect of gangliosides on the morphology of neurons

The present data suggest two conclusions: that gangliosides, after their insertion in the membrane, may have an effect on the cellular proliferation and/or on the cellular attachment to the substrate or to other cells and that an increase in the amount of gangliosides in the membrane may have important trophic effects on the neuron. The first conclusion is supported by the increase in the number of cell bodies per microscopic field which may, at first sight, correspond to an increase in cell attachment rather than to an increase in cell proliferation. Gangliosides confer, in fact, a negative charge on the membrane; polylysine,

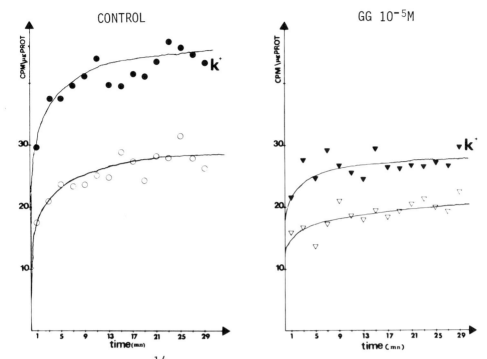

Fig. 5. Efflux of [^{14}C]dopamine. 7 day-old neurons were preincubated 4 hours with 4 µCi of [^{14}C]dopamine (Amersham, 57 mCi/mmole) in 4 ml of culture medium containing 17.6 mg/ml ascorbic acid and 19.6 mg/ml pargyline. After washing with NaCl 0.147 M, cells were incubated in 4 ml Krebs Ringer solution, pH 7.4, at 37°C with or without 50 mM potassium. Every two minutes, the radioactivity contained in 100 µl of the incubation medium was determined in an Intertechnique scintillation spectrometer by adding in a vial 0.4 ml bidistilled water and 10 ml of Rotiszint 22.

the support for the growth of neurons, has essentially positive charges, and thus an increase of negative charges in the neuronal membrane should have an effect similar to the one which has been observed.

The trophic effects were also rapidly observed already after 1 day of treatment; this pertained to the size of the cell bodies, the number of secondary processes, and the length of the primary processes. These results are a reflection of the results on the fluxes of metabolites such as choline and GABA.

Moreover, it was also shown, for other cell types, that the addition of gangliosides to culture medium stimulated morphological differentiation.[11,12]

Effect of added gangliosides on the endogenous metabolism of glycolipids

The results show that the interaction(s) between neurons and added gangliosides engender modifications at the membrane level which result firstly in important changes of the gangliosidic metabolism. This is shown by: a) after an intensive washing of the treated cells with a medium supplemented with fetal calf serum the amount of endogenous gangliosides increased considerably, thus suggesting an incorporation of the added glycolipids into the neuronal membrane (as has been shown) and/or a strong binding to membranes; b) the incorporated gangliosides produce changes in the metabolism of sialic acid terminal as can be judged after incubation with N-Ac-[^{14}C]Mann. The observed decrease in radioactivity may suggest a difference in the accessibility of endogenous and exogenous sialic acid groups. Similarly the micelles containing the different species of gangliosides may not be equally incorporated in the neuronal membranes.

Effect of gangliosides on fluxes of metabolites across the membrane

The trophic role of gangliosides might also be deduced by the increase which has been observed in the intracellular incorporation of choline into choline containing compounds. Such an increase was observed from the 30th min of incubation and further indicates that the effect of the glycolipids may well be on the metabolism of choline rather than only on its transport. A similar increase had already been seen in GABA influx and we have already reported the indication of an increase of the number of sites of transport of GABA after ganglioside treatment.[10] The observed effect moreover appears not to be due to a general modification of the membrane properties since the influx of dopamine did not change.

Similarly the different effects observed on the efflux of choline and the release of dopamine and GABA may indicate a differentiation effect of gangliosides upon these efflux or release mechanisms. However, the possibility cannot be completely excluded that an increase in gangliosidic concentration at the neuronal surface leads to an increased probability of complexation of the NeuAc terminals with Ca^{++} ions. Even if the observed effects were due to such an increased binding of Ca^{++} (thus reducing its entry into the neuron) it should be concluded that the three release mechanisms which have been studied are differently influenced by Ca^{++}. This is true for choline efflux[13] but not so for dopamine and GABA.

In conclusion, the present results show:

a) that gangliosides added exogenously to a growth medium are incorporated and metabolized by neurons,
b) that such an incorporation produces sprouting of neuronal expansions and a possible effect on cell proliferation and/or cell to cell contacts,
c) that there is a trophic effect on the neuronal metabolism and finally,
d) that gangliosides may directly or indirectly intervene on nerve transmission and on the overall dynamics of nerve membranes.

REFERENCES

1. D. R. Critchley, and M. G. Vicker, Glycolipid as membrane receptors important in growth regulation and cell-cell interactions, in: "Dynamic Aspects of Cell Surface Organization," G. Poste and G. L. Nicolson, eds., Elsevier North Holland Biomedical Press, Amsterdam (1977).
2. S. Roseman, The synthesis of complex carbohydrates by multi-glycosyltransferase systems and their potential function in intercellular adhesion, Chem. Phys. Lipids 5:270 (1970).
3. H. Dreyfus, J. C. Louis, S. Harth, and P. Mandel, Gangliosides in cultured neurons, Neuroscience 5:1647 (1980).
4. G. Lee, S. M. Aloj, and L. D. Kohn, The structure and function of glycoprotein hormone receptors: ganglioside interactions with luteinizing hormone, Biochem. Biophys. Res. Commun. 77:434 (1977).
5. H. Dreyfus, J. C. Louis, S. Harth, M. Durand, and R. Massarelli, Role of sialoglycoconjugates in the transport of neurotransmitters, in: "Neural Transmission, Learning and Memory," R. Caputto and C. Ajmone Marsan, eds., Raven Press, New York (1983).
6. A. Gorio, G. Carmignoto, L. Facci, and M. Finesso, Motor nerve sprouting induced by ganglioside treatment. Possible implications for gangliosides on neuronal growth, Brain Res. 197:236 (1980).
7. B. Pettmann, J. C. Louis, and M. Sensenbrenner, Morphological and biochemical maturation of neurons cultured in the absence of glial cells, Nature 281:378 (1979).
8. J. C. Louis, B. Pettmann, J. Courageot, J. F. Rumigny, P. Mandel, and M. Sensenbrenner, Developmental changes in cultured neurons. An ultrastructural and neurochemical study, Exp. Brain Res. 42:63 (1981).
9. J. C. Louis, H. Dreyfus, T. Y. Wong, G. Vincendon, and R. Massarelli, Uptake, transport and metabolism of neurotransmitters in pure neuronal cultures, in: "Neural Transmission, Learning and Memory," R. Caputto and C.

Ajmone Marsan, eds., Raven Press, New York (1983).
10. J. C. Louis, A. Gorio, R. Massarelli, S. Harth, and H. Dreyfus, Effect of gangliosides on the development of neurons in cell cultures, in: "Nervous System Regeneration," B. Haber, J. Regino Perez-Polo, G. A. Hashim and A. M. Guiffrida Stella, eds., Alan R. Liss, Inc., New York (1983).
11. F. J. Roisen, H. Bartfeld, R. Nagele, and G. Yorke, Ganglioside stimulation of axonal sprouting in vitro, Science 214:577 (1981).
12. A. Leon, L. Facci, D. Benvegnu, and G. Toffano, Morphological and biochemical effects of gangliosides in neuroblastoma cells, Develop. Neurosci. 5:108 (1982).
13. T. Y. Wong, D. Hoffmann, H. Dreyfus, J. C. Louis, and R. Massarelli, Efflux of choline from neurons and glia in culture, Neurosci. Lett. 29:293 (1982).

STUDIES OF GANGLIOSIDES IN DIVERSE NERVE CELL CULTURES

W. Dimpfel and U. Otten*

Rudolph Buchheim Inst. f. Pharmakologie
Justus Liebig-Universitat GieBen
6300 Giessen, GFR
*Abtlg. Pharmakologie
Biozentrum, Klingelbergstr
4056 Basel, CH

INTRODUCTION

Gangliosides have been reported to be involved in a great number of biological features.[1] When working with these compounds in tissue cultures a remarkable phenomenon has been observed. The neuroblastoma cell line "neuro 2a" could be induced in the presence of gangliosides to create excessive neurite outgrowth.[2] Therefore the question arose whether this so-called sprouting might be accompanied by changes in ganglioside synthesis or if gangliosides would only provide the receptor for a yet unknown trophic factor.

Two models were chosen to address this question. Firstly primary cell cultures known to produce extensive growth of neurites during the first two weeks of culturing were analysed with respect to their synthesis of different gangliosides. Secondly two phaeochromocytoma cell lines known to respond to the exposure of nerve growth factor (NGF) with neurite growth were analysed in the presence and absence of NGF.

The phaeochromocytoma cell line PC12 was introduced by Greene and Tischler for studying peripheral noradrenergic actions under the well controlled conditions of tissue culture.[3] The cell line ZPH is a subclone of these PC12 cells which shows an induction of tyrosine hydroxylase in response to the exposure to nerve growth factor (to be published).

MATERIAL AND METHODS

For production of primary nerve cell cultures the standard procedure as originally developed in Dr. P. Nelson's laboratory at the National Institutes of Health in Bethesda, U.S.A., was followed.[4] During the course of the reported experiments only 100 mm diameter collagen coated dishes were taken in order to have enough material for analysis. A total of 18 dishes from one dissection (= sister cultures) were grown in the presence of 10% pig serum and analysed by taking three of them at a time every 4 days. No antibiotics were used. During the last 24 hours before analysis they were exposed each to 3 µCi of D-[C^{14}]glucosamine-hydrochloride (Radiochemical Centre Amersham) in the presence of complete medium. Analytic procedures for determination of the ganglioside pattern and its quantitation were performed exactly as described earlier.[5] Thin-layer chromatography was performed exactly as described by Ando et al.[6] Reference gangliosides were obtained from Dr. Madaus, Cologne, GFR. The tumor cell lines PC12 and ZPH were grown on 100 mm diameter collagen coated Falcon plastic dishes in the presence of 5% fetal calf and 10% adult horse serum but without antibiotics. Precursor labelling was done with 3 µCi of radioactive glucosamine-hydrochloride per dish. Remaining procedures were the same as given for the primary cultures. Tumor cells were grown for 6 days in the presence or absence of NGF before addition of the labelled ganglioside precursor glucosamine. The concentration of NGF was 50 ng/ml during these 6 days. For other experimental details see the preceding paper.[5] Nomenclature of gangliosides is given according to Svennerholm.[8]

RESULTS

Primary nerve cell cultures disclose approximately the same ganglioside pattern as has been observed in vivo.[7] Quantitation of the rate of synthesis has been reported for primary cultures at day 15 but nothing is known with respect to changes during the culturing period. In Fig. 1 the time course of the rate of synthesis is documented for a three week period. Whereas the rate of synthesis of the gangliosides GM_1 and GD_{1a} remains rather stable, a continuous drop is observed with respect to GD_{1b} and GT_{1b}, most prominently for GT_{1b}. The monosialo-ganglioside GM_3 shows a sharp drop in synthesis but recovers to a major degree.

The phaeochromocytoma cells synthesize nearly the same types of gangliosides as primary cultures. Fig. 2 documents the difference with respect to the ganglioside GD_3 which is produced instead of the expected GD_{1a}. This semiquantitative result as obtained by the thin-layer chromatogram scan already points to a

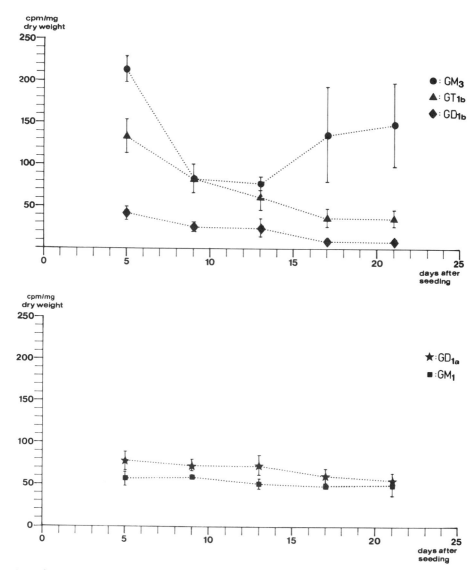

Fig. 1. Synthesis of different gangliosides in primary cultures. The cultures were exposed for 24 hours to radioactive glucosamine. Lipid soluble radioactivity of the cells was analyzed by thin-layer chromatography and quantitated by determination of scraped off material after identification by means of a radioactivity scanner. Abscissa: time of analysis. Ordinate: amount of radioactivity confined to different gangliosides (symbols: mean ± s.d. of N = 3 cultures.

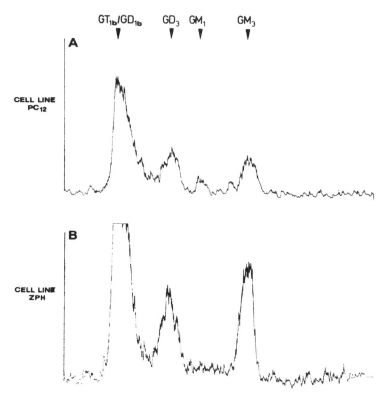

Fig. 2. Separation of different gangliosides by thin-layer chromatography. Comparison of two phaeochromocytoma cell lines PC12 and ZPH. Identification of gangliosides was achieved by running standards in a parallel trace and visualization of the spots by Cu-resorcinol spray.

remarkably large amount of a ganglioside with the mobility of GD_{1b}/GT_{1b}. As expected no clear differences can be observed between the two similar phaeochromocytoma cell clones PC12 and ZPH. Fig. 3 gives the quantitative relationship between the synthesis rate of the different ganglioside types on a dry weight basis and as percentage of total synthesis. The similarity of these two cell lines with respect to ganglioside synthesis is striking. Both cell types synthesize more than double the amount of the most polar gangliosides corresponding to the spot occupied by GT_{1b} and GD_{1b} in primary cultures.

Pretreatment of the tumor cells with nerve growth factor for 6 days did not have any influence on the synthesis rate of ganglioside. Fig. 4 shows a comparison of the two cell lines in response to NGF pretreatment. Despite extensive morphological

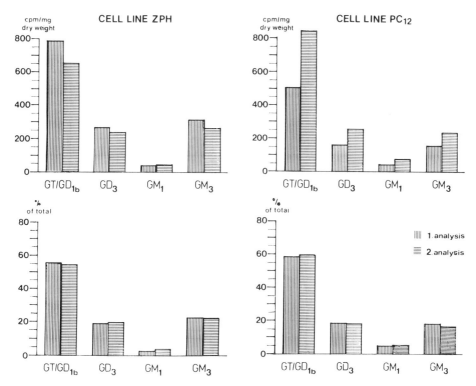

Fig. 3. Synthesis rate of different gangliosides. Identified spots of radioactivity were scraped off from the chromatogram and counted by means of a Beckmann scintillation counter for content of ^{14}C. Two independent culture dishes of each cell type were analyzed separately (note different symbols). Same experiment is shown twice with two different ordinates.

differences as documented in Fig. 5 between NGF-treated and control cultures, no change in ganglioside synthesis could be observed.

DISCUSSION

Originally it was hoped to correlate fiber outgrowth of neurons with changes in ganglioside synthesis. With respect to primary cultures the reported results remain inconclusive, as during the suspected period of extensive neurite formation and synaptogenesis (first two weeks after seeding) no increase of synthesis of any of the different gangliosides took place. On the contrary, polar gangliosides (GD_{1b} and GT_{1b}) dropped from period to

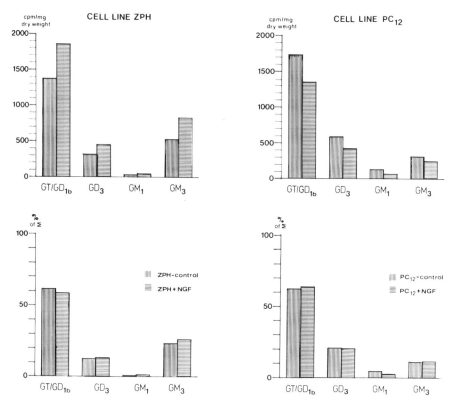

Fig. 4. Synthesis rate of gangliosides in presence of NGF. Similar analysis as in Fig. 3 but comparing cultures treated for 6 days with 50 ng/ml of NGF against controls. Neither of the two cell lines showed a difference in ganglioside synthesis after the differentiation process.

period, whereas GD_{1a} and GM_1 remained rather stable. As these four gangliosides have been associated with neuronal cells,[5] it remains to be asked if we were monitoring a loss of certain neurons carrying especially GD_{1b} and GT_{1b}. In favor of this explanation was the absence of morphologically recognizable neurons in the phase contrast picture at day 21 (not shown). The drop in the synthesis rate of GM_3 can be explained on the basis that we used a mitosis inhibitor between day 5 and 9 aiming at a reduction of the amount of nonneuronal cells in order to prevent overgrowth of fibroblasts. The consecutive recovery of the synthesis of GM_3 is in line with this interpretation. As we did not evaluate fiber outgrowth quantitatively, the data cannot be used to argue for or against the involvement of gangliosides during the process of neurite extension.

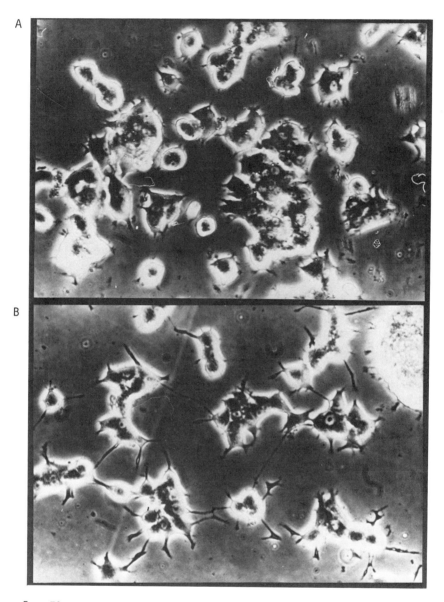

Fig. 5. Phase contrast pictures of PC12 cells in the presence and absence of NGF-treatment for 6 days.
A: control culture. B: NGF treated culture.

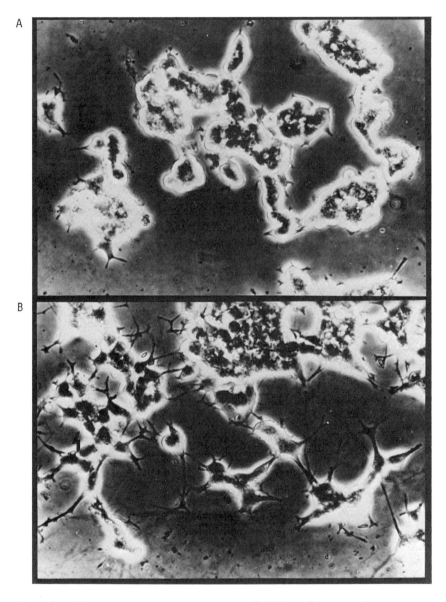

Fig. 6. Phase contrast pictures of ZPH cells in the presence and absence of NGF-treatment for 6 days.
A: control culture. B: NGF treated culture.

The phaeochromocytoma cells are able to synthesize a rather similar pattern of gangliosides in comparison to primary cultures with the exception of producing GD_3 instead of GD_{1a}. But as these two gangliosides do not separate very well with the applied method, we cannot exclude that these cells might synthesize GD_{1a} in small amounts. With regard to the most polar ganglioside resembling GD_{1b}/GT_{1b} according to superficial chromatographic analysis, it has recently been shown by Margolis et al. that this ganglioside is different from GT_{1b} or GD_{1b}, but a more detailed analysis seems to be in progress.[10] However, the synthesis of such a polar ganglioside seems to be rather unique for a tumor cell as all tumor cells checked so far disclose a "simpler" ganglioside pattern. They lack these particular gangliosides. It is tempting to speculate that these two gangliosides might have something to do with catecholaminergic transmission, because the production and release of catecholamines is one of the prominent features of these cells. It would also be interesting to learn more about the stationary concentrations of these gangliosides in order to calculate turn-over rates of these important membrane constituents. Pretreatment of phaeochromocytoma cells with NGF leads to a clearly visible outgrowth of cell processes (Figs. 5 and 6). Thus in this model a possible relationship between gangliosides and neurite outgrowth could be checked. Despite the fact that both cell lines responded to the exposure with NGF with extensive sprouting, no measurable differences were recorded with respect to ganglioside synthesis. It can therefore be concluded that changes in ganglioside synthesis are not needed for neurite outgrowth, but that gangliosides could well function as target molecules for NGF. This suggestion is in line with the results reported by Schwartz and Spirman[9] who found that an anti GM_1-antibody, purified by affinity chromatography, was able to specifically inhibit the NGF induced sprouting in cultured chicken dorsal root ganglion cells. However, a general increase of ganglioside synthesis - along with other changes in glycoprotein synthesis - under treatment of PC12 cells with NGF has also been reported.[10] This result is at variance with ours unless the difference can be explained on the basis of a different reference taken by these authors (total protein instead of nondialysable dry weight as in our experiments).

SUMMARY

Primary nerve cell cultures and tumor cells have been analysed with respect to synthesis of gangliosides. Neither developmental aspects of primary cultures nor nerve growth factor-induced fiber extension in PC12 or ZPH tumor cells gave any evidence of an involvement of gangliosides. Thus, it is concluded that neurite sprouting and ganglioside synthesis may not be interrelated, but that gangliosides might well serve as receptors for trophic factors.

ACKNOWLEDGEMENT

We appreciate the excellent technical assistance of Mr. H. Muller. This work was supported by the DFG (Sonderforschungsbereich 47).

REFERENCES

1. L. Svennerholm, P. Mandel, H. Dreyfus and P.-F. Urban, "Structure and function of gangliosides," Advances in experimental medicine and biology" 125, Plenum Press, New York (1980).
2. W. Dimpfel, W. Moeller and U. Mengs, Ganglioside induced fiber outgrowth in cultured neuroblastoma cells, Naunyn Schmiedebergs Arch. Pharmacol. 308:Suppl. K46 (1979).
3. L. A. Greene and A. S. Tischler, Establishment of a noradrenergic clonal line of rat adrenal phaeochromocytoma cells which respond to nerve growth factor, Proc. Nat. Acad. Sci. USA 73:2424 (1976).
4. W. Dimpfel, Mammalian nerve cell culture, Meth. Enzym. 77:162 (1981).
5. W. Dimpfel, Differential distribution of gangliosides in nerve cell cultures, Cellular and Molec. Neurobiol. 2:105 (1982).
6. S. Ando, N.-C. Chang and R. K. Yu, High performance thin-layer chromatography and densitometric determination of brain ganglioside compositions of several species, Anal. Biochem. 89:437 (1978).
7. W. Dimpfel, R. T. C. Huang and E. Habermann, Gangliosides in nervous tissue cultures and binding of ^{125}I-labeled tetanus toxin, a neuronal marker, J. Neurochem. 29:329 (1977).
8. L. Svennerholm, Chromatographic separation of human brain gangliosides, J. Neurochem. 10:613 (1963).
9. M. Schwartz and N. Spirman, Sprouting from chicken embryo dorsal root ganglia induced by nerve growth factor is specifically inhibited by affinity purified antiganglioside antibodies, Proc. Nat. Acad. Sci. (USA) 79:6080 (1982).
10. R. K. Margolis, S. R. J. Salton and R. U. Margolis, Complex Carbohydrates of cultured PC12 phaeochromocytoma cells, J. Biol. Chem. 258:4110 (1983).

IN-VITRO AND IN-VIVO STUDIES ON GANGLIOSIDES IN THE DEVELOPING AND

REGENERATING HIPPOCAMPUS OF THE RAT

Wilfried Seifert and Hans-Joachim Fink

Max-Planck-Institut fur biophysik. Chemie
Department of Neurobiology
Gottingen, W. Germany, BRD

INTRODUCTION

Gangliosides may have important biological functions in cell-cell recognition during development, for regenerative processes in the nervous system, as receptors or receptor modulators, and possibly in the process of synaptic transmission. In order to investigate some of these aspects we have used in our laboratory nerve cell cultures as model systems for studies on ganglioside functions.[1,2] In spite of their obvious advantages, neuroblastoma cell lines are inadequate models for such studies due to their lack of the higher polysialo-gangliosides[2] and due to their transformed properties as tumor derivatives. We have therefore developed in recent years a primary culture system of the developing rat hippocampus which allows in-vitro studies on development and synapse formation of pyramidal neurons.[3] We report here some results on the ganglioside pattern of these hippocampal neurons. The hippocampus is known for its functional and morphological plasticity. Following entorhinal lesions and resulting degeneration in the area dentata, a regenerative process of axonal sprouting and renewed synapse formation is induced. In a parallel in-vitro study we have investigated changes in ganglioside biosynthesis during this process of regenerative synaptogenesis in the hippocampus of the adult rat.

METHODS

Cell Culture

As a source of hippocampal neurons we have used the hippocampus of the 18-19 day old rat fetus. At this developmental time the pyramidal neurons have just become postmitotic and will develop under appropriate cell culture conditions.[3-5] The serum-free neuronal culture system developed in our laboratory[3] uses rather low cell densities (about $3-4 \times 10^4$ cells per cm^2), polylysine-coated glass coverslips, and a hormone supplemented medium containing insulin, transferrin, hydrocortison and triodothyronine. Under such conditions these cultures contain about 90% neurons and only about 10% glial cells (mostly astrocytes). Cell types have been characterized by immunofluorescence and neurotransmitter uptake studies. Synapse formation starts after 10 days of in-vitro culture as demonstrated by electromicroscopy and electrophysiology.[3] Survival and neurite outgrowth of these neurons depend on the presence of an astrocyte-produced neurotrophic factor (NTF) which we have recently characterized.[3-6] Labeling of cell cultures (initial cell number 1×10^5 cells) was done in a fructose-containing DMEM medium, supplemented with hormones, by applying 10 Ci/ml of ^{3}H-glucosamine for 8 hrs before harvest.

Entorphinal Cortex Lesion

Partial deafferentation of the dentate gyrus of male young adult rats (Sprague-Dawley, 280-320 g) was induced by electrolytic lesions of the entorhinal cortex. The electrode consisted of a teflon insulated nichrome wire. Current (1 mA) was applied for 45 sec at six different sites. Six days later, after injecting 500 Ci of ^{3}H-glucosamine into the brain's right ventricle, the gangliosides were labelled for 20 hrs and the rats were killed by cervical dislocation.

Ganglioside Analysis

Gangliosides were extracted as described by Svennerholm and Fredman.[13] For ganglioside extraction from cell cultures, carrier gangliosides (Sigma, type III) were added to the extraction medium at a concentration of 15 g per 4 - 5 coverslips. The ganglioside fractions were freed from low molecular weight material by passing the extracts through a Sephadex G-50 medium column.[14] Individual gangliosides were separated on HPTLC plates (Merck, Darmstadt) using C/M/0.2% aq. $CaCl_2$ 55:45:10 (v/v/v) as running solvent. Thin layer chromatograms were scanned for radioactivity with a TLC scanner (Berthold, Wildbad). Alternatively, radioactive spots were localized using 3-H Ultrofilm (LKB, Bromma, Sweden). Reference gangliosides were visualized by spraying with resorcinol/HCl.[15]

RESULTS AND DISCUSSION

Cell Culture Studies

Under certain conditions relatively homogenous cell populations can be obtained in culture. This is a considerable advantage of in-vitro systems for ganglioside studies as compared to in-vitro studies dealing with nervous tissue consisting of varying amounts of different cell types. Once the ganglioside pattern of certain classes of nerve cells are known, the results of in-vivo studies can be more clearly interpreted. Therefore we have first determined the ganglioside pattern of glial cells in culture and then of neuronal cells, both obtained from the brain of the developing rat.

An almost pure glial cell culture can be obtained from the cerebral hemispheres of a 1-week postnatal rat. As shown in Fig. 1a, these cultures are dominated by protoplasmic astrocytes which continue to proliferate in-vitro in the presence of serum. Immunofluorescent staining with Anti-GFAP (not shown here) demonstrates that these cultures contain about 80-90% astrocytes.

The ganglioside pattern (Fig. 1b) shows a dominant GM3 peak together with a small peak of GM2. The higher gangliosides are virtually absent. This is in agreement with a previous study from the Strasbourg group[7] on gangliosides of glial cells. Thus astrocytes in culture have a peculiar ganglioside pattern quite different from neuronal cells. The high GM3 content in their plasmamembrane could be used for identification of these cells at postnatal stages in a similar way as the unique galactocerebroside content of oligodendrocytes where a monoclonal antibody against this marker has been used for developmental studies.[8] According to studies by Irwin et al.[9] on the ganglioside pattern of fetal rat and mouse brain, the GM3 content is rather high at early embryonic stages (22% on day 15) but decreases markedly towards the end of prenatal life (4% on day 20). Since the glial cell number is greatly increasing during this time, it seems likely that the remaining GM3 of the postnatal brain is mostly localized on the astrocytes.

For the ganglioside pattern of neuronal cells we have used the serum-free hippocampal cell culture system established in our laboratory.[3] These cultures consist mostly of neurons (about 90% of total cells) as demonstrated by immunofluorescence studies. Autoradiographic GABA-uptake experiments indicate that about 30% of these cells are gabaergic, while the other 60% are pyramidal neurons utilizing the excitatory neurotransmitter glutamate. In presence of the neurotrophic factor NTF[3,5,6] provided by cerebral astrocytes or their conditioned medium these neurons develop in-vitro from round postmitotic cells (taken on day 18 prenatally) to differentiated neurons with long processes which form a neuronal network as shown in Fig. 2a. Labeling of these cultures with ^{3}H-glucosamine for 8 hrs and separation of the labeled gangliosides on HPTLC plates (as described in Methods) reveals the ganglioside pattern of these neurons. We have used a rather sensitive method which allows performance of this analysis on 1×10^5 neurons.

As shown in Fig. 2b, the neurons exhibit the full complement of gangliosides including the higher polysialo-gangliosides. The comparison of the different time points (cultures labeled on day 0, 2, 4 and 6) shows a significant increase in a band running between GD1a and GD1b on days 2 and 4. This band is identified tentatively as GD2. On day 6 there is an increase of GM3 which is probably due

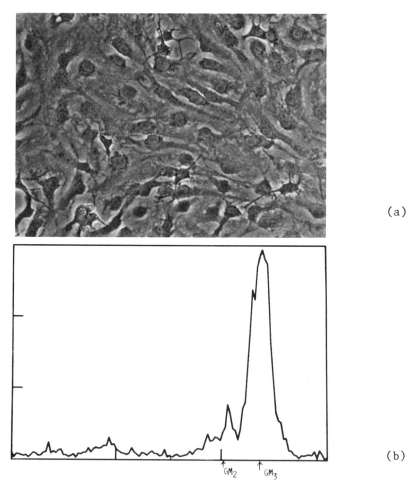

Fig. 1. Glial culture of postnatal rat cerebrum (a). Scanning of radioactive gangliosides from glial culture (b).

to slowly proliferating astrocytes. The neurons in these serum-free cultures and under these labeling conditions are mostly dying around this time. We cannot exclude the possibility that our neurons may contain small amounts of GM3, but it seems likely that most of the GM3 content in these "neuronal cultures" is due to the residual 10% of astrocytes which contain predominantly GM3 (see Fig. 1b).

The presence of individual gangliosides on the surface of these hippocampal neurons can also be demonstrated by immunofluorescence with specific antisera. Unfortunately antisera against the more interesting polysialo-gangliosides are difficult to obtain. Fig. 3 shows an immunofluorescence with a monoclonal antiserum against the ganglioside GQ which was originally developed in Nirenberg's laboratory.[10] In this case we used double-immunofluorescence to stain the neurons and their processes with the anti-GQ (3b) and at the same time eventually present astrocytes with anti-GFAP (3c).

Regeneration Studies

Lesion-induced synaptogenesis in the hippocampal formation, especially in the dentate gyrus, has been studied extensively in recent years as a particular model of CNS synapse plasticity.[11,12] Following lesions in the entorhinal cortex, the afferent fibers from this area to the dentate gyrus (perforant path fibers) degenerate and as a result their synaptic areas in the dendritic molecular layer of granular neurons become vacant. This event then induces axonal sprouting and concomitant formation of new synapses by neighboring fibers, in particular associational and commissural fibers from the CA4 area as well as fiber terminals from the septum.

Since the start of synaptogenesis on day 4 - 5 is rather synchronized following the lesion, this system allows to investigate biochemical mechanisms of regenerative synapse formation in the mammalian CNS. Therefore we have used this model system for studying possible changes of ganglioside biosynthesis under these conditions. Fig. 4 outlines these events in a schematic way. An actual lesion of the entorhinal cortex and the laminar terminal fields of the major afferent pathways are shown in a Timm stained section in Fig. 5.

In our experiments ^{3}H-glucosamine was injected into the lateral ventricle and the gangliosides were labeled for 20 hrs before sacrificing the animal. The ganglioside pattern was analyzed as described in Methods. Fig. 6 shows the resulting scannings of radioactive gangliosides after HPTLC separation. The first lane represents the control (hippocampus from unlesioned rat), the second lane the left hippocampus (unlesioned site) and the third lane the right hippocampus (lesioned site).

The results can be summarized as follows: Compared with normal rat hippocampus, the hippocampus of the lesioned animals showed a striking large amount of incorporated label in one band running between GD1a and GD1b in our HPTLC system. We therefore tentatively identify this band as GD2. This increased labeling was observed both in the ipsilateral and contralateral hippocampus of the lesioned animal, on day 4 and at later time points. Other

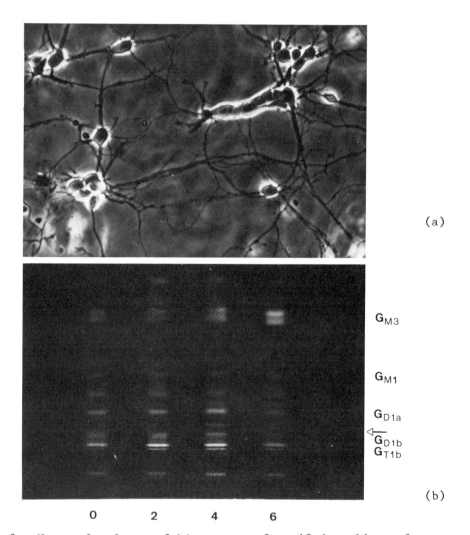

Fig. 2. Neuronal culture of hippocampus from 18 day old rat fetus after 4 days in-vitro (a). Autoradiography of radioactive gangliosides from neuronal cultures on days 0, 2, 4, 6 (b).

a) Phase contrast

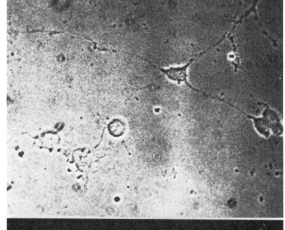

b) Anti-GQ stain
    with monoclonal
    antibody A2B5
    for neurons

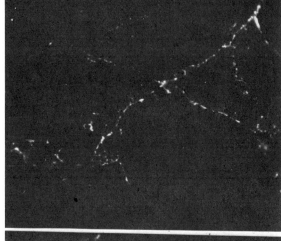

c) Anti-GFAP stain
    for astrocytes

Fig. 3.
Double immunofluo-
rescence of neuro-
nal cell cultures

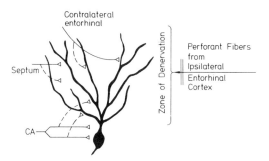

Fig. 4. Reactive synaptogenesis in the dentate gyrus according to Cotman. Broken lines indicate sprouting of nerve endings after lesion of the entorhinal cortex.

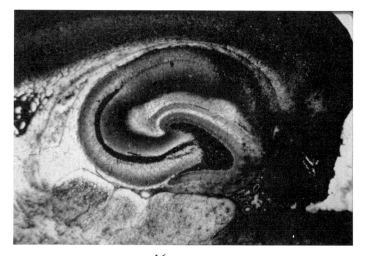

Fig. 5. Timm stained section[16] showing the entorhinal lesion and the hippocampus with laminar terminal fields of the major afferent pathways.

changes in individual gangliosides were also observed but appeared less striking than the GD2 increase (see Fig. 6). Another important observation (not shown here) is a significant increase in total incorporation of label into the ganglioside fraction about 4 days after lesion. This corresponds with the known time course of reactive synaptogenesis following entorhinal lesions.[12]

Both the increased ganglioside biosynthesis and the specific increase in labeled GD2 seem to be biochemical expressions of regenerative synapse formation as a consequence of the lesion stimulus. The fact that the increased synthesis of gangliosides was observed to a lesser extent in the contralateral side, is in agreement with the observation that the entorhinal lesion also induces turnover of synapses in the contralateral side.[12]

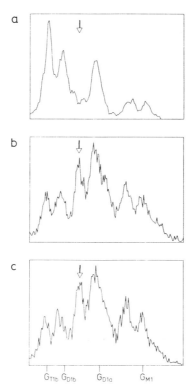

Fig. 6. Scannings of radioactive gangliosides from lesioned animals: a) Control; b) Left hippocampus; c) Right hippocampus.

CONCLUSIONS

Our cell culture studies have demonstrated by using a newly developed neuronal culture system of the developing rat hippocampus that defined populations of CNS neurons can be investigated during in-vitro development with respect to their ganglioside pattern and biosynthesis under controlled conditions.

In contrast to the almost exlcusive GM3 content of astrocytes in culture (Fig. 1a, 1b), hippocampal neurons (mostly of pyramidal type) exhibit the full complement of the higher polysialo-gangliosides (Fig. 2a, 2b). Under the influence of the neurotropic factor NTF (3), these neurons - originally round postmitotic cells - form neurites and develop into neuronal networks with synapse formation starting after about 10 days of in-vitro culture. The most prominent change observed in these cultures was a significant increase in the ganglioside GD2 on day 2 and 4 after plating. Since this is the most active phase of neurite outgrowth in this culture system, we assume that the increase in GD2 reflects or is coupled with the stimulated synthesis of neurite-related membrane molecules. The reason for the particular increase in GD2 biosynthesis, cannot be answered at this time.

Interestingly our studies on ganglioside biosynthesis in the hippocampal formation during lesion-induced synaptogenesis also show a striking increase in GD2 (Fig. 6), together with a general increase of incorporated radioactive precursor into the ganglioside fraction. It seems that under these conditions of lesion-induced axonal sprouting and synapse formation biosynthesis of gangliosides is greatly stimulated and for unknown reasons GD2 biosynthesis is of particular prominence. Thus our studies seem to indicate a striking parallel with regard to ganglioside GD2 biosynthesis between the in-vitro developing hippocampal neurons and the in-vivo regenerative situation of lesion-induced synaptogenesis.

ACKNOWLEDGEMENT

We thank Dr. Greg Rose for his advice and contribution in the lesion work, and Mr. Carlos Pascual for skillful technical assistance.

REFERENCES

1. J. Morgan and W. Seifert, Growth factors and gangliosides: a possible new perspective in neuronal growth control, J. Supramol. Struct. 10:111 (1979).

2. W. Seifert, Gangliosides in nerve cell cultures, in: "Gangliosides in Neurological and Neuromuscular Function, Development and Repair", M. Rapport and A. Gorio, eds., Raven Press, New York (1981).
3. W. Seifert, B. Ranscht, H. J. Fink, F. Forster, S. Beckh, and H. W Muller, Development of hippocampal neurons in cell culture: A molecular approach, in: "Neurobiology of the Hippocampus", W. Seifert, ed., Academic Press, London (1983).
4. H. W. Muller and W. Seifert, Neurotrophic factor (NTF), J. Neurosci. Res. 8:195 (1982).
5. W. Seifert and H. W. Muller, Neuronal-glial interaction: A trophic factor (NTF) for central neurons, in: "Cell Culture Methods for Molecular and Cell Biology", D. Barnes, D. Sirbasku, and G. Sato, eds. (1984)
6. H. W. Muller, S. Beckh, and W. Seifert, Neurotrophic factor (NTF) for central neurons, Proc. Natl. Acad. Sci. USA, submitted (1983).
7. J. Robert, L. Freysz, M. Sensenbrenner, P. Mandel, and G. Rebel, Gangliosides of glial cells: A comparative study of normal astroblasts in tissue culture and glial cells isolated on sucrose-Ficoll gradients, FEBS Lett. 50:144 (1975).
8. B. Ranscht, P. A. Clapshaw, J. Price, M. Noble, and W. Seifert, Development of oligodendrocytes and Schwann cells studied with a monoclonal antibody against galactocerebroside, Proc. Natl. Acad. Sci. USA 79:2709 (1982).
9. L. N. Irwin, D. B. Michael, and C. C. Irwin, Ganglioside patterns of fetal rat and mouse brain, J. Neurochem. 34: 1527 (1980).
10. G. S. Eisenbarth, F. S. Walsh, and M. Nirenberg, Monoclonal antibody to a plasma membrane antigen of neurons, Proc. Natl. Acad. Sci. USA 76:4913 (1979).
11. G. Lynch, Extrinsic influences on the development of afferent topographies in mammalian brain, in: "The Neurosciences", F. O. Schmitt and F. G. Worden, eds., The MIT Press, Cambridge, Mass., USA (1979).
12. C. W. Cotman and M. Nieto-Sampedro, Brain function, synapse renewal and plasticity, Ann. Rev. Psychol. 33:371 (1982).
13. L. Svennerholm and P. Fredman, A procedure for the quantitative isolation of brain gangliosides, Biochim. Biophys. Acta 617:97 (1980).
14. K. Ueno, S. Ando, and R. U. Yu, Gangliosides of human, cat and rabbit spinal cords and cord myelin, J. Lipid Res. 19:863 (1978).
15. L. Svennerholm, Quantitative estimation of sialic acids, Biochim. Biophys. Acta 24:604 (1957).
16. F. Timm, Zur Histochemie des ammonshorngebietes, Z. Zellforschung mikroskop. Anat. 48:548 (1958).

Biomedical Potential of Gangliosides

GANGLIOSIDE TREATMENT OF GENETIC AND ALLOXAN-INDUCED DIABETIC NEUROPATHY

Alfredo Gorio, Ferrante Aporti,* Franco Di Gregorio, Antonella Schiavinato, Renata Siliprandi, and Maurizio Vitadello

Fidia Research Laboratories, Dept. of Cytopharmacology and *Dept. of Biochemistry; Via Ponte della Fabbrica 3/A, 35031 Abano Terme, Italy

INTRODUCTION

Spontaneous diabetes is quite common in animals and man, and it can also be induced by a single injection of alloxan or streptozotocin in experimental animals. The discovery and consequent use of insulin has increased the ability to treat the disease but often its dramatic side effects are unaffected. The neuropathy, developed in diabetes, affects motor, sensory and autonomic components of the nervous system. It is characterized by decreased nerve conduction velocity,[1] axonal degeneration,[2] segmental demyelination and axonal atrophy.[3,4] Its development may be correlated with the degree of duration and severity of hyperglycemia, but often it may not develop even after several years of diabetes.[1,5]

We have decided to investigate the development of such a disease using both the mutant diabetic mouse C57BL/Ks (db/db) and rats 30 days after alloxan injection. The autosomal recessive mutation of the db/db mouse was discovered at the Jackson Laboratory. The earliest abnormality observed is a hyperinsulinemia, occurring already 10 days after birth accompanied by degranulation of beta-cells. A few weeks later the animals are hyperphagic, obese, hyperglycemic and hyperinsulinemic. By four to five months of age they show hypoinsulinemia and hyperglycemia.[6-10] The db/db mouse also develops several symptoms which are common to human diabetic neuropathy. The sensory and motor systems are symmetrically affected with a loss of conduction velocity, axonal atrophy and degeneration.[10-18] In addition, there are changes in slow axonal transport while fast transport seems unaffected.[19,20]

Alloxan induces diabetes by interacting with pancreatic beta-cells and inducing their necrosis.[10,21,22] The animal then develops a neuropathy with several similarities to the genetic one (db/db) although of toxin origin.[23] We decided to investigate the ability of gangliosides to counteract the development of such a pathology. Previous results had shown that these molecules are capable of stimulating axonal sprouting both in cultures and in peripheral and central nervous system during axonal regeneration,[24-32] while in toxic and diabetic peripheral neuropathies[15,33,34] they improve axonal morphometry and conduction velocity.

MATERIALS AND METHODS

Electrophysiology

This part of the study was performed on 153 diabetic mice (db/db) and 58 non-diabetic mice (+/?).

Animals were anaesthetized with thiopental (30 mg/kg i.p.), then the sciatic nerves were dissected and, to eliminate spontaneous firing, bathed in Hanseleit bicarbonate solution, bubbled with 95% O_2 and 5% CO_2, for 10 mins.

Recordings were performed by placing the nerve into a moist chamber onto platinum wire electrodes. The chamber was kept at 37°C in a thermostatically controlled water bath. The proximal end of the nerve was stimulated with supramaximal pulses of 0.1 msec duration and a frequency of 1 per sec delivered from a Digit 3T stimulator (Romagnoli Elettronica, Livorno, Italy). Two recording points were used with the fixed distance of 10 mm. Action potentials were recorded and monitored on a 5115 Tektronix oscilloscope.

Axonal morphometry

Sciatic nerves were quickly dissected and fixed with 2% paraformaldehyde and 2% glutaraldehyde in 0.1 M phosphate buffer for four hours at 4°C; post fixation was carried out with OsO_4 for two hours at 4°C. Semi-thin sections were cut from Agar 1000 included specimens, stained with toluidine blue and photographed. Photographs at x 1500 magnification were used to measure axon diameter with a 624004 A.S.M. particle analyzed (Leitz, Wetzlar, West Germany).

Acoustical cortical evoked potentials

Acoustical Cortical Evoked Potentials (CER) were recorded by means of two Aesculap Michel electrodes, they were placed one on the median line of the cranium and the other on the mastoid zone. The animals were blocked in a holder placed in a sound proof chamber and presented to stimuli of unfiltered clicks, entered at 4 KHz, at a rate of 3/sec. The loud-speaker (Philips AD 0140/T4) was placed 15 cm away and facing the experimental animal. CER were preamplified and evaluated by means of a Nicolett model 1170 neuro-averager. We monitored the threshold, which was considered the minimal sound intensity necessary to evoke an auditory potential and the latencies of the waves 2 and 3 in the brain stem potentials.

Myelin freeze-fracture

Age matched diabetic and control mice were anaesthetized with Pentothal and perfused via the left ventricle with 100 ml Krebs' solution, which contained heparin (200 units/ml) and 0.1% xylocaine. After the Krebs' rinse, fixative constituted by 3% glutaraldehyde and 1% paraformaldehyde solution in 0.1 M phosphate buffer pH 7.4 was perfused. Sciatic nerve was then removed, cut in small bits and placed in the same aldehyde mixture at $4°C$ for a few hours. The tissue was then immersed in glycerol at an increasing concentration up to 30% in 0.1 M phosphate buffer pH 7.4 for at least 30 min. Then the nerve bits were rapidly frozen in freon 22, cooled in liquid nitrogen, fractured and shadowed in a Balzers BAF 400 D apparatus (Balzers, Liechtenstein).

The freeze-fracture replicas were cleaned in a sodium hypoclorite solution, rinsed in distilled water, mounted on a copper grid and examined in a Philips 400 EM. The quantitative evaluation was carried out on photographs 150,000 X of randomly chosen axons.

The number of particles was estimated in a 0.25 μm^2 area and then expressed as number of particles per μm^2.

AChE transport

db/db mice were anaesthetized, the sciatic nerve cut 5 mm from the ischiatic foramen and a 2 mm segment was resected to assay the basal levels of AChE molecular forms. AChE accumulation was allowed to proceed for 24 hours, then 2 mm of the extremity of the proximal stump were resected and the assay carried out.[35]

Specimens of 4 animals were pooled and homogenized in 300 µl of buffer constituted by 1 M NaCl, 0.1 M phosphate buffer pH 7.0, 1% triton X-100; 1 mg/ml bacitracin and 0.16 mg/ml benzamidine were added as protease inhibitors. The homogenate was centrifuged at

20,000 g and tested for AChE activity. The remainder of the supernatant was then layered on a 5-20% linear concentration sucrose gradient and centrifuged at 250,000 g for 17 hours at 4°C in a SW41 Beckman or TST 41 Kontron rotor. The gradients were then divided into 60-70 fractions and assayed for AChE activity. The amounts of each AChE form were then determined and corrected for the concentration normally present within intact nerve.

Muscle reinnervation

Rats were injected with 100 mg/kg of alloxan to induce diabetes. Thirty days later EDL muscle denervation was obtained by crushing the sciatic nerve and the time course of reinnervation monitored by means of standard intracellular recordings, light and electron microscopy. The details of the above technical procedures have been recently published.[24-26]

Collateral reinnervation of soleus muscle was stimulated by partially denervating the muscle with surgical resection of L5 root and monitored by measuring the isometric tension induced by nerve stimulation, as previously described in detail.[27]

Ganglioside treatment

For the NCV and morphometric studies, db/db mice were treated for 30 days with daily i.p. injections of either 1 or 10 mg/kg of gangliosides (GM_1 21%, GD_{1a} 39.7%; GD_{1b} 16% and GT_{1b} 19%). One group was treated at 80 days of age, then a part of it subjected to morphological and NCV analysis; the remaining ones received a second 30 days treatment from 125-155 days of age and were subsequently analyzed as above. Later stages of diabetic neuropathy were analyzed by treating two groups of animals for 30 days, one at 150 days of age and a second at 250. Measurements were performed at 180 and 280 days, respectively. In the collateral reinnervation experiments, rats were treated daily with 5 mg/kg gangliosides from the day of surgery. Gangliosides were purified by HPLC from bovine brain cortex to high purity; no phospholipids or amino acids or peptides were present.

RESULTS

Axonal transport, morphometry and NCV

From as early as 80 days of age up to 220 days the amount of AChE activity per unit length of sciatic nerves from diabetic mice remained constant ranging from 0.50 ± 0.10 units/2 mm of nerve length at 80 days to 0.42 ± 0.04 at 220 days. The absolute values were not different from the control heterozygotes.

Also the distribution of AChE molecular forms (G_1, G_2, G_4 and A_{12}) showed no alterations as expressed by the sedimentation analysis.[19,20] When AChE transport was tested throughout this period of life, it was found that up to 130 days of age the db/db mouse showed no obvious difference from controls. However, at 180 days there was a 10-20% reduction of AChE accumulation in the diabetic animal. This fall was mainly due to reduced accumulation of G_1 and G_2 forms, while G_4 and A_{12} were unaffected. Between 130 and 180 days of age there was significant increase in the transport of G_1 and G_2 forms in control mice, while in db/db mice there was an 85% decrease.[19,20]

Since the two larger forms G_4 and A_{12} are transported at a fast rate, it can be concluded that in such an experimental model fast axonal transport is not impaired, while the flow rate of G_1 and G_2 is clearly affected at 180 days of life.

It must be noted that db/db mice show alterations in nerve conduction velocity (NCV) as early as 80 days after birth, while changes in AChE transport are observed only at 180 days of life. These data suggest that some pathophysiological changes occurred between 130 and 180 days of life and that the disease may have several stages of evolution.

NCV, as mentioned above, and axonal size seem to be affected by the disease at a much earlier stage. We have monitored the development of axonal size throughout the life of both diabetic and heterozygote non-diabetic mice. As shown in Fig. 1, db/db mice show a reduced axonal size compared to the controls, which is due to a higher percentage of small fibers.[36] Both +/? and db/db mice show a progressive increase in axonal size during the animal's growth, diabetic animals exhibiting smaller axons and the difference between the two becoming wider with time and reaching the maximum at 180 days of age. However, the difference at 280 days is hardly significant. This would indicate that at each single age examined the growth rate of db/db mice is slower than controls, as other authors have suggested.[37] Fig. 1 also shows that at 400 days of age there is a large drop in axonal size of both diabetic and heterozygote animals, perhaps due to aging. The extent of this decrease is very similar and there is no difference between the two groups of animals.

We showed in the past that if 150 day old db/db mice were treated with 10 mg/kg gangliosides for 30 days, an improvement of both NCV and axonal morphometry was observed.[15,23] However, the effect of this treatment on axonal morphometry was not significant at 110 days of life, but only at 180.

To evaluate the effect of gangliosides on NCV, we treated the

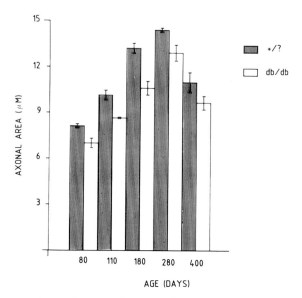

Fig. 1. Mean value of axonal area in three sciatic nerves (about 10,000 myelinated axons) of 80, 110, 180, 280 and 400 day old db/db and +/? mice. Note the slower growth of the db/db mice axons towards maximum mean diameter. The drop observed at 400 day is however similar.

animals for 30 or 60 days prior to electrophysiological testing, as described in Materials and Methods. db/db and +/? mice at 80 days of age were treated daily with either saline or gangliosides for 30 days. At 110 days of age some animals were used to measure NCV, while others, after 15 days of standard rearing, were treated in the same way for another 30 days. Then at the age of 155 days the animals were examined. NCV of +/? mice was 47.9 ± 1.8 at 110 days and 48.0 ± 1.3 at 155 days, while NCV of db/db was 40.5 ± 1.2 at 110 days and 40.4 ± 1.5 at 155. At both stages of treatment, 110 and 155 days of life, no difference was observed between saline and ganglioside treated mice.[36] These results indicate that gangliosides failed to prevent the establishment of neuropathy, while insulin was reportedly active in improving NCV at this stage of db/db mouse life.[16] However, we observed the opposite result at later stages; if animals were treated with 1 mg/kg gangliosides either from 150 to 180 days of life or from 250 to 280 days, NCV improved respectively from 37.5 ± 0.6 to 41.3 ± 0.8 m/sec (180 day old) and from 40.9 ± 1.1 to 44.9 ± 1.0 m/sec (280 day old). The improvement was highly significant at the 0.01 level. NCV of heterozygote was 48.3 ± 0.9 at 180 days and 50.4 ± 1.1 at 280 days of life. At this stage of the disease the animal was reported to be insensitive to insulin.

Myelin freeze-fracture

Diabetic and non-diabetic mice were processed for freeze-fracture analysis at 110 and 180 days of life after 30 days treatment with either 10 mg/kg of gangliosides or saline.

As reported by Fukuma et al.,[38] myelin P face of streptozotocin treated rats shows fewer particles then controls at both ages (Figs. 2, 3). In particular the P face particle density per μm^2 is 480 in the +/? and 445 for db/db mice at 110 days, while at 180 days it is 505 for +/? and 460 for db/db. This deficiency is counteracted by ganglioside treatment at both stages but reaches significance only at 180 days; at this stage the treatment brought back the number of particles to 495 per μm^2 (Fig. 4). Again also in this case the treatment was particularly effective at 180 days of age for the db/db mouse.

Hearing loss in db/db mouse

Hearing loss is one of the complications of diabetes; it is present also in the db/db mouse and is age related. Data obtained by comparison with the heterozygote used as control show that the loss is about 12% at 60 days of life, 22% at 100 days and 64% at 180 days. This loss is indicative of a lesion at the receptive site. In addition to the threshold of auditory evoked response, we found that the latency of the waves 2 and 3 of the Brain Stem Evoked Potentials (BSEP) were delayed, as shown in Table 1. These data suggest also an alteration of central acoustic pathways. If the animals were treated for 30 days with gangliosides and then examined, all the affected parameters recovered significantly (Table 1). Therefore, the treatment improved both the alterations of the receptive system and the conduction velocity of CNS pathways involved in auditory processes.

Alloxan induced diabetic neuropathy

We have previously shown that in alloxan induced diabetes, there is a loss in sciatic nerve conduction velocity accompanied by axonal atrophy; in addition the animals show a loss in auditory response.[23] Ganglioside treatment improved all the above parameters in a significant manner.[23] Since the pharmacologically induced diabetic neuropathy shows a decreased slow axonal transport, it was of interest to see whether regeneration processes were affected. Lasek and Hoffman[39] had proposed that slow transport, in particular the slow component B, determined the rate of axonal elongation; indeed our results indicate that the rate of axonal elongation in diabetes is decreased by about 30%.

As shown in Materials and Methods, we lesioned normal and diabetic rats crushing the sciatic nerve in the same manner and at

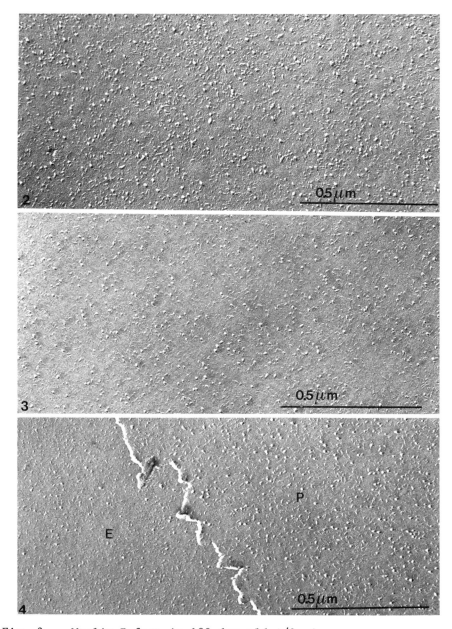

Fig. 2. Meylin P face in 180 day old +/? mice.
Fig. 3. Myelin P face in 180 day old db/db mice.
Fig. 4. Myelin P and E face in 180 day old db/db mice after 30 day ganglioside treatment.

Table 1. Effect of Gangliosides on Brainstem-Evoked Potentials of Diabetic Mice.

B.S.E.P.	% increase from values for db/m		
	Th	L_2	L_3
Sal	63.0 ± 7.0	13.5 ± 3.3	13.5 ± 3.8
G_{10}	37.0 ± 6.0	3.0 ± 2.2	1.7 ± 3.0

Brainstem-evoked potentials (BSEP) of db/db mice are reported as percentage increase from db/m values, which are: Threshold (Th) = 43.3 ± 1.67 dBSPL, latency of the second wave (L_2) = 2.46 ± 0.045 msec; latency of the third wave (L_3) = 3.61 ± 0.090 msec. Data are reported as mean ± SE from 6 animals for db/db, and from 9 and 10 animals, respectively for db/db treated with saline (Sal) and with gangliosides 10 mg/kg (G_{10}). dBSPL = decibel sound pressure level. Reprinted with kind permission of authors and editors, A. Gorio, F. Aporti, and F. Norido, from Ganglioside Treatment in Experimental Diabetic Neuropathy, in: "Gangliosides in Neurological and Neuromuscular Function, Development and Repair", M. M. Rapport and A. Gorio, eds., Raven Press, New York (1981).

the same position. EDL muscle reinnervation occurred between days 13 and 14 in normal animals and around day 20 in the diabetic ones. The rate of synaptic maturation was greatly affected by the disease since resting neurotransmitter release, measured as m.e.p.p. frequency, was only 30% of control 50 days after crush while normal animals had completely recovered.[25] The time course of muscle polyinnervation was however identical between normal and diabetics. The difference was only due to the fact that reinnervation occurs 1 week later in the latter and consequently the whole process is delayed by about 1 week. These results suggest that sprouting capacity of regenerating motoneurons is normal in spite of the slower rate of elongation (Figs. 5, 6). However, collateral reinnervation of the soleus muscle proceeded in a different manner in diabetic animals. In fact, 30 days after partial denervation, muscles of normal rats were fully reinnervated if 8 or more motor units survived surgery. However, alloxan-treated animals failed to do so; some muscles, even with 15 or 20 surviving motor units, were not fully reinnervated. This phenomenon was not observed if the animals were treated daily with 5 mg/kg of gangliosides, since each muscle examined was fully reinnervated if 7 or more motor units survived surgery.

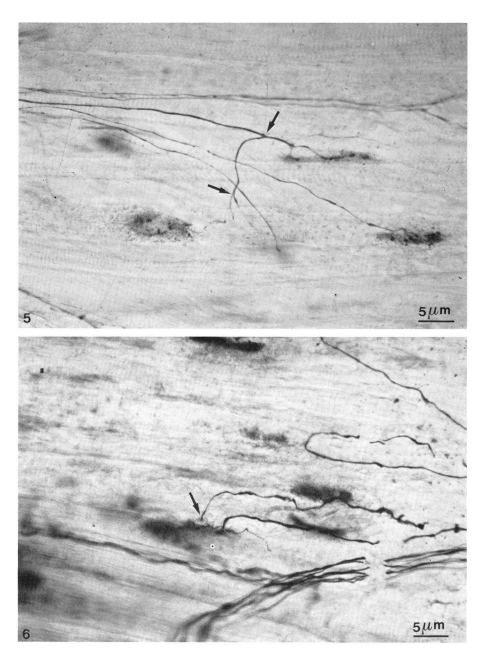

Fig. 5. EDL muscle from alloxan treated rats. 3 days after sciatic nerve crush, muscle was dissected, electrophysiologically monitored, frozen and stained as reported in ref. 26. The picture shows axon sprouting (arrows). Bar = 5 μm.

Fig. 6. EDL muscle as in Fig. 5. The picture shows sprouting and polyinnervation (arrow). Bar = 5 μm.

DISCUSSION

Alloxan and genetic models of diabetes have allowed us to investigate the development and characteristics of diabetic neuropathy and its treatment with gangliosides throughout the life span of the animal. Diabetic neuropathy can be induced by a single injection of alloxan or streptozotocin, the induced axonal damage being reversible with insulin[40,41] as in human juvenile diabetic neuropathy.[1] In such a case, the decreased excitability is accompanied by a drop of axonal transport.[42-44] Similar results were also reported for the db/db mouse.[45,46]

In the latter model, we showed that fast transport of AChE is unaffected up to 220 days of age of the mouse, while the flow rate of G_1 and G_2, which are probably transported at a slow rate, is reduced by 85% at 180 days of life. These results suggest a modification in neuronal cell body metabolism. The axonal size is also reduced at the early stages of the db/db life but the maximal difference is reached at 180 days of life (Fig. 1). However, these changes are not correlated with the establishment of the neuropathy as defined by the NCV, which is already reduced as early as 80 days after birth and the difference is significant at 110 days.

These early changes in NCV are sensitive to brief treatments with insulin. In fact, it was shown that if db/db mice were treated with insulin at the age of 90 days, they showed a recovery in NCV; however, if the treatment was performed later, i.e., at the age of 160 days, a lack of response was observed.[16] These authors called "metabolic" the early stage of the disease. Both our results and those of Robertson and Sima[16] indicate that this second stage is characterized by resistance to insulin treatment and is preindicated by a drop in the transport of the G_1 and G_2 molecular forms of AChE. A further suggestion about the shift in nature of the disease is that now the peripheral nerve becomes sensitive to ganglioside treatment. Prior to 150 days of age, single or double 30 day cycles of treatment failed to change NCV; however, at later stages the effect was remarkable. The improvement of NCV was correlated with recovery in axonal size which, also at previous stages (110 days), was unaffected by the treatment. The fact that this late stage, at which insulin becomes inactive, is marked by a drop in AChE slow transport, would indicate that the disease is now of "neuronal" nature, while previously it was more generally metabolic.

In addition to axonal parameters, ganglioside treatment also affects myelin. Diabetic mice exhibit a lower density of myelin lamellae particles as evidenced by freeze-fracture. Their significance is not very clear but since their number decreases in pathological conditions characterized by NCV alteration, there must be a relationship between their density and nerve function. The efficacy of ganglioside treatment correlates with this statement since at 180 days of age particle density improves significantly along with NCV, while at 110 days both NCV and particle density are not affected. A decreased particle density may indicate a less efficient myelin insulation, which consequently would cause a drop in conduction velocity. In such a case the recovery of NCV and particle density would correlate very well. The effect of the treatment is not restricted to the motor system since also the sensory is affected, as shown by the protective effect on the progressive hearing loss developed by the db/db mouse.

The alloxan induced neuropathy may be different from the one developed by the db/db mouse for two reasons: a) it is sensitive to insulin in spite of the drop in slow transport;[42-44] b) one month after treatment with alloxan the animal is already sensitive to ganglioside treatment, while the db/db mouse must have the neuropathy a long time to become sensitive to ganglioside. Perhaps the clue to understanding the process is the drop in AChE slow transport, which is immediately affected in alloxan treated rats and drastically drops between 130-180 days of life in the db/db mouse. This change may be intrinsically neuronal and independent of both insulin efficacy and glycemic levels. This indicates that when axonal transport is affected the system may be sensitive to gangliosides, while insulin, which is probably active via other mechanisms, is not necessarily pharmacologically active in improving neuronal parameters. Indeed, there are correlations between changes in slow transport and ganglioside efficacy; slow axonal transport is the back-bone of neuronal regeneration and we showed that gangliosides enhance regenerative capacity of neurons in normal and in pathological conditions.[24,27] This change in transport, due to axonal trauma or biochemical disturbances, may be the clue to an understanding of the biological and pharmacological actions of gangliosides. On the other hand it has been reported that GM_1 is present only at the nodes of Ranvier in peripheral nerves;[47] if this is the site of action of gangliosides it is easy to understand how a local alteration may affect NCV and nodal sprouting. Local interactions may enhance enzymatic activities,[48] change permeabilities to important ions such as Ca^{++}[29] or adjuvate the action of trophic factors.[5,24] It has been reported that phosphatidylinositol metabolism is affected in diabetic neuropathy and that a decreased PI turnover may impair (Na^+-K^+) ATPase functions.[49-51] However, if NCV decrease is due to lower activity of the Na-pump, we do not understand why gangliosides would be active only from 180 days of age and not earlier since they have been reported to enhance pump activity in vitro and in vivo.[5,52] Therefore, either the Na-pump is affected only in the second period of the animal's life or the action of gangliosides is not via this mechanism.

SUMMARY

Peripheral neuropathy is a common complication of diabetes. Using the mutant diabetic mouse C57BL/ks (db/db) and alloxan-treated rats, 30 days after intoxication, we investigated development and treatment with gangliosides of such a disease. The db/db mouse develops a neuropathy characterized by a loss in conduction velocity shown as early as 80-90 days after birth and maintained throughout life. At later stages (5-6 months of age) there is a drop in slow transport and myelin particle density. These changes are correlated by a lack of response to insulin treatment, which, prior to this stage, is capable of improving nerve conduction velocity (NCV). On the other hand gangliosides became effective, improving NCV, myelin particle density and sensory perception (auditory deficit) at 5 months of age in the db/db mouse. We presume that this differential neuronal response to insulin and gangliosides indicates a change of the neuropathy from a metabolic stage to neuronal. Alloxan induced diabetic neuropathy is treatable with gangliosides even 30 days after intoxication.

REFERENCES

1. G. Gregerson, Diabetic neuropathy; influence of age, sex, metabolic control and duration of diabetes on motor conduction velocity, Neurology 17:972 (1967).
2. M. J. Brown, J. R. Martin, and A. K. Asbury, Painful diabetic neuropathy. A morphometric study, Arch. Neurol. 33:164 (1976).
3. M. J. Brown, A. J. Sumner, D. A. Greene, S. M. Diamond, and A. K. Asbury, Distal neuropathy in experimental diabetes mellitus, Ann. Neurol. 8:168 (1980).
4. P. J. Dyck, The causes, classification and treatment of peripheral neuropathy, New Engl. J. Med. 307:283 (1982).
5. J. Pirart, Diabetes and its degenerative complications: a prospective study of 4,400 patients observed between 1947 and 1973, Diabetes Care 1:168 (1978).
6. A. A. Like, Spontaneous diabetes in animals, in: "The Diabetic Pancreas", B. W. Volk and K. F. Wellman, eds., Plenum Press, New York (1977).
7. A. E. Renold, Spontaneous diabetes and/or obesity in laboratory rodents, in: "Advances in Metabolic Disorders", R. Levine and R. Luft, eds., Academic Press, New York (1968).
8. L. Herberg and D. L. Coleman, Laboratory animals exhibiting obesity and diabetes syndromes, Metabolism 26:59 (1977).
9. D. L. Coleman, Obese and diabetes: two mutant genes causing diabetes obesity syndromes in mice, Diabetologia 14:141 (1978).
10. J. P. Mordes and A. A. Rossini, Animal models of diabetes, Am. J. Med. 70:353 (1981).

11. D. L. Coleman and K. P. Hummel, Studies with the mutation diabets in the mouse, Diabetologia 3:238 (1967).
12. D. L. Coleman and K. P. Hummel, Lesions physiologique et morphologiques caracterisant le diabete par mutation (db) chez la souris, J. Ann. Diabet. Hotel Dieu. 9:19 (1968).
13. L. Herberg and D. L. Coleman, Laboratory animals exhibiting obesity and diabetes syndromes, Metabolism 26:59 (1977).
14. S. A. Moore, R. G. Peterson, D. L. Felten, T. R. Cartwright, and B. L. O'Connor, Reduced sensory and motor conduction velocity in 25-week-old diabetic C57BL/Ks (db/db) mice, Exp. Neurol. 70:548 (1980).
15. F. Norido, R. Canella, and A. Gorio, Ganglioside treatment of neuropathy in diabetic mice, Muscle & Nerve 5:107 (1982).
16. D. M. Robertson and A. A. F. Sima, Diabetic neuropathy in the mutant mouse C57BL/Ks (db/db). A morphometric study, Diabetes 29:60 (1980).
17. A. A. F. Sima and D. M. Robertson, Peripheral neuropathy in mutant diabetic mouse C57BL/Ks (db/db), Acta Neuropathol. 41:85 (1978).
18. A. A. F. Sima and D. M. Robertson, Peripheral neuropathy in the diabetic mutant mouse. A ultrastructural study, Lab. Invest. 40:627 (1979).
19. M. Vitadello, J. Y. Couraud, R. Hassig, A. Gorio, and L. Di Giamberardino, Axonal transport of acetylcholinesterase molecular forms in sciatic nerve of genetically diabetic mice, in: "Axoplasmic Transport in Physiology and Pathology", D. G. Weiss and A. Gorio, eds., Springer-Verlag, Berlin (1982).
20. M. Vitadello, J. Y. Couraud, R. Hassig, A. Gorio, and L. Di Giamberardino, Axonal transport of acetylcholinesterase in the diabetic mutant mouse, Exper. Neurol., in press (1983).
21. L. J. Fischer and D. E. Rickert, Pancreatic islet-cell toxicity, CRC Crit. Rev. Toxicol. 231 (1975).
22. W. E. Dulin and M. G. Soret, Chemically and hormonally induced diabetes, in: "The Diabetic Pancreas", B. W. Volk and K. F. Wellman, eds., Plenum Press, New York (1977).
23. A. Gorio, F. Aporti, and F. Norido, Ganglioside treatment in experimental diabetic neuropathy, in: "Gangliosides in Neurological and Neuromuscular Function, Development and Repair", M. M. Rapport and A. Gorio, eds., Raven Press, New York (1981).
24. A. Gorio, G. Carmignoto, L. Facci, and M. Finesso, Motor nerve sprouting induced by ganglioside treatment. Possible implications for gangliosides on neuronal growth, Brain Res. 197:236 (1980).
25. G. Carmignoto, M. Finesso, R. Siliprandi, and A. Gorio, Muscle reinnervation. I. Restoration of transmitter release mechanisms, Neuroscience 3:393 (1983).
26. A. Gorio, G. Carmignoto, M. Finesso, P. Polato, and A. Gorio, Muscle reinnervation, II. Sprouting, synapse formation and

repression, Neuroscience 3:403 (1983).
27. A. Gorio, P. Marini, and R. Zanoni, Muscle reinnervation. III. Motoneuron sprouting capacity, enhancement by exogenous gangliosides, Neuroscience 3:427 (1983).
28. A. Gorio and G. Carmignoto, Reformation maturation and stabilization of neuromuscular junctions in peripheral nerve regeneration: the possible role of exogenous gangliosides on determining motoneuron sprouting, in: "Post-Traumatic Peripheral Nerve Regeneration", A. Gorio, H. Millesi, and S. Mingrino, eds., Raven Press, New York (1981).
29. A. Gorio, G. Carmignoto, and G. Ferrari, Axon sprouting stimulated by gangliosides: a new model for elongation and sprouting, in: "Gangliosides in Neurological and Neuromuscular Function, Development and Repair", M. M. Rapport and A. Gorio, eds., Raven Press, New York (1981).
30. G. Ferrari, M. Fabris, and A. Gorio, Gangliosides enhance neurite outgrowth in PC12 cells, Develop. Brain Res. 8:215 (1983).
31. G. Jonsson, H. Kojima, and A. Gorio, GM_1 ganglioside has a counteracting effect on neurotoxin induced alteration of the postnatal development of central serotonin (5 HT) neurons, 7th European Neuroscience Congress, Hamburg 12-16 September (1983).
32. B. Ceccarelli, F. Aporti, and M. Finesso, Effects of brain gangliosides on functional recovery in experimental regeneration and reinnervation, in: "Advance in Experimental Medicine and Biology", G. Porcellati, B. Ceccarelli, and G. Tettamanti, eds., Plenum Press, New York (1976).
33. A. Gorio, G. Carmignoto, G. Ferrari, F. Norido, M. G. Nunzi, R. Rubini, and R. Zanoni, Pharmacological aspects of experimental peripheral neuropathy, in: "International Conference on Peripheral Neuropathies", S. Refsum, C. L. Bolis, and A. Portera-Sanchez, eds., Excerpta Medica, Amsterdam (1981).
34. F. Aporti and M. Finesso, Effetto dei gangliosidi nelle polinevriti tossiche e sperimentali, La Medicina del Lavoro 68:296 (1977).
35. J. Y. Couraud and L. Di Giamberardino, Axonal transport of the molecular forms of acetylcholinesterase in chick sciatic nerve, J. Neurochem. 35:1035 (1980).
36. F. Norido, R. Canella, R. Zanoni, and A. Gorio, The development of diabetic neuropathy in the C57BL/Ks (db/db) mouse and its treatment with gangliosides, submitted to Exptl. Neurology.
37. G. R. Jefferys and P. K. Thomas, Diabetic neuropathy update, Trends in Neurosci. 8 (1981).
38. M. Fukuma, J. L. Carpentier, L. Orci, D. A. Greene, and A. I. Winegrad, An alteration in internodal myelin membrane structure in large sciatic nerve fibres in rats with acute streptozotocin diabetes and impaired nerve conduction

velocity, Diabetologia 15:65 (1978).
39. R. J. Lasek and P. M. Hoffman, The neuronal cytoskeleton, axonal transport and axonal growth, in: "Microtubules and Related Proteins", R. Goldman, T. Pollard, and J. Rosenbaum, eds., Cold Spring Harbor Laboratory, New York (1976).
40. A. K. Sharma and P. K. Thomas, Peripheral nerve structure and function in experimental diabetes, J. Neurol. Sci. 23:1 (1974).
41. D. A. Greene, P. V. Dejesus, and A. I. Winegrad, Effects of insulin and dietary myoinositol on impaired peripheral motor nerve conduction velocity in acute streptozotocin diabetes, J. Clin. Invest. 55:1326 (1975).
42. J. Jakobsen and P. Sidenius, Decreased axonal transport of structural proteins in streptozotocin diabetic rats, J. Clin. Invest. 66:292 (1980).
43. R. E. Schmidt, F. M. Matschinsky, D. A. Godfrey, A. D. Williams, and D. B. McDougal Jr., Fast and slow axoplasmic flow in sciatic nerve of diabetic rats, Diabetes 24:1081 (1975).
44. P. Sidenius and J. Jacobsen, Axonal transport in early experimental diabetes, Brain Res. 173:315 (1979).
45. A. Giachetti, Axoplasmic transport of noradrenaline in the sciatic nerves of spontaneously diabetic mice, Diabetologia 16:191 (1979).
46. S. A. Moore, R. G. Peterson, D. L. Felten, T. R. Cartwright, and N. L. O'Connor, Reduced sensory and motor conduction velocity in 25-week-old diabetic C57BL/Ks (db/db) mice, Exp. Neurol. 70:548 (1980).
47. A. L. Ganser, D. A. Kirschner, and M. Willinger, Localization of gangliosides on the surfaces of peripheral nerve fibers by cholera toxin binding, Accepted to J. Neurocytol. (1983).
48. M. M. Rapport and A. Gorio, "Gangliosides in Neurological and Neuromuscular Function, Development, and Repair", Raven Press, New York (1981).
49. M. E. Bell, R. G. Peterson, and J. Eichberg, Metabolism of phospholipids in peripheral nerve from rats with chronic streptozotocin induced diabetes: Increased turnover of phosphatidylinositol-4,5-biphosphate, J. Neurochem. 39:192 (1982).
50. V. Natarajan, P. J. Dyck, and H. H. O. Schmid, Alterations of inositol lipid metabolism of rat sciatic nerve in streptozotocin induced diabetes, J. Neurochem. 36:413 (1981).
51. P. A. Simmons, A. I. Winegrad, and D. B. Martin, Significance of tissue myo-inositol concentrations in metabolic regulation in nerve, Science 217:848 (1982).
52. F. Aporti, L. Facci, A. Pastorello, R. Siliprandi, M. Savastano, and G. Molinari, Brain cortex gangliosides and (Na^+-K^+) ATPase system of stria vascularis in guinea pig, Acta Otolaryngol. 92:433 (1981).

DOUBLE-BLIND CONTROLLED TRIAL OF PURIFIED BRAIN GANGLIOSIDES

IN AMYOTROPHIC LATERAL SCLEROSIS AND EXPERIENCE WITH PERIPHERAL

NEUROPATHIES

Walter G. Bradley

Department of Neurology
University of Vermont College of Medicine
Burlington, Vermont 05405

INTRODUCTION

As fully reviewed in earlier chapters in this volume, the gangliosides are a complex series of acidic glycolipids, which are concentrated in neuronal cell membranes, and appear to play a role in neuronal excitability, enzyme activity and receptor function. In tissue culture of spinal cord and muscle, exogenous gangliosides enhance the development of neuromuscular junctions.[1] In the human and feline gangliosidoses, the neurons have excessive proliferation of dendritic trees.[2-4] These observations led to the suggestion that exogenous gangliosides might increase regeneration in diseases caused by neuronal degeneration. Bovine brain gangliosides have been reported to enhance axonal regeneration and reinnervation of denervated skeletal muscle following nerve section,[5-9] and in neuropathies in animals[10,11] and in humans.[12-14] It is suggested that ganglioside treatment causes faster reinnervation by stimulating axonal sprouting.[15]

Purified bovine brain gangliosides (Cronassial) has recently been released by the Food and Drug Administration for experimental human trials in the United States. We report here the results of a double-blind controlled trial in 40 patients with amyotrophic lateral sclerosis treated for six months, and of ten patients with a variety of chronic peripheral neuropathies treated in an open trial for a similar period.

MATERIALS AND METHODS

Amyotrophic lateral sclerosis (ALS) trial

For incorporation into the trial, patients were required to have the clinical diagnosis of ALS with both upper and lower motor neuron disease, and no other known etiology. The patients were judged to be likely to survive for the six months of the trial.

Quantitated neurological parameters. A large number of parameters of neurological function were quantitated every month in patients, and these data provided the basis for the analysis of the response to treatment. The degree of upper and lower motor neuron involvement in each limb was separately assessed. Bulbar involvement was graded, as were the tendon reflexes and plantar responses. A research physical therapist performed 70 tests of neurological function in the areas of respiratory function, bulbar function, upper and lower extremity activities of daily living, time motor activities and isometric strength. Respiratory function was quantitated with a respirometer. Forced cough force was rated. Bulbar function was assessed by the time taken to repeat a number of syllables, and swallowing was measured by the time it took to drink four fluid ounces of water through a straw. Ten upper extremity and nine lower extremity activities of daily living were rated by interview. Timed activities of repetition of upper and lower limb function were measured. The maximal isometric strength of 18 muscle groups on each side were measured with a strain gauge tensiometer, and grip strength was measured by a dynamometer. Electromyography and nerve conduction studies were performed in a standard fashion.

Since the rate of progression of ALS varies greatly from patient to patient, great care was taken to balance patients in the two treatment groups. This was achieved by a computer algorithm balancing for four prognostically significant parameters. These were age, combined upper and lower motor neuron grade in all four limbs, the rate of progression of the disease, defined as the percentage of deterioration of upper and lower limb functional assessment composite score divided by the time in days since the onset of the first symptoms, and the presence of bulbar involvement. The computer was programmed to allot each case on a random basis but in such a fashion as to ensure that the intergroup difference was minimized by the additional insertion.

Forty patients were incorporated into the ALS trial, 19 of whom received gangliosides and 21 placebo. Ganglioside patients received 40mg of Cronassial by daily intramuscular injection, and control patients received 2ml of phosphate buffer solution alone.

Data were analyzed with the aid of a computer in a number of ways. Comparison was made between the ganglioside and the placebo groups at each month of the trial for each of the approximately 120 neuromuscular parameters studied using the one-tailed Student's t test. The best straight line fit for the 7 monthly data points for each parameter was derived by the least mean squares method. Comparison was made of the mean slope for each parameter for the two groups. A number of other manipulations were undertaken to try to improve the power of the study, including eliminating patients showing no change during the study period in any one particular parameter, and the grouping of similar functions (bulbar, muscle strength, lower motor neuron score, upper motor neuron score, etc.).

Peripheral neuropathy study

Eleven patients with sensory and/or motor peripheral nerve degenerations were incorporated into a non-blinded study. All had a clinical pattern which was chronically progressive or had remained static for more than 9 months prior to the commencement of the trial. The diagnostic categories are shown in Table 1. Patients received 40 mg of purified brain gangliosides in 2 ml (Cronassial) daily by intramuscular injection. Qualitative and quantitative assessments of neuromuscular function were undertaken monthly as described in the ALS study above.

RESULTS

ALS trial

In the patient's and physician's assessments, three patients of the 40 in the trial showed some improvement. All three were receiving gangliosides, and no patient receiving placebo reported improvement. One patient ceased to deteriorate and began to improve at the third month of the trial. Two patients noted a decrease in the rate of deterioration of their disease. The one patient who noted improvement began to deteriorate again about 6 weeks after withdrawal of the gangliosides, and has subsequently died.

Quantitative neuromuscular parameters showed no systematic statistically significant difference between the two treatment groups in terms of neurological assessment, physical therapy assessment or electrophysiological data. A small number of results achieved statistical significance at the 0.05 level, but these observations were approximately equally divided between the two treatment groups with the exception noted below, and the total number did not exceed the expected 5% level. However, there was a difference between the groups in the slope of deterioration of the 70 neuromuscular assessments performed by the physical therapist. This slope was less in the ganglioside group than in the placebo in 49 parameters, and greater in the ganglioside group than in the placebo in 21 parameters. If these data were independent variables, this observation would be statistically significant ($p<0.001$) by the Mann-Whitney Nonparametric Test. However since these data derive from the same group of patients, they are not independent variables. A computer generated Monte Carlo analysis indicated that the 49-21 separation was due to linkage of data.

Eight patients failed to complete the trial, four in each treatment group. Five patients died and three developed respiratory failure requiring ventilatory support. No toxic side effects of the drug were noted.

Peripheral neuropathy study

Ten of the eleven patients completed three to six months of ganglioside treatment, and one patient withdrew from the study. Physician's assessment indicated improvement in six of the ten patients, no change in one patient, and continuing deterioration in three patients. Of the six assessed to have improved by the physician, five assessed their own status as improved, and one considered himself unchanged. Improvement was seen in several different disease categories (Table 1). Greatest improvement was seen in one case of adrenomyeloleukodystrophy, and two cases of idiopathic sensorimotor polyneuropathy. Improvement appeared in the first month in two patients, in the second month in a further two patients, and in the third month in the remaining two.

An attempt is made in Table 2 to provide an indication of the extent of improvement in each of the categories of parameters. The results are separated into those patients assessed to have improved by their physician, and those assessed to have remained static or deteriorated. Improvement in any parameter was defined as having less abnormality in the final determination than at the initial determination. In the six patients assessed to have improved, this improvement extended over all the parameters. The extent of the improvement however was relatively slight as indicated by the only minor improvement in functional grades. No overall improvement was noted in electromyographic findings. It should be noted that the units in Table 2 are arbitrary, and can only be taken as an indication of the extent of improvement.

Table 1. Diagnosis, Clinical Features, Physician's & Patient's Assessment in Open Trial Cronassial Therapy.

Patient No.	Diagnosis	Age Yrs.	Sex	Duration of disease (yrs.)	Progression[a]	Physician Assessment %[b]	Patient Assessment %[b]
1	Spinal muscular atrophy	19	M	12	C.P.	110	125
2	Post-polio progressive muscular atrophy	47	M	5	C.P.	75	75
3	Recurrent dorsal root ganglion degeneration	40	M	2	22 months	130	130
4	Idiopathic sensorimotor polyneuropathy	66	M	2	C.P.	80	80
5	Idiopathic sensorimotor polyneuropathy	67	M	2	C.P.	150	150
6	Idiopathic sensorimotor polyneuropathy	54	M	4	C.P.	90	90
7	Familial amyloid neuropathy	62	M	8	C.P.	100	100
8	Idiopathic sensorimotor polyneuropathy	75	F	2	C.P.	200	180
9	X-linked adrenomyeloleuko-dystrophy	30	M	8	C.P.	125	160
10	Idiopathic sensory polyneuropathy	51	M	3	C.P.	withdrew	
11	Asymmetric dorsal root ganglion degeneration	45	M	7	12 months	125	100

a: Progression: C.P. = Chronic Progressive, S = static.
b: Assessment as overall state at end of treatment period as percent of state immediately prior to treatment.

Table 2. Clinical & Electrophysiological Response to Cronassial Treatment in 10 Patients With Motor and/or Sensory Neuronal Degenerations.

Parameter	Improved (6 patients)			Static or Deteriorated (4 patients)		
	N	Proportion Improved[a]	Mean ± S.D.[b]	N	Proportion Improved[a]	Mean ± S.D.[b]
Physician Assessment	6	100	143 ± 33	4	0	86 ± 11
Patient Assessment	6	83	149 ± 38	4	0	86 ± 11
Functional Grades	50	12	105 ± 44	40	0	94 ± 14
Functional Measurements	50	90	136 ± 35	35	49	106 ± 16
Strength Measurements	114	61	111 ± 22	84	32	92 ± 24
Sensory Measurements	259	50	155 ± 96	171	8	92 ± 28
Distal Latency and Nerve Conduction Velocity	44	40	195 ± 62	43	37	114 ± 86
EMG	51	4	99 ± 8	51	14	109 ± 47

N = No. of determinations.
a = Proportion of determinations which improved from beginning to end of treatment period, expressed as percentage.
b = Mean of the ratio (last determination): (first determination), expressed as percentage ± S.D.

DISCUSSION

ALS trial

Though there was some qualitative indication that the progression of the ALS was affected by ganglioside treatment in three patients, no statistically significant improvement was demonstrated in the 19 ALS patients receiving 40mg per day of gangliosides for six months compared with 21 computer-matched patients receiving placebo. It is however too early to conclude that gangliosides are ineffective in treating ALS. The dose of gangliosides used in the ALS trial (40 mg per day) was chosen as the maximum available at the initiation of the trial. This dose (approximately 0.5 mg/kg), or less, has been widely used in man, though it is below that shown to be therapeutically effective in animal models (1-50 mg/kg).[10,12] Also it transpired that the power of the study[16,17] may not have been adequate to exclude that the drug had a therapeutic effect. The decision to incorporate 40 ALS patients into this trial was based upon the availability of patients and testing resources. Our data on the rates of deterioration in ALS revealed the clinically well-known variability between patients with this disease. This variance greatly reduces the power of the study, and requires a large number of cases to demonstrate a small therapeutic effect. We calculate that at best 80 patients would be required to exclude a 25% slowing of the rate of deterioration produced by the drug with a p value of less than 0.05. Our experience indicates that a definitive trial would use each patients as his/her own control in a double-blind placebo-controlled sequential study of the rate of deterioration, using a dose of gangliosides of greater than 75 mg per day.

Peripheral neuropathy study

The open study of gangliosides in peripheral nerve sensory and motor neuronal degenerations was undertaken to determine whether any particular disease category was especially responsive to ganglioside therapy. Such an open trial has inevitable drawbacks, including the placebo effect and observer bias. Objective measurements of strength, sensation, function and electrophysiological parameters were used in an attempt to circumvent these drawbacks. Of the ten patients who completed three to six months therapy, six improved on the basis of the physician's assessment of neurological function. Though some learning effect occurs with functional strength testing, the proportion of tests improving by greater than 10% in these six patients exceeded the proportion showing a similar learning effect in six age-matched controls (46% cf 23%; p<0.05). The improvement was however relatively modest in most patients, though one patient with a sensorimotor neuropathy recovered all sensory loss, and one patient with a dorsal root ganglion degeneration went from requiring two canes to walk to being able to run. All the patients have been chosen to have a chronically progressive or static picture such that it would not have been expected that they would show any spontaneous improvement. Several different categories of disease showed improvement, which was therefore not specifically related to any particular diagnosis. In the patient with adrenomyeloleukodystrophy, motor, sensory, autonomic and cerebellar function all improved significantly.

These results in peripheral nervous system degenerations are sufficiently encouraging to warrant extended double-blind controlled trials of gangliosides in the treatment of many diseases of the peripheral nervous system.

SUMMARY

A double-blind control trial of 40 mg purified brain gangliosides in 40 patients with ALS for six months showed no significant improvement with the drug. An open trial of the same dose of ganglioside for the same period showed measurable improvement in six out of ten patients with a variety of peripheral motor and sensory neuronal degenerations. These studies indicate the need for continuing clinical investigation of the therapeutic usefulness of the purified brain gangliosides.

REFERENCES

1. R. Obata, M. Oide, and S. Handa, Effects of glycolipids on in vitro development of neuromuscular junction, Nature 266:369 (1977).
2. D. P. Purpura and H. J. Baker, Neurite induction in mature cortical neurons in feline GM1 ganglioside storage disease, Nature 266:553 (1977).
3. D. P. Purpura, G. D. Pappas, and H. J. Baker, Fine structure of meganeurites and secondary growth processes in feline GM1 gangliosidosis, Brain Res. 143:13 (1978).
4. D. P. Purpura and S. V. Walkley, Aberrant neurite and spine generation in mature neurons in the gangliosidoses, in: "Gangliosides in Neurological and Neuromuscular Function, Development and Repair," M. M. Rapport and A. Gorio, eds., Raven Press, New York (1981).
5. B. Ceccarelli, F. Aporti, and M. Finesso, Effects of brain gangliosides on functional recovery in experimental regeneration and reinnervation, in: "Ganglioside Function: Biochemical and Pharmacological Implications," G.

Porcellati, B. Ceccarelli, and G. Tettamanti, eds., Plenum Publ., New York (1976).
6. A. Gorio, G. Carmignoto, L. Facci, and M. Finesso, Motor Nerve sprouting induced by ganglioside treatment. Possible implications for gangliosides on neuronal growth, Brain Res. 197:236 (1980).
7. B. Ceccarelli, F. Aporti, and M. Finesso, Effects of brain gangliosides on reinnervation of fast-twitch rat skeletal muscle, Adv. Exp. Med. Biol. 83:283 (1977).
8. M. R. Caccia, G. Meola, C. Cerri, L. Frattola, G. Scarlato, and F. Aporti, Treatment of denervated muscle by gangliosides, Muscle & Nerve 2:382 (1979).
9. F. Norido, R. Canella, and F. Aporti, Acceleration of nerve regeneration by gangliosides estimated by the somatosensory evoked potentials, Experientia 37:301 (1981).
10. C. Bulgheroni, M. Maroni, A. Colombi, O. L. Cotrone, V. Foa, R. Gilioli, E. Rota, D. Pellucchette, E. Bottacchi, and A. Volta, Study of the effect of gangliosides on experimental carbon disulfide neuropathy, in: "Peripheral Neuropathies. Devel. Neurol.," N. Canal, et al, eds., Elsevier-North Holland, Amsterdam (1978).
11. F. Norido, R. Canella, and A. Gorio, Ganglioside treatment of neuropathy in diabetic mice, Muscle & Nerve 5:107 (1982).
12. G. Pozza, V. Saibene, V. G. Comi, and N. Canal, The effect of ganglioside administration in human diabetic peripheral neuropathy in: "Gangliosides in Neurological and Neuromuscular Function, Development and Repair," M. M. Rapport and A. Gorio, eds., Raven Press, New York (1981).
13. S. Bassi, M. G. Albizzati, E. Calloni, and L. Frattonla, Electromyographic study of diabetic and alcoholic polyneuropathic patients treated with gangliosides, Muscle & Nerve 5:351 (1982).
14. A. Mingioni, M. Monteleone, A. Pruzzi, O. Soragni, G. Cristiani, C. Moretti, W. Mega, and F. Scanabissi, Research in the use of cerebral gangliosides in neurolysis of the upper limbs, Electromyog. Clin. Neurophysiol. 19:353 (1979).
15. A. Gorio, G. Carmignoto, and G. Ferrari, Axon sprouting stimulated by gangliosides: A new model for elongation and sprouting, in:"Gangliosides in Neurological and Neuromuscular Function, Development and Repair," M. M. Rapport and A. Gorio, eds., Raven Press, New York (1981).
16. J. Cohen, "Statistical Power Analysis for the Behavioral Sciences," Academic Press, New York (1977).
17. M. H. Brooke, G. M. Fenichel, R. C. Griggs, J. R. Mendall, R. Moxley, J. P. Miller, and M. A. Province, Clinical investigation in Duchenne dystrophy: 2. Determination of the "power" of therapeutic trials based on the natural history, Muscle & Nerve In press (1983).

TRIALS OF GANGLIOSIDE THERAPY FOR AMYOTROPHIC LATERAL SCLEROSIS
AND DIABETIC NEUROPATHY

Mark Hallett, Hugh Harrington, H. Richard Tyler,
Thomas Flood,* and Norma Slater*

Section of Neurology, Department of Medicine
Brigham and Women's Hospital
*Joslin Clinic; and Harvard Medical School
Boston, MA, USA

INTRODUCTION

In a number of experimental situations exogenously administered gangliosides have been demonstrated to speed recovery from axonotmesis. First demonstrated for the pre- and post-ganglionic sympathetic fibers of the cat nictitating membrane,[1] it has now been demonstrated also for the rat sciatic nerve[2] and rat tail nerve.[3] The mechanism of action seems to be stimulation of the sprouting process.[3,4] Hence, gangliosides would be expected to be beneficial in conditions such as traumatic nerve injury and mononeuritis from focal nerve infarction. Therapy would reduce the time of disability. Additionally, if regrowth of nerve is sometimes limited by fibrosis in the region of injury, then rapid growth out of this region would improve the quality of recovery.

For types of peripheral nerve disease other than axonotmesis there is no clear expectation that gangliosides would reverse the pathophysiological process. In most conditions, however, the body does make spontaneous attempts at regeneration; to the extent that gangliosides will aid this process the neuropathy will be improved. In any event, some reports have appeared showing efficacy of ganglioside therapy in experimental neuropathies such as that seen in diabetic mice,[5] for symptoms of human neuropathies such as alcoholic neuropathy,[6] and for improvement of electrophysiological parameters in human neuropathies such as diabetic neuropathy.[7,8]

In this chapter we summarize our experience to date with two clinical trials.

TRIAL FOR AMYOTROPHIC LATERAL SCLEROSIS[9]

ALS is a devastating disease, the course of which has never been influenced by any therapy. There are no animal models of this illness which are satisfactory; hence to test the possible influence of ganglioside therapy it is necessary to proceed directly with human trials.

Forty patients with ALS were treated in a double-blind fashion with daily intramuscular injection of either 40 mg of mixed gangliosides (Cronassial) or placebo for six months. Patients were evaluated with manual muscle testing by a neurologist and a physical therapist and with objective muscle testing of certain muscles using dynamometers. Pulmonary function tests were obtained which included forced vital capacity, one-second forced vital capacity, maximum expiratory flow rate, maximum mid expiratory flow rate, maximum inspiratory pressure and maximum expiratory pressure. These pulmonary function tests are measures of the strength of muscles of respiration and have been demonstrated to be useful in evaluating patients with ALS.

Eighteen patients in the ganglioside group and 14 patients in the placebo group finished the entire study and were compared. There were 2 deaths in the ganglioside group and 3 in the placebo group. Both groups declined on all parameters. The analysis was complicated by the fact that the ganglioside group was (by chance) slightly stronger at the beginning and the end of the study. Hence it was necessary to compare rates of decline. Since the scale for manual muscle testing is non-linear, non-parametric methods were needed to analyze these data. The rates of decline were statistically indistinguishable for the two groups on all the tests. No significant toxicity of the gangliosides was noted.

Ganglioside therapy of 40 mg daily for 6 months has no demonstrable influence on the course of ALS.

TRIAL FOR DIABETIC NEUROPATHY

There are several forms of diabetic neuropathy. The most common is a generalized, symmetrical, sensorimotor neuropathy which can be seen in patients with either type I (insulin dependent) or type II (insulin independent) diabetes. Less common is a mononeuropathy, or mononeuropathy multiplex, where individual nerves are affected. Autonomic neuropathies are also seen. The pathophysiology is not known for any of these, but in the mononeuropathies infarcts of nerve have been demonstrated presumably caused by the vascular disease in diabetes. The prognosis of the mononeuropathies is often good since the nerve can regrow after the axonotmestic injury. As noted above gangliosides should be useful in speeding up the process of repair. At the time of writing this chapter we are engaged in a double-blind study of treatment of different types of diabetic neuropathy: generalized neuropathy in type I diabetes, generalized neuropathy in type II diabetes and mononeuropathy multiplex. A preliminary interim analysis of data was performed lumping all the patients together and looking at their features of generalized neuropathy.

Patients were divided into groups by type of neuropathy, and each group was separately randomized to either 40 mg of mixed gangliosides or placebo for daily intramuscular injection for 12 weeks. Patients were selected for study only if symptomatic for their neuropathy, and symptoms of paresthesia, pain and cramps were followed. Motor and sensory nerve conduction of the median nerve, motor conduction of the peroneal nerve and sensory conduction of the sural nerve were carried out periodically, and glycosylated hemoglobin was monitored. Of the forty patients who will eventually be enrolled in the study, the present analysis concerns the first twenty-five to finish the entire three months. There were 13 on drug and 12 on placebo. The average age was similar in the two groups, about 52 years, and while there were about the same number of men and women in the drug group, the placebo group had nine men and three women. The initial glycosylated hemoglobin values were slightly higher in the drug group.

While symptoms in both groups generally improved, patients treated with gangliosides appeared to do somewhat better. Both the frequency and intensity of paresthesias, pains and cramps all improved more in the ganglioside treated group. For example, in relation to frequency of paresthesias, the ganglioside group's symptoms declined 31% to the placebo group's 25%. For intensity of paresthesias, the ganglioside group's symptoms declined 14% to the placebo group's increase of 5%.

There were no discernable changes in the nerve conduction parameters, either in motor or sensory conduction velocity or in the amplitudes of the sensory action potentials or compound muscle action potentials. Perhaps it was overly optimistic to expect to see changes in nerve conduction with this short time period. A warning about the use of nerve conduction in this circumstance has appeared in the literature.[10]

These results, while favorable, are preliminary and are subject to change with more extensive analysis of the complete body of data.

The positive effect in diabetic neuropathy, which is similar to that shown in previous trials noted above, is slight. There has never been a clear idea of what dose of gangliosides is necessary in humans for a good pharmaceutical effect. Perhaps doses of more than 40 mg daily, taken for more than three or six months, would have greater efficacy in this disease or ALS.

SUMMARY

Double blind placebo controlled trials of daily intramuscular injections of 40 mg of mixed gangliosides were carried out in ALS and diabetic neuropathy. Forty patients with ALS were treated for six months and monitored with tests of strength and pulmonary function. No effect of gangliosides was found. Forty patients with symptomatic diabetic neuropathy will be treated for three months. Some of the data from the first 25 patients to complete the study were analyzed in a preliminary fashion. While no difference in nerve conduction studies was found in the treated group compared to the control group, there was greater symptomatic improvement in the patients treated with gangliosides.

ACKNOWLEDGEMENT

These studies were supported by grants from Fidia; important guidance came from Bio/Basics International Corporation.

REFERENCES

1. B. Ceccarelli, F. Aporti, and M. Finesso, Effects of brain gangliosides on functional recovery in experimental regeneration and reinnervation, in: "Advances in Experimental Medicine and Biology," G. Porcellati, B. Ceccarelli, and G. Tettamanti, eds., Plenum Press, New York (1975).
2. A. Gorio, G. Carmignoto, L. Facci, and M. Finesso, Motor nerve sprouting induced by ganglioside treatment: possible implications for gangliosides on neuronal growth, Brain Res. 197:236 (1980).
3. F. Norido, R. Canella, and F. Aporti, Acceleration of nerve regeneration by gangliosides estimated by the somatosensory evoked potentials (SEP), Experientia 37:301 (1981).
4. A. Gorio, G. Carmignoto, and A. Ferrari, Axon sprouting stimulated by gangliosides: a new model for elongation

and sprouting, in: "Gangliosides in Neurological and Neuromuscular Function, Development and Repair," M. M. Rapport, A. Gorio, eds., Raven Press, New York (1981).

5. F. Norido, R. Canella, and A. Gorio, Ganglioside treatment of neuropathy in diabetic mice, Muscle and Nerve 5:107 (1982).

6. B. Mamoli, G. Brunner, R. Mader, and H. Schanda, Effects of cerebral gangliosides in the alcoholic polyneuropathies, Eur. Neurol. 19:320 (1980).

7. G. Pozza, V. Saibene, G. Comi, and N. Canal, The effect of ganglioside administration in human diabetic peripheral neuropathy, in: "Gangliosides in Neurological and Neuromuscular Function, Development and Repair," M. M. Rapport, A. Gorio, eds., Raven Press, New York (1981).

8. S. Bassi, M. G. Albizzati, E. Calloni, and L. Brattola, Electromyographic study of diabetic and alcoholic polyneuropathic patients treated with gangliosides, Muscle and Nerve 5:351 (1982).

9. H. Harrington, M. Hallett, and H. R. Tyler, Trial of ganglioside therapy for amyotrophic lateral sclerosis, Neurology 33(Suppl 2):101 (1983).

10. D. A. Greene, M. J. Brown, S. N. Braunstein, S. S. Schwartz, A. K. Asbury, and A. I. Winegrad, Comparison of clinical course and sequential electrophysiological tests in diabetics with symptomatic polyneuropathy and its implications for clinical trials, Diabetes 30:139 (1981).

TREATMENT OF PAINFUL DIABETIC POLYNEUROPATHY WITH MIXED GANGLIO-

SIDES

>Allan Naarden, Jaime Davidson, Leslie Harris, Jeanne
>Moore, and Stephen DeFelice
>
>Texas Neurological Institute at Dallas
>7777 Forest Lane
>Dallas, Texas 75230

INTRODUCTION

Peripheral neuropathy is a common complication of diabetes mellitus[1] with an incidence of from 15% to 60%.[2,3,4] While estimates vary, as many as 20% of these patients may suffer from considerable painful discomfort.[5] Control of blood glucose has been demonstrated to improve motor nerve conduction in diabetic rats[6] and humans,[7,8] though this may not improve peripheral nerve function.[9] Treatment of painful diabetic neuropathy has had variable success, and as an attempt to deal with this serious problem, this study was initiated.

METHODS

A double-blind study was initiated amongst 20 insulin-dependent diabetic patients whose signs and symptoms were consistent with painful neuropathy. Table 1 outlines the patient population. Of the original 20 patients nerve conduction data from 18 were available for study. One patient's data were technically unsatisfactory (take off points could not be determined) and one patient was lost to follow-up. All patients were evaluated by a neurologist (A.L.N.) at entry and exit from the study. A diabetologist (J.D.) supervised diabetes control which was monitored by serum glucose and glycosylated hemoglobin (Table 2). Bilateral median sensory and motor and peroneal motor nerve conductions were performed by one technician (J.M.) utilizing a TECA TE4 unit (surface temperature monitored by thermistor at 37° C.) at entrance and exit from the study. Laboratory studies

Table 1. Patient Population.

			Treatment Group	Placebo Group
1.	Number:		10	10
		Men	6	5
		Women	4	5
2.	Age:	Mean (yrs)	50.1	44.5
		Range (yrs)	(29-71)	(19-61)
3.	Duration of Diabetes	Mean (yrs)	16.6	20.2
		Range (yrs)	(5-30)	(10-34)

Table 2. Monitoring of Diabetes Control.

Patient #	Age	Duration of Diabetes (Yrs.)	HbAlC(%) Start	End
1	59	11	9.4	9.6
2	38	20	10.1	10.4
3	42	10	7.7	10.8
4	19	17.5	6.5	7.4
5	59	5	5.3	6.9
6	31	17	10.5	10.8
7	55	19	6.6	8.3
8	46	11	9.7	10.5
9	71	14	8.7	8.2
10	40	22	9.1	10.2
11	43	28	9.6	10.2
12	61	17	11.9	12.2
13	52	30	13.2	13.8
14	55	20	8.5	9.4
15	36	34	7.0	8.3
16	46	27	8.8	11.3
17	67	15	13.2	11.1
18	29	12	14.3	10.2
19	56	18	---	---*
20	41	21	9.2	9.3

*Did not complete study

included CBC, SMA-20 and urinalysis. Patients and/or family members were instructed in the daily parenteral administration of mixed gangliosides (40 mg per 2 cc) and were seen at bimonthly intervals in followup for compliance and registration of any side effects.

The duration of the study was three months. At the completion all patients remaining in the study were re-evaluated clinically and electromyographically after a washout period of from three weeks to three months.

RESULTS

Diabetes control was monitored by measurement of glycosylated hemoglobin levels done before, during and at the end of the study (Fig. 1). The glycosylated hemoglobin levels did not change significantly throughout the clinical trial indicating stable diabetes control. During the study patients utilized home blood sugar monitoring as a means of glycemic control and no significant changes were observed. Overall there was no statistically significant alteration of physical examination, though there were several individual patients who had dramatic changes while receiving mixed gangliosides:

J.O. Marked gait improvement, dramatic improvement in pain; return of potency.

J.T. Marked improvement in pain relief; return of potency.

The most significant changes occurred in nerve conduction studies. Median sensory conductions improved an average of 4 m/s (0 to -8 m/s) during treatment and declined during washout on

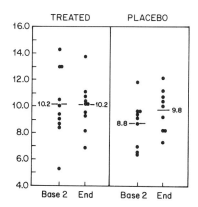

Fig. 1. Content (%) of glycosylated hemoglobin (HbA1C).

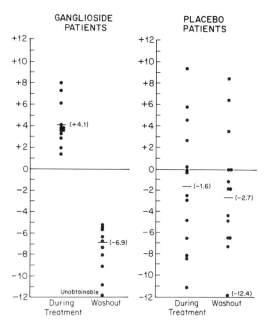

Fig. 2. Median sensory nerve conduction velocity change from baseline during treatment and then washout in meters per second.

average by 6.9 m/s (+5 to -11 m/s). The placebo treated group declined by an average of 1.6 m/s (+9.4 to -11 m/s) during treatment and during washout, fell further by an average of 2.7 m/s (+8.6 to -12.4 m/s) (Fig. 2). Median motor conduction in the treated group improved on average by 5.2 m/s (-5 to +12.2 m/s) and fell after a washout period by 5.5 m/s (0 to -12.4 m/s) while the placebo group essentially showed no change during treatment (Fig. 3). The peroneal nerve conductions revealed an improvement during treatment of 0.8 m/s (+4.8 to -3.4 m/s) which fell during washout by 3.9 m/s (+5.8 to -15.7 m/s) while the placebo group fell by 3.2 m/s during treatment (+4.6 to -15.6 m/s) and for unknown reasons improved after washout by 2.7 m/s (+8.4 to -6 m/s).

These results were subjected to statistical analysis (Table 3A). Analysis of this table indicates clear evidence of an improvement in conduction in the median sensory and peroneal motor nerve conductions and a significant drop during washout in conduction of all nerves (Figs. 2-4). Changes in latencies (Table 3B) paralleled the nerve conduction changes except for a significant change in median motor latency (though no concomitant change in nerve conduction was seen) and there was no significant change in peroneal motor latency (Fig. 5). In examining the data we

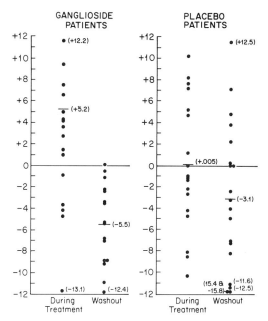

Fig. 3. Median motor nerve conduction velocity change from baseline during treatment and then washout in meters per second.

noted that when changes occurred they were, in general, more pronounced on the right side than the left (Fig. 6). Hence, we decided to also present the results for analyses done using right side measurements only (Tables 4A and B). Had the right side only been examined, the effect of drug would have been even more pronounced.

Side effects were minimal. Half of the patients receiving gangliosides, and 20% of patients receiving placebo complained of an increase in the pain during the first two weeks of treatment, though no patient wished to stop treatment because of this. This complaint dissipated within one or two weeks of onset. One patient receiving gangliosides developed hematuria but a urological evaluation determined that this was related to prostatism and not drug treatment. One patient receiving placebo had a change in her menstrual cycle.

DISCUSSION

While the etiology of diabetic polyneuropathy remains to be completely elucidated, evidence is increasingly mounting that a

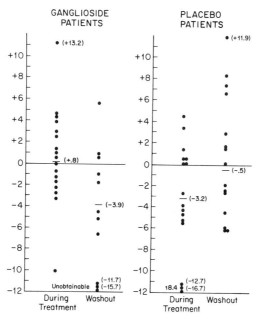

Fig. 4. Peroneal nerve conduction velocities (m/sec).

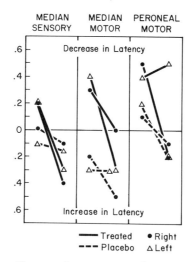

Fig. 5. Change in average latency (msec).

Table 3A. Summary of Conduction Velocity Analysis (Average of Right and Left Side by Nerve).

	Main Effect		Washout	
Parameter	Change in velocity m/sec	p-Value	Change in velocity m/sec	p-Value
Median sensory	5.0	NS	4.7	NS
Median motor	1.7	NS	3.6	0.06
Peroneal	4.9	0.07	4.5	0.10

Table 3B. Summary of the Latency Analysis (Average of Right and Left Side by Nerve).

	Main Effect		Washout	
Parameter	Latency (msec)	p-Value	Latency (msec)	p-Value
Median sensory	0.27	0.02	0.25	0.05
Median motor	0.57	0.02	0.29	0.08
Peroneal	0.07	NS	0.01	NS

metabolic disturbance underlies this abnormality.[2] Spencer et al.[10] have suggested that there is a disturbance in energy metabolism underlying polyneuropathies, while Winegrad et al.[11] have summarized evidence that this may be true for diabetes as well. They have speculated that underlying this is a reduction in Na^+,K^+)-ATPase activity.

These changes in energy metabolism have been associated with alterations in levels of myoinositol, sorbitol and fructose.[12] Controversy surrounds the role of these sugars since Dyck et al. found variable levels of sorbitol, fructose and myoinositol in endoneural studies, and could not relate levels of these substances to clinical, neurophysiological or pathological severity of the neuropathology.[13] Winegrad et al. have found results that are quite opposite to this, while attempts at increasing endoneural myoinositol by oral myoinositol[14] or by using an aldose reductase inhibitor[15] have also left opposite results.

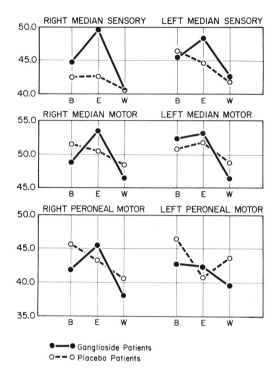

Fig. 6. Average nerve conduction velocities (m/sec). B, beginning; E, end; W, washout.

The pathological correlate of these biochemical changes is most prominent in the distal part of the peripheral nerve and involves a loss of myelinated and unmyelinated axons and segmental demyelination.[1] These changes may in fact be a secondary phenomenon and axonal degeneration may be the earlier and more pertinent abnormality.[16] Brown et al. found, in three patients with painful diabetic neuropathy, that the most striking abnormalities were to be found in unmyelinated and small myelinated fibers, and that unmyelinated nerve fiber sprouting was evident.[16]

Experimental parenteral administration of gangliosides in diabetic mice improved nerve conduction and returned fiber size to control values.[18]

Gangliosides have been demonstrated to have a direct effect on axonal sprouting[17,18] as well as on cellular differentiation,[19] neuronal growth,[20] regeneration and transformation,[21] neuronal maturation[22] and axonal transport.[23] The mechanism by which gangliosides improve nerve conduction remains obscure, but

Table 4A. Summary of Conduction Velocity Analysis (Right Side Only).

Parameter	Main Effect		Washout	
	Change in velocity (m/sec)	p-Value	Change in velocity (m/sec)	p-Value
Median sensory	6.0	0.005	6.5	0.005
Median motor	4.7	0.05	4.7	0.05
Peroneal	5.6	0.07	5.7	0.06

Table 4B. Summary of Latency Analysis (Right Side Only).

Parameter	Main Effect		Washout	
	Latency (msec)	p-Value	Latency (msec)	p-Value
Median sensory	0.93	0.005	0.87	0.005
Median motor	0.52	0.05	0.02	NS
Peroneal	0.26	NS	0.12	NS

indirect evidence suggests a modification of neuronal membrane structure.[24]

Our results extend the observations of Pozza et al.[25] on the effect of gangliosides on improving nerve conduction in diabetic patients. In addition, we have once more documented an asymmetry in nerve conductions in what is often thought to be a clinically symmetrical illness. This ought to alert investigators to the need to do bilateral studies.

SUMMARY

1. We studied 18 patients with painful diabetic neuropathy in a double-blind study of 40 mg per day of mixed gangliosides.*

---

* Cronassial, Fidia Research Laboratories

2. Diabetes control was maintained throughout by analysis of serum glucose and glycosylated hemoglobin levels.

3. Median motor and sensory, and peroneal motor conductions were evaluated in placebo and treated groups before and after a treatment period of three months. All conductions were performed by one technician on a TECA-4 EMG machine with surface temperature controlled at 37°C.

4. There was a definite improvement in nerve conductions in the treated group, particularly noted in the median sensory conductions.

5. We have demonstrated a difference between right and left-sided conductions in the same patients confirming that this illness, at least from an electrophysiological point of view, is asymmetric. Clinical improvement was variable but when present was dramatic.

6. Side effects of this drug were minimal. Half of the patients complained of a transient increase in pain during the first two weeks of treatment. No patient stopped the drug because of this complaint.

We conclude that in this three-month study mixed gangliosides caused a significant improvement in some nerve conductions without significant side effects. Further studies seem warranted to determine the nature and extent of this effect.

REFERENCES

1. P. J. Dyck, The causes, classification and treatment of peripheral neuropathy, New Engl. J. Med. 307:283 (1982).
2. P. K. Thomas and S. G. Eliasson, Diabetic neuropathy, in: "Peripheral Neuropathy," Dyck, Thomas and Lambert, eds., W. B. Saunders, (1972).
3. W. G. Oakley, D. A. Pyke, R. B. Tattersall, and D. J. Watkins, Longterm diabetes. A clinical study of 92 patients after 40 years, Quarterly J. Med. 196:145 (1974).
4. A. T. Paz-Guevera, T. H. Hsu, and P. White, Juvenile diabetes after 40 years, Diabetes 24:559 (175).
5. J. Pirart, Diabetes mellitus and its degenerative complications, Diabetes Care 1:168 (1978).
6. A. Winegrad and D. A. Greene, Diabetic polyneuropathy New Engl. J. Med. 295:1416 (1976).
7. R. Graf, J. B. Halter, M. A. Pfeifer, E. Harlar, F. Brozovich, and D. Porte, Glycemic control and nerve conduction abnormalities in non-insulin dependent diabetic subjects, Ann. Int. Med. 94:307 (1981).

8. E. Harlar, R. Graf, J. Halter, F. Brozovich and T. Soine, Diabetic neuropathy: A clinical, laboratory and electrodiagnostic study, Arch. Phys. Med. Rehab. 63:298 (1982).
9. F. J. Service, J. Daube, P. C. O´Brien, B. R. Zimmerman, C. J. Swanson, M. D. Brennan, and P. J. Dyck, Effect of blood glucose control on peripheral nerve function in diabetic patients, Mayo Clin. Proc. 58:283 (1983).
10. P. S. Spencer, M. I. Sabris, H. Schaumberg, and C. L. Moore, Does a defect of energy metabolism in the nerve fibers underlie axonal degeneration in polyneuropathies? Ann. Neurol. 5:501 (1979).
11. A. Winegrad, D. Simmons, and D. B. Martin, Has one diabetic complication been explained? Editorial, New Eng. J. Med. 308:152 (1983).
12. A. A. F. Sima, Structural and functional characterization of the neuropathy in the spontaneously diabetic BB-Wistar rat, in: "Excerpta Medica," Y. Goto, A. Horuchi, and K. Kogue, eds., Amsterdam, (1982).
13. P. J. Dyck, W. B. Sherman, L. M. Hallcher, J. Service, P. O´Brien, L. Grina, P. Palumbo, and C. Swanson, Human diabetic endoneural sorbitol, fructose, and myoinositol related to sural nerve morphometry, Ann. Neurol. 8:590 (1980).
14. G. Gregerson, B. Bertelsen, H. Harbo, E. Larsen, J. Rud Andersen, A. Helles, M. Schmiegelow, J. Just Christensen, Oral supplementation of myoinositol: Effects on peripheral nerve functions in human diabetics and on concentration in plasma, erythrocytes, urine and muscle tissue in human diabetics and normals, Acta Neurol. Scand. 67:164 (1983).
15. R. G. Judzewitsch, J. B. Jaspan, K. S. Polonsky, C. R. Weinberg, J. B. Halter, E. Halar, M. A. Pfeifer, C. Vukadinovic, L. Bernstein, M. Schneider, K.-Y. Liang, K. H. Gabbay, A. H. Rubenstein, D. Porte, Jr., Aldose reductase inhibition improves nerve conduction velocity in diabetic patients, New Engl. J. Med. 308:119 (1983).
16. M. Brown, J. R. Martin, A. K. Asbury, Painful diabetic neuropathy, Arch. Neurol. 33:164 (1976).
17. F. Roisen, H. Bartfeld, R. Nagele, and G. Yorke, Ganglioside stimulation of axonal sprouting in vitro, Science 214:577 (1981).
18. A. Gorio, G. Carmignoto, L. Facci, M. Finesso, Motor nerve sprouting induced by ganglioside treatment. Possible implications for gangliosides on neuronal growth, Brain Res. 197:236 (1980).
19. S. Ando, N.-C. Chang, and R. K. Yu, High performance thin-layer chromatography and densitometric determination of brain ganglioside compositions of several species, Anal. Biochem. 89:437 (1978).

20. P. H. Fishman, J. Moss and V. C. Manganiello, Synthesis and uptake of gangliosides by choleragen responsive human fibroblasts, Biochem. 16:1871 (1977).
21. P. H. Fishman, R. O. Brady, and S. A. Aronson, A comparison of membrane glycoconjugates from mouse cells transformed by murine and primate RNA sarcoma virus, Biochem. 15:201 (1976).
22. F. Roisen, H. Bartfeld, and M. Rapport, Ganglioside mediation of in vitro neuronal maturation, in: "Gangliosides in Neurological and Neuromuscular Function, Development and Repair," M. M. Rapport and A. Gorio, eds., Raven Press, New York, (1981).
23. M. Kalra and J. DePalma, Ganglioside induced acceleration of axonal transport following nerve crush injury in the rat, Neuroscience Letters 34:1 (1982).
24. G. Tettamanti, Ganglioside and receptor activity (Abstract), Muscle and Nerve Nov. (1978).
25. G. Pozza, V. Saibene, G. Comi, and N. Canal, The effect of ganglioside administration in human diabetic peripheral neuropathy, in: "Gangliosides in Neurological and Neuromuscular Function, Development and Repair," M. M. Rapport and A. Gorio, eds., Raven Press, New York (1981).

GANGLIOSIDE (CRONASSIAL) THERAPY IN DIABETIC NEUROPATHY

Steven H. Horowitz

Department of Medicine
Long Island Jewish-Hillside Medical Center
New Hyde Park, New York 11042

INTRODUCTION

Preliminary experimental and clinical studies suggest that cerebral gangliosides promote peripheral nerve regeneration and produce symptomatic and electrophysiological improvement in patients with diabetic neuropathy. However, in view of the chronicity of this neuropathy and the limitation of those clinical studies to 40 days duration,[1,2] more extensive investigations are desirable to sustain the efficacy of ganglioside therapy. To this end Cronassial[R], a combination of GM1, GD1a, GD1b, and GT1b (Fidia Research Laboratories, Italy), was administered with a prolonged double blind controlled protocol.

PATIENTS

Forty patients with symptomatic diabetic distal symmetrical neuropathy are participating in this study. At its completion 20 will have received 40 mg of Cronassial and 20 a saline placebo, intramuscularly on a daily basis, for the first six months and each patient can elect to receive 40 mg of Cronassial daily in an open drug trial for a minimum of another six months. At present 25 patients (12 receiving Cronassial and 13 receiving placebo) have completed the double blind controlled part of the study and it is their data that are the subject of this presentation. The pertinent personal data on these patients are presented in Table 1.

ADMISSION CRITERIA

The following criteria were employed for admission into this study: 1) patients aged 18 to 75 excepting fertile females

Table 1a. Patient Clinical Data

	Sex	Age	Duration of Diabetes (yr)	Duration of Neuropathic Symptoms (yr)	Insulin Dependence	HgbA$_1$C (%) Start	End
				CRONASSIAL GROUP			
1.	F	36	18.5	5.0	+	3.9	6.4
2.	M	46	9.5	2.0	0	4.4	5.0
3.	M	72	12.5	1.0	0	3.7	2.9
4.	M	52	30.0	6.0	+	6.5	6.2
5.	F	65	6.0	4.0	0	7.3	7.5
6.	M	54	28.0	11.0	+	5.7	6.7
7.	M	50	48.0	5.0	+	7.5	6.7
8.	F	53	20.0	6.0	+	5.2	6.2
9.	M	57	12.0	5.0	0	9.1	9.5
10.	M	29	25.0	2.0	+	4.5	4.9
11.	M	61	14.0	3.0	0	8.3	8.1
12.	M	45	4.0	4.0	+	5.5	3.2
		51.7	19.0	4.5	7	6.0	6.1

Table 1b. Patient Clinical Data

	Sex	Age	Duration of Diabetes (yr)	Duration of Neuropathic Symptoms (yr)	Insulin Dependence	HgbA$_1$C (%) Start	End
				PLACEBO GROUP			
1.	M	74	14.0	1.0	+	5.3	4.0
2.	M	46	7.0	10.0	+	3.1	4.1
3.	M	74	2.0	1.0	0	3.6	4.2
4.	F	61	25.0	12.0	+	5.5	5.5
5.	F	53	7.0	1.0	+	7.9	3.4
6.	M	56	4.0	1.5	+	3.3	3.6
7.	M	57	5.0	1.5	+	3.0	4.2
8.	F	62	19.0	10.0	+	6.0	9.1
9.	M	42	14.0	6.0	+	8.7	6.6
10.	F	48	29.0	1.0	+	7.2	4.7
11.	F	46	12.0	1.0	+	9.3	7.8
12.	F	62	2.0	1.0	0	4.2	4.3
13.	F	52	4.0	1.0	0	5.0	4.6
		56.4	11.1	3.7	10	5.5	5.1

in the child-bearing years; 2) stable diabetes mellitus of at least two years duration; 3) symptoms of peripheral nerve dysfunction (paresthesias, dysesthesias, numbness, weakness) for at least one year; 4) no other etiology for the neuropathy; 5) electrophysiological evidence of peripheral nerve dysfunction. However, once into the study, it was not deemed ethically proper to allow patients in poor control (HbA1C levels >5.9% or 2 hour post-prandial glucose levels >300 mg%) to continue as such, and these patients were encouraged to improve their control. Occasional patients were able to do so.

Informed consent was obtained from all patients.

METHODS

The clinical evaluation consisted of a 22-point neurological symptom score and a 1000-point semi-quantitative neurological examination score. This evaluation plus routine blood studies, urinanalysis, two hour post-prandial glucose and HbA1C levels (Koenig, et al.[3] modification of the Trivelli, et al.[4] method) were performed at 0, 3, 6, 9, 12, 18 and 26 weeks.

The following electrophysiological studies employing the DISA 1500 EMG system were conducted at 0, 6, 12, and 26 weeks. Electromyography was performed in two to four distal and proximal muscles in the arms and legs. Sensory conduction studies, utilizing near-nerve electrode techniques were performed on the right sural nerve from dorsum pedis to lateral malleolus and from lateral malleolus to midcalf, on the left superficial peroneal nerve from ankle to capitulum fibulae, and on the left median nerve from Digits I and III to wrist and elbow. Left deep peroneal and left median nerve motor conduction velocities and distal latencies were obtained. The results were considered abnormal if they fell outside two standard deviations from the means of controls. The median nerve sensory evoked potential responses during 30 minutes of ischemia were also measured. Stimulation occurred at Digit I and recordings were made from wrist and elbow.

RESULTS

Clinical evaluations

The neurological symptom score, indicating the presence of 22 different symptoms, was reduced by two points (symptoms) or more in six of the 12 (50%) patients on placebo. The improvements reflected the loss of dysesthesias, paresthesias and occa-

Table 2 Neurological Examination Score

	CRONASSIAL				PLACEBO		
Patient	Start	End	Change	Patient	Start	End	Change
1	896	934	+ 38	1	918	939	+ 21
2	942	966	+ 24	2	889	892	+ 3
3	943	963	+ 20	3	789	784	- 5
4	820	891	+ 71	4	891	904	+ 13
5	838	969	+131	5	908	964	+ 56
6	869	867	- 2	6	859	859	0
7	798	816	+ 18	7	822	893	+ 71
8	646	710	+ 64	8	777	861	+ 84
9	892	912	+ 20	9	887	898	+ 11
10	888	943	+ 55	10	948	959	+ 11
11	926	984	+ 58	11	877	874	- 3
12	888	925	+ 37	12	915	914	- 1
				13	906	880	- 26
Mean	862.2	906.7	+ 44.5	Mean	875.8	893.9	+ 18.1

sionally numbness, most often in the lower extremities.

The mean change in the neurological examination score was +44.5 points for the Cronassial patients and +18.1 points for the placebo patients ($p<0.04$). Ten of the 12 (83%) Cronassial patients experienced an improvement of 20 points or more whereas four of the 13 (31%) placebo patients had similar improvement (Table 2). These improvements reflected increased perception of the sensory stimuli in the lower extremities, often in the upper extremities as well, with occasional return of lower extremity reflexes. Little change in muscle strength occurred.

Electrophysiological studies

Sensory and motor conduction analysis in the Cronassial patients revealed small increases in sensory velocities (1.1 to 4.4 meters/sec) at the sural, peroneal and median nerves, representing changes of +2.9 to +12.9%. Most of these changes are within the range of error of the technique and presently are not deemed clinically significant. However, the increase of 4.4m/s (+12.9%) occurring at the distal sural nerve may represent definite improvement. No significant changes were seen in the amplitudes, configurations or synchronization of the sensory and motor evoked potentials, nor in the motor velocities or distal latencies. No significant change occurred in any electrophysiologic parameter in the placebo patients.

Ischemic responses

The ischemic responses in this study have been compared to those of 20 normal controls, 40 patients with non-metabolic peripheral neuropathies and 161 patients with diabetes mellitus without or with peripheral nerve dysfunction. In the normal subjects, patients with non-metabolic neuropathies and diabetics in excellent control, the sensory evoked potential at the elbow disappears within the first 24 minutes of ischemia. The persistence of the potential after 30 minutes is definitely abnormal. Previous studies indicate highly significant correlations between diabetic control (HbAlC levels) and the potential amplitudes at the elbow at 30 minutes ($n=161$, $r=0.5722$, $p<0.001$).

The 12 patients receiving Cronassial exhibited the typical correlation between HbAlC levels and the elbow sensory potential amplitude ($r=0.6799$, $p<0.02$) at the start. As the study continued the correlation progressively decreased, its statistical significance disappearing at 6 weeks and remaining so thereafter. At the end no relationship appeared to exist between HbAlC levels and elbow sensory potential amplitudes (Fig. 1). The 13 patients receiving placebo did not exhibit the loss of correlation between these two parameters.

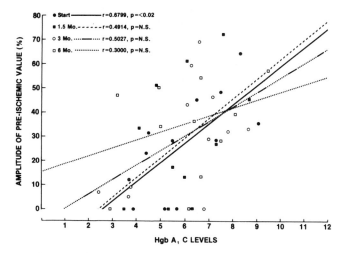

Fig. 1. Correlation between sensory evoked potential amplitudes at 30 minutes of ischemia recorded at the elbow and HgbA1C levels in patients receiving Cronassial.

DISCUSSION

The findings reported here suggest that cerebral gangliosides may have beneficial clinical effects on the neuropathy associated with diabetes mellitus. Clinical improvement approaching standard statistical significance is demonstrated in the patients receiving Cronassial when compared to the patients receiving placebo. Positive changes exist in the sensory conduction velocities of the Cronassial patients but they are small and not significant and their relevance remains to be seen. The causes and significance of the alterations in the ischemic responses in the Cronassial patients are obscure. A ganglioside effect on neuronal anaerobic metabolism can be postulated, but not proven.

Given the chronicity of diabetic neuropathy its multiplicity of pathological findings and the proposed mechanisms of cerebral ganglioside effect on peripheral nerves, it is to be assumed that a slow and gradual course of improvement would exist if gangliosides are an effective therapy. The encouraging results have prompted continued administration of Cronassial and evaluation of these patients for periods of longer than six months. These studies are now in progress.

SUMMARY

Cronassial^R, a combination of four different cerebral gangliosides, was evaluated in a six month double blind controlled study involving 25 patients with symptomatic diabetic neuropathy (12 receiving Cronassial, 13 receiving placebo). Clinically there was mild but definite improvement in the Cronassial patients. Small increases in sensory conduction velocities in the Cronassial patients were also seen; their significance requires further clarification. Measurements of neuronal responses during ischemia suggest that Cronassial may enhance anaerobic metabolism.

REFERENCES

1. G. Pozza, V. Saibene, G. Comi, and N. Canal, The effect of ganglioside administration in human diabetic peripheral neuropathy, in: "Gangliosides in Neurological and Neuromuscular Function, Development and Repair," M. M. Rapport and A. Gorio, eds., Raven Press, New York (1981).
2. S. Bassi, M. G. Albizzati, E. Calloni and L. Frattola, Electromyographic study of diabetic and alcoholic polyneuropathic patients treated with gangliosides, Muscle & Nerve 5:351 (1982).
3. R. J. Koenig, C. N. Peterson, R. L. Jones, C. Saudek, M. Lehrman, and A. Cerami, Correlation of glucose regulation and hemoglobin A_1C in diabetes mellitus, N. Engl. J. Med. 295:417 (1976).
4. L. A. Trivelli, H. M. Ranney and M. T. Lai, Hemoglobin components in patients with diabetes mellitus, N. Engl. J. Med. 284:353 (1971).

MULTICENTRE TRIAL ON GANGLIOSIDES IN DIABETIC PERIPHERAL

NEUROPATHY[¶]

Domenico Fedele, Gaetano Crepaldi and Leontino Battistin*

Institute of Internal Medicine
*Institute of Neurology
University of Padova, Italy

INTRODUCTION

In spite of the reports on the therapy with pyridoxine,[1] aldose reductase inhibitors,[2-4] and myoinositol,[5,6] at present no specific treatment is available for diabetic peripheral neuropathy, the most frequent long-term complication of diabetes mellitus.

Recent data have suggested that brain gangliosides, a class of glycosphingolipids pharmacologically active on different forms of peripheral neuropathies,[7,8] enhance the physiological reinnervation processes by stimulating sprouting mechanisms.[9-13] Thus, they can improve peripheral neuropathy both in animals,[14-16] and in man.[8,17]

So, we investigated the efficacy of the ganglioside treatment on patients with diabetic peripheral neuropathy with a multicentre, randomized, cross-over, double-blind, controlled vs placebo study design.

[¶] The trial has been carried out in collaboration with: A. Tiengo, P. Negrin (Padova), G. Pozza, N. Canal, G. C. Comi (Milano), G. F. Lenti, G. F. Pagano, L. Bergamini, W. Troni (Torino), F. Frigato, C. Ravenna, C. Mezzina (Mestre-Venezia), D. Massari, R. Gallato, M. Massarotti, R. Matano (Abano Terme), F. Grigoletto (Padova), H. Davis (Rochester), and M. Klein (New Jersey).

SUBJECTS AND METHODS

Four centres in North Italy (Padova, Milano, Torino, Mestre) participated in the trial after a preliminary 14 month period, necessary for the standardization of the neurological and metabolic tests used in the protocol. Diabetic subjects included in the study were insulin-treated for at least one year, aged between 18 and 55 years, had impaired motor or sensory conduction velocity in at least two nerves, and had no other causes of peripheral neuropathy. No subject was pregnant; the alcohol intake had to be less than 500 ml of wine a day and cigarette smoking less than 10 a day.

One hundred fifty one diabetic subjects agreed to enter the study, but only 140 completed it: the drop-outs were 11, 5 of them due to family reasons and 6 due to the onset of other diseases. The patients were divided into two protocols according to the severity of their neurological symptoms. Prot. I included 97 diabetic subjects without or with mild symptoms, and Prot. II grouped 43 patients with severe and frequent symptoms. Mean ($\pm$S.D.) age was 34 ± 11 in subjects of Prot. I and 43 ± 9 ($p < 0.001$) in Prot. II. Male subjects were 65 in the first group and 26 in the second one; females were 32 and 17 respectively. Duration of diabetes was shorter ($p < 0.001$) in Prot. I (9.4 ± 4 yrs) than in Prot. II (16 ± 8.6 yrs). No differences existed between the two groups for mean values of metabolic parameters. Insulin requirement was 51 ± 16 vs 45 ± 20 U/day, glycosylated hemoglobin was 11.1 ± 2.2 vs $11.1 \pm 23\%$ and mean daily plasma glucose was 215 ± 72 vs 232 ± 79 mg/dl in Prot. I and II respectively.

These metabolic parameters were evaluated in all centres with the same methods, standardized before the study began.

Neurological investigations included assessment of both symptoms and electrophysiological parameters. Paresthesias and nocturnal pain were scored from 1 to 4, according to their severity and frequency. Score 1 represented absence of symptoms, 2 mild and rare symptoms, 3 frequent ones, and 4 constant and severe symptoms. The 97 patients included in Prot. I scored 1 to 2, while all the 43 subjects of Prot. II rated 3 or 4. Sensory Conduction Velocity (SCV) of median (finger-wrist and wrist-elbow) and sural nerves, Motor Conduction Velocity (MCV), mean Distal Latency (DL), and Motor Action Potential (MAP) amplitude of ulnar and peroneal nerves were recorded in all subjects. Stimulation with surface electrodes was always supramaximal. A double evaluation (A and B) was performed at baseline.

The study design of both protocols was multicentre, randomized, cross-over, double-blind, controlled vs placebo. Randomization was performed separately for each centre. The study period lasted 16 weeks: 6 weeks of drug or placebo treatment and 4 intermediate weeks of washout. Each patient was assigned at random to one of the two treatment sequences, drug-washout-placebo or placebo-washout-drug. Both placebo and drug were administered i.m. The ganglioside dose was 20 mg/day in a single administration. Clinical and metabolic data were assessed at the beginning and at 3, 6, 10, 13, 16 weeks for Prot. I, and at the beginning and at 2, 4, 6, 10, 12, 14, 16 weeks for Prot. II. Electrophysiological data were recorded at the beginning and at the end of each treatment period for both protocols.

Statistical analyses were performed by means of covariance analysis to compare the response to drug and to placebo, and by means of MacNemar's and Gart's tests to evaluate the qualitative variables related to the symptoms.

RESULTS

No changes in the metabolic parameters were observed. Neither insulin requirement nor glycosylated hemoglobin nor mean daily plasma glucose levels showed significant variations during the study period.

The evaluation of the neurological symptom scores showed that paresthesias improved significantly ($p < 0.02$) in Prot. II. At the end of the study period, only 10 out of the 43 patients who initially had neurological symptoms still reported paresthesias, while 17 subjects did not present them and 14 reported them rarely.

Table 1 reports the non-parametric analysis of main and carry-over ganglioside effects on electrophysiological parameters.

It can be noted that the drug treatment induced a significant main effect in DL of ulnar nerve ($p < 0.01$) and in MCV of peroneal nerve ($p < 0.03$) in subjects of Prot. I, and in MCV of ulnar nerve ($p < 0.002$) and in SCV of median nerve ($p < 0.06$) in patients of Prot. II.

No side effects related to the treatment were reported by the patients and the physicians during the study period.

DISCUSSION

This trial has demonstrated that ganglioside treatment improves some symptoms and signs of diabetic peripheral neuropathy, and has confirmed previous reports on mice with congenital and

Table 1. Non-parametric Analysis (Two Tailed Test) of the Electrophysiological Data Obtained in 140 Diabetic Patients With Peripheral Neuropathy Treated With Gangliosides.

		p value			
	Electroneurological	Prot. I (97)		Prot. II (43)	
NERVE	parameters	Main	Carry-over	Main	Carry-over
MEDIAN	SCV finger-wrist	n.s.	n.s.	n.s.	n.s.
	wrist-elbow	n.s.	0.02	0.06	n.s.
SURAL	SCV	0.02*	n.s.	n.s.	0.05
ULNAR	MCV	n.s.	n.s.	0.002	0.002
	DL	0.01	n.s.	n.s.	0.1
	MAP	n.s.	n.s.	n.s.	n.s.
PERONEAL	MCV	0.03	0.1	n.s.	0.3
	DL	n.s.	n.s.	n.s.	n.s.
	MAP	n.s.	n.s.	n.s.	0.06

* placebo

alloxan-induced diabetes[14-16] and on man.[8,17] Together with the improvement of paresthesias in Prot. II, the treatment improved significantly MCV of peroneal (Prot. I) and ulnar nerves (Prot. II), and SCV of median nerve (Prot. II). The lack of significant changes in metabolic parameters underlines that this improvement is related to the drug effect rather than to an amelioration of metabolic control. Moreover, the carry-over effect detected in some electrophysiological findings suggests that gangliosides could have a delayed or persistent action after discontinuing therapy.

The experimental data on the function and properties of gangliosides could explain the positive results obtained in the treatment of diabetic peripheral neuropathy. Gangliosides have been proved to be incorporated at the neuronal membrane level in peripheral nerves,[10] to increase the physiological process of reinnervation by stimulating sprouting mechanisms,[9,10,13] and to activate some enzyme activities.[18,19] Activation of (Na^+,K^+)ATPase may lead to a proper intra- and extra-axonal ion balance, improving nerve conduction velocities.

The partiality of the results obtained with this study may be due to the short-term treatment period and to the low ganglioside doses administered. This suggests that next studies about the ganglioside effects on peripheral neuropathy should be carried out with higher doses and for a longer period of time.

REFERENCES

1. E. R. Levin, T. A. Hanscom, M. Fisher, W. A. Laustad, A. Lui, A. Ryan, D. Glokner, and S. R. Levin, The influence of pyridoxine in diabetic peripheral neuropathy, Diabetes Care 4:606 (1981).
2. K. H. Gabbay, N. Spack, S. Loo, J. L. Hirsh, and A. A. Ackil, Aldose reductase inhibition: studies with alrestatin, Metabolism 28 (Suppl):471 (1979).
3. A. Culebras, J. Alio, and J. L. Herrera, Effect of aldose reductase inhibitor on diabetic peripheral neuropathy, Arch. Neurol. 38:133 (1981).
4. R. G. Judzerwitsch, J. B. Jaspan, K. S. Polonsky et al., Aldose reductase inhibition improves nerve conduction velocity in diabetic patients, New Engl. J. Med. 308:119 (1983).
5. D. A. Greene, P. V. De Jesus, and A. I. Winegrad, Effects of insulin and dietary myoinositol on impaired peripheral motor nerve conduction velocity in acute streptozotocin diabetes, J. Clin. Invest. 55:1326 (1975).
6. R. S. Clements, Jr., B. Vaurganti, T. Kuba, S. J. Oh, and B. Darnell, Dietary myoinositol intake and peripheral nerve function in diabetic neuropathy, Metabolism 28:477 (1979).
7. A. Mingione, M. Monteleone, G. Paruzzi, O. Soragni, G. Cristiani, C. Moretti, W. Mega, and F. Scanabissi, Research in the cerebral gangliosides in neurolyses of the upper limb, Clin. Neurophysiol. 19:353 (1979).
8. S. Bassi, M. G. Albizzati, and L. Frattola, Electromyographic study of diabetic and alcoholic polyneuropathic patients treated with gangliosides, Muscle & Nerve 5:351 (1982).
9. B. Ceccarelli, F. Aporti, and M. Finesso, Effects of brain gangliosides on functional recovery in experimental regeneration and reinnervation, in: "Ganglioside Function," G. Porcellati, B. Ceccarelli, and G. Tettamanti, eds., Plenum Press, New York and London (1976).
10. A. Gorio, G. Carmignoto, L. Facci, and M. Finesso, Motor nerve sprouting induced by gangliosides treatment. Possible indications for gangliosides on neuronal growth, Brain Res. 197:236 (1980).
11. A. Gorio, G. Carmignoto, and G. Ferrari, Axon sprouting

stimulated by gangliosides. A new model for elongation and sprouting, in: "Gangliosides in Neurological and Neuromuscular Function, Development and Repair," M. M. Rapport and A. Gorio, eds., Raven Press, New York (1981).
12. F. Norido, R. Canella, and F. Aporti, Acceleration of nerve regeneration by gangliosides estimated by the somatosensory evoked potentials, Experientia (Basel) 37:301 (1981).
13. A. Gorio, P. Marini, and R. Zanoni, Muscle reinnervation III. Motoneuron sprouting capacity, enhancement by exogenous gangliosides, Neuroscience 8:417 (1983).
14. A. Gorio, F. Aporti, and F. Norido, Ganglioside treatment in experimental diabetic neuropathy, in: "Gangliosides in Neurological and Neuromuscular Function, Development and Repair," M. M. Rapport and A. Gorio, eds., Raven Press, New York (1981).
15. A. Gorio, G. Carmignoto, G. Ferrari, F. Norido, M. G. Nunzi, R. Rubini, and R. Zanoni, Pharmacological aspects of experimental peripheral neuropathy, in: "International Conference on Peripheral Neuropathies," S. Refsum, C. L. Bolis, and Portera-Sanchez, eds., Excerpta Medica, Amsterdam, Oxford-Princeton (1982).
16. F. Norido, R. Canella, and A. Gorio, Ganglioside treatment of neuropathy in diabetic mice, Muscle & Nerve 5:107 (1982).
17. G. Pozza, V. Saibene, G. Comi, and N. Canal, The effect of ganglioside administration in human diabetic peripheral neuropathy, in: "Gangliosides in Neurological and Neuromuscular Function, Development and Repair," M. M. Rapport and A. Gorio, eds., Raven Press, New York (1981).
18. J. W. Daly, The effect of gangliosides on activity of adenylate cyclase and phosphodiesterase from rat cerebral cortex, in: "Gangliosides in Neurological and Neuromuscular Function, Development and Repair," M. M. Rapport and A. Gorio, eds., Raven Press, New York (1981).
19. A. Leon, L. Facci, G. Toffano, S. Sonnino, and G. Tettamanti, Activation of $(Na^+ K^+)$ATPase by nanomolar concentrations of GM1 ganglioside, J. Neurochem. 37:350 (1981).

A DOUBLE BLIND PLACEBO CONTROLLED TRIAL OF MIXED GANGLIOSIDES

IN DIABETIC PERIPHERAL AND AUTONOMIC NEUROPATHY

R. R. Abraham, R. M. Abraham* and V. Wynn

Alexander Simpson Laboratory for Metabolic Research
Mint Wing, St. Mary's Hospital
Praed Street, London, W.2.
*Dept. of Neurophysiology
Central Middlesex Hospital
Park Royal, London, W.3.

INTRODUCTION

Diabetic neuropathy is a common complication of diabetes that is incurable at present. The likelihood of clinical manifestations increases with the duration and severity of the hyperglycemia.[1-7] Apart from the beneficial effects of insulin by continuous infusion[8,9] or the alcohol myoinositol,[10-13] the only other drugs tried with questionable benefit in clinical diabetic neuropathy are isaxonine[14] and sorbinil.[15-18]

Gangliosides are a chemical family of plasma membrane-located complex sialoglycolipids found in high concentration in nerve tissue. They have been shown to be biologically effective initiators of dendritogenesis when added to nerve cells in culture.[19] In central denervation experiments in rats,[20,21] mixed gangliosides have been shown to result in an acceleration of spontaneous functional recovery, with biochemical evidence of increased regeneration of synapses (increased acetylcholine esterase and choline acetyl transferase activities and reduced serotonin denervation supersensitivity). These findings have led to the suggestion by some investigators that gangliosides act as receptor molecules for endogenous trophic nerve growth factors.[22]

We used a preparation of 4 gangliosides ('Cronassial' manufactured by Fidia Farmaceutici, Abano Terme, Italy) in a double-blind placebo controlled trial in diabetic peripheral and autonomic neuropathy. This mixed preparation of gangliosides has been shown to result in improvement in nerve conduction velocity and axonal morphometry in the genetically diabetic (db/db) mouse.[23]

PATIENTS AND METHODS

Patients

Twenty-six patients were studied and their clinical data are shown in Table 1. The criteria for inclusion in the trial were that the patients had a symptom or clinical sign of diabetic neuropathy (including simple loss of ankle reflexes or postural hypotension with a systolic blood pressure drop of >20 mm Hg) and at least one abnormality on neurophysiological investigation (viz. a compound motor nerve action potential <2mV, motor nerve conduction velocity <35 m/sec, sural nerve action potential <2.5 µV and sural nerve conduction velocity <33 m/sec). Patients were excluded from the trial if they had other conditions associated with neuropathy (declared excess alcohol consumption or MCV >100, untreated pernicious anemia, rheumatoid arthritis, etc.) or were treated with drugs recognized as being a potential cause of neuropathy (e.g. perhexilene, nitrofurantoin). In addition, patients with general debility or at risk of pregnancy or with moderately severe abnormalities of liver function were also excluded from the trial.

The placebo and Cronassial groups were not matched prior to randomization. Each patient was investigated before and immediately after 6 weeks' continuous daily treatment with either 20 mg (4 ml) Cronassial or placebo (buffer without the Cronassial) (supplied by Fidia Farmaceutici, Abano Terme, Italy) given intramuscularly by a nurse at the patient's home. Four weeks following cessation of treatment, a further assessment was carried out to record any persistent effect of the drug. At each consultation, patients were interviewed for symptom complaints and had a clinical, neurophysiological and laboratory evaluation. The tests of autonomic dysfunction were done in the pre-treatment and immediate post-treatment periods.

A symptom history was obtained using a standard questionnaire (which included enquiry on the occurrence of paraesthesiae, numbness, pain, cramps, restless legs, faintness, diarrhea and hesitancy).

The laboratory tests included full hematology, routine biochemistry, fasting cholesterol, triglycerides and high density lipoproteins, and the glycosylated hemoglobin HbA1C.

Table 1. Clinical Data of Cronassial and Placebo Groups.

	Total No.	Age (years)	Number of patients that were symptomatic	Height (cm)	Weight (kg)	Duration of diabetes (years)	Number not on insulin	Years on insulin
Placebo	7 M, 7 F	62.9±2.3 (14)	10	167.8±2.6 (14)	70.9±4.3 (14)	13.0±3.5 (14)	5	5.9±3.5 (14)
Cronassial	7 M, 5 F	55.2±5.3 (12)	10	168.9±3.7 (12)	67.9±3.8 (12)	15.9±2.7 (12)	1	9.9±2.6 (12)

Electrophysiological measurements

A single operator (R.M.A.), unaware of the treatment category of each patient, carried out all the neurophysiological tests using a Medelec MS6 machine. Electrophysiological measurements were made in a warm room (>22°C), on the (R) leg of all patients. Skin surface temperatures were measured on the lateral border of the right foot. The proximal and distal compound motor action potentials of the common peroneal nerve were measured with the stimulating electrode at the head of the fibula and the ankle, and recordings were made after maximal stimulation, using saddle electrodes with an interelectrode distance of 5 cm. The conduction velocity was calculated from the measured latencies. The motor action potential (DMNAP) value presented is that obtained after stimulation distally and was measured as the maximum deflection from the baseline. When present, the minimum f wave latency was measured as the shortest of 20 successive f latencies. In one case on placebo treatment, the common peroneal nerve compound action potential was absent and the anterior tibial nerve was used instead. The results are not shown as there was no significant change during or after the course of treatment. The sural action potential was measured with "near nerve" needle electrodes by taking the maximum difference between the initial negative deflection and any positive overshoot.

Autonomic neuropathy measurement

(a) The orthostatic test. The blood pressure and heart rate response to standing was measured in 19 patients. All heart rate measurements were made from measurements of ECG R-R intervals. Patients were allowed to rest supine for up to 30 minutes or when a basal heart rate was established continuously for at least 5 minutes. After a basal blood pressure was recorded, (the diastolic pressure was taken at the point of disappearance of the Korotkov sounds), the patient was asked to voluntarily stand up rapidly (within 5 seconds). Continuous ECG records were made for a minute and then after 5 and 10 minutes, with blood pressure recordings at 1 minute, 5 minutes and 10 minutes after standing. Various indices were calculated from these data:
    (1) The heart rate at beat 15 over that at beat 30;[24]
    (2) An acceleration index[25] (expressed here as the fastest heart rate in the first minute after standing - resting heart rate/resting heart rate, x 100).
    (3) A brake index[25] (expressed here as the fastest-slowest heart rate in the first minute after tilt/resting heart rate, x 100).

(b) The vital capacity test. Each patient was asked to take a relaxed deep single breath during continuous ECG recording. The inspiratory and expiratory phases lasted at least 5 seconds. The results are presented as the difference in heart rate between the fastest heart rate during inspiration and the slowest heart rate during expiration and as the E/1 ratio (slowest heart rate during expiration) : (fastest heart rate during inspiration).[26]

(c) <u>The Valsalva Maneuver</u>. Each patient was asked to blow into a mouthpiece attached to a sphygmomanometer through a tube with a 2 mm hole in it, in order to introduce a constant leak. The patient was required to blow hard to a pressure of 40 mm Hg for as long as possible, or 15 seconds (whichever was the shorter). All patients were not included in this test as some of our elderly patients were not able to perform this test satisfactorily. Results were calculated as the fastest heart rate during the blow/slowest heart rate after the blow and also as the slowest heart rate during the test/resting heart rate.[26]

Review of diabetic control

All patients were encouraged to do regular daytime diurnals of capillary blood glucose at home using either commercial meters or blood testing sticks. Between 5 and 7 readings were taken at intervals throughout the day and evening. 24-hour profiles of plasma glucose were obtained on some patients who were partially sighted, and these consisted of venous samples taken every 2 hours in hospital. All glucose readings were corrected for comparable plasma values. Where 3 or 4 successive daily profiles were available, the average was taken. The glycosylated hemoglobin HbAlC was measured before treatment, after 6 weeks' treatment, and 4 weeks after the wash-out period.

Plasma glucose was measured by an autoanalyzer method using glucose-oxidase. Total glycosylated hemoglobin was measured by a micro-electrophoretic method (Corning).

No patients withdrew from the study and there were no serious adverse reactions to Cronassial. The randomization code was broken for one patient who was admitted to hospital in left ventricular failure the day after starting his injections. As he was in the placebo group, his participation in the trial was continued following discharge from hospital. One patient conceived while on treatment with Cronassial. Permission for the use of Cronassial in diabetic neuropathy in the U.K. was obtained from the C.C.S.M. and subsequently from the Ethical Committee of St. Mary's Hospital. Signed informed consent for participation in the trial was obtained from all patients.

RESULTS

Electrophysiological evaluation

Tables 2 and 3 give the mean values of paired comparisons in the two groups after 6 weeks' treatment (Table 2) and 4 weeks later in the wash-out period (Table 3). Tables 4 and 5 give the paired increases from the pre-treatment values in individual patients in the placebo (Table 4) and Cronassial (Table 5) groups.

(a) The compound motor action potential of the common peroneal nerve (DMNAP). Both placebo and Cronassial groups had comparable pre-treatment values [1.57 ± 0.38 (n=11) mV and 1.27 ± 0.25 (n=12) mV respectively]. There was no significant difference in the compound motor action potential in the placebo group either after 6 weeks' treatment or in the wash-out period. In the Cronassial group, a slight improvement was noted after 6 weeks' treatment [pre-treatment value 1.27 ± 0.25 (n=12) mV, post-treatment value 1.93 ± 0.39 (n=12) mV] and this increased after the wash-out period, but did not reach statistical significance [pre-treatment paired value 1.14 ± 0.24 (n=11) mV vs. 4 weeks post-treatment paired value 2.33 ± 0.67 (n=11) mV].

Four patients showed increases >1.0 mV in their DMNAP measurements after Cronassial (+2.83, +2.80, +1.20, +2.00 mV) while none of the placebo group did so ($p<0.047$, one-tailed Fisher exact probability test). The increases >1.0 mV in the Cronassial group after the wash-out period were +1.49, +3.90, +5.50, +3.5 mV compared to the only increase >1.0 mV in the placebo group of 1.07 mV (N.S. at $p<0.05$). Concomitant surface temperature increases were recorded in two patients, RA and RS, of $1.5-2.9°C$. No patient in the placebo group showed any change greater than 1.0 mV in the three successive observations made over 10-14 weeks.

(b) The motor nerve conduction velocity (MCV). Both placebo and Cronassial groups had comparable values for the motor nerve conduction velocities before the start of treatment [placebo 37.7 ± 1.4 (n=11) m/sec and Cronassial 36.9 ± 2.3 (n=11) m/sec. - Table 2]. Although increases in mean conduction velocity of 1.5 m/sec (after 6 weeks of treatment) and 2.0 m/sec (after the wash-out period) were observed in the Cronassial group, without any similar changes in the placebo group, these did not reach statistical significance at the $p<0.05$ level (Tables 2 and 3). The higher mean values in the Cronassial group after treatment and wash-out periods resulted from big improvements in five patients. They showed individual changes in motor nerve conduction velocity of +4.5, +4.3, +4.9, +3.7, +4.0 (after 5-7 weeks of treatment) and +2.8, +2.1, +0.1, +7.6 and +2.75 m/sec (after the wash-out period of 4 weeks) (Table 5). Three of these patients also showed large improvements in the size of their motor action potentials. Allowing for an increase in conduction velocity of

Table 2. Paired Comparisons of Neurophysiological Data After 6 Weeks Treatment.

	Placebo		Cronassial	
	Pre-treatment	Post-treatment	Pre-treatment	Post-treatment
DMNAP (mV)	1.57±0.38 (11)	1.73±0.43 (11)	1.27±0.25 (12)	1.93±0.39 (12)
MCV (m/sec)	37.7±1.4 (11)	37.5±1.6 (11)	36.9±2.3 (11)	38.4±2.3 (11)
f latency (msec)	65.3±3.3 (6)	66.2±7.2 (6)	57.0±2.3 (7)	56.9±2.8 (7)
Sural action potential (µV)	6.7±0.5 (3)	6.1±0.3 (3)	7.7±1.0 (5)	5.7±0.9 (5)
Sural NCV (m/sec)	41.6±2.4 (3)	35.0±2.0 (3)	41.6±2.6 (5)	41.7±3.0 (5)

All comparisons are not significantly different from each other

Table 3. Paired Comparisons of Neurophysiological Data 4 Weeks After Completion of Treatment.

	Placebo		Cronassial	
	Pre-treatment	4 weeks Post-treatment	Pre-treatment	4 weeks Post-treatment
DMNAP (mV)	1.37±0.36 (12)	1.38±0.26 (12)	1.14±0.24 (11)	2.33±0.67 (11)
MCV (m/sec)	37.3±1.4 (12)	37.7±1.7 (12)	35.8±2.3 (10)	37.8±1.6 (10)
f latency (msec)	62.3±3.8 (6)	57.6±4.0 (6)*	58.0±2.5 (6)	59.0±2.4 (6)
Sural action potential (μV)	5.7±1.1 (4)	7.5±1.3 (4)	7.4±1.4 (8)	6.7±1.4 (8)
Sural NCV (m/sec)	40.9±1.8 (4)	38.6±2.2 (4)*	38.5±2.7 (7)	36.4±1.9 (7)

*$p \leq 0.05$ Students t test

DMNAP not significant using: Students t test
 Wilcoxon Matched Pairs
 Friedman 2-Way Analysis of Variance

Table 4. Change in Measurement from Pre-Treatment Values - Placebo Group.

Initials	D MNAV 6	D MNAV 10	MNCV 6	MNCV 10	f latency 6	f latency 10	Sural AP 6	Sural AP 10	Sural NCV 6	Sural NCV 10
F.B.	+0.37	-0.18	-3.6	+1.2	-	-	-0.9	+3.5	-6.8	-3.6
R.B.	+0.3	-1.8	+6.8	+5.9	-	-	NF	NF	NF	NF
I.B.	-	+0.08	-	+0.5	-	-	NF	NF	NF	NF
J.C.	+0.18	+0.27	-0.4	+7.2	-27	-8	NF	NF	NF	NF
S.D.	+0.3	-0.2	-2.5	-2.8	-4	-3.5	NF	NF	NF	NF
G.M.	+0.1	-0.4	-1.06	-0.06	-	-	-1.4	+1.6	-9.1	-1.4
E.P.	+1.0 *	+0.30	+1.0 *	-0.4	-	-3	NF	-2.1	NF	-
C.R.	+0.4	+0.6	+0.7	-1.3	+31	-	NF	NF	NF	NF
H.S.	+0.12	+0.08	+2.9	-3.0	-	-	NF	NF	NF	NF
R.Sp.	-0.13 t	-0.36 tt	+9.9 t	+2.9 tt	+1 t	+2 tt	NF	NF	NF	NF
J.S.	-0.77	-	-6.3	-	+6	-	NF	NF	NF	NF
C.T.	+0.15	+0.70	+1.0	-2.0	0	-1	+0.5	0	-3.7	-3.2
D.T.	-0.40	-0.40	-1.5	+1.1	-1	-1	NF	NF	NF	NF
H.Y.	-	+1.07	-	-2.1	-	-12	-	+2.0	-	-1.2

NF = not found
6 - refers to values 5-7 weeks after start of treatment
10 - refers to values 4 weeks after completion of treatment
R.Sp. values refer to tibial nerve not common peroneal nerve

* temperature increase of +1.6°C
t temperature increase of +3.8°C
tt temperature increase of +4.1°C

Table 5. Change in Measurement from Pre-Treatment Values - Cronassial Group.

Initials	D MNAV 6	D MNAV 10	MNCV 6	MNCV 10	f latency 6	f latency 10	Sural AP 6	Sural AP 10	Sural NCV 6	Sural NCV 10
R.A.	+2.83*	+1.49	-5*	-4.7	-2*	+4	-1.0*	-0.5	+7.2*	-2.2
G.A.	+0.01	-0.01	-0.2	+3.4	-	-	NF	NF	NF	NF
S.G.	0	-0.03	-	-	-	-	NF	NF	NF	NF
J.L.	-0.8	-0.5	+2.6	-1.1	+3	+1	+1.3	-1.4	-16.1	-7.5
M.Pks.	-0.1	-	+2.4	-	-2	-	NF	NF	NF	NF
M.Pt.	+2.8	+3.9	+4.5	+2.8	-	-	-1.0	-0.2	+0.3	-3.8
E.S.	+1.2	+5.5	+4.3	+2.1	-	-	-	-0.8	-	+4.4
R.S.	+2.03 †	+3.52 ††	+4.9 †	+0.1 ††	-5 †	-3 ††	-6.7 †	+0.5 ††	+5.5 †	-4.1 ††
A.V.S.	-0.32	-0.34	-0.8	+1.3	-	-	-	-0.6	-	-0.9
R.T.	-0.29 Δ	-0.4 Δ	+3.7 Δ	+7.6 Δ	+4 Δ	+1 Δ	NF Δ	NF Δ	NF Δ	NF Δ
T.T.	-0.37	+0.06	+4.0	+2.75	0	0	NF	NF	NF	NF
C.W.	+1.0	-0.03	+0.77	+5.67	+1	+3	-2.3	-2.7	+3.3	-0.9

NF = not found
6 - refers to values 5-7 weeks after start of treatment
10 - refers to values 4 weeks after completion of treatment

* temperature increase of +2.9°C
† temperature increase of +1.5°C
†† temperature increase of +2.0°C
Δ temperature increase of +3.3°C

0.53 m/sec per °C increase in surface temperature,[27] the temperature increases noted in RS and RT (Table 5) of 1.5 and 3.3°C respectively, would only have accounted for conduction velocity increases of up to 1.5 m/sec, whereas the measured improvements were 4.9 and 3.7 m/sec respectively. Two patients in the placebo group also showed large improvements in motor nerve conduction velocity of 6.8, 9.9 m/sec (5-7 weeks after the start of treatment) and 5.9, 2.9 m/sec (after the 4 week wash-out period) but these improvements were not associated with parallel changes in the size of their motor nerve compound action potentials. Here too, the temperature change of 3.8°C with R.Sp. would only account for a conduction velocity increase of 2.0 m/sec of the 9.9 m/sec measured.

(c) Minimum f wave latency. The f wave was not consistently obtained in all our patients. In only 11/14 of the placebo group and 8/12 of the Cronassial group were f waves reliably obtained and Cronassial treatment made no significant difference at any time to the minimum latencies of the f wave when compared to placebo values (Tables 2 and 3).

(d) Sural nerve action potential. No detectable sural action potential was obtained in 9 patients in the placebo group and in 5 patients of the Cronassial group. There was a small fall in the paired means in the Cronassial group [7.7 ± 1.0 μV pre-treatment vs. 5.7 ± 0.9 μV (n=5) post-treatment] but this was not significant at the $p<0.05$ level and the numbers in this comparison are too small.

(e) The sural nerve conduction velocity. Small numbers in this group again made comparisons between the placebo and Cronassial groups impossible but in the 5 patients who had sural nerve action potentials in the Cronassial group, no significant maintained improvement was seen (Table 2).

Autonomic neuropathy tests

The results of all the measurements made in the 3 tests of autonomic function are shown in Table 6. There was no significant change in any of the values after treatment with either placebo or Cronassial. The two groups were different before the start of the trial with respect to the acceleration, brake and 15:30 indices of the orthostatic tilt test, suggesting a much more severe sympathetic and parasympathetic denervation in the placebo group than in the Cronassial group.

Changes in symptoms and signs

There were minor and comparable changes in symptomatology in both placebo and Cronassial groups. Four patients in the Cronassial group and one in the placebo group reported a complete disappearance of symptoms (pain, numbness, weakness, autonomic hypotensive symptoms and vomiting) reported as severe, intoler-

Table 6. Effect of Cronassial on Three Tests for Autonomic Neuropathy.

	Placebo		Cronassial	
	Pre-treatment	Post-treatment	Pre-treatment	Post-treatment
Orthostatic Test				
Change in systolic BP (1 min) (mmHg)	-11.4±5.7 (7)	-13.3±4.4 (7)	-17.3±12.2 (11)	-25.0±9.8 (10)
Change in systolic BP (10 min) (mmHg)	-12.5±7.9 (8)	-13.8±6.5 (8)	-23.2±10.3 (11)	-23.2±7.3 (11)
Change in diastolic BP (1 min) (mmHg)	-1.7±3.8 (6)	+4.2±6.2 (6)	-4.3±5.1 (11)	-6.5±5.7 (11)
Change in diastolic BP (10 min) (mmHg)	-2.1±3.8 (7)	-1.4±2.4 (7)	+0.5±5.7 (11)	-8.2±3.7 (11)*
Resting heart rate (b.p.m.)	77.4±4.4 (9)	78.8±4.6 (9)	70.6±3.5 (11)	74.6±4.1 (11)
Acceleration index (%)	14.1±2.8 (7)	15.2±3.4 (7)	30.4±6.7 (10) †	29.6±5.2 (10) †
Brake Index (%)	7.06±2.3 (7)	8.06±2.8 (7)	20.9±4.4 (10) ††	21.7±3.5 (10) †††
15/30 index	1.01±0.01 (7)	0.93±0.07 (7)	1.03±0.01 (10)	1.00±0.02 (10)*
Vital Capacity Test				
Change in heart rate during single breath	10.2±1.8 (9)	8.9±1.8 (9)	11.0±2.5 (11)	12.45±1.52 (11)
E/1 ratio	0.88±0.02 (9)	0.90±0.02 (9)	0.87±0.03 (11)	0.86±0.02 (11)
Valsalva Test				
Max. HR/ min. HR	1.15±0.07 (5)	1.17±0.06 (5)	1.30±0.21 (5)	1.31±0.12 (5)
Min. HR/ RHR	1.05±0.02 (5)	1.00±0.04 (5)	1.06±0.07 (5)	0.99±0.02 (5)

All comparisons not significant except * = $p \leq 0.05$ Students t test
† $p \leq 0.05$ from equivalent placebo value
†† $p \leq 0.02$ from equivalent placebo value
††† $p \leq 0.01$ from equivalent placebo value

able and persistent in nature before treatment. However, the occurrence of fresh, severe and persistent symptoms (numbness, sweating, dizziness) occurred following treatment in one patient in the placebo group and three in the Cronassial group.

Changes in neurological examination were variable even in the placebo group. A comparable number of patients reported an improvement in two or more signs in the placebo (3/14) and Cronassial groups (4/12) although 8/14 of the placebo group and 1/12 of the Cronassial group reported signs that signified a deterioration when compared with the pre-treatment examination. The four patients who showed marked improvements in the size of the compound motor action potential, all had minor clinical neurological deficits. Two had no neurological abnormalities at all and the remaining two had absent ankle reflexes, with impaired vibration sense in one case 68 years of age.

Hematological, biochemical and diabetic assessment

Apart from small differences in pre-treatment fasting triglyceride levels [placebo 148.1 ± 16.6 (n=11) mg% vs. Cronassial 90.4 ± 11.2 (n=9) mg% p<0.05] and serum total protein levels [placebo 73.0 ± 1.7 (n=11) g/l vs. Cronassial 68.5 ± 1.1 (n=11) g/l p<0.05] the placebo and Cronassial groups had similar hematology and biochemistry. There were no other significant changes in Hb, WBC, ESR, urea, Na, K, creatinine, albumin, total protein, Ca^{+2}, PO_4^{-3} bilirubin, aspartate transaminase, alkaline phosphatase, GTP, fasting cholesterol, triglycerides and HDL cholesterol.

Diabetic control

There was no significant change in the mean HbA1C concentrations of both the placebo and Cronassial groups throughout the study and this suggests a constant degree of diabetic control in the two groups. The mean HbA1C of the Cronassial group was always higher than the values of the placebo group and this achieved significance in the 5-7 week values [placebo group 9.8 0.5 (n=9) g% vs. Cronassial group 12.6 ± 0.7 (n=10) g% p<0.05].

Mean plasma glucoses calculated from values obtained during 12-24 hour diurnal profiles were not significantly different in the two groups and did not deteriorate significantly during the trial period. Individually, there was a deterioration in the mean plasma glucose of the diurnal profile in E.S. (not confirmed by HbA1C values) but her neurophysiology improved over the trial period. The three other patients on Cronassial who showed improvements in compound motor nerve action potentials showed no significant changes in diabetic control.

DISCUSSION

Most animal work on neural regeneration using mixed gangliosides has confirmed the biological efficacy of this preparation at doses of 1 mg/kg body weight. The dose permitted for the present study, 20 mg, may therefore have been insufficient and may account for the failure to obtain a greater number of patients showing definite improvements after treatment with Cronassial. Some investigators[13] have suggested that at least six months' treatment is required before any improvements occur but this length of treatment is less acceptable to patients receiving their treatment by daily intramuscular injections. Clinical trials in diabetic neuropathy have rarely been conducted for long enough to take into account the variability of many of the parameters measured. The only objective measurements made in this trial were the electrophysiological and autonomic function tests as the symptom history and clinical signs were dependent on inconsistent and unreliable patient responses. The DMNAP in our placebo group never varied by more than 2.1 mV in the entire study period, though conduction velocity measurements, which are more susceptible to temperature changes, varied by as much as 9.9 m/sec in two successive measurements in one of the placebo treated subjects.

Although mean values of the DMNAP of the common peroneal nerve were higher after treatment with Cronassial when compared to the placebo group, this result was not significant at the $p < 0.05$ level. Inspection of the individual data showed that the higher mean of the Cronassial group was elevated by large increases in the motor nerve action potentials of four patients. Temperature increases may have been a factor in the increases observed in two of these patients. The same two patients were also relatively young, aged 27 and 29 years, and it is significant that all four patients who improved were relatively free of abnormal neurological clinical signs [absent ankle jerks in two patients (one aged 68) and no abnormal clinical signs in two patients]. Although their symptomatology was also mild and of infrequent occurrence, their pre-treatment compound motor action potentials were very low (0.81, 1.7, 1.2 and 0.57 mV in the four responders). Three of these four patients also showed significant (>4 m/sec) improvements in conduction velocity. This implies improved function of the fastest and largest myelinated fibres and also possibly of a decrease in the extent of segmental demyelination.[28] The increase in compound motor action potentials in four subjects suggests that the mixed gangliosides have also had some effect in increasing the number of functioning fibres as well as in improving the function of the largest myelinated fibres in three of these four patients.

The five patients who showed motor nerve improvements (either in action potential size or conduction velocity) all had diabetes for less than 18 years, whereas all the non-responders, except two, had diabetes for more than 18 years. It is well recognized that diabetic neuropathy has a higher incidence in older age groups[29] and this may reflect a more severe neuropathy in those with long-standing diabetes. Diabetic neuropathy has a mixed pathology; there is loss of myelinated and unmyelinated axons, a proportion of the damaged nerves show segmental demyelination and remyelination, and some others show axonal degeneration.[28,30] It is possible that Cronassial benefited a subgroup that possessed early, reversible pathology with less axonal degeneration. The patient population in this study had a modal age of between 50-60 years with a duration of about 15 years from the initial diagnosis of their diabetes. A high proportion had abnormalities on autonomic testing; the placebo group, in particular, had a greater number affected with poor autonomic function and absent sural sensory action potentials.

Nerve conduction velocity can be improved by reducing the mean plasma glucose levels in diabetic patients.[5,6,8] The diabetic control in the two groups was unchanged throughout the study discounting the possibility that improved control of the hyperglycemia was responsible for any of the clinical or electrophysiological improvements.[4,8]

There is no information on the persistence of neurological benefit after treatment with Cronassial. Exogenously administered gangliosides appear to be incorporated into neural plasma membranes and synaptosomes and can be shown to travel down the axon. After the 4 week wash-out period, we have shown a persisting increase in the motor action potentials of all four responders and in two of the three who showed improved conduction velocities.

SUMMARY

Twenty-six patients with signs or symptoms of diabetic neuropathy who also had motor or sensory neurophysiological abnormalities, were intensively studied in the first double-blind, placebo controlled trial of a ganglioside mixture in the U.K. (Cronassial, Fidia Farmaceutici, Abano Terme, Italy; 6 weeks of 20 mg daily intramuscularly). Diabetes control was good in both groups and there was no deterioration in either group during the course of the trial. Of the five neurophysiological parameters measured - the motor nerve compound action potential, motor nerve conduction velocity and minimum f wave latency of the common peroneal nerve and the sensory action potential and conduction velocity of the sural nerve - only the mean motor nerve action potential and conduction velocity increased though the changes were not significant. Four patients showed improvements of >1.0 mV after treatment with Cronassial while none of the placebo group did so ($p<0.05$). Of the four patients who showed dramatic improvements in motor nerve action potentials, three also showed improvements in motor nerve conduction velocity. There were no significant changes in three tests of autonomic function after Cronassial therapy (orthostatic tilt reactions, single deep breath and Valsalva maneuver). Four patients in the Cronassial group (not those showing neurophysiological improvement) but only one in the placebo group reported marked improvements in their symptomatic complaints after treatment. We conclude that Cronassial is a promising new drug for the treatment of diabetic peripheral neuropathy with clear benefits for some patients.

ACKNOWLEDGEMENTS

We are grateful to Sir Roger Bannister, Director of the Department of Electrophysiology at St. Mary´s Hospital, London, W.2., for the use of a Medelec MS6 machine. Our clinical colleagues, Dr. M. W. J. Davie, Dr. J. H. Parr and Dr. T. Lockington, made important contributions and M. Williams contributed to the data analysis. We also wish to thank Mrs. C. V. Abraham for giving all the injections. Dr. R. Kohn, Dr. J. R. Whittington (Advisory Services (Clinical and General) Ltd., U.K.), and Dr. M. Massarotti (Fidia Farmaceutici, Abano Terme, Italy) gave helpful advice.

REFERENCES

1. J. D. Ward, Diabetic neuropathy, Clin. Endocrinol. Metab. 1:809 (1972).
2. J. Pirart, Diabetes and its degenerative complications: a prospective study of 4,400 patients observed between 1947 and 1973, Diabetes Care 1:168 (1978).
3. R. J. Graf, J. B. Halter, E. Halar, and D. Porte, Nerve conduction abnormalities in untreated maturity-onset diabetes: relation to levels of fasting plasma glucose and glycosylated hemoglobin, Ann. Intern. Med. 90:298 (1979).
4. D. Porte, R. J. Graf, J. B. Halter, M. A. Pfeifer, and E. Halar, Diabetic neuropathy and plasma glucose control, Am. J. Med. 70:195 (1981).
5. R. J. Graf, J. B. Halter, M. A. Pfeifer, E. Halar, F. Brozovich, and D. Porte, Jr., Glycemic control and nerve conduction abnormalities in non-insulin-dependent diabetic subjects, Ann. Intern. Med. 94:307 (1981).
6. D. J. Ward, C. G. Barnes, D. J. Fisher, J. D. Jessop, and R. W. R. Baker, Improvement in nerve conduction following

treatment in newly diagnosed diabetics, Lancet 1:428 (1971).
7. G. Gregersen, Diabetic neuropathy: Influence of age, sex, metabolic control and duration of diabetes on motor conduction velocity, Neurol. 17:972 (1967).
8. A. Pietri, A. L. Ehle, and P. Raskin, Changes in nerve conduction velocity after six weeks of glucoregulation with portable insulin infusion pumps, Diabetes 29:668 (1980).
9. J. R. Warmolts, J. R. Mendell, T. M. O'Dorisio, and S. Cataland, (abstract) Evaluation of portable insulin infusion pump on peripheral nerve function in diabetes mellitus, Neurology 31(II):129 (1981).
10. R. S. Clements, Jr., B. Vourganti, T. Kuba, S. J. Oh, and B. Darnell, Dietary myoinositol intake and peripheral nerve function in diabetic neuropathy, Metabolism 28:477 (1979).
11. G. Gregersen, H. Borsting, P. Theil, and C. Servo, Myoinositol and function of peripheral nerves in human diabetics, Acta Neurol. Scan. 58:241 (1978).
12. J. G. Salway, L. Whitehead, J. A. Finnegan, A. Karunana-Yaka, D. Barnett, and R. B. Payne, Effect of myoinositol on peripheral nerve function in diabetes, Lancet 2:1282 (1978).
13. D. A. Greene, M. J. Brown, S. N. Braunstein, S. C. Schwartz, A. K. Asbury, and A. I. Winegrad, Comparison of clinical course and sequential electrophysiological tests in diabetics with symptomatic polyneuropathy and its implications for clinical trials, Diabetes 30:139 (1981).
14. P. Augustin and M. Rathery, Clinical trial of isaxonine in diabetic neuropathies La Nouvelle Presse Medicale 11(6):1265 (1982).
15. R. G. Judzewitsch, J. B. Jaspan, K. S. Polonsky, C. R. Weinberg, J. B. Halter, E. Halar, et al., Aldose reductase inhibition improves nerve conduction velocity in diabetic patients, N. Eng. J. Med. 308:119 (1983).
16. D. J. Handelsman and J. R. Turtle, Clinical trial of an aldose reductase inhibitor in diabetic neuropathy, Diabetes 30:459 (1981).
17. K. H. Gabbay, N. Spack, S. Loo, H. J. Hirsch, and A. A. Ackil, Aldose-reductase inhibition: studies with Alrestatin, Metabolism 28:471 (1979).
18. A. Culebras, J. Alio, J-L. Herrera, and I. P. Lopez-Fraile, Effect of an aldose reductase inhibitor on diabetic peripheral neuropathy: preliminary report, Arch. Neurol. 38:133 (1981).
19. F. J. Roisen, H. Bartfeld, R. Nagele, and G. Yorke, Ganglioside stimulation of axonal sprouting in vitro, Science 214:577 (1981).
20. S. E. Karpiak, (abstract) Recovery of function after CNS damage enhanced by gangliosides, International Society

for Neurochemistry Satellite Meeting "Ganglioside Structure, Function and Biomedical Potential", Parksville, BC, Canada, (1983).
21. G. Toffano, (abstract) (v.s.) Effect of gangliosides on the regeneration of central nervous system, (1983).
22. W. Dimpfel and U. Otten, (abstract) (v.s.) Studies of gangliosides in diverse nerve cell cultures, (1983)
23. A. Gorio, F. Norido, R. Canella, and R. Zanoni, (abstract) (v.s.) The development of diabetic neuropathy in the C57BL/KS (db/db) mouse and its treatment with gangliosides, (1983).
24. D. J. Ewing, I. W. Campbell, A. Murray, J. M. M. Neilson, and B. F. Clarke, Immediate heart-rate response to standing: simple test for automatic neuropathy in diabetes, B.M.J. 1:145 (1978).
25. G. Sundqvist, B. Lilja, and L-O. Almer, Abnormal diastolic blood pressure and heart rate reactions to tilting in diabetes mellitus, Diabetologia 19:433 (1980).
26. T. Bennett, I. K. Farquhar, D. J. Hosking, and J. R. Hampton, Assessment of methods for estimating autonomic nervous control of the heart in patients with diabetes mellitus, Diabetes 27:1167 (1978).
27. M. Kato, The conduction velocity of the ulnar nerve and the spinal reflex time measured by means of the H wave in average adults and athletes, Tohoku. J. Exper. Med. 73:74 (1960).
28. F. Behse, F. Buchthal, and F. Carlsen, Nerve biopsy and conduction studies in diabetic neuropathy, J. Neurol. Neurosurg. Psych. 40:1072 (1977).
29. M. M. Martin, Diabetic neuropathy. A clinical study of 150 cases, Brain 76:594 (1953).
30. P. K. Thomas and R. G. Lascelles, The pathology of diabetic neuropathy, Quart. J. Med. 35:489 (1966).

GANGLIOSIDES--CLINICAL OVERVIEW

Stephen L. DeFelice and Max Ellenberg*

Bio/Basics International Corporation
Cranford, New Jersey 07016
*Mt. Sinai School of Medicine
New York, New York 10029

The preclinical ganglioside data hold great promise for these substances as potential therapy for a variety of neurologic conditions ranging from acute stroke and spinal cord injuries to chronic degenerative diseases. These data clearly indicate striking activity in both in vitro and in vivo models including the peripheral and central nervous systems. Unlike quick-acting drugs such as insulin and antibiotics, time and patience are required to establish the clinical utility of this class of substances. The following is a brief historical overview describing the rationale behind the clinical program including a discussion of preliminary results.

Early in the research program, the positive effect of gangliosides on the repair of crushed peripheral nerves was observed in animals utilizing doses of 50 mg/kg of the mixture chronically administered via the intraperitoneal route. Concurrently, the phenomenon of "ganglioside-induced" sprouting was discovered, which involves an accelerated reconnection of damaged nerves. Based on the preclinical data, it was believed at that time that higher doses given over a longer period were necessary to insure the delivery of sufficient ganglioside concentration to the receptor area for a particular clinical indication.

Further developments presented a more challenging picture. Gangliosides were evaluated in rodents with genetic diabetic neuropathy. It was discovered that nerve conduction velocity increased at doses of 1 mg/kg and above. The necessity of high dosing to achieve activity was now questioned. This was also supported by some preclinical data showing activity at low concentrations. In addition, it appeared likely that "nerve sprouting" was not entirely responsible for the effect on diabetic neuropathy, but that additional mechanisms were probably involved.

Subsequent to this, experiments in the CNS were initiated. It was noted that in physically damaged but otherwise normal CNS tissue, animal brain gangliosides could accelerate healing. In addition, an increase in certain enzyme activity after 14 days of administration was observed. It became apparent, therefore, that chronic administration may not always be necessary. This position was supported by in vitro models such as tissue cultures, where the activity of gangliosides can be detected within hours.

There was also another factor involved in planning the clinical development of gangliosides--that of cost. Most drugs used in medicine are produced synthetically at relatively low cost. Gangliosides must be carefully and painstakingly isolated from bovine brains and purified in an expert manner. The result is the very high cost of sophisticated raw material. In order for it to be generally available in sufficient quantity, it would be helpful to obtain the minimally effective clinical dose. With insulin a clear handle on dose-response curves for various types of hyperglycemia can be determined within 30 days. Regrettably, at this stage of knowledge it is almost impossible to achieve precision with ganglioside dosing for the clinical models are complicated, the studies long, and the end points imprecise.

There are additional problems. Most of the animal work done used intraperitoneal administration (i.p.). One cannot, however, chronically administer a drug intraperitoneally or intravenously (i.v.) as a practical matter. In addition, because gangliosides are destroyed in the digestive tract, the only route remaining is intramuscular (i.m.). The problem with this is the uncertainty regarding the rate and degree of i.m. absorption, since clinical distribution studies with gangliosides cannot be done due to technical limitations. Subsequent to this, however, biological activity has been demonstrated in animals utilizing the i.m. route--which was welcome news indeed.

With the preceding in mind, the approach of the ongoing present clinical program was divided into two categories.

1. Acute damage to normal tissue such as spinal cord injury and stroke: These studies would utilize a high dose and short-term i.v. administration.

2. Chronic peripheral neuropathies and certain chronic CNS diseases: These studies would utilize a lower dose, with chronic i.m. administration daily.

The clinical papers presented at the meeting represent the early findings of an extensive program to evaluate the effects of gangliosides in chronic neurologic disease. The data are still incomplete since several key clinical trials are ongoing. The present evidence available demonstrates clinical activity of gangliosides in man, the nature and importance of which should become clearer as clinical trials are completed and data analyzed.

The clinical presentations at this meeting dealt with three categories of diseases:

1. Amyotrophic lateral sclerosis: In two trials no statistical differences were noted between the ganglioside versus placebo-treated group, though one of the investigators believed he observed unexpected arrest or improvement of the clinical condition in three out of the twenty ganglioside-treated patients during the last month of the study. No such phenomenon was noted in the placebo-treated group. Because of the suggested possibility that higher dose--longer duration treatment may have some beneficial activity, we have initiated another clinical trial in ALS.

2. Diabetic neuropathy: The initial trial evaluated the effect of gangliosides on electrophysiologic parameters in insulin-dependent diabetics.[1] It was an open clinical trial with an untreated control group. A significant effect was noted on nerve sensory action potential latency and mixed nerve conduction velocity of certain nerves.

   Subsequent to this, all clinical trials were double blinded; one in the United Kingdom, one multicentered trial in Italy, and three in the U.S. (all three of which are ongoing.) Both symptoms and electrophysiologic changes were evaluated. The preliminary results, to date, do not demonstrate a consistent effect among all trials. This could be due to the unexpected variability of certain parameters evaluated in certain studies. What is clear, however, is that gangliosides appear to have an effect on both the improvement of symptoms and of electrophysiologic parameters, the nature and magnitude of which cannot be fully characterized until the clinical trials are completed and the data more thoroughly analyzed.

3. Chronic idiopathic polyneuropathy: The rationale behind this study was to search for certain uncommon neurologic diseases where gangliosides might have an obvious effect.

For this reason patients who were not improving or who were worsening with serious, rare diseases were selected. Though the clinical trial was not double-blind, it was believed that historical controls had value, i.e., if patients with these particular diseases began to improve on gangliosides, it would be improbable that this happened by chance. To date, fifty percent of the patients have shown both clinical and objective improvement.

It is clear that there are a number of unanswered questions ranging from the proper dose-duration regimen to the choice of the proper clinical indications for gangliosides. In addition to the clinical results presented at this meeting, we are conducting additional, and hopefully, definitive trials in the U.S. for the following indications:

1. Chronic idiopathic polyneuropathy

2. Friedreich's Ataxia

3. Charcot-Marie-Tooth

4. Adrenoleukodystrophy

5. ALS

6. Nonhemorrhagic stroke**

7. Spinal cord injury**

At the conclusion of these studies, we expect to have many answers regarding the clinical promise of gangliosides in a variety of neurologic indications--both GM1 and ganglioside mixture-by the end of 1984.

**We are awaiting FDA clearance to initiate these trials.

REFERENCES

1. G. Pozza, V. Saibene, G. Comi, and N. Canal, The Effect of Ganglioside Administration in Human Diabetic Peripheral Neuropathy, in: "Gangliosides in Neurological and Neuromuscular Function, Development and Repair," M. M. Rapport, and A. Gorio, eds., Raven Press, New York, (1981).

PARTICIPANTS

ABRAHAM, RALPH
   Metabolic University, Mint Wing, St. Mary's Hospital, Praed
   Street, London, W2, ENGLAND

ANDO, SUSUMU
   Department of Biochemistry, Tokyo Metropolitan Institute of
   Gerontology, 35-2 Sakaecho, Itabashiku, Tokyo 173, JAPAN

BASU, SUBHASH
   Department of Chemistry, University of Notre Dame, Biochemistry
   and Biophysics Program, Notre Dame, IN 46556, U.S.A.

BAUMANN, NICOLE
   Laboratoire de Neurochimie, Inserum U 134, CNRS ERA 421,
   Hospital de la Salpetriere, 75651 Paris Cedex 13, FRANCE

BRADLEY, WALTER
   Dept. of Neurology, University of Vermont College of Medicine,
   University Health Center, Burlington, Vermont 05405, U.S.A.

CAPUTTO, RANWEL
   Departmento de Quimica Biology, Facultad de Ciencias Quimicas,
   University Nacional de Cordoba, Cordoba, 5000, ARGENTINA

CREPALDI, GAETANO
   Istituto di Medicina Clinica, Cattedra di Patologia Medica
   Policlinico, Via Giustiniani. 2,35100 Padova, ITALY

DAIN, JOEL
   Department of Biochemistry & Biophysics, University of Rhode
   Island, Kingston, RI 02881, U.S.A.

DAWSON, GLYN
   University of Chicago, Department of Pediatrics, Box 82,
   950 East 59th Street, Chicago, IL 60637, U.S.A.

DeFELICE, STEPHEN
  Bio/Basics International Corp., 411 North Avenue, East Cranford,
  N.J. 07016, U.S.A.

DIMPFEL, W.
  AM PFAD 8,6301 Linden c/o Giessen, WEST GERMANY

DREYFUS, HENRI
  Centre de Neurochimie du CNRS, 5 Rue Blaise Pascal, 67084
  Strasbourg Cedex, FRANCE

EGGE, HEINZ
  Physiologische Chemie Institut der Universitat, Nussalle D-5300
  Bonn, WEST GERMANY

ETO, YOSHIKATSU
  Department of Pediatrics, Tokyo Jikei University School of
  Medicine, Minato-ku, 3-25-8 Nishi-shinbashi, Tokyo 105, JAPAN

FREDMAN, PAM
  Dept. of Neurochemistry, Psychiatric Research Centre, Univ. of
  Goteborg, St. Jorgen Hospital, S-422 03 Hisings Backa, SWEDEN

GATT, SHIMON
  Laboratory of Neurochemistry, Department of Biochemistry, Hebrew
  University, Hadassah Medical School, Jerusalem, ISRAEL

GHIDONI, RICCARDO
  Instituto di chimica Biologica, Facolta di Medicina e Chirurgie
  dell Universitadi Milano, Via C, Saldini 50, Milano 20133, ITALY

GORIO, ALFREDO
  Department of Cytopharmacology Fidia Research Laboratories,
  35031 Abano Terme, ITALY

GRANT, CHRIS
  Department of Biochemistry, University of Western Ontario,
  London, Ontario, CANADA NGA 5c1

HAKOMORI, SEN-ITIROH
  Biochemistry, Oncology Division, Fred Hutchinson Cancer Research
  Center, 1124 Columbia Street, Seattle WA 98104, U.S.A.

HALLETT, MARK
  Division of Neurophysiology, Brigham & Woman's Hospital, 75
  Francis Street, Boston, Mass 02115, U.S.A.

HANDA, SHIZUO
  Dept. of Biochemistry, Faculty of Medicine, Tokyo Medical and
  Dental Univ., Yushima, Bunkyo-ku, Tokyo 113, JAPAN

PARTICIPANTS

HOGAN, EDWARD
　Dept. of Neurology, Medical University of South Carolina, 171 Ashley Ave., Charleston, SC 29403, U.S.A.

HOROWITZ, STEVEN
　Long Island Jewish-Hillside Medical Center, 270-05 76th Ave., New Hyde Park, NY 11042, U.S.A.

IRWIN, LOUIS
　Dept. of Biochemistry, Eunice Kennedy Shriver Center, 200 Trapelo Road, Waltham, Mass 02254, U.S.A.

KARPIAK, STEPHEN
　New York State Psychiatric Institute, Division of Neuroscience, 722 West 168th Street, New York, NY 10032, U.S.A.

KOHN, LEONARD
　Building 4, Room B1-32, National Institutes of Health, Bethesda, MD 20205, U.S.A.

LEDEEN, ROBERT
　Dept. of Neurology, Albert Einstein College of Medicine, 1300 Morris Park Avenue, Bronx, NY 10461, U.S.A.

LI, YU-TEH
　Dept. of Biochemistry, Tulane Medical School 1430 Tulane Avenue, New Orleans, LA 70112, U.S.A.

MAKITA, AKIRA
　Biochemistry Laboratory, Cancer Institute, Hokkaido Univ. School of Medicine, Sapporo 060, JAPAN

MANDEL, PAUL
　Centre de Neurochimie du CNRS, 5 Rue Blaise Pascal, 67084 Strasbourg Cedex, FRANCE

MARCUS, DONALD
　Dept. of Medicine, Baylor College of Medicine 1200 Moursund Avenue, Houston, TX 77030, U.S.A.

MARKWELL, MARY ANN
　Dept. of Microbiology, Molecular Biology Institute, Univ. of California, Los Angeles, CA 90024, U.S.A.

McCLUER, ROBERT
　E.K. Shriver Center for Medical Retardation, 200 Trapelo Road, Waltham, Mass. 02154, U.S.A.

MIYATAKE, TADASHI
　Dept. of Neurology, Brain Research Institute, Niigata University, Niigata 951 Asahimachi 1, JAPAN

MOSER, HUGO
  John F. Kennedy Institute, 707 North Broadway, Baltimore, MD 21203, U.S.A.

NAARDEN, ALLAN
  Texas Neurological Institute at Dallas, 7777 Forest Lane, Dallas, TX 75230, U.S.A.

NAGAI, YOSHITAKA
  Dept. of Biochemistry, Faculty of Medicine, Univ. of Tokyo, Hongo, Bunkyo-ku, Tokyo 113, JAPAN

ODERFELD-NOWAK, BARBARA
  Nencki Institute of Experimental Biology, Polish Academy of Sciences, 3 Pasteur Str. 02 093, Warsaw, POLAND

RAHMANN, HINRICH
  Zoological Institute, Univ. Stuttgart-Hohenheim, 7000 Stuttgart 70 (Hohenheim), Garbenstrabe 30-BIO II, WEST GERMANY

RAPPORT, MAURICE
  Dept. of Neuroscience, New York State Psychiatric Institute, 772 West 168th Street, New York, NY 10032, U.S.A.

RATTAZZI, MARIO
  Children's Hospital of Buffalo, Dept. of Pediatrics, Div. of Human Genetics, 86 Hodge Avenue, Buffalo, NY 14222, U.S.A.

ROISEN, FRED
  Dept. of Anatomy, UMDNY-Rutgers Medical School, P. O. Box 101, Piscataway, NJ 08854, U.S.A.

ROSENBERG, ABRAHAM
  Dept. of Biochemistry & Biophysics, Stritch School of Medicine, Loyola University, Room 6648, Maywood, IL 60153, U.S.A.

SANDHOFF, KONRAD
  Institut fur Organische Chemie und Biochemie, der Universitat Bonn, Gerhard-Domagk-Strasse 1, D-5300 Bonn 1, WEST GERMANY

SCHAUER, ROLAND
  Biochemisches Institut der Christian-Albrechts University, Olshausenstrasse 40/60, D-2300 Kiel, WEST GERMANY

SCHENGRUND, CARA-LYNNE
  Dept. of Biological Chemistry, Pennsylvania State University, M.S. Hershey Medical Center, Hershey, PA 17033, U.S.A.

PARTICIPANTS

SEIFERT, WILFRIED
  Max-Planck-Institut fur Biophys, Chemie, P. O. Box 968, 3400 Gottingen, WEST GERMANY

SELA, BEN-AMI
  Department of Biophysics, Weizmann Institute of Science, Rehovot, ISRAEL

SEYFRIED, THOMAS
  Dept. of Neurology, Yale University School of Medicine, 333 Cedar Street, New Haven, CT 06510, U.S.A.

SKRIVANEK, JOSEPH
  Division Natural Science, State University of New York, College at Purchase, Purchase, NY 10577, U.S.A.

SONNINO, SANDRO
  Dept. of Biological Chemistry, The Medical School, University of Milano, Via Saldini 50, Milano 20133, ITALY

SUZUKI, AKEMI
  Tokyo Metropolitan Institute of Medical Science, 3-18 Honkomagome, Bunkyo-ku, Tokyo 113, JAPAN

SUZUKI, KUNIHIKO
  Dept. of Neurology and Neuroscience, Rose F. Kennedy Center for Research at Albert Einstein College of Medicine, 1300 Morris Park Avenue, Bronx, NY 10461, U.S.A.

TAKETOMI, TAMOTSU
  Department of Biochemistry, Shinshu University School of Medicine, Matsumoto, Nagano 390, JAPAN

TETTAMANTI, GUIDO
  Instituto di Chimia Biologica, Facolta di Medicina e Chirurgia, dell' Universitadi Milan, Via Saldini 50, Milano 20133, ITALY

TOFFANO, GINO
  Department of Biochemistry, Fidia Research Laboratories, 35031 Abano Terme, ITALY

WALKLEY, STEPHEN
  Dept. of Neuroscience, Rose F. Kennedy Center for Research at Albert Einstein College of Medicine, 1300 Morris Park Avenue, Bronx, NY 10461, U.S.A.

WOLFE, LEONHARD
  Dept. of Neurology and Neurosurgery, Montreal Neurological Institute, McGill Univ. 3801 University St., Montreal, Quebec H3A 2B4, CANADA

YAMAKAWA, TAMIO
   Tokyo Metropolitan Institute of Medical Science, 3-18
   Honkomagome, Bunkyo-ku, Tokyo 113, JAPAN

YATES, ALLAN
   Division of Neuropathology, Ohio State University, 473 West 12th
   Avenue, Columbus, OH 43210, U.S.A.

YOHE, HERBERT
   Building 7 - Room 203, VA Medical Center, West Haven, CT 06516,
   U.S.A.

YU, ROBERT
   Department of Neurology, Yale University School of Medicine, 333
   Cedar Street, New Haven, CT 06510, U.S.A.

AUTHOR INDEX

A

Abe M., 263
Abraham R.M., 607
Abraham R.R., 607
Agnati L.F., 475
Aldinio C., 475
Aloj S., 355
Ando S., 241
Aporti F., 549
Ariga T., 103
Autilio-Gambetti L., 431

B

Bartfeld H., 499
Basu M., 249
Basu S., 249
Battistin L., 601
Baumann N., 297
Berry-Kravis E., 341
Bradley W.G., 565
Bremer E.G., 381

C

Calza L., 475
Caputto R., 147
Chien J.-L., 287
Chigorno V., 273, 307
Chon H.-C., 249
Cohen I.R., 441
Cohen O., 441
Crepaldi G., 601

D

Dal Tosso R., 475

Dasgupta S., 287
Davidson J., 581
Dawson G., 341
DeFelice S., 581, 625
Demou P.C., 87
DiGregorio F., 549
Dimpfel W., 525
Dreyfus H., 27, 513

E

Egge H., 55
Ellenberg M., 625
Endo T., 455
Eto Y., 431

F

Fedele D., 601
Ferret B., 513
Fink H.-J., 535
Fiorilli A., 273
Flood T., 575
Fredman P., 369
Freysz L., 513
Fuxe K., 475

G

Gasa S., 111
Ghidoni R., 273, 307
Gierliani A., 273
Gorio A., 465, 513, 549
Grant C.W.M., 119
Grollman E.F., 355

H

Hakomori S.-I., 333, 381
Hallett M., 575
Handa S., 65
Hanfland P., 55
Hara A., 419
Harpin M.-L., 297
Harrington H., 575
Harris L., 581
Harth S., 513
Hashimoto Y., 263
Held S.J., 499
Hilbig R., 395
Hofleig J.H., 155
Hogan E.L., 287
Horowitz S.H., 593
Huang Y.-Y., 15

I

Irwin L.N., 319
Iwamori M., 135

J

Janigro D., 465

K

Karpiak S.E., 489
Kasama T., 419
Kiuchi Y., 263
Koerner T.A.W., Jr., 87
Kohn L.D., 355
Kon K., 241
Konat G., 441
Kundu S.K., 455
Kushi Y., 65
Kyle J.W., 249

L

Lacetti P., 355
Leon A., 475
Lev-Ram V., 441
Li S.-C., 213
Li Y.-T., 213

M

Makita A., 111
Malesci A., 307
Mandel P., 27
Marcus D.M., 455
Markwell, M.A.K., 369
Massarelli R., 513
Matsui Y., 27
McGrath J.T., 431
Miyatake T., 103
Moore J., 581
Muhleisen M., 395

N

Naarden A., 581
Nagai Y., 135, 183
Nakamura M., 111

O

Offner H., 441
Ono Y, 241
Otten U., 525

P

Peter-Katalinic J., 55
Peters M.W., 119
Pohlentz G., 227
Portoukalian J., 297
Prestegard J.H., 87
Probst W., 395

R

Rahmann H., 395
Rapport M.M., 15
Rebel G., 27
Roisen F.J., 499

S

Sanai Y., 183
Sandhoff K., 227
Savoini G., 475
Scarsdale J.N., 87
Schauer R., 75
Schiavinato A., 549

AUTHOR INDEX

Schröder C., 75
Schwarzmann G., 227
Scott D.D., 455
Seifert W., 535
Sekine M., 103
Sela B.-A., 441
Seyfried T.N., 169
Shukla A.K., 75
Siliprandi R., 549
Slater N., 575
Sonnino S., 273, 307
Spero D.A., 499
Stewart S.S., 455
Suzuki A., 263
Suzuki K., 407
Svennerholm L., 369

T

Taketomi T., 419
Tanaka Y., 241
Tettamanti G., 197, 273, 307
Tipnis U.R., 155
Toffano G., 475
Tombaccini D., 355
Tsuji S., 183
Tyler H.R., 575

V

Valenti G., 475
Venerando B., 273
Vitadella M., 549

W

Warner J.K., 155
Wynn V., 607

Y

Yamakawa T., 3, 263
Yanagisawa K., 111
Yates A.J., 155
Yorke G., 499
Yu R.K., 39, 87, 169
Yusuf H.K.M., 227

Z

Zalc B., 297
Zanoni R., 465
Zini I., 475

SUBJECT INDEX

A2B5 monoclonal antibody
  ganglioside GQ 539, 541
  production, 16
  specificity, 18
ABO blood group, 9, 10
O-acetylated sialic acid,
    307-317
N-acetylgalactosaminyl-
    transferase, 206
  genetic regulation, 263
  opioid peptides, 206
β-N-acetylhexosaminidase,
    see Hexosaminidase A
N-acetylneuraminate lyase,
    79, 81
N-acetylneuraminic acid
    (see also sialic acid)
  capillary GLC, 76, 80
  HPLC-MS, 81, 83
  membrane surface, 127
  NMR, 92
AChE transport, diabetic
    neuropathy, 551
Activator proteins
  and genetic defects, 410
  glycosphingolipid
    catabolism, 213
  GM2-gangliosidosis, 217
  normal urine presence, 224
Active E-rosette test, 444
Acylneuraminate cytidylyl-
    transferase, 79-80
Adenylate cyclase
  neuritogenesis, 508
  serotonin receptor, 341
  TSH antibodies, 355
Adrenal gland, 103-109

Adrenoleukodystropy, 412
    568, 569
Affinity chromatography, 19
Aging effects, GM1, 241-247
Alkali labile gangliosides,
    307-317
"Allosteric Regulator"
    model, 390-391
Alloxan diabetic neuropathy,
    549-561
Alternation behavior, CNS
    damage recovery, 490
"Aminoglycolipid route", 199
Amphibians, ganglioside pattern
    320-323
Amyotrophic lateral sclerosis
    treatment, 565, 575, 627
2,7-Anhydroneuraminic acid, 84
Animal models, genetic diseases,
    411, 412
Anomeric configurations, 45
Anti-asialo-GM1 antibody, 184-187
Anti-Forssman antibody, 141
Anti-ganglioside antibodies
  pathology, 413
  regeneration inhibition, 477
Anti-GM1 antibodies
  behavioral dysfunction, 493
  MS and SLE, 456
  neurite outgrowth, 469, 493
  NGF-induced sprouting inhibition,
    533
  Trembler mutant, 300, 303
Anti-GM2(NeuGc) antibodies,
    266, 267
Anti-GM3, Trembler, 300
Anti-GQ, 539

Anti-GSL antibodies, 455
Antibodies
  asialo-GM1, 183
  function, 19-22
  specificity, 16-20, 455-461
  raising, 16
  glycosphingolipids, 455-461
Antigens
  erythrocyte glycolipids, 10
  multiple sclerosis, 441-451
Antiidiotype theory, 365
Arylsulfatase A, catabolism,
      activator, 214
Asialo-GM1
  antibody specificity, 19, 20
  and cellular activity,
      antibodies, 183
  in MS and SLE, antibodies,
      456-461
  GM1-gangliosidosis, 424
  NK cell relationship, 186
Asialo-GM2
  canine gangliosidosis, 433
  GM1-activator, 216-217
Astrocytes
  in culture, 29, 30, 32, 33,
      537
  and GD3, 172
Autoimmune thyroid disease,
      355-366
Avian gangliosides, 321, 322
Axonal regeneration, 465-470
Axonal size, diabetic neuro-
      pathy, 553
Axonal transport (see also
      slow transport)
  brain gangliosides, 204
  diabetic neuropathy, 552
  peripheral nerve, 155-162

Bacterial toxins, 334-336
BALB/c mice, GM2(NeuGc)
      expression, 264-270
Bass fish, alkaline labile
      gangliosides, 308,
      310, 311
Batten disease, 412
Birds
  phylogeny, 321-323
  thermal adaptation, 399

Bone marrow gangliosides, 140, 141
Brain, chick
  neuron cultures, 28-29
  ontogeny, 324-326
Brain, dog, GM2-gangliosidosis,
      432, 433
Brain gangliosides
  aging and hormone effects,
      241-247
  alkali labile type, 307-317
  bovine, 499-509
  degradation, 202
  distribution, 136, 137, 140
  incorporation rate, myelin,
      241-247
  metabolism, and axon flow, 204
  molecular diversity, 136
  phylogeny and ontogeny, 319
  rat, neuron culture, 28, 29
  Trembler mutant, 300-304
Brain, lamprey, alkali labile
      gangliosides, 308
Brain, mouse
  alkali labile gangliosides, 310
  myelin GM1, 241-247
Brain, rabbit, alkali labile
      gangliosides, 314, 315
Brain, rat, gangliosides, 140
Buffy coat gangliosides, 140
Butyrate, metabolic effects, 206

Calcium complexes, thermal
      adaptation, 399-403
Canine GM2-gangliosidosis, 431-439
Capillary GLC, sialic acids, 76-84
Carcinoma treatment, 22
Cat GM2-gangliosidosis, 439
Cell cultures, 27-34, 513-523,
      525-533, 535-544
Cell growth, ganglioside
      inhibition, 381-391
Cell proliferation, GQ1b
      promotion, 187-190
Cell surface
  gangliosides, topography,
      141-143
  glycosphingolipid markers,
      183-187
Cerebellar gangliosides
  alkaline labile form, 314

Cerebellar gangliosides (continued)
  distribution, rat, 140
  neurological mutants, 169
  thermal adaptation, 315, 316
Ceramide
  field-desorption, 67, 69
  ganglioside precursor, 202
Cerebroside sulfate (see also Sulfatide)
  discovery, 5
Chemical ionization mass spectrometry, 65-72
  N-acetylneuraminic acid, 83
  carbohydrate sequencing, 43
  versus electron impact, 243
Chicken gangliosides
  adrenal gland, 106-108
  in culture, 28, 29
  ontogeny, 324-326
  phylogeny, 322
  skeletal muscle, 287-294
"Chol-1" antibodies, 16, 21
Cholera toxin
  GM1 localization, 21
  receptors, overview, 334
  TSH receptor expression, 363-366
Cholinergic damage, recovery, 484
Chromaffin granules, gangliosides, 106
Chromium trioxide method, 45
Chromosome markers, glycolipid expression, 269
Chronic idiopathic polyneuropathy, 568, 627
Classification, gangliosides, 40
CMP-sialic acid, as contaminant of gangliosides, 166
Conduction velocity, ganglioside treatment, 552, 570, 577, 583, 603, 610
Cronassial therapy, 565, 575, 593, 607
"Cryptic glycolipids, 122

Cyclic AMP
  metabolic regulation, 206
  neuritogenesis, 508
  serotonin mediation, 343
Cytochalasin D, neurite growth, 504, 507
Cytoskeleton, neurite formation, 500-509
Cytoskeleton-associated glycolipid, 144

Demyelinating disease, immunology, 441-451
Dentritic development, 325
Dentritic spines, GT1a, 176
2-Deoxy-2,3-dehydroneuraminic acid derivatives, 76
Deuterium labeling, advantages, 242
Diabetic neuropathy treatment, 549, 575, 581, 593, 601, 607, 627
Dopaminergic neurons, recovery, GM1, 477
Dormouse, gangliosides, thermal adaptation, 315, 399, 401
Dorsal root ganglia, neurite formation, 500

E-rosetting, demyelinating disease, 441-451
Ectoglycosyl transferase, neuronal membrane, culture, 32-33
Ectotherms, 395-403
EEG spiking, GM1 antibody, 19
Elasmobranchs, 320
Electron impact mode MS
  carbohydrate sequencing, 43
  versus chemical ionization, 243, 244
Enkephalin
  ganglioside effects, 5HT, 342, 351
  receptors, 348
Entorhinal lesions, recovery, 490-494, 535-544
Enzyme defects, catabolism, 408-410
Epidermal growth factor
  cell growth, 382-391

Epidermal growth factor
(continued)
  neuritogenesis, 508
Epilepsy, anti-ganglioside
    antiserum, 413
Erythrocytes
  equine, 111-117
  FAB spectra, 59, 61, 62
  ganglioside discovery, 8-10
  ganglioside distribution,
    rat, 140, 141
  ganglioside topography,
    surface, 142, 143
Ester linkages, gangliosides,
    307-317
Evoked potentials
  Cronassial effect, 597
  diabetic neuropathy, 555
Evolution, 319-327
Experimental allergic
    encephalomyelitis
  immunology 447
  pathology, 413
Extraction procedure, radio-
    labelled contaminants,
    164

Fast-atom bombardment
  carbohydrate sequencing, 43
  glycosphingolipid structure,
    55-62
  method, 55-56
Fetal gangliosides, 325
Fibroblast growth factor, 382
Fibronectin receptors, 334
Field-desorption MS
  carbohydrate sequencing, 43
  glycolipid analysis, 65-72
  versus secondary ion MS, 71
Fish gangliosides
  phylogeny, 320-323
  thermal adaptation, 395-403
Freeze-etching, 124-126
Frog gangliosides
  ontogeny, 323, 324
  phylogeny, 322
Fucoganglioside
  equine erythrocytes, 111
  NMR, 112-115
  structure, 117

Fucolipids, GM1-ganglio-
    sidosis, 426
Fucosyl-GM1, 142

GABA release
  GM1 antibody effect, 20
  neuron cultures, 520
Galactocerebrosides, 143-144
α-Galactosaminidase, 293
α-Galactosidase, 293
β-Galactosidase
  activator protein, 216
  developing muscle, 293
  GM1 gangliosidosis, 409
Galactosylceramide (see also
    Galactocerebroside)
  discovery, 3
  distribution, 3-4
  GM1 gangliosidosis, 421
Galactosyltransferase, chick
    brain, 250-255
Ganglio-N-tetraose, tissue,
    140, 141
Ganglio series
  biosynthesis, 249-258
  metabolism, 197-207
Ganglioside biosynthesis, 249-258
  golgi apparatus, 227-238
  new approaches, 273-282
  pathway, 198-204
Ganglioside degradation, 199-204,
    273-282
Ganglioside isolation procedures,
    163-166
Ganglioside metabolism, 197-207,
  new approaches, 273-282
Ganglioside receptors, 333
  lectin binding, 125, 126
  overview, 333-336
  and regeneration, 493
  Sendai virus, 369-378
Ganglioside structure, 39-47, 55,
    65, 75, 87, 103, 111
"Ganglioside syndrome", 413, 449
Ganglioside therapy
  ALS, 565-572
  autonomic neuropathy, 607
  diabetic neuropathy, 549, 575,
    581, 593, 601, 607
  asymmetry effect, 585

SUBJECT INDEX 643

Ganglioside therapy (continued)
  overview, 414, 625-628
Ganglioside transport (see axonal transport)
GD1a
  erythrocyte surface topography, 142
  FAB spectra, 57, 58
  FD-MS and SI-MS, 67, 69, 70
  golgi apparatus synthesis, 228-238
  in cell cultures, 30
  in MS, immunology, 445
  in mutants, cerebellum, 170-172, 174-178
  in nerve cell culture, 514, 526, 538
  in rat tissues, 141
  Sendai virus receptor, 371
  stepwise synthesis, 249-252
  thermal adaptation, 401
  Trembler mutant, 300-304
GD1b
  in cell cultures, 31
  immunologic cross-reactions, 16
  in mutants, cerebellum, 170
  in nerve cell cultures, 514, 526, 538
  in neuron cultures, 29
  rat tissues, 141
  thermal adaptation, 401
  Trembler mutant, 300-304
GD2
  in neuron cultures, 29
  in regeneration, 540-544
GD3
  in adrenal gland, 106-108
  antibody specificity, 16
  in cell cultures, 29-32
  developmental changes, 289
  genetic dystrophy, 290
  golgi apparatus synthesis, 228-238
  mutant studies, cerebellum, 171
  ontophyletics, 325
  phosphatidylglycerol stimulation, 228

GD3 (continued)
  in rat tissues, 140
  reactive gliosis, 172
  serotonin binding protein, 350
GM1
  activator protein, 215
  aging and hormone effects, 241-247
  antibody function, 19-22
  antibody specificity, 20
  brain damage recovery, 477-485, 491-494
  cell cultures, 29-32
  cell growth inhibition, 382
  erythrocytes, 141-143
  FAB spectra, 57, 58
  FD-MS and SI-MS, 66
  in GM1 gangliosidosis, 424
  genetic regulation, 265
  golgi apparatus synthesis, 228-238
  and growth factors, 382
  mutant studies, cerebellum, 171
  in MS, immunology 445
  myelin incorporation, 241
  ontophyletics, 325, 326
  thermal adaptation, 401
  Trembler mutant, 303
  tunicamycin inhibition, 230-232
GM1-activator, 215, 222
GM1-gangliosidosis
  activator protein, 216
  enzyme defect, 409-410
  organ gangliosides, 419
GM1(NeuGc), genetic regulation, 265, 269
GM1 uptake, NCB-20 cells, 349
GM2
  activator protein, 216-224
  antibody specificity, 16
  genetic dystrophy, 291
  genetic expression, 263
  golgi apparatus synthesis, 228-238
  muscle development, 291
  1-D and 2-D NMR, 88-95
  rat bone marrow, 141
  structure, 88
  tunicamycin inhibition, 229-232

GM2-gangliosidosis, 216, 224
  canine, 431-440
  enzyme defect, 408-411
  GM2-activator, 217, 410
  juvenile, 408, 409
GM2(NeuGc), genetic
    expression, 263
GM3
  adrenal gland, 106-108
    subcellular fractions,
      107-108
  antibody specificity, 16
  in cell cultures, 30, 537
  cell growth inhibition,
    382-391
  developmental changes,
    289, 290
  genetic dystrophy, 290
  genetic regulation, 263
  and growth factors, 382
  ontophyletics, 325, 326
  peripheral nerve, 300
  in rat tissues, 140, 141
  Trembler mutant, 300-304
GM4
  antibody specificity, 16
  rat tissues, 141
GQ1b
  MS, immunology 445
  mutants, cerebellum, 170
  nerve growth promoter, 187
  Sendai virus receptor, 371
  thermal adaptation, 401
GT1a
  in mutants, cerebellum, 170
  development, 174-178
  Purkinje cell enrichment,
    176
GT1b
  antibody specificity, 16-20
  FAB spectra, 57-58
  in mutants, cerebellum, 170
  Sendai virus receptor, 372
  thermal adaptation, 401
  transferase inhibition and
    light, 151, 152
  Trembler mutant, 301, 303
Genetic abnormalities, 408
Genetic diabetic neuropathy,
  549-561

Genetic dystrophy, 287-294
GLC-mass spectrometry, 42
Glial cell patterns
  cell culture, 29-33, 537
  glioblastoma, 32
Gliosis, see Reactive gliosis
Globoside
  developmental changes, 291
  erythrocyte membranes, 9-10
α-Glucosidase, 293
Glucosyl ceramide
  activator protein, 215
  distribution, 3-4
  ganglioside precursor, 202
Glycoproteins
  carbohydrate structures, 10
  light stimulus effects, 148
  receptors, overview, 334
  TSH receptor, 361-366
Glycosidases
  carbohydrate sequencing, 42
  development, chicks, 292-294
  genetic dystrophy, 289-292
Glycosidic substitution, 43-45
Glycosphingolipids
  anomeric configuration, 45
  in membranes, 119-126
  structure analysis, FAB, 55
  structure, NMR, 87
  subcellular distribution, 143
  surface antigens, 183-187
Glycosyl transferases
  golgi vesicles, purified, 227-238
  metabolic regulation, 205
  nerve cell culture, 32-33
Golgi complex gangliosides
  new approaches to study, 273-282
  rat liver, 227-238
  synthesis, 160, 161, 202-204
GOTO cells, 188-190
Granule cells, 169-178
Graves' disease, 356, 365, 366

Headgroup dynamics, glycolipids,
  119
Hematopoietic system
  ganglioside diversity, 136
HeLa cells, Sendai virus receptors,
  372
Hematoside, erythrocytes, 8

SUBJECT INDEX

Hepatocytes, GM1 metabolism, 280-282
Hexosaminidase
　animal models, defects, 411, 412
　canine GM2-gangliosidosis, 435
　catabolism, activator, 214
　developing muscle, 293
　in gangliosidoses, 218-224, 408, 409
　Sandhoff disease, 409
Hibernation
　alkali labile gangliosides, 315, 316
　dormouse gangliosides, 399
Hippocampus, 535-544
Homeotherms, 395-403
HPLC
　vs capillary GLC-MS, 83
　and mass spectrometry, 80
　sialic acids, 77-84
Hypertrophic neuropathy, 297

Idiopathic sensorimotor polyneuropathy, 568, 627
Immunology
　gangliosides, present status, 15-22
　multiple sclerosis, 441
Interferon
　anti-asialo-GM1, 185
　ganglioside receptors, 334
Intestine, ganglioside distribution, 140
Isoelectric focusing
　glycosyltransferases, 256
　sialytransferase, 257

Juvenile GM2-gangliosidosis
　and canine gangliosidosis, 437
　enzyme defect, 408, 409

Kidney
　ganglioside distribution, 140
　GM1 gangliosidosis, 419-428

Lactonic gangliosides, 314

Lactosylceramide
　canine gangliosidosis, 433
　dystropic muscle, 292, 294
　ganglioside synthesis, 198
　GM1-gangliosidosis, 424
Lafora disease, 412
Lamprey
　alkaline labile gangliosides, 308-313
　phylogeny, 320
　thermal adaptation, 396-399
Learned avoidance, 413
Lectins
　freeze-etching, 125, 126
　ganglioside receptor binding, 125, 126
Light exposure, ganglioside labeling, 147
Liposomes
　freeze-etching, 125, 126
　immunology studies, 458
Liver
　ganglioside distribution, 140-144
　GM1 metabolism 273-282
　GM1-gangliosidosis, 419
　GM2-gangliosidosis, 432
　genetic expression, 263
LM1 ganglioside
　GM1-gangliosidosis, 426
　stepwise biosynthesis, 249
Lung, ganglioside distribution, 140
Lurcher mutation, 169-178
Lymphocytes, MS, 441-451
Lysosomal enzymes
　ganglioside degradation, 202
　GM1 metabolism, liver, 276
　plasma membrane relationship, 203, 204
　study approaches, 273-282

Mammary tumors, anti-asialo GM1 serum, 185
α-Mannosidase, muscle, 293
Mass-fragmentography, 41
Mass spectrometry
　and capillary GLC, 80-84
　carbohydrate sequencing, 43
　CI versus EI, 243, 244
　glycolipid analysis, 55-63, 65-72
　and HPLC, 80-84

MDBK cells, Sendai virus, 372
MDCK cells, Sendai virus, 372
Median nerve conduction, 597, 603
Membrane fusion, 373-378
Metachromatic leuko-
    dystrophy, 412
Microsomes
  adrenal gangliosides, 106
  ganglioside synthesis, 160
Monkey, adrenal gland
    gangliosides, 106-108
Monoclonal antibodies
  anti-GQ, culture, 541
  carbohydrate sequencing, 43
  thyrotropin receptor, 355
Mononeuropathy, treatment, 577
Morphine analgesia, inhibition
    by GM1 antibody, 19
Motoneurons, sprouting, 469
Motor action potentials, 610
Motor nerve conduction, 577, 584, 597, 610
Mouse
  adrenal gland gangliosides, 106-108
  ontophyletics, 325, 326
Mouse hepatitis virus, 185
Mucolipidosis type IV, 410
Mucopolysaccharidosis, 412
Multiple sclerosis
  anti-asialo-GM1 titer, 186
  GSL antibodies, 455-461
  secondary abnormalities, 413
Muscular dystrophy, 287-294
Myelin
  diabetic neuropathy, 555
  galactosyl ceramide
      localization, 3
  ganglioside immunization, 449
  GM1 incorporation, 241-247
Myelination, 241, 324, 421

Natural killer cells, 186
NB-1 cells, GQ1b effect, 188
NCB-20 cells, 342-351
Negative ion mode, FAB, 56

Nerve conduction velocity, diabetic
    neuropathy, 570, 577, 583, 603, 610
Nerve growth factor
  in cell cultures, 528-533
  ganglioside interaction, 469
  GM1 antibody interaction, 20
  neuritogenesis, 508
Nerve regeneration, see
    Regeneration
Neuraminic acid (see also
    Sialic acid)
  ganglioside chromogen, 8
  myelin incorporation, aging, 244, 245
Neuraminic acid derivatives
  capillary GLC, 76
  HPLC, 77-84
Neuraminidase (see also Sialidase)
  for structure determination, 42
Neuritogenesis
  in culture, 529-533
  embryos, GM1, 481-485
  ganglioside treatment, 469, 470, 489
  and GQ1b, 188-190
  hippocampal neurons, 544
  and microfilaments, 500
  and microtubules, 500
  and microvilli, 500
  N-2A cells, 499-509
  overview, 465-470
Neuron cultures, see Cell cultures
Neurotransmitters, 341-351
N-glycolylneuraminic acid (see also
    Sialic acid)
  adrenal medula, 107
  HPLC, 79
Niemann-Pick disease, 412
Nuclear Magnetic Resonance
  anomeric configuration, 45
  applications, 87-98
  fucoganglioside, 111-117
  glycosidic substitution sites, 44

Oligodendroglioma, 32
Opiate receptors, sulfatide, 334, 336
Opioids, metabolism effect, 206

SUBJECT INDEX

Optic nerve, ganglioside transport, 150, 153
Optic neuritis, 446
Optic tectum
  ganglioside ontogeny, 324
  ganglioside transport to, 147-154
Orthostatic test, 610
Ox, adrenal gland gangliosides, 106

P-face, myelin, diabetic neuropathy, 555
Painful diabetic neuropathy, 581
Paresthesias, ganglioside treatment, 577, 604
Passive avoidance learning, 19
PC12 cells
  neuritogenesis, 508
  sprouting study, 469, 525
Periodate oxidation technique, 43, 44
Peripheral nerve gangliosides, 155, 300
Permethylated gangliosides
  FAB, 57, 58, 60-62
  glycosidic substitution, 42-44
  mass spectrometry, 67, 69, 70
Peroneal nerve conduction velocities, 586, 597, 603, 610
Phosphatidylglycerol stimulation
  ganglioside synthesis, 207
  rat liver golgi complex, 227
Phosphatidylinositol
  diabetic neuropathy, 560
  serotonin receptor, 346
  synthesis stimulation, 228
Phosphorylation, glycosyl-transferases, 206
Phylogeny, 319-327, 395-403
Plasma membrane
  in culture, exogenous gangliosides, 514
  freeze-etching, 124-126
  GM1 metabolism, 277, 278

Plasma membrane (continued)
  glycolipid behavior, 119
  lysosome relationship, 203
  metabolism site, 201, 273
  neuritogenesis, 499, 508
  permeability, GD3, 177
  regeneration effects, 476, 493
Platelet-derived growth factor, 382-391
Positive ion mode, FAB, 56
Primary neuronal cultures, 28, 525, 535
6-$\underline{n}$-propyl-2-thiouracil, 245
Purkinje cell degeneration mutant, 169-178
Puromycin, metabolism changes, 205

Rabbit gangliosides
  adrenal gland, 106-108
  alkali labile type, 314
  antibodies, 16-18
  demyelinating disease model, 447, 460
  peripheral nerve, 155-166
Rainbow trout, ontogeny, 323
Rat gangliosides
  adrenal gland, 106-108
  erythrocyte topography, 142, 143
  phylogeny, 322
  tissue distribution, 140
Reactive gliosis
  cerebellar gangliosides, 169-178
  GD3, marker, 172
Receptor modulation, 381-391 (see also Ganglioside receptors)
Regeneration
  brain damage, 475, 489
  damage protection, 470
  diabetic neuropathy, 557
  embryo model, 481-485
  ganglioside treatment, 469, 475, 489
  hippocampus, 535-544
  overview, 465-470, 476
  and slow transport, 560
Regio quadrigemina
  alkali labile gangliosides, 315
Reptiles, phylogeny, 321-323
Reserpine sedation block, GM1 antibody effect, 19

Retina
  ganglioside labeling, 147
  ganglioside ontogeny, 324

Sandhoff's disease
  and cat GM2-gangliosidosis, 439
  enzyme defect, 409, 410
Sciatic nerve
  diabetic neuropathy, 549
  ganglioside synthesis and transport, 155-166
  Trembler mouse, 297-304
Secondary ion mass spectrometry, 65
Seminolipid
  discovery, 5-6
  FD-spectra and SI-spectra, 69-71
Sendai virus, 334, 336, 369
Sensory nerve conduction, 583, 597
Serotonin binding protein, and GD3, 350
Serotonin receptor
  adenylate cyclase, 341-351
  clonidine, 344-345
Serotonin release, GM1 antibody effect, 20
Serum albumin, and ganglioside binding, 126-129
Sialate O-acetyl esterase, 80, 83
Sialic acid (see also Neuraminic acid)
  in alkaline labile gangliosides, 307, 311
  capillary GLC, 75-84
  developmental changes, 289
  FAB analysis, 57, 59-60
  ganglioside metabolism, 198-201
  HPLC, 75-84
  membrane release, 127-129
  myelin incorporation, 244
  plasma membrane activity, 205
  proposed structures, 7-8
  Sendai virus receptor, 371
  thermal adaptation, 397-403

Sialidase (see also Neuraminidase)
  degradation pathway, 199, 201
  mucolipidosis IV, defect, 409, 410
  receptor destruction 370
  site of action, 202-205
Sialoglycoproteins, 375, 376
Sialosylparagloboside
  antibody specificity, 17
  developmental changes, 289
  peripheral nerve, 300
Sialyl transferase
  developmental changes, 292
  ganglioside synthesis, 252, 257
  genetic dystrophy, 289-294
  and light exposure, 151
  neuronal membrane, 32, 33
  plasma membrane, 205
  regulating factors, 206
"Sialylation-desialylation" cycle, 205
Side effects, ganglioside treatment, 585
Slow transport (see also Axonal transport)
  diabetic neuropathy, 555-560
  sprouting, 467
Sodium pump, diabetic neuropathy, 560
"Soft ionization techniques", 65
Solid phase antibody methods, 18
Spinal cord
  ganglioside distribution, rat, 140
  thermal adaptation, 315
Spleen gangliosides, 140, 419
75-84 Sprouting
  brain damage, GM1, 478-485
  cell cultures, 513-523, 525-533
  entorhinal lesions, 535
  ganglioside treatment, 469
  overview, 467
Staggerer mutation, 169-178
Starfish, sialic acid, 76
Stomach, ganglioside distribution, 140
Striatal neurons
  embryo development, GM1, 481, 482
  recovery, 478-480

SUBJECT INDEX

Subacute sclerosing panencephalitis, 412
Substantia nigra, damage recovery, 478-481, 484
Sulfated glycolipids
  discovery, 4-7
  FD-MS and SI-MS, 69
Sural nerve conduction, 597, 603, 610
Swiss mouse 3T3 cell growth, ganglioside inhibition, 382-391
Synaptogenesis
  cell cultures, 529, 535-544
  diabetic neuropathy, 557
  entorhinal lesions, 535
  GD2, 544
  weaver mutant, 174-178
Synaptosomes, ganglioside synthesis, 160
Systemic lupus erythematosus
  anti-asialo-GM1 titer, 186
  GSL antibodies, 455-461

Tay-Sachs disease, see GM2-gangliosidosis
Teleosts, 320, 397
Testis, ganglioside distribution, 140
Thermal adaptation
  alkali labile gangliosides, 315, 316
  vertebrates, 395-403
Thymus, ganglioside distribution, 141
Thyroid gangliosides, antibodies, 360
Thyroid-stimulating antibodies, 355-366

Thyrotropin receptor
  and autoimmune thyroid disease, 355-366
  monoclonal antibodies, 355-366, 390, 391
Thyroxine
  GM1 incorporation, myelin, 245-247
  TSH antibodies, 357
Transformed cells, ganglioside patterns, 30-32
Trembler mutant, 297-304
Triiodothyronine, TSH antibodies, 357
Trimethylsilyl ethers, mass spectra, 76, 81, 82
Tumor growth, anti-asialo-GM1, 184, 185
Tunicamycin inhibition
  ganglioside synthesis, 206
  liver golgi complex, 227
Two-dimensional NMR, 88-98

Ulnar nerve conduction, 603

Valsalva maneuver, 611
Vesicles (see also Golgi complex)
  Golgi apparatus, 203
  neurite growth, 501
  receptor marking, 126
Visual system, ganglioside transport, 161, 162

Weaver mutant
  cerebellar development, 173-178
  cerebellum gangliosides, 169-178
WHT/Ht strain
  brain gangliosides, 266
  GM2(NeuGc) expression, liver, 263-270

ZPH cell line, 525-533